Microporous and Mesoporous Materials for Catalytic Applications

Microporous and Mesoporous Materials for Catalytic Applications

Guest Editor

Narendra Kumar

MDPI

Basel • Beijing • Wuhan • Barcelona • Belgrade • Novi Sad • Cluj • Manchester

Guest Editor
Narendra Kumar
Laboratory of Industrial
Chemistry and Reaction
Engineering
Åbo Akademi University
Turku
Finland

Editorial Office
MDPI AG
Grosspeteranlage 5
4052 Basel, Switzerland

This is a reprint of the Special Issue, published open access by the journal *Catalysts* (ISSN 2073-4344), freely accessible at: www.mdpi.com/journal/catalysts/special_issues/Microporous_Mesoporous_Catalytic.

For citation purposes, cite each article independently as indicated on the article page online and using the guide below:

Lastname, A.A.; Lastname, B.B. Article Title. *Journal Name* **Year**, *Volume Number*, Page Range.

ISBN 978-3-7258-3598-0 (Hbk)
ISBN 978-3-7258-3597-3 (PDF)
https://doi.org/10.3390/books978-3-7258-3597-3

Cover image courtesy of Narendra Kumar
Transmission Electron Micrograph of Ru-MCM-41 Mesoporous Material Catalyst.

Contents

About the Editor

Narendra Kumar

Prof. Dr. Narendra Kumar obtained his Dr. Tech. (Chemical Engineering) from Åbo Akademi University, Turku/Åbo, Finland, in 1996. He has worked at the Laboratory of Industrial Chemistry and Reaction Engineering, Åbo Akademi University, since 1990. His duties are research, education, and the supervision of PhD, MSc., and B. Sc. students. He has published 498 peer-reviewed research papers in reputed, high-impact factor international scientific journals in the field of catalysis, catalytic materials, catalyst synthesis and characterizations, microporous and mesoporous materials, chemical engineering, petro-chemicals, oil refining, biofuels, environmental catalysis, fine chemicals, pharmaceuticals, and speciality chemical synthesis. His H-index is 43. He has co-supervised more than 75 doctoral, licentiate, and M. Sc. theses.

Dr. Narendra Kumar was awarded the Åbo Akademi University Bronz and Silver medals for continuous services at the university for more than 30 years. He has evaluated doctoral theses and acted as an opponent and examiner for universities in Finland, Sweden, and Spain. He is a member of several scientific committees and acts as a reviewer for the Federation of European Zeolite Association, International Zeolite Association, International Conference on Engineering for Waste and Biomass Valorization, International Conference on Catalysis and Nordic Symposium on Catalysis.

Dr. Narendra Kumar is an expert for the evaluation of European Union Projects, Horizon Europe (2021–2027). He is a member of professional and scientific organizations: Finnish Association for Graduate Engineers, Finnish Catalysis Society, European Federation of Chemical Engineers, British Zeolite Association, and International Zeolite Association. He acts as a reviewer for more than 35 international journals for publishers such as Elsevier, American Chemical Society, Royal Society of Chemistry, Springer, MDPI, Lidsen Publisher, and Universal Wiser Publisher.

Preface

I would like to express my gratitude toward the authors of the published articles in this Special Issue Microporous and Mesoporous Materials for Catalytic Applications. These research papers and review articles were peer-reviewed by selected prominent international researchers in the field of microporous and mesoporous materials. In considering that these catalytic materials are of immense importance for the production of green fuels and chemicals, intensive research is being carried out in academic and industrial research laboratories around the globe. Furthermore, for the sustainable development of our society, it is of immense significance that environment-friendly technologies, materials, and methods are applied to the manurfaturing of fine and speciality chemicals and renewable energy production. Microporous and mesoporous materials will play an important role in the development of green process technology, the mitigation of climate change, water purification, and thge synthesis of pharmaceutical components. Articles published in this Special Issue with novel research results, valuable experimental data, and creative discussions will help upcoming and established researchers, scientists, and academics inthe development of new microporous and mesoporous materials for catalytic applications in refinery processes, fine chemical synthesis, and renewable energy production.

Narendra Kumar
Guest Editor

catalysts MDPI

Editorial

Microporous and Mesoporous Materials for Catalytic Applications

Narendra Kumar

Laboratory of Industrial Chemistry and Reaction Engineering, Faculty of Science and Engineering, Åbo Akademi University, Aurum, Henriksgatan 2, FI-20500 Turku, Finland; narendra.kumar@abo.fi

The Special Issue "Microporous and Mesoporous Materials for Catalytic Applications has twelve peer-reviewed articles (Contributions 1–12), out of which there are eight research papers (Contributions 1–8) and four review papers (Contributions 9–12). The main research topics of the published peer review articles are synthesis and physico-chemical characterizations of structured acidic and metal-modified microporous and mesoporous materials. The prepared acidic, transition, and noble metal-modified microporous and mesoporous materials have been utilized in the production of fuel components such as gasoline, diesel, and jet fuels, pharmaceuticals, fine chemicals, and specialty and medicinal drug molecules. Catalytic reaction mechanism, kinetic studies, and deactivation of catalysts have been reported for these applications. Furthermore, reactions of immense interest for academic and industrial researchers have been studied in the published research papers. These important reactions are methanol to aromatics on hybrid zeolite catalysts (Contribution 1), synthesis of CBO (CO_3O_4/Bi_2O_3) degradation of fipronil and acetochlor pesticides in aqueous medium (Contribution 2), hydrodesulfurization of thiophene in an n-heptane stream (Contribution 3), dimethyl ether to olefins (Contribution 4), catalytic distillation of atmospheric residue of petroleum (Contribution 5), effect of basic production on porous supported alumina for production of acetins (Contribution 6), Ni and Ce grafted mesoporous silica KIT-6 for adsorption of CO_2 (Contribution 7), Ordered mesoporous nZVI/Zr-Ce-SBA-15 catalysts used for nitrate reduction (Contribution 8), a comprehensive review of production of fine chemicals (Contribution 9), porous aerogel structures as promising matrials for photocatalysis (Contribution 10), Hierarchical zeolite by alkaline treatment (Contribution 11) and catalytic ozonation of textile waste water (Contribution 12).

The synthesis of aluminosilicate microporous and ordered mesoporous materials using hydrothermal synthesis has been the focus of research in these published research papers. Preparation of transition and noble metal-modified microporous and mesoporous materials has been carried out using evaporation impregnation, ion-exchange, deposition precipitation, co-precipitation, chemical vapor deposition, and in situ synthesis methods. It is noteworthy to mention that authors in the published peer-reviewed research papers have also utilized novel synthesis methods, such as the introduction of transition and noble metals during the hydrothermal synthesis and post-synthesis bi-metallic modifications.

In-depth physico-chemical characterizations of the as-synthesized pristine and metal-modified microporous and mesoporous materials have been performed using characterization methods such as X-ray powder diffraction for determination of the structural features and phase purity and scanning electron microscopy for the measurements of the crystal morphology, shape, and size distributions (Contributions 1,3,4). The amounts, strengths, and types of Brønsted and Lewis acid sites have been analyzed by FT-IR equipment using pyridine as a probe molecule. The authors of some research papers have also carried out measurements of the amounts of Brønsted and Lewis acid sites and total acidity using temperature-programmed desorption of ammonia (Contributions 3,5,6). Furthermore, the characterization of basic sites in pristine and metal-modified has been carried out using

Citation: Kumar, N. Microporous and Mesoporous Materials for Catalytic Applications. *Catalysts* **2024**, *14*, 723. https://doi.org/10.3390/catal14100723

Received: 30 September 2024
Accepted: 14 October 2024
Published: 16 October 2024

temperature-programmed desorption of CO_2. Transmission electron microscopy and high resolution transmission electron microscopy have been applied for the determination of the size and dispersion of noble and transition metals (Contributions 2,3,5). The oxidation states of metal nanoparticles have been measured using X-ray photoelectron spectroscopy. The chemical compositions of the synthesized acidic and metal-modified microporous and mesoporous materials have been measured using energy dispersive X-ray microanalyses and inductively coupled plasma spectroscopy (Contributions 9,11,12). Surface area, pore volume, and pore size distributions have been measured using nitrogen physisorption (Contributions 9–10).

The novel research results published in the research papers of the Special Issue titled "Microporous and Mesoporous Materials for Catalytic Applications" will enhance the scientific knowledge in the research fields of the synthesis, characterization, and applications of these industrially important catalytic materials for the development of green process technology and environmentally friendly industrial processes for the production of renewable energy, green fuels, and chemicals. Furthermore, the published research papers and reviews will contribute to the understanding of the fundamental, theoretical and practical aspects of reaction mechanisms, adsorption phenomena, shape selectivity diffusion of reactants and products, catalyst deactivation, and regenerations.

Funding: This research received no external funding.

Acknowledgments: I would like to thank all the authors who shared their research and the reviewers for their invaluable contributions.

Conflicts of Interest: The author declares no conflicts of interest.

List of Contributions:

1. Magomedova, M.V.; Galanova, E.G.; Starozhitskaya, A.V.; Afokin, M.I.; Matevosyan, D.V.; Egazaryants, S.V.; Tsaplin, D.E.; Maximov, A.L. Methanol to aromatics on hybrid structure zeolite catalysts. *Catalysts* **2024**, *14*, 461. https://doi.org/10.3390/Catal14070461.
2. Saeed, M.; Panchal, S.; Bajaber, M.A.; Alalwiat, A.A.; Ahmed, A.E.; Razzaq, U.; Rab Nawaz, H.Z.; Hussain, F. Synthesis of CBO (CO_3O_4-Bi_2O_3) heterogeneous photocatalysts for degradation of fipronil and acetochlor pesticides in aqueous medium. *Catalysts* **2024**, *14*, 392. https://doi.org/10.3390/Catal14060392.
3. Coutinho, A.C.S.L.S.; Barros, J.M.F.; Araujo, M.D.S.; Silva, J.B.; Souza, M.J.B.; Delgado, R.C.O.B.; Fernandes, V.J., Jr.; Araujo, A.S. Hydrodesulfurization of thiophene in n-heptane stream using CoMo/SBA-15 and CoMo/AlSBA-15 mesoporous catalysts. *Catalysts* **2024**, *14*, 198. https://doi.org/10.3390/Catal14030198.
4. Magomedova, M.V.; Starozhitskaya, A.V.; Davidov, I.A.; Tsaplin, D.E.; Maximov, A.L. Dimethyl Ether to Olefins on Hybrid Intergrowth Structure Zeolites. *Catalysts* **2023**, *13*, 570. https://doi.org/10.3390/Catal13030570.
5. Morais, C.G.D.P.; Silva, J.B.; Almeida, J.S.; Oliveira, R.R.; Araujo, M.D.S.; Fernandes, G.J.T.; Delgado, R.C.O.B.; Coriolano, A.C.F.; Fernandes, V.J.; Araujo, A.S. Catalytic distillation of atmospheric residue of petroleum over HY-MCM-41 micro-mesoporous materials. *Catalysts* **2023**, *13*, 296. https://doi.org/10.3390/Catal13020296.
6. Bezerra, R.d.C.F.; Mota, G.; Vidal, R.M.B.; Carmo, J.V.D.; Saraiva, G.D.; Campos, A.; Oliveira, A.C.; Lang, R.; Otubo, L.; Jiménez, J.J.; et al. Effect of Basic Promoters on Porous Supported Alumina Catalysts for Acetins Production. *Catalysts* **2022**, *12*, 1616. https://doi.org/10.3390/Catal12121616.
7. Suba, M.; Popa, A.; Verdeș, O.; Borcănescu, S.; Barvinschi, P. Ni and Ce grafted mesoporous silica KIT-6 for CO_2 Adsorption. *Catalysts* **2022**, *12*, 1339. https://doi.org/10.3390/Catal12111339.
8. Zhang, R.; Liu, H.; Jiang, W.; Liu, W. Ordered mesoporous nZVI/Zr-Ce-SBA-15 Catalystsused for nitrate reduction: Synthesis, optimization and mechanism. *Catalysts* **2022**, *12*, 797. https://doi.org/10.3390/Catal12070797.
9. Lantos, J.; Kumar, N.; Saha, B. A comprehensive review of fine chemical production using metal modified and acidic microporous and mesoporous catalytic materials. *Catalysts* **2024**, *14*, 317. https://doi.org/10.3390/Catal14050317.

10. Lee, K.H.; Arshad, Z.; Dahshan, A.; Alshareef, M.; Alsulami, Q.A.; Bibi, A.; Lee, E.-J.; Nawaz, M.; Zubair, U.; Javid, A. Porous Aerogel Structures as Promising Materials for Photocatalysis, Thermal Insulation Textiles, and Technical Applications: A Review. *Catalysts* **2023**, *13*, 1286. https://doi.org/10.3390/Catal13091286.
11. Oliveira, D.S.; Lima, R.B.; Pergher, S.B.C.; Caldeira, V.P.S. Hierarchical zeolite by alkaline treatment: Advantages and applications. *Catalysts* **2023**, *13*, 316. https://doi.org/10.3390/catal13020316.
12. Bilińska, M.; Bilińska, L.; Gmurek, M. Homogeneous and Heterogeneous Catalytic Ozonation of Textile Wastewater: Application and Mechanism. *Catalysts* **2023**, *13*, 6. https://doi.org/10.3390/catal13010006.

catalysts

Article

Methanol to Aromatics on Hybrid Structure Zeolite Catalysts

Maria V. Magomedova [1,2], Ekaterina G. Galanova [1], Anastasia V. Starozhitskaya [1,*], Mikhail I. Afokin [1], David V. Matevosyan [1], Sergey V. Egazaryants [3], Dmitry E. Tsaplin [1,3] and Anton L. Maximov [1]

1 A.V. Topchiev Institute of Petrochemical Synthesis, Russian Academy of Sciences (TIPS RAS), 29 Leninsky Prospekt, 119991 Moscow, Russia
2 Lomonosov Institute of Fine Chemical Technologies, Russian Technological University (MIREA), 86 Vernadsky Prospekt, 119454 Moscow, Russia
3 Faculty of Chemistry, Lomonosov Moscow State University, 1 Leninskie Gory Str., 119234 Moscow, Russia
* Correspondence: av-star@ipc.as.ru

Abstract: A study on the reaction of methanol to aromatic hydrocarbons using catalysts based on hybrid zeolites MFI-MEL, MFI-MTW, and MFI-MCM-41 at a temperature of 340 °C and a pressure of 10.0 MPa was carried out. It is shown that in the synthesis of hydrocarbons under pressure, the activity of the studied samples is similar and does not have a linear correlation with their total acidity. It was found that the catalyst's activity is primarily determined by the rate of the initial methanol conversion reaction, which is related to the volume of micropores—more micropores lead to higher activity. Additionally, increasing the volume of mesopores results in the formation of heavier aromatic compounds, specifically C_{10}–C_{11}.

Keywords: methanol to aromatics; hybrid zeolites; intergrowth structure; MEL; MTW; MCM-41; MFI

Citation: Magomedova, M.V.; Galanova, E.G.; Starozhitskaya, A.V.; Afokin, M.I.; Matevosyan, D.V.; Egazaryants, S.V.; Tsaplin, D.E.; Maximov, A.L. Methanol to Aromatics on Hybrid Structure Zeolite Catalysts. *Catalysts* **2024**, *14*, 461. https://doi.org/10.3390/catal14070461

Academic Editor: Javier Ereña

Received: 19 June 2024
Revised: 11 July 2024
Accepted: 16 July 2024
Published: 18 July 2024

1. Introduction

The process of converting methanol into hydrocarbons on a zeolite catalyst, referred to as the MTG (Methanol to Gasoline) process, has been known since the second half of the 20th century. However, with the advancement of decarbonization in the petrochemical industry and the development of methanol synthesis technologies using CO_2 as a reagent, research into the process of synthesizing hydrocarbons from methanol remains relevant.

In the synthesis of hydrocarbons from oxygenates, a catalyst based on the MFI zeolite structure is most commonly used. The most well-known industrial brand of this zeolite is ZSM-5 (Zeolite Socony Mobil–5) from Mobil (Spring, TX, USA). It is characterized by high productivity for liquid hydrocarbons and a low deactivation rate. Until 2010, the most common methods for modifying the catalyst were its promotion with various metals and zeolite desilication/dealumination. For instance, the introduction of Ga, Zn, and Ag into the catalyst increases the yield of aromatic compounds [1,2]. Desilication/dealumination leads to the formation of randomly oriented mesopores surrounded by micropores, which slightly extends the catalyst's lifetime and increases the propylene/ethylene ratio due to its open mesoporosity [3].

Currently, alternative zeolite structures such as zeolites with a two-dimensional channel structure (MEL) or a one-dimensional channel structure (MTW) are being considered for the synthesis of hydrocarbons from oxygenates [4–6]. For example, in [7], it has been shown that the MEL zeolite can compete with MFI zeolite in terms of catalyst deactivation rate and propylene selectivity (35% versus 27%). Catalysts based on the MTW zeolite structure are also a good alternative to traditional zeolites in terms of propylene and butene formation, with a combined yield reaching 65 mol% C [8].

Additionally, to increase the yield of target products, the synthesis of zeolites with a hybrid structure, by combining different zeolite structures such as MFI and MEL, MFI and MTW, MFI and BEA, MFI and AEL, and MFI and FER, is actively being developed.

Hybrid materials can be obtained through the co-crystallization of zeolites with different structures or in the synthesis of "core-shell" type structures [9–14]. For example, in [11], samples based on the MFI/MTW structure were synthesized by various methods. It has been shown that compared to the standard MFI-based sample, these samples exhibit high selectivity for lower olefins (28% versus 19%) with a high proportion of propylene in the product (propylene/ethylene ratio of 0.79 versus 0.58) during methanol conversion. High results were also achieved with the hybrid co-crystallized MFI/MEL zeolite, which showed a 47% propylene formation selectivity at 100% methanol conversion, compared to 27% for MFI zeolite [9].

In a previous study [15], we investigated catalyst samples based on co-crystallized MFI-MEL, MFI-MTW, and "core-shell" MFI-MCM-41 structures in the reaction of synthesizing olefins from dimethyl ether (DME). It has been shown that at high DME conversions for catalysts based on MFI and MFI-MEL zeolites, the molar ratios of ethylene/propylene are close and range from 1.1 to 1.2. For the MFI-MTW-based catalyst, the lowest ethylene/propylene ratio of 0.8 was observed, while the "core-shell" MFI-MCM-41 structure showed the highest ratio of 1.45. It was suggested that under atmospheric pressure, the presence of micropores influences the selectivity of product formation, particularly ethylene, by stabilizing key intermediates like alkylcyclopentyl cations and polymethyl-substituted aromatic hydrocarbons. Moreover, it was shown that the topology of the hybrid zeolite determines the hydrogen transfer reaction rates but does not affect the isomerization activity of the catalyst. The next step in the research is to study how the hybrid structure of zeolites influences the synthesis of liquid hydrocarbons, particularly aromatic compounds, under increased pressure. The aim of this study is to establish the relationship between the physicochemical characteristics of the zeolite component of the catalyst and the composition of the resulting gaseous and liquid products in the reaction of methanol conversion to hydrocarbons at elevated pressure.

2. Results and Discussion

Under the conditions of catalyst testing, T = 340 °C and P = 10.0 MPa, methanol conversion increases in the following order: MFI-MTW/Al_2O_3 < MFI-MCM-41/Al_2O_3 < MFI/Al_2O_3 < MFI-MEL/Al_2O_3 (Figure 1a). For the three samples based on MFI, MFI-MEL, and MFI-MCM-41 zeolites, the pattern of methanol conversion on contact time can be described by a single dependence (Figure 1b).

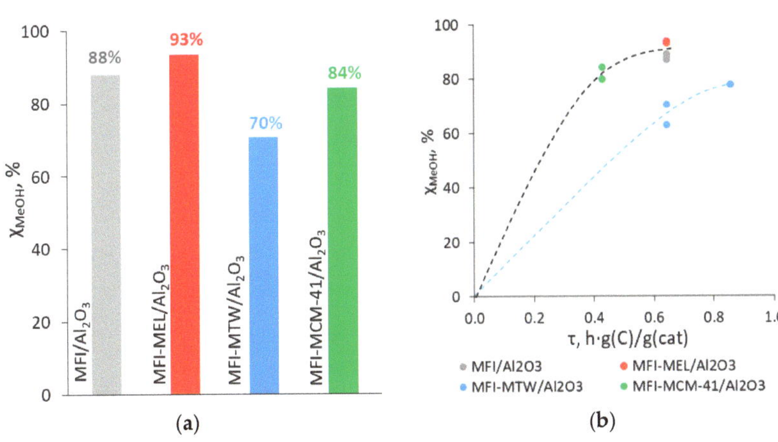

Figure 1. (**a**) Methanol conversion on hybrid zeolite catalysts. (**b**) Dependence of methanol conversion on specified contact time. T = 340 °C, P = 10.0 MPa.

When testing these catalysts in the reaction of converting dimethyl ether to lower olefins at atmospheric pressure, we observed a linear correlation between the activity of the samples and their total acidity: the MFI/Al_2O_3 sample, with a total acidity of 495 µmol NH_3/g(cat)×h, demonstrated the highest activity, while the MFI-MTW/Al_2O_3 sample with a total acidity of 266 µmol NH_3/g(cat)×h, showed the lowest activity [15]. However, such a correlation was not observed when the reaction was conducted under pressure. The observed effect can be explained by the decrease in the mean free path as pressure increases, which results in a shift diffusion from Knudsen to Fickian. Since the Fickian diffusion coefficient is inversely proportional to pressure, increasing the pressure to 10 MPa significantly reduces the diffusion coefficient [16,17]. Consequently, under increased pressure, the observed reaction rate is mainly determined by internal diffusion. Although the intrinsic reaction rate increases with the catalyst's acidity, this is not observable due to the limiting diffusion constraints. Therefore, the observed reaction rate under increased pressure is determined by the total volume of micropores, where the primary reactions of the MTO process occur. Therefore, a correlation was observed: with an increase in the volume of micropores of the catalyst, its activity increases (Table 1, Supplementary Materials: Figure S6)

Table 1. Product yields of the methanol-to-hydrocarbon reaction on hybrid zeolite catalysts.

	MFI/Al_2O_3	**MFI-MEL/Al_2O_3**	**MFI-MTW/Al_2O_3**	**MFI-MCM-41/Al_2O_3**
Micropore volume, cm^3/g	0.057	0.088	0.012	0.04
Micropore diameter, nm	0.9	0.91	0.69	0.63
Mesopore volume, cm^3/g	0.142	0.198	0.431	0.300
Mesopore diameter, nm	8.27	5.79	19.67	7.58
Yield, wt.C%:				
DME	0.4	0.4	15.1	0.5
Lower olefins C_2–C_4	0.2	0.6	3.1	0.5
Gaseous hydrocarbons C_1–C_4	35.7	24.3	18.8	26.6
Liquid hydrocarbons C_5–C_{11}	36.2	52.4	34.6	54.3
Mass ratio gas/liquid	1.0	0.5	0.6	0.5

The confirmation of the proposed relationship between the catalyst activity and micropore volume is the high yield of dimethyl ether (15.1 wt.C%) and the presence of lower olefins C_2–C_4 (yield = 3.1 wt.C%) in the gas phase on the MFI-MTW/Al_2O_3 catalyst with the smallest volume of micropores (Table 1, Figure 2).

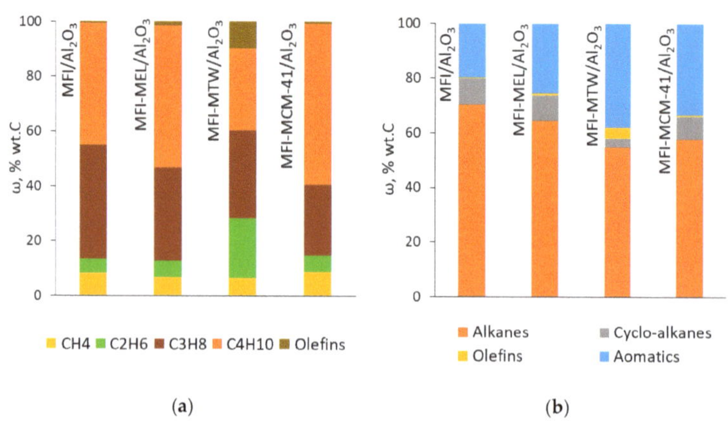

Figure 2. Distribution of individual compounds in methanol conversion on hybrid zeolite catalysts. (**a**) gaseous hydrocarbons C_1–C_4; (**b**) liquid hydrocarbons C_5–C_{11}.

The gaseous hydrocarbons C_1–C_4 for the MFI/Al_2O_3, MFI-MEL/Al_2O_3, and MFI-MCM-41/Al_2O_3 catalysts were mainly composed of propane and butanes, with a total proportion exceeding 80 wt.% (Figure 2). For the MFI-MTW/Al_2O_3 catalyst, a significant amount of olefins C_2–C_4 (9.5 wt.%) was present among the gaseous hydrocarbons, directly confirming the primary formation of lower olefins in the synthesis of liquid hydrocarbons.

For the MFI/Al_2O_3, MFI-MEL/Al_2O_3, and MFI-MCM-41/Al_2O_3 catalysts, the major constituent of the liquid phase was isoalkanes C_5–C_{11}, comprising 52 to 66 wt%. The highest content of aromatic hydrocarbons was observed for MFI-MTW/Al_2O_3 at 38.2 wt.%.

For the MFI/Al_2O_3 and MFI-MEL/Al_2O_3 samples, within the aromatic compounds, the predominant constituents were C_9–C_{10} compounds, with their total share reaching 72–75 wt%. (Figure 3). For the MFI-MTW/Al_2O_3 and MFI-MCM-41/Al_2O_3 catalysts, the focus shifted towards the formation of C_{10}–C_{11} aromatic hydrocarbons (Figure 3, Supplementary Materials: Figure S7). This can be linked to the volume of mesopores in the catalyst, which was 0.142 and 0.198 cm^3/g for MFI/Al_2O_3 and MFI-MEL/Al_2O_3, respectively, and 0.431 and 0.300 cm^3/g for MFI-MTW/Al_2O_3 and MFI-MCM-41/Al_2O_3, respectively. The increased volume of mesopores in catalysts based on hybrid zeolite promotes the release of bulky aromatic compounds from the zeolite pores into the gas phase. Consequently, we can make another important conclusion linking the textural characteristics of the catalytic material and the distribution of reaction products: the larger the volume of mesopores in the catalyst structure, the faster the hydrogen transfer reactions, leading predominantly to the formation of aromatic compounds.

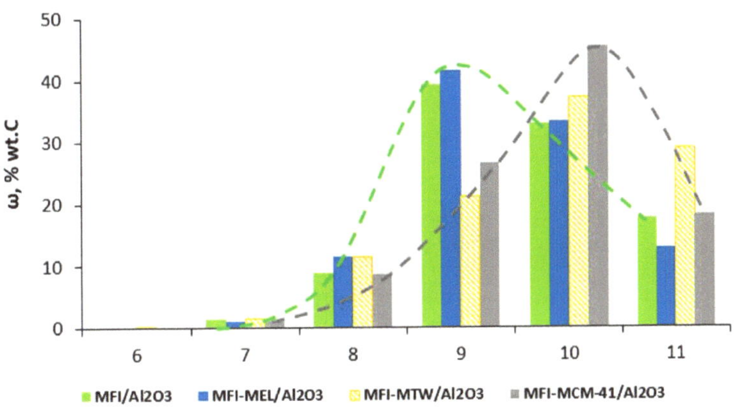

Figure 3. Distribution of individual aromatic compounds in methanol conversion on hybrid zeolite catalysts. Green dotted line: interpolation for MFI/Al_2O_3 and MFI-MEL/Al_2O_3, grey dotted line: interpolation for MFI-MCM-41/Al_2O_3.

3. Experimental Section

3.1. Zeolite Synthesis

The MFI zeolite and hybrid co-crystallized zeolites (MFI-MEL, MFI-MTW, MFI-MCM-41) were synthesized at the Department of Chemistry, Moscow State University [15]. Detailed studies on the physicochemical properties of zeolites and catalysts based on were conducted in [15] and the patents [18,19]. The most important physicochemical properties of zeolites and catalysts are presented in the Supplementary Materials (Supplementary Materials: Tables S1–S4, Figures S1–S5). The Si/Al ratio for the standard MFI-based sample is 48; for the hybrid zeolite samples (MFI-MEL, MFI-MTW, MFI-MCM-41), the ratios are similar, ranging from 53 to 55. The phase ratios in the hybrid zeolites are as follows: MFI-MEL = 55/45 wt.%, MFI-MTW = 60/40 wt.%, and MFI-MCM-41 = 80/20 wt.%.

3.2. Catalyst Preparation

42.7 g of AlO(OH) were placed in a mixer and 20 mL of a peptizing solution (5 mL of 1.0 M aqueous HNO_3 + 15 mL of H_2O) was added. The mixture was stirred for 5 min. Then, 76.5 g of dry zeolite was added. The mass was stirred for 20 min at 60 °C. After that, the catalytic mass was passed through an extruder with a die diameter of 2.5 mm. The extrudates are dried in an oven at 80 °C, 90 °C, 100 °C, 110 °C, and 120 °C for 3 h at each temperature, then calcined in a muffle furnace at 550 °C for 7 h. During calcination, AlO(OH) is converted to Al_2O_3. The final content of Al_2O_3 in the material obtained was 30 wt%.

3.3. Experimental Procedure

The experimental unit for the synthesis of liquid hydrocarbons from methanol has been described in several of our works, such as [20,21]. The methanol-to-hydrocarbon conversion was carried out in a flow-circulation mode at a pressure of 10.0 MPa and a temperature of 340 °C, using hydrogen as a carrier gas at a flow rate of 5 L/h. The reactor was loaded with 2–3 g of catalyst. Methanol was fed into the system using a dosing pump at a rate of 6 mL/h. The circulating gas flow rate was 180–185 L/h, with a vent gas flow rate of 5 L/h.

Liquid reaction products (organic and aqueous phases) were collected after 24 h of operation, and their quantity and composition were analyzed. The volume of dissolved gases in the liquid phase was determined using a special burette. The composition of the liquid phase, vent gases, and degassing products was determined chromatographically.

For each experiment, a material balance was calculated based on the composition and quantity of the resulting products. The analysis of the gas phase (C_1–C_5) was performed using a Chrom-5 gas chromatograph (Laboratory Instruments, Prague, Czech Republic) with a combined packed column containing the sorbent Polysorb 1, modified with Carbowax 3000, a flame ionization detector, and the carrier gas was helium, and the column temperature was programmed to rise from 50 to 180 °C.

The composition of the liquid organic phase was analyzed using a Crystallux 4000M chromatograph (Production Company Meta-Chrom, Yoshkar-Ola, Russia) with a flame ionization detector and a capillary column with a non-polar phase Petrocol (100 m × 0.325 mm × 0.5 μm) in a temperature-programmed mode (35–250 °C, heating rate 2 °C/min), with helium as the carrier gas (flow rate 2 mL/min).

3.4. Data Processing

The main indicators used to evaluate the process were methanol conversion, the product yields based on carbon, and the distribution of individual components or groups.

Methanol conversion was calculated using the following formula:

$$X_{MeOH} = \frac{m_{\text{in}(MeOH)} - m_{\text{out}(MeOH)}}{m_{\text{in}(MeOH)}} \cdot 100, \% \quad (1)$$

where $m_{\text{in}(MeOH)}$ and $m_{\text{out}(MeOH)}$ are the masses of methanol inlet and outlet, respectively, in g/h.

The yield of carbon-containing products was calculated using the following formula:

$$Y_i = \frac{m_{\text{out}}(\text{C})_{\text{i}}}{m(\text{C})_{\text{in}(MeOH)}} \cdot 100,\ \text{wt.C\%} \quad (2)$$

where $m_{\text{out}}(\text{C})_{\text{i}}$ is the mass of carbon of the i-th product, in g/h, and $m(\text{C})_{\text{in}(MeOH)}$ is the mass of carbon in the inlet methanol, in g/h.

The yields of gaseous and liquid hydrocarbons were calculated using Equations (3) and (4):

$$Y_{GHC} = \sum\nolimits_{k=1}^{4} Y_k,\ \text{wt.C\%} \quad (3)$$

$$Y_{LHC} = \sum_{k=5}^{11} Y_k, \text{ wt.C\%} \quad (4)$$

where Y_k is the yield of an individual hydrocarbon, in wt.C%.

The distribution of individual substances and component groups was calculated using the following formula:

$$\omega_i = \frac{m(C)_i}{\sum_i m(C)_i} \cdot 100, \text{ wt.C\%} \quad (5)$$

where $m(C)_i$ is the mass of carbon in the i-th product, in g/h, and $\sum_i m(C)_i$ is the total mass of carbon in the group, in g/h.

4. Conclusions

The study investigated the conversion of methanol to aromatic hydrocarbons using co-crystallized hybrid zeolite catalytic systems in a flow-circulation reactor unit at a pressure of 10.0 MPa and a temperature of 340 °C.

The results showed that the catalytic systems based on MFI, MFI-MEL, and MFI-MCM-41 zeolites showed similar activity, achieving methanol conversions of 84–94%. The catalyst based on MFI-MTW zeolite showed the lowest activity with a methanol conversion of 70%. For this catalyst, the gaseous hydrocarbons contained dimethyl ether and C_2-C_4 olefins.

It was demonstrated that under increased pressure, the acidic properties of the samples play a secondary role. The activity of the catalyst is mostly determined by the diffusion restrictions and by the total volume of micropores. Therefore, at high pressure, the catalyst activity and the distribution of reaction products are determined by the textural properties of the catalyst: the volume of micropores is responsible for the observed rate of primary MTO-reactions, while the volume of mesopores is responsible for secondary reactions such as H-transfer and aromatization. The use of hybrid co-crystalline zeolites MFI-MCM-41 and MFI-MTW leads to an increase in the content of heavier aromatic compounds in liquid hydrocarbons due to the enlargement of the catalyst's mesopore volume.

Supplementary Materials: The following supporting information can be downloaded at: https://www.mdpi.com/article/10.3390/catal14070461/s1, Table S1: Composition of hybrid zeolites; Figure S1: XRD patterns of the catalysts; Table S2: Textural properties of the catalysts and indexed hierarchy factor of the catalysts [22]; Figure S2: BJH plots (desorption curve) for the catalysts, Figure S3: t-plots for nitrogen adsorbed in the catalysts; Figure S4 NH_3-TPD profiles of the catalysts; Table S3: Acidic properties of catalysts; Figure S5: IR spectra of pyridine adsorbed on hybrid zeolites; Table S4: Acid characteristics of hybrid zeolites; Figure S6: Dependence of methanol conversion on the volume of micropores in the catalyst; Figure S7: Dependence of average number of carbon atoms in aromatic compounds on the volume of mesopores in the catalyst.

Author Contributions: Conceptualization, M.V.M. and E.G.G.; methodology, E.G.G., D.V.M. and D.E.T.; formal analysis, E.G.G., M.I.A. and A.V.S.; resources, A.L.M.; writing—original draft preparation, M.V.M., E.G.G. and A.V.S.; writing—review and editing, M.V.M. and A.V.S.; visualization, M.V.M. and E.G.G.; supervision, M.V.M.; project administration, A.L.M. and S.V.E. All authors have read and agreed to the published version of the manuscript.

Funding: This work was financially supported by the Russian Science Foundation (grant No. 24-13-00242).

Data Availability Statement: The data presented in this study are available on request from the corresponding author.

Conflicts of Interest: The authors declare no conflict of interest.

References

1. Freeman, D.; Wells, R.P.K.; Hutchings, G.J. Conversion of Methanol to Hydrocarbons over Ga_2O_3/H-ZSM-5 and Ga_2O_3/WO_3 Catalysts. *J. Catal.* **2002**, *205*, 358–365. [CrossRef]
2. Yiwen, F.; Ji, T.; Xiaochang, H.; Weibin, S.; Yibing, S.; Changyong, S. Aromatization of Dimethyl Ether over Zn/H-ZSM-5 Catalyst. *Chin. J. Catal.* **2010**, *31*, 264–266.
3. Hartmann, M.; Machoke, A.G.; Schwieger, W. Catalytic test reactions for the evaluation of hierarchical zeolites. *Chem. Soc. Rev.* **2016**, *45*, 3313–3330. [CrossRef] [PubMed]

4. Ji, Z.; Jiao, C.; Jiao, Q.; Wang, Q.; Dai, W.; Zheng, J.; Wang, Y.; Li, W.; Li, R. Synthesis of hierarchical ZSM-11 and its catalytic performances during methanol to propylene. *Microporous Mesoporous Mater.* **2024**, *374*, 113142. [CrossRef]
5. Liu, C.; Uslamin, E.A.; van Vreeswijk, S.H.; Yarulina, I.; Ganapathy, S.; Weckhuysen, B.M.; Kapteijn, F.; Pidko, E.A. An integrated approach to the key parameters in methanol-to-olefins reaction catalyzed by MFI/MEL zeolite materials. *Chin. J. Catal.* **2022**, *43* 1879–1893. [CrossRef]
6. Jang, H.-G.; Ha, K.; Seo, G. Active intermediates formed on phosphorous-modified MTW zeolites in methanol to olefin conversion An ESR study. *Appl. Catal. A Gen.* **2015**, *499*, 168–176. [CrossRef]
7. Kai, Y.; Jia, X.Y.; Sen, W.; Sheng, F.; He, S.P.; Wang, P.F.; Dong, M.; Qin, Z.F.; Fan, W.B.; Wang, J.G. Effect of framework structure of ZSM-11 and ZSM-5 zeolites on their catalytic performance in the conversion of methanol to olefins. *J. Fuel Chem. Technol.* **2023**, *51* 1652–1662.
8. Taniguchi, T.; Nakasaka, Y.; Yoneta, K.; Tago, T.; Masuda, T. Size-Controlled Synthesis of Metallosilicates with MTW Structure and Catalytic Performance for Methanol-to-Propylene Reaction. *Catal. Lett.* **2016**, *146*, 666–676. [CrossRef]
9. Conte, M.; Xu, B.; Davies, T.E.; Bartley, J.K.; Carley, A.F.; Taylor, S.H.; Khalid, K.; Hutchings, G.J. Enhanced selectivity to propene in the methanol to hydrocarbons reaction by use of ZSM-5/11 intergrowth zeolite. *Microporous Mesoporous Mater.* **2012**, *164* 207–213. [CrossRef]
10. Wang, X.; Meng, F.; Chen, H.; Gao, F.; Wang, Y.; Han, X.; Wang, L. Synthesis of a hierarchical ZSM-11/5 composite zeolite of high SiO_2/Al_2O_3 ratio and catalytic performance in the methanol-to-olefins reaction. *Comptes Rendus Chim.* **2017**, *20*, 1083–1092 [CrossRef]
11. Aghdam, M.S.; Askari, S.; Halladj, R.; Tajar, A.F. Facile and selective approach towards synthesis of a series ZSM-5/ZSM-12 catalysts for methanol to hydrocarbons reactions: Applying different synthesis driving force and conditions. *Adv. Powder Technol* **2022**, *33*, 103502. [CrossRef]
12. Mirshafiee, F.; Khoshbin, R.; Karimzadeh, R. Steam-assisted methanol conversion to green fuel over highly efficient hierarchical structured MFI/BEA composite zeolite synthesized by incorporation method. *Renew. Energy* **2022**, *197*, 1061–1068. [CrossRef]
13. Wang, X.; Guo, S.; Niu, Z. Synthesis of a ZSM-5 (core)/SAPO-11 (shell) composite zeolite and its catalytic performance in the methylation of naphthalene with methanol. *RSC Adv.* **2023**, *13*, 2081–2089. [CrossRef] [PubMed]
14. Kim, J.J.; Jeong, D.J.; Jung, H.S.; Hur, Y.G.; Choung, I.W.; Baik, J.H.; Park, M.J.; Chung, C.-H.; Bae, J.W. Dimethyl ether conversion to hydrocarbons on the closely interconnected FER@ZSM-5 nanostructures. *Microporous Mesoporous Mater.* **2022**, *340*, 112034 [CrossRef]
15. Magomedova, M.V.; Starozhitskaya, A.V.; Davidov, I.A.; Tsaplin, D.E.; Maximov, A.L. Dimethyl Ether to Olefins on Hybrid Intergrowth Structure Zeolites. *Catalysts* **2023**, *13*, 570. [CrossRef]
16. Xiao, J.; Wei, J. Diffusion mechanism of hydrocarbons in zeolites—I. Theory. *Chem. Eng. Sci.* **1992**, *47*, 1123–1141. [CrossRef]
17. Froment, G.F.; Bischoff, K.B.; De Wilde, J. *Chemical Reactor Analysis and Design*; John Wiley & Sons: Hoboken, NJ, USA, 2010.
18. Maksimov, A.L.; Magomedova, M.V.; Afokin, M.I.; Tsaplin, D.E.; Kulikov, L.A.; Ionin, D.A. Method for Producing an HZSM-Type Zeolite (Variants) and Method for Producing Aromatic Hydrocarbons of the C_6-C_{11} Fraction. RUS Patent RU2753263C1, 12 August 2021.
19. Tsaplin, D.E.; Kulikov, L.A.; Maksimov, A.L.; Karakhanov, E.A. Method for Obtaining a Composite Material with a Hierarchical Structure. RUS Patent RU2773945C1, 14 June 2022.
20. Magomedova, M.V.; Peresypkina, E.G.; Ionin, D.A.; Afokin, M.I.; Golubev, K.B.; Khadzhiev, S.N. Effect of Feedstock and Gas Atmosphere Composition on Selectivity and Distribution of Hydrocarbon Groups in Gasoline Synthesis from Oxygenates. *Pet. Chem.* **2017**, *57*, 1052–1057. [CrossRef]
21. Maximov, A.L.; Magomedova, M.V.; Galanova, E.G.; Afokin, M.I.; Ionin, D.A. Primary and secondary reactions in the synthesis of hydrocarbons from dimethyl ether over a Pd-Zn-HZSM-5/Al_2O_3 catalyst. *Fuel Process. Technol.* **2020**, *199*, 106281. [CrossRef]
22. Verboekend, D.; Mitchell, S.; Milina, M.; Groen, J.C.; Pérez-Ramírez, J. Full Compositional Flexibility in the Preparation of Mesoporous MFI Zeolites by Desilication. *J. Phys. Chem. C* **2011**, *115*, 14193–14203. [CrossRef]

catalysts

MDPI

Article

Synthesis of CBO (Co_3O_4-Bi_2O_3) Heterogeneous Photocatalyst for Degradation of Fipronil and Acetochlor Pesticides in Aqueous Medium

Muhammad Saeed [1,*], Sandeep Panchal [2], Majed A. Bajaber [3], Ahlam A. Alalwiat [3], Ahmed Ezzat Ahmed [4,5], Ujala Razzaq [1], Hafiza Zahra Rab Nawaz [1] and Farhat Hussain [1]

1 Department of Chemistry, Government College University Faisalabad, Faisalabad 38000, Pakistan
2 Department of Civil Engineering, Government Polytechnic Mankeda, Agra 283102, UP, India
3 Chemistry Department, Faculty of Science, King Khalid University, P.O. Box 9004, Abha 61413, Saudi Arabia
4 Department of Biology, College of Science, King Khalid University, Abha 61413, Saudi Arabia
5 Prince Sultan Bin Abdelaziz for Environmental Research and Natural Resources Sustainability Center, King Khalid University, Abha 61421, Saudi Arabia
* Correspondence: msaeed@gcuf.edu.pk; Tel.: +92-346-9010903

Abstract: The excessive use of pesticides has led to the harmful contamination of water reservoirs. Visible-light-driven photocatalysis is one of the suitable methods for the removal of pesticides from water. Herein, the development of CBO (Co_3O_4-Bi_2O_3) as a heterogeneous catalyst for the visible light-assisted degradation of Fipronil and Acetochlor pesticides is reported. After synthesis via coprecipitation using cobalt (II) nitrate hexahydrate ($Co(NO_3)_2{\cdot}6H_2O$), bismuth (III) nitrate pentahydrate ($Bi(NO_3)_3{\cdot}5H_2O$) and sodium hydroxide (NaOH) as precursor materials, the prepared CBO was characterized using advanced techniques including XRD, EDS, TEM, SEM, FTIR, and surface area and pore size analysis. Then, it was employed as a photocatalyst for the degradation of Fipronil and Acetochlor pesticides under visible light irradiation. The complete removal of Fipronil and Acetochlor pesticides was observed over CBO photocatalyst using 50 mL (100 mg/L) of each pesticide separately within 120 min of reaction. The reaction kinetics was investigated using a non-linear method of analysis using the Solver add-in. The prepared CBO exhibited a 2.8-fold and 2-fold catalytic performance in the photodegradation of selected pesticides than Co_3O_4 and Bi_2O_3 did, respectively.

Keywords: CBO; pesticides; Fipronil; Acetochlor; photocatalysis

Citation: Saeed, M.; Panchal, S.; Bajaber, M.A.; Alalwiat, A.A.; Ahmed, A.E.; Razzaq, U.; Rab Nawaz, H.Z.; Hussain, F. Synthesis of CBO (Co_3O_4-Bi_2O_3) Heterogeneous Photocatalyst for Degradation of Fipronil and Acetochlor Pesticides in Aqueous Medium. *Catalysts* **2024**, *14*, 392. https://doi.org/10.3390/catal14060392

Academic Editor: Narendra Kumar

Received: 26 May 2024
Revised: 14 June 2024
Accepted: 17 June 2024
Published: 19 June 2024

1. Introduction

The intense growth in the human population has led to a significant increase in the demand for food and clean water. Additionally, rapid expansion in industries has accelerated the deterioration of the fresh water supply [1–4] Various strategies have been adopted for the increase in the production of food. One of the strategies is the employment of agrochemicals called pesticides. Pesticides are agrochemicals commonly employed for the eradication of useless organisms and the preservation of agricultural products. Pesticides are employed for the abatement of pests and weeds to promote the growth of crops and increase the production of food [5,6]. About 2 million tons of pesticides are used annually around the world. However, the excessive use of pesticides causes severe environmental pollution because most of the pesticides disseminate to the environment. It is assumed that less than 15% of the pesticides employed meet their targets [7] (As most of these pesticides are stable and persistent, they therefore remain in soil for a long time. Ultimately, they leach to water bodies as persistent organic pollutants (POPs). Several studies have shown that the toxicological effects of these micropollutants persist even at low concentrations due to their endocrine disruptor potential and become further complicated by their presence as a mixture [8–10].

Fipronil and Acetochlor are commonly used pesticides in Pakistan. Fipronil, which is generally employed for pest control in veterinary applications, household applications, and agriculture, is a wide-ranging phenylpyrazole insecticide. Its IUPA name is [5-amino-1-[2,6-dichloro-4-(trifluoromethyl) phenyl]-4-[(trifluoromethyl)sulfinyl]-1H-pyrazole-3-carbonitrile] [11]. It is efficacious in the control of insects that have resistance to other insecticides like carbamate, organophosphorus, cyclodiene, and pyrethroids. It disrupts the central nervous system through interfering with the flow of Cl^{1-} through the γ-aminobutyric acid-regulated chloride channel [12]. Acetochlor, which is widely used in agriculture applications for the control of annual grasses and broad-leaf weeds, is an effective chloroacetanilide herbicide. Its IPAC name is [2-chloro-N-(ethoxymethyl)-*N*-(2-ethyl-6-methylphenyl) acetamide] [13,14]. The excessive use of these pesticides causes the contamination of water resources due to water runoff in agricultural fields. Additionally, the industries of pesticide production and domestic wastewater also contribute to the formation of the pesticide-contaminated wastewater. These pesticides are considered as emerging micropollutants. These micropollutants are stable over a long time; therefore, they are transported in water and cause pollution far away from their point source. The existence of these pesticides in surface water and drinking water is a big concern because they are extremely hazardous due to their endocrine disruptor potential. The sporadic use of pesticides has affected ecosystems, which threaten fish, domestic animals, wildlife, birds, and livestock.

Therefore, the treatment of pesticide-contaminated wastewater is a crucial research domain. The conventional methods, including chemical methods (such as electrochemical and chemical precipitation), physical methods (such as adsorption), and biological methods are not effective in the eradication of the pesticide pollutants from the environment. Therefore, advanced methods are needed for the treatment of pesticide-contaminated water [7,15]. Techniques known as advanced oxidation processes (AOPs) are considered effective strategies for the complete remediation of the agrochemicals present in water. These techniques have abilities to mineralize the pesticides to non-toxic inorganic molecules. AOPs comprise various methods including photo-Fenton, Fenton, UV/O_3, O_3, UV/H_2O_2, photocatalyst/UV-A, photocatalyst/UV-B, and photocatalyst/Vis. All these methods involve the in situ production of hydroxyl radicals ($OH^{\bullet}$) [16–18]. The semiconductor metal oxides-based visible-light-driven photocatalytic method is superior to other methods owing to a low processing cost, the complete decomposition of the pesticide molecules, and mild reaction conditions [19,20]. Only the narrow-band gap semiconductor metal oxides can be employed as visible-light-driven photocatalysts for the degradation of pesticide molecules in the aqueous medium. However, the narrow-band gap semiconductor metal oxides have the drawback of low photocatalytic performance due to the fast recombination of photoinduced charges. Various approaches have been adopted for the improvement of photocatalytic performances of narrow-band gap semiconductor metal oxides. The construction of heterostructure/heterojunction through coupling two or more semiconductor metal oxides is one of the approaches used for the enhancement of photocatalytic performance under visible light irradiation due to its effectiveness and feasibility for the spatial separation of positive holes and electrons [21–23]. The coupling of two semiconductors of different work functions creates a built-in electric field at the interface of the two semiconductors. The creation of a built-in electric field is due to the flow of electrons from the semiconductor having a higher work function compared to the one with a lower work function. As a result, the positive holes accumulate in one component of the heterojunction while the electrons in accumulate other components, and consequently, the recombination is inhibited [24]. The cobalt oxide (Co_3O_4) and bismuth oxide (Bi_2O_3) are typical narrow-band semiconductors with a band gap energy of 1.2–2.1 eV and 2.5–2.9 eV, respectively. Although both have a response to the absorption of visible light, they have limited photocatalytic performances due to the quick recombination of positive holes and electrons. The construction of the heterojunction between cobalt oxide (Co_3O_4) and bismuth oxide (Bi_2O_3) creates a driving force for the separation of positive holes and electrons. Hence, the construction of a cost-effective CBO (Co_3O_4-Bi_2O_3) composite for the efficient photodegradation of Fipronil and Acetochlor pesticides under sunlight is reported in this study.

2. Results and Discussion

2.1. Characterization

XRD analysis is a powerful technique to estimate the crystalline nature and crystal structure of the synthesized catalyst. Therefore, the successful synthesis of CBO was confirmed using XRD analysis. Figure 1 shows the XRD spectrum of CBO and standard XRD spectra of Co_3O_4 and Bi_2O_3. According to JCPDS No. 42-1476, the diffraction peaks observed in the XRD of CBO at 2θ ~ 18, 30, 37, 43, and 58° indicate the existence of a cubic structure of Co_3O_4 in fabricated CBO. These peaks have been indexed to 111, 200, 311, 200, and 511 planes of the cubic structure of Co_3O_4, respectively [25].). Similarly, the diffraction peaks observed at 2θ ~ 19°, 22°, 25°, 26°, 27°, 28°, 33°, 35°, 38°, 46°, 52°, 53°, 54°, and 58° correspond to Bi_2O_3 according to JCPDS No 41-1449 [26–28]. It can also be observed that the characteristics diffraction peaks of Bi_2O_3 in CBO have been slightly shifted. Ding et al. [29] have reported that this slight shift is due to the incorporation of Co^{3+}, which has a smaller ion radius (0.063 nm) than Bi^{3+} (0.117 nm).

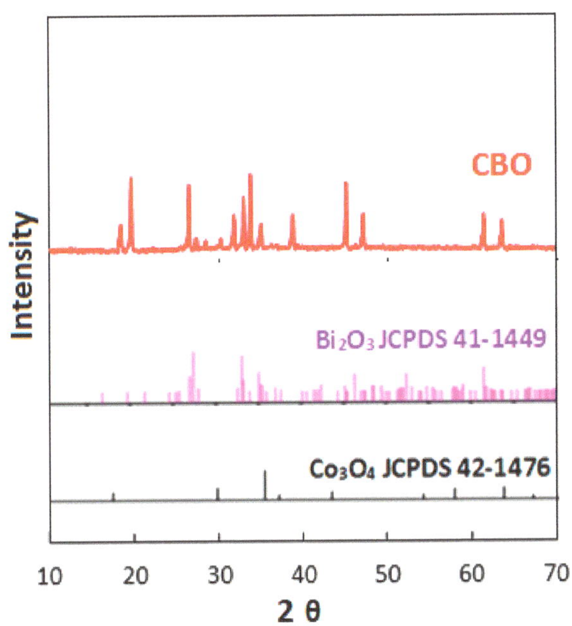

Figure 1. XRD analyses.

Energy dispersive X-ray spectroscopy (EDS) was utilized for the determination of the elemental composition of CBO. The EDS spectrum of CBO is given in Figure 2. The spectroscopic analysis showed that the synthesized heterostructure is composed of cobalt, bismuth, and oxygen. The percentages of these elements are 11.1, 70.8, and 13.9, respectively. The EDS spectrum shows the existence of 4.2% carbon as well in the synthesized CBO. The exact source of this carbon is not known; however, it is considered an impurity in the precursor materials.

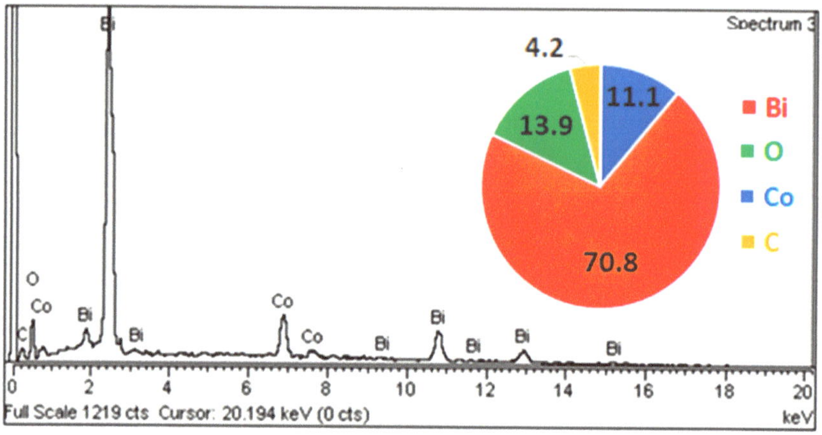

Figure 2. EDS spectrum and percent composition of CBO.

The shape and size of the particles of CBO were studied using TEM and SEM analyses. Figure 3 shows the TEM and SEM images of CBO. TEM indicates that the particle size of CBO is less than 50 nm. The observable lattice fringes in HR-TEM confirm the crystalline nature of CBO. Both the XRD and HR-TEM results support each other. The results of SEM show the dispersed, non-agglomerated, CBO particles and their irregularity in their morphology. The dispersed, non-agglomerated particles have higher catalytic performance because the active centers of such particles are easily accessible to reactant molecules [30]. The TEM image was analyzed for particle size distribution using ImageJ software (https://imagej.net/ij/). The particle size distribution is depicted in Figure 3.

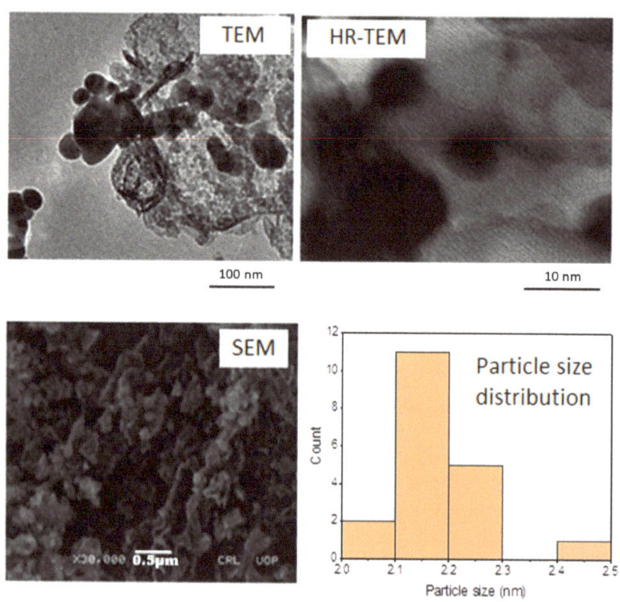

Figure 3. TEM, SEM analyses, and particle size distribution.

The functional groups and bonds associated with synthesized CBO were examined with FTIR spectroscopy. Figure 4 depicts the FTIR spectrum of CBO. Several absorption bands can be observed in the FTIR spectrum of CBO. The presence of an absorption band at 1634 cm^{-1} indicates the existence of OH groups on the CBO heterostructure. The existence of Bi-O and Bi-O-Bi linkages can be confirmed using absorption bands at ~445 cm^{-1} and 846 cm^{-1}, respectively. The absorption bands at ~571 and 668 cm^{-1} are representative peaks of Co_3O_4 due to Co-O stretching vibrations [31–33].

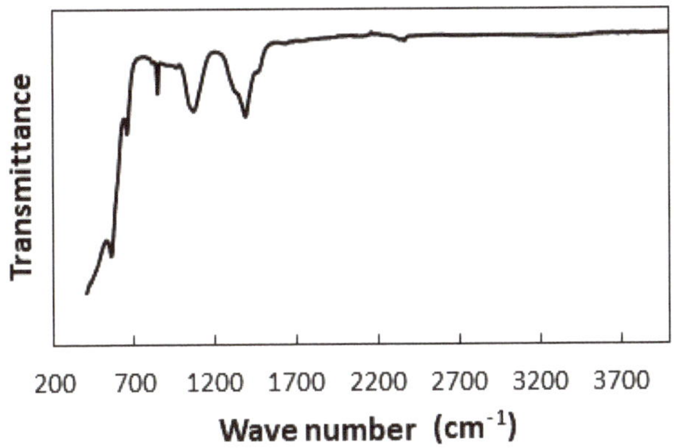

Figure 4. FTIR spectrum of CBO.

The surface area and pore size analyses were carried out by investigating the adsorption of nitrogen at 77.4 K. The surface area calculated through analyzing the nitrogen adsorption data according to the BET equation was found as 38.53 m^2/g (Figure 5). Similarly, the pore surface area and pore volume were determined as 11.13 m^2/g and 15.342 cc/g using the BJH method, respectively (Figure 5).

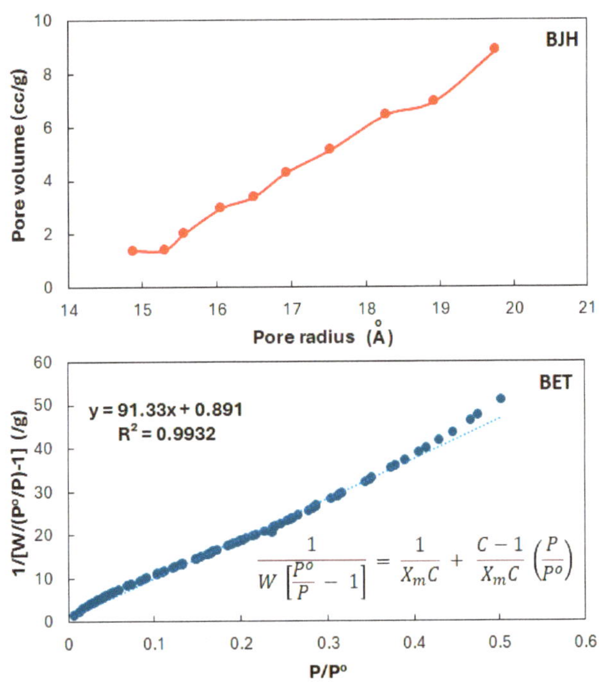

$$\frac{1}{W\left[\frac{P^o}{P}-1\right]} = \frac{1}{X_m C} + \frac{C-1}{X_m C}\left(\frac{P}{P^o}\right)$$

Figure 5. Surface area and pore size analysis.

2.2. Photocatalysis

Fipronil and Acetochlor were selected as model pollutants for the evaluation of photocatalytic activities. In the first step, 50 mL of 100 mg/L solution of Fipronil and Acetochlor were stirred under the irradiation of light for 1 h separately. The analysis of the reaction mixture after one hour of stirring showed that there was no loss in the concentration of Fipronil and Acetochlor after stirring the solutions. It confirms that there is no degradation of Fipronil and Acetochlor due to photolysis. In the second step, 0.05 g CBO was added to the solution of Fipronil and Acetochlor in separate experiments. The reaction mixtures were stirred under dark conditions for half an hour. The analyses of reaction mixtures showed about a 20% decrease in the concentration of Fipronil and Acetochlor. This decrease in concentration is attributed to the adsorption of Fipronil and Acetochlor onto the surface of CBO. In the third step, the separate solutions of Fipronil and Acetochlor containing CBO were stirred under sunlight for the evaluation of photocatalytic degradation activities. The analyses showed the complete mineralization of Fipronil and Acetochlor within two hours of the reaction duration. For comparison, the photocatalytic degradation of Fipronil and Acetochlor was carried out with Bi_2O_3 and Co_3O_4 under similar reaction conditions. More than a 70 and 60% mineralization of Fipronil and Acetochlor were achieved using Bi_2O_3 and Co_3O_4 as photocatalysts, respectively. Figure 6 depicts the obtained data.

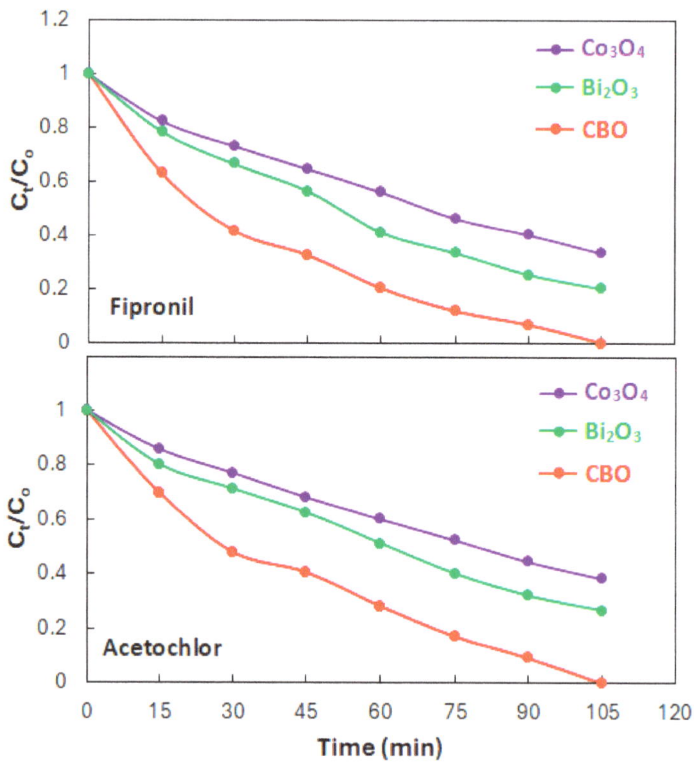

Figure 6. Photocatalytic degradation of Fipronil and Acetochlor.

The optimization of catalyst dose is also an important parameter in catalysis. The use of an optimized catalyst dose avoids the unnecessary utilization of a catalyst in a catalytic reaction. Therefore, the optimum catalyst doses were determined for the photocatalytic degradation of Fipronil and Acetochlor pesticides. This study was accomplished by performing separate photodegradation experiments of Fipronil and Acetochlor with different dosages of CBO in the range of 0.03 to 0.1 g under the same experimental conditions. Separate 50 mL aqueous solutions of Fipronil and Acetochlor (100 mg/L) were used in this investigation. The reaction duration was 60 min. The obtained results are given in Figure 7. The given data indicate that initially both the percent degradation of Fipronil and Acetochlor and their rate of reaction increased with a catalyst dose from 0.03 to 0.05 g and then a decrease was observed with a higher catalyst dose. Hence, a 0.05 g dose was declared as the optimum catalyst dose in this study. Initially, the catalytic activity increases with the catalyst dose because the number of active centers increases with the catalyst dose. However, the higher catalyst dose causes the scattering of light resulting in a decrease in catalytic activity [3].

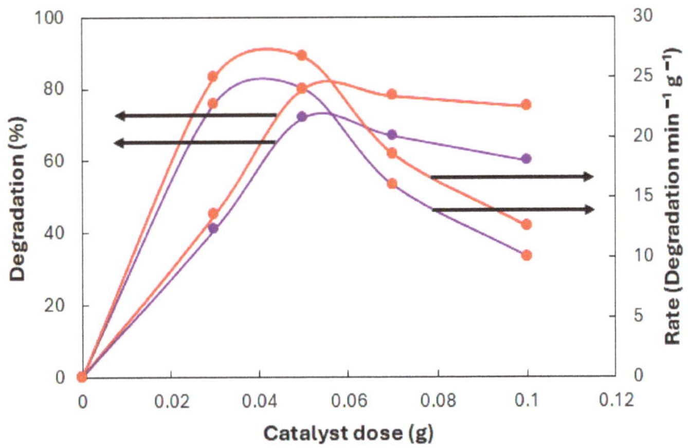

Figure 7. Effect of catalyst dose on photodegradation of Fipronil and Acetochlor.

The best photocatalyst is one that can be recycled many times without any loss in its photocatalytic activities. Hence, the reusability and stability of the CBO heterogeneous photocatalyst were also studied. The spent CBO samples were collected from the reaction mixture using filtration after photocatalytic degradation experiments. The collected spent sample was first washed many times with ethanol and then with distilled water. After washing, it was dried at 100 °C for 12 h. Then, it was reused as a heterogeneous photocatalyst for the degradation of acetochlor under similar experimental conditions as used earlier. The analyses of samples taken from the reaction mixture showed that the recycled CBO catalyst has almost the same photocatalytic activity as the fresh sample. Hence, it is concluded that fabricated CBO is a stable and reusable heterogeneous photocatalyst for the degradation of pesticides.

2.3. Mechanism and Kinetics

The prime reactive species involved in the degradation of selected pesticides were identified through trapping experiments. Separate experiments were performed for the CBO-catalyzed photodegradation of Fipronil in the presence of EDTA and BQ. It was observed that photodegradation of Fipronil decreased from 80% to 68% and 53% in the presence of EDTA and BQ, respectively. The EDTA and BQ have been used as scavengers for h^+ and superoxide anion radicals, respectively. These results confirm that h^+ and superoxide anion radicals are the primary reactive species involved in the degradation of pesticide molecules [34–37]. Hence, the enhanced photocatalytic activity of CBO is attributed to the formation and separation of h^+ and e^-. As the XRD suggests, synthesized CBO is a mixture of the Co_3O_4-Bi_2O_3 heterojunction and a composite of these two components. Hence, the enhanced photocatalytic activity of synthesized CBO can be explained by two mechanisms.

The CBO is a second-generation photocatalyst. It is also known as a heterojunction. The second-generation photocatalysts were designed to overcome the limitations or drawbacks associated with first-generation photocatalysts. The fast recombination of positive holes and electrons is the most significant drawback associated with first-generation photocatalysts. In second-generation photocatalysts, these positive holes and electrons are confined to the valence band of one component and the conduction band of other components. Hence, the separation between positive holes and electrons impedes their recombination. Consequently, these species result in the production of highly reactive hydroxyl radicals that take part in the degradation of organic molecules [38,39]. The band edge positions of the components of the CBO heterojunction have been estimated through Mott–Schottky measurement using the theory of electronegativity as reported by Pradhan et al. [40]. Accordingly, the band gap energy, VB, and the CB of Co_3O_4, have been calculated as 2.4 eV, 2.63 eV, and 0.23 eV, respectively. On the other hand, the band gap energy, VB, and CB for Bi_2O_3, have been calculated as 2.7 eV, 2.78 eV, and 0.08 eV, respectively. In a heterojunction, the excited electrons (e^-) move from the lower CB of Co_3O_4 to the higher CB of Bi_2O_3. The higher CB of Bi_2O_3 acts as a barrier to charge recombination. The transfer of e^- from one component to the other component of heterojunction reduces the recombination of h^+ and e^-. The separated charges (h^+ and e^-) yield hydroxyl radicals in reaction with water and oxygen, respectively. Hydroxyl radicals are powerful oxidizing agents that degrade pollutant molecules [41–43]. Figure 8a explains the possible mechanism.

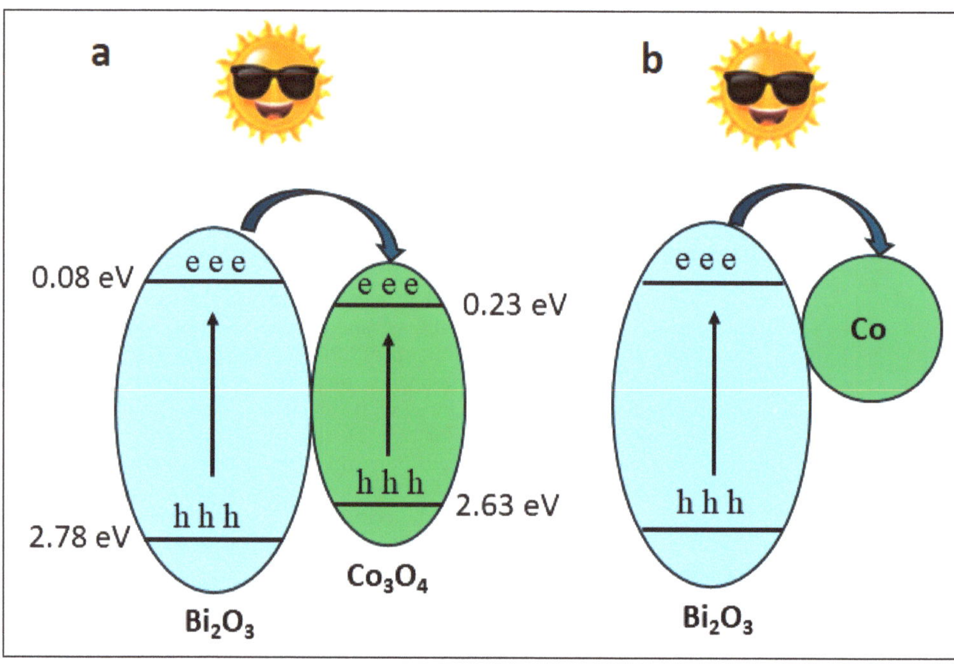

Figure 8. Mechanism of photocatalytic process (**a**) heterojunction (**b**) doped composite.

The enhanced photocatalytic performance of synthesized CBO can also be explained by another mechanism considering the composite or doped nature of synthesized CBO. As it has been reported in the discussion of XRD analysis, the characteristics diffraction peaks of Bi_2O_3 in CBO have been slightly shifted. Ding et al. [29] have reported that this slight shift is due to the incorporation of Co^{3+} which has a smaller ion radius (0.063 nm) than Bi^{3+} (0.117 nm). The incorporation of Co^{3+} in Bi_2O_3 acts as a sink for photoexcited electrons and obtains the electrons which restrict the recombination of electrons and holes, resulting in the enhancement of the photocatalytic activity of the CBO [44,45]. Figure 8b explains this mechanism.

The following reactions describe the photocatalytic process:

$$CBO + Visible\ irradiation \rightarrow h^{+}(CBO) + e^{-}(CBO) \quad (1)$$

$$h^{+}(CBO) + H_2O \rightarrow OH^{\bullet} \quad (2)$$

$$e^{-}(CBO) + (O_2)_{ads} \rightarrow O_2^{\bullet -} \quad (3)$$

$$O_2^{\bullet -} + H_2O \rightarrow OH^{\bullet} \quad (4)$$

$$Pesticide\ molecules + OH^{\bullet} \rightarrow Degradation\ products \quad (5)$$

Based on the above mechanism, the rate of reaction is proposed as (hv: Light energy, P: Concentration of pesticide)

$$-\frac{dP}{dt} = k(hv)(O_2)_{ads}(P) \quad (6)$$

The following experimental conditions make the rate of reaction independent of the first two terms in the above rate expression.

The reaction mixture is continuously irradiated.

The reaction mixture is open to the atmosphere.

Therefore, the rate expression becomes as follows:

$$-\frac{dP}{dt} = k(P) \quad (7)$$

$$-\frac{dP}{P} = k\ dt \quad (8)$$

$$P = P_o\, e^{-kt} \quad (9)$$

where P and P_o represent the concentration at a given time and the initial concentration of pesticides in mg/L, respectively. The photocatalytic degradation data of Fipronil and Acetochlor were analyzed according to the kinetic model given in Equation (9) via the non-linear method of analysis using the Solver add-in. Figure 9 shows the analyses of degradation data of Fipronil and Acetochlor via a non-linear method of analysis. The rate constants (k) for the degradation of Fipronil and Acetochlor were determined as given in Table 1. The higher values of R^2 indicate the fitness of the kinetics model to experimental data. The observed rate constants show that CBO exhibited 2.8-fold and 2-fold catalytic performance in the photodegradation of selected pesticides than Co_3O_4 and Bi_2O_3, respectively.

Table 1. Rate constants for degradation of Fipronil and Acetochlor.

Catalyst	Fipronil			Acetochlor		
	k (min^{-1})	R^2	Half Life (min)	k (min^{-1})	R^2	Half Life (min)
Co_3O_4	0.0101	0.998	68.6	0.0088	0.999	78.8
Bi_2O_3	0.0145	0.997	47.8	0.0119	0.995	58.2
CBO	0.0283	0.997	24.5	0.0233	0.992	29.7

The results of photodegradation of Fipronil and Acetochlor over CBO reported in this study were compared to already reported data as well. For example, Tenri and co-workers [46] have employed 5% AC/Chitosan-$Ca_2Fe_2O_5$ for the photodegradation of Fipronil. They irradiated a 25 mg/L solution of Fipronil containing 0.3 mg/L catalyst for 1 h. They obtained a 77% degradation of Fipronil. Hirashima et al. [47] have investigated the photodegradation of Fipronil using tandem mass spectrometry-liquid chromatography for the detection of degradation products. They detected many degradation products including dechlorinated compounds. Similarly, Mianjy and Niknafs [48] reported fipronil-desulfinyl as the main degradation product in the photolysis of Fipronil under the irradiation of UV light. Fu et al. [13] studied the H_2O_2/UV-assisted photocatalytic degradation of acetochlor using α-Fe_2O_3 as a catalyst. They investigated the effect of various parameters such as pH, concentration, and α-Fe_2O_3 dose on the photocatalytic degradation of acetochlor. The results showed a 91% photocatalytic degradation efficiency with a 50 m/L solution of acetochlor. Similarly, the TiO_2/UV system has been employed for the successful degradation of metolachlor, acetochlor, and alachlor pesticides in aqueous medium. The maximum degradation efficiency of these pesticides was reported as 98%, 93%, and 97%, respectively. A 1000 mL solution of 10 mg/L concentration was used for each pesticide. Furthermore, the toxicity of these pesticides was also investigated using freshwater alga Chlorella kessleri [49]. Wang and co-workers [50] reported the removal of Acetochlor via adsorption and degradation using $MnFe_2O_4$@AC as the adsorbent and catalyst. The removal efficiency through adsorption using 0.2 g/L $MnFe_2O_4$@AC as the adsorbent was found as 226 mg/g at 25 °C after 10 h of agitation. The removal efficiency of Acetochlor through catalytic degradation was found as >90% through heat-activated peroxymonosulfate (PMS) oxidation. Another research group reported BiOBr as a catalyst for the degradation of acetochlor under visible light irradiation. They mainly focused on the dependence of degradation efficiency on catalyst dosage. They found that degradation efficiency increased exponentially with increasing catalyst dosage [51].

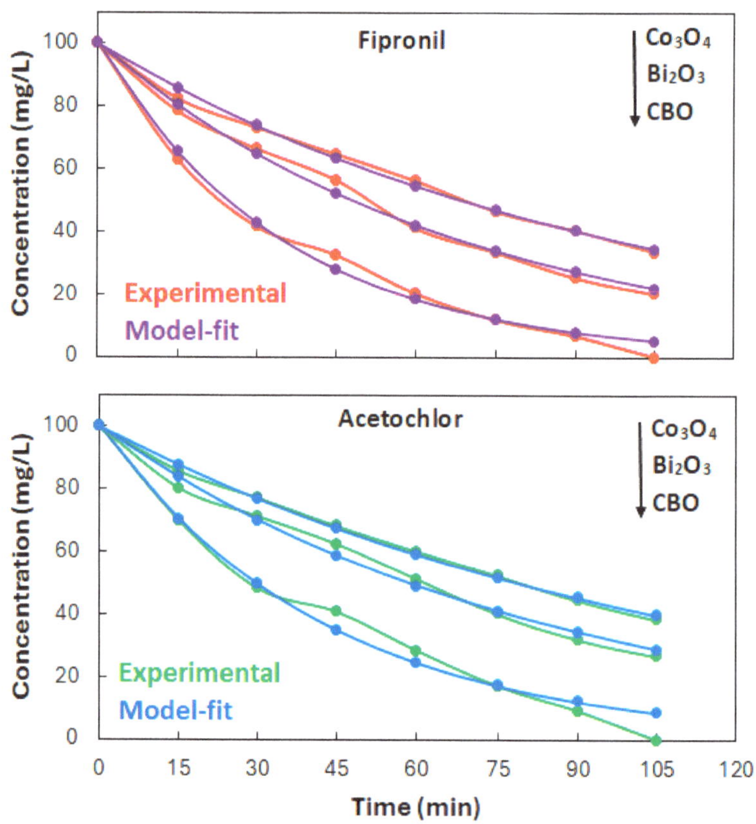

Figure 9. Kinetics analyses.

3. Experimental Section

3.1. Synthesis of CBO Heterogeneous Photocatalyst

The photocatalyst CBO was synthesized through the coprecipitation method according to the procedure reported earlier [30]. Typically, a 10 mL solution was prepared by dissolving 10 mmol (2.9 g) cobalt (II) nitrate hexahydrate ($Co(NO_3)_2{\cdot}6H_2O$) in de-ionized water. Another 10 mL solution was prepared by dissolving 10 mmol (4.85 g) bismuth (III) nitrate in 1 M nitric acid. Then, the solutions of cobalt (II) nitrate hexahydrate and bismuth (III) nitrate were mixed in a beaker. The pH of the resultant mixed solution was adjusted at 12 by dropwise addition of 1 M sodium hydroxide solution from the burette. The resultant mixed solution was continuously stirred for 3 h at a speed of 120 rpm while keeping the temperature constant at 60 °C. As a result of stirring for 3 h, green precipitates of hydroxides of Co and Bi were formed. The resultant precipitates were separated through filtration. The precipitates were washed for the removal of unreacted precursor materials. The washed precipitates were dried at 100 °C for 24 h. Finally, the dried substrate was ground and annealed at 500 °C for 3 h. After cooling to room temperature in the desiccator, the resultant black CBO was ground and sieved via 200 mesh. It was then stored in a glass vial for further study.

3.2. Characterization

The synthesized CBO was characterized by various advanced techniques. Various techniques used for the characterization of CBO include X-ray diffraction spectroscopy, Energy dispersive X-ray spectroscopy, Scanning electron microscopy, Transmission electron microscopy, Infrared spectroscopy and surface area and pore size analysis using JOEL-JDX-3532 (made in Tokyo, Japan), JSM5910 (made in Tokyo, Japan), JEOL-JSM 5910 (made in Tokyo, Japan), JEM 2100 JEOL (made in Tokyo, Japan), and Bruker VRTEX70 (made in Billerica, MA, USA), and NOVA 2200e (made in Connersville, IN, USA), respectively.

3.3. Photocatalysis

The photocatalytic activities of synthesized CBO were explored through the degradation of pesticides Fipronil and Acetochlor, which were selected as model pollutants. The evaluation of photocatalytic activities of CBO was explored in three steps.

A 50 mL solution of Fipronil and/or Acetochlor having a specific concentration was taken in a 500 mL Pyrex glass beaker. The solution of Fipronil and/or Acetochlor was stirred continually under irradiation of ultraviolet or visible light for 1 h. After stirring for one hour, a sample was extracted, and its absorbance was measured. This experiment was performed to know whether Fipronil and Acetochlor degrade under irradiation of ultraviolet or visible light. The analysis showed that there was no degradation of ultraviolet or visible light due to photolysis.

A 0.05 g of CBO was added to the solution of Fipronil and/or Acetochlor after the photolysis experiment. The Fipronil and/or Acetochlor solution containing CBO was stirred continuously under dark conditions for half an hour. After stirring under the dark, a sample was extracted, and its absorbance was measured with a UV-visible spectrophotometer. This experiment was performed to know the removal of Fipronil and Acetochlor due to adsorption on CBO.

The Fipronil and/or Acetochlor solution containing CBO was stirred continuously under irradiation of ultraviolet or visible light. Reaction mixture samples were extracted at different time intervals. The extracted samples were analyzed via measurement of their absorbance values. This experiment was performed to know the removal of Fipronil and Acetochlor due to photocatalytic degradation.

3.4. Analysis of Reaction Mixture

For estimation of photocatalytic activity, the reaction mixture was analyzed via measurement of absorbance of reaction mixture samples extracted at different time intervals at λ_{max} of Fipronil and Acetochlor. A UV-visible double-beam spectrophotometer (Lambda 25,

Perkin Elmer, Waltham, MA, USA) was used for the analyses of the reaction mixture. For the estimation of photocatalytic degradation of selected pesticides, 10, 15, 20, 25, and 30 mg/L standard solutions of Fipronil and Acetochlor were prepared. The absorbance of each standard solution was measured in the range of 200–250 nm; 350 and 219 nm were found as λ_{max} of Fipronil and Acetochlor, respectively. The absorbance of each standard solution at λ_{max} was plotted against concentration. Figure 10 shows these results. The resultant calibration curves were used for the determination of the centration of Fipronil and Acetochlor in the reaction samples extracted at different time intervals. Then, the following equation was used for the estimation of photocatalytic activity in terms of percent photodegradation of the Fipronil and Acetochlor:

$$Percent\ degradation = \frac{Co - C}{Co} \times 100 \quad (10)$$

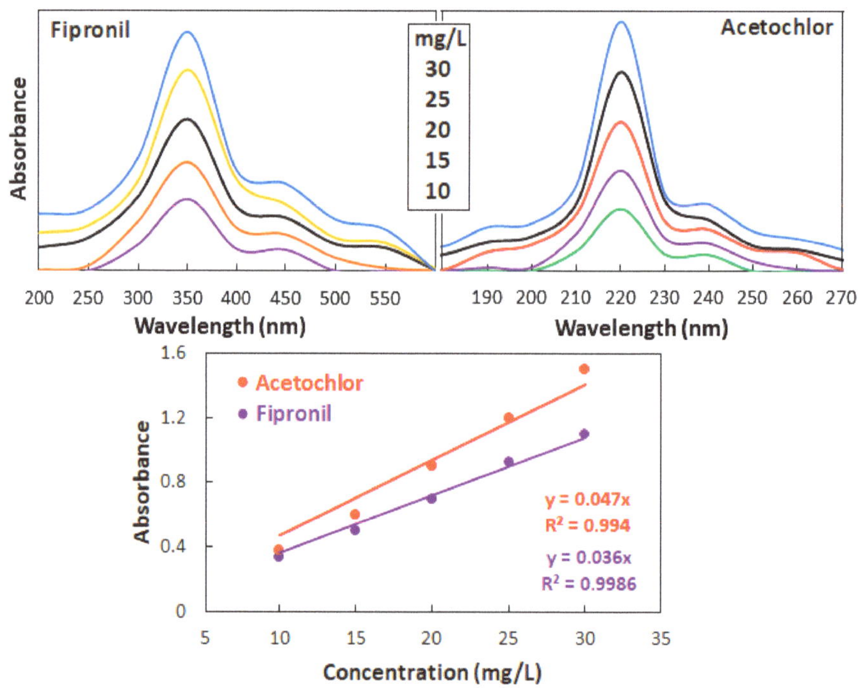

Figure 10. Calibration curves.

4. Conclusions

The photocatalyst CBO was synthesized via coprecipitation using cobalt (II) nitrate hexahydrate, bismuth (III) nitrate, nitric acid, and sodium hydroxide as the precursor materials. The fabrication of $Co_3Bi_2O_7$ was confirmed with XRD, EDS, TEM, SEM, TGA, FTIR, and surface area analysis. The fabricated Co_3O_4-Bi_2O_3 was employed as a photocatalyst for the degradation of Fipronil and Acetochlor under visible light irradiation. An almost complete ~75 and ~62% removal of both pesticides was observed with the CBO, Bi_2O_3, and Co_3O_4 photocatalysts, respectively. The photodegradation data were analyzed according to a first-order kinetics model using a non-linear method of analysis. The fabricated CBO was found to be an efficient, stable, and reusable photocatalyst for the photodegradation of Fipronil and Acetochlor under visible light irradiation.

Author Contributions: Conceptualization, M.S.; methodology, M.S., U.R.; software, S.P.; validation, M.S., and S.P.; formal analysis, M.S., and U.R.; investigation, M.S., U.R., H.Z.R.N., and F.H.; resources, A.E.A.; data curation, M.S.; writing—original draft preparation, M.S.; writing—review and editing, S.P.; visualization, M.S.; supervision, M.S.; project administration, M.S.; funding acquisition, M.A.B. and A.A.A. All authors have read and agreed to the published version of the manuscript.

Funding: The authors express their appreciation to the Deanship of Scientific Research at King Khalid Uni-versity for funding this work through the large group research project under grant number (R.G.P. 2/589/45).

Data Availability Statement: Data are contained within the article.

Acknowledgments: The authors express their appreciation to the Deanship of Scientific Research at King Khalid University for funding this work through the large group research project under grant number (R.G.P. 2/589/45).

Conflicts of Interest: The authors declare no conflicts of interest.

References

1. Bruckmann, F.S.; Schnorr, C.; Oviedo, L.R.; Knani, S.; Silva, L.F.O.; Silva, W.L.; Dotto, G.L.; Bohn Rhoden, C.R. Adsorption and Photocatalytic Degradation of Pesticides into Nanocomposites: A Review. *Molecules* **2022**, *27*, 6261. [CrossRef]
2. Rout, P.R.; Zhang, T.C.; Bhunia, P.; Surampalli, R.Y. Treatment technologies for emerging contaminants in wastewater treatment plants: A review. *Sci. Total. Environ.* **2021**, *753*, 141990. [CrossRef]
3. Saeed, M.; Akram, N.; Atta-Ul-Haq Naqvi, S.A.R.; Usman, M.; Abbas, M.A.; Adeel, M.; Nisar, A. Green and eco-friendly synthesis of Co_3O_4 and Ag-Co_3O_4: Characterization and photo-catalytic activity. *Green Process. Synth.* **2019**, *8*, 382–390. [CrossRef]
4. Younis, S.; Usman, M.; Atta ul Haq Akram, N.; Saeed, M.; Raza, S.; Siddiq, M.; Bukhtawar, F. Solubilization of reactive dyes by mixed micellar system: Synergistic effect of nonionic surfactant on solubilizing power of cationic surfactant. *Chem. Phys. Lett.* **2020**, *738*, 136890. [CrossRef]
5. Lian, L.; Jiang, B.; Xing, Y.; Zhang, N. Identification of photodegradation product of organophosphorus pesticides and elucidation of transformation mechanism under simulated sunlight irradiation. *Ecotoxicol. Environ. Saf.* **2021**, *224*, 112655. [CrossRef]
6. Wang, Z.; Zhang, B.; He, C.; Shi, J.; Wu, M.; Guo, J. Sulfur-based Mixotrophic Vanadium (V) Bio-reduction towards Lower Organic Requirement and Sulfate Accumulation. *Water Res.* **2021**, *189*, 116655. [CrossRef]
7. Yeganeh, M.; Charkhloo, E.; Sobhi, H.R.; Esrafili, A.; Gholami, M. Photocatalytic processes associated with degradation of pesticides in aqueous solutions: Systematic review and meta-analysis. *Chem. Eng. J.* **2022**, *428*, 130081. [CrossRef]
8. Ferhi, S.; Vieillard, J.; Garau, C.; Poultier, O.; Demey, L.; Beaulieu, R.; Penalva, P.; Gobert, V.; Portet-Koltalo, F. Pilot-scale direct UV-C photodegradation of pesticides in groundwater and recycled wastewater for agricultural use. *J. Environ. Chem. Eng.* **2021**, *9*, 106120. [CrossRef]
9. González, V.J.; Vázquez, E.; Villajos, B.; Tolosana-Moranchel, A.; Duran-Valle, C.; Faraldos, M.; Bahamonde, A. Eco-friendly mechanochemical synthesis of titania-graphene nanocomposites for pesticide photodegradation. *Sep. Purif. Technol.* **2022**, *289*, 120638. [CrossRef]
10. Luna-Sanguino, G.; Ruíz-Delgado, A.; Duran-Valle, C.J.; Malato, S.; Faraldos, M.; Bahamonde, A. Impact of water matrix and oxidant agent on the solar assisted photodegradation of a complex mix of pesticides over titania-reduced graphene oxide nanocomposites. *Catal. Today* **2021**, *380*, 114–124. [CrossRef]
11. Gomes Júnior, O.; Borges Neto, W.; Machado, A.E.H.; Daniel, D.; Trovó, A.G. Optimization of fipronil degradation by heterogeneous photocatalysis: Identification of transformation products and toxicity assessment. *Water Res.* **2017**, *110*, 133–140. [CrossRef]
12. de Lima, O.J.; de Araújo, D.T.; Marçal, L.; Crotti, A.E.M.; Machado, G.S.; Nakagaki, S.; de Faria, E.H.; Ciuffi, K.J. Photodegradation of Fipronil by Zn-$AlPO_4$ Materials Synthesized by Non-Hydrolytic Sol–Gel Method. *Chemengineering* **2022**, *6*, 55. [CrossRef]
13. Fu, Y.; Li, Y.; Hu, J.; Li, S.; Qin, G. Photocatalytic degradation of acetochlor by α-Fe_2O_3 nanoparticles with different morphologies in aqueous solution system. *Optik* **2019**, *178*, 36–44. [CrossRef]
14. Li, H.; Wang, Y.; Fu, J.; Hu, S.; Qu, J. Degradation of acetochlor and beneficial effect of phosphate-solubilizing Bacillus sp. ACD-9 on maize seedlings. *3 Biotech* **2020**, *10*, 67. [CrossRef]
15. Bokhari, T.H.; Ahmad, N.; Jilani, M.I.; Saeed, M.; Usman, M.; Haq, A.U.; Rehman, R.; Iqbal, M.; Nazir, A.; Javed, T. UV/H_2O_2, UV/H_2O_2/SnO_2 and Fe/H_2O_2 based advanced oxidation processes for the degradation of disperse violet 63 in aqueous medium. *Mater. Res. Express* **2020**, *7*, 015531. [CrossRef]
16. Dewil, R.; Mantzavinos, D.; Poulios, I.; Rodrigo, M.A. New perspectives for Advanced Oxidation Processes. *J. Environ. Manag.* **2017**, *195*, 93–99. [CrossRef]
17. Miklos, D.B.; Remy, C.; Jekel, M.; Linden, K.G.; Drewes, J.E.; Hübner, U. Evaluation of advanced oxidation processes for water and wastewater treatment—A critical review. *Water Res.* **2018**, *139*, 118–131. [CrossRef]
18. Ribeiro, A.R.; Nunes, O.C.; Pereira, M.F.R.; Silva, A.M.T. An overview on the advanced oxidation processes applied for the treatment of water pollutants defined in the recently launched Directive 2013/39/EU. *Environ. Int.* **2015**, *75*, 33–51. [CrossRef]

19. Xue, Y.; Liu, X.; Zhang, N.; Shao, Y.; Xu, C.C. Enhanced photocatalytic performance of iron oxides@HTCC fabricated from zinc extraction tailings for methylene blue degradation: Investigation of the photocatalytic mechanism. *Int. J. Miner. Met. Mater.* **2023**, *30*, 2364–2374. [CrossRef]
20. Yu, F.; Li, C.; Li, W.; Yu, Z.; Xu, Z.; Liu, Y.; Wang, B.; Na, B.; Qiu, J. Π-Skeleton Tailoring of Olefin-Linked Covalent Organic Frameworks Achieving Low Exciton Binding Energy for Photo-Enhanced Uranium Extraction from Seawater. *Adv. Funct. Mater.* **2024**, *34*, 2307230. [CrossRef]
21. Li, R.; Gao, T.; Wang Yao Chen, Y.; Luo, W.; Wu, Y.; Xie, Y.; Wang Yong Zhang, Y. Engineering of bimetallic Au–Pd alloyed particles on nitrogen defects riched g-C_3N_4 for efficient photocatalytic hydrogen production. *Int. J. Hydrogen Energy* **2024**, *63*, 1116–1127. [CrossRef]
22. Liu, Y.; Liu, N.; Lin, M.; Huang, C.; Lei, Z.; Cao, H.; Qi, F.; Ouyang, X.; Zhou, Y. Efficient visible-light-driven S-scheme $AgVO_3/Ag_2S$ heterojunction photocatalyst for boosting degradation of organic pollutants. *Environ. Pollut.* **2023**, *325*, 121436. [CrossRef]
23. Soni, V.; Sonu Sudhaik, A.; Singh, P.; Thakur, S.; Ahamad, T.; Nguyen, V.H.; Thi, L.A.P.; Quang, H.H.P.; Chaudhary, V.; Raizada, P.; et al. Visible-light-driven photodegradation of methylene blue and doxycycline hydrochloride by waste-based S-scheme heterojunction photocatalyst Bi_5O_7I/PCN/tea waste biochar. *Chemosphere* **2024**, *347*, 140694. [CrossRef]
24. Yang, H. A short review on heterojunction photocatalysts: Carrier transfer behavior and photocatalytic mechanisms. *Mater. Res. Bull.* **2021**, *142*, 111406. [CrossRef]
25. Malefane, M.E. $Co_3O_4/Bi_4O_5I_2/Bi_5O_7I$ C-scheme heterojunction for degradation of organic pollutants by light-emitting diode irradiation. *ACS Omega* **2020**, *5*, 26829–26844. [CrossRef]
26. Yasin, M.; Saeed, M.; Muneer, M.; Usman, M.; ul Haq, A.; Sadia, M.; Altaf, M. Development of Bi_2O_3-ZnO heterostructure for enhanced photodegradation of rhodamine B and reactive yellow dyes. *Surf. Interfaces* **2022**, *30*, 101846. [CrossRef]
27. Yi, S.; Yue, X.; Xu, D.; Liu, Z.; Zhao, F.; Wang, D.; Lin, Y. Study on photogenerated charge transfer properties and enhanced visible-light photocatalytic activity of p-type Bi_2O_3/n-type ZnO heterojunctions. *New J. Chem.* **2015**, *39*, 2917–2924. [CrossRef]
28. Hu, L.; Zhang, G.; Liu, M.; Wang, Q.; Dong, S.; Wang, P. Application of nickel foam-supported Co3O4-Bi2O3 as a heterogeneous catalyst for BPA removal by peroxymonosulfate activation. *Sci. Total. Environ.* **2019**, *647*, 352–361. [CrossRef]
29. Ding, Y.; Zhu, L.; Huang, A.; Zhao, X.; Zhang, X.; Tang, H. A heterogeneous Co_3O_4-Bi_2O_3 composite catalyst for oxidative degradation of organic pollutants in the presence of peroxymonosulfate. *Catal. Sci. Technol.* **2012**, *2*, 1977–1984. [CrossRef]
30. Saeed, M.; Alwadai, N.; Ben Farhat, L.; Baig, A.; Nabgan, W.; Iqbal, M. Co_3O_4-Bi_2O_3 heterojunction: An effective photocatalyst for photodegradation of rhodamine B dye. *Arab. J. Chem.* **2022**, *15*, 103732. [CrossRef]
31. Hammad, A.H.; Marzouk, M.A.; ElBatal, H.A. The Effects of Bi_2O_3 on Optical, FTIR and Thermal Properties of SrO-B_2O_3 Glasses. *Silicon* **2016**, *8*, 123–131. [CrossRef]
32. Li, R.; Zhou, D.; Luo, J.; Xu, W.; Li, J.; Li, S.; Cheng, P.; Yuan, D. The urchin-like sphere arrays Co_3O_4 as a bifunctional catalyst for hydrogen evolution reaction and oxygen evolution reaction. *J. Power Sources* **2017**, *341*, 250–256. [CrossRef]
33. Tang, X.; Feng, Q.; Huang, J.; Liu, K.; Luo, X.; Peng, Q. Carbon-coated cobalt oxide porous spheres with improved kinetics and good structural stability for long-life lithium-ion batteries. *J. Colloid Interface Sci.* **2018**, *510*, 368–375. [CrossRef]
34. Khan, I.; Luo, M.; Guo, L.; Khan, S.; Wang, C.; Khan, A.; Saeed, M.; Zaman, S.; Qi, K.; Liu, Q.L. Enhanced visible-light photoactivities of porous $LaFeO_3$ by synchronously doping Ni^{2+} and coupling TS-1 for CO_2 reduction and 2,4,6-trinitrophenol degradation. *Catal. Sci. Technol.* **2021**, *11*, 6793–6803. [CrossRef]
35. Khan, I.; Luo, M.; Khan Sohail Asghar, H.; Saeed, M.; Khan Shoaib Khan, A.; Humayun, M.; Guo, L.; Shi, B. Green synthesis of SrO bridged $LaFeO_3$/g-C_3N_4 nanocomposites for CO2 conversion and bisphenol A degradation with new insights into mechanism. *Environ. Res.* **2022**, *207*, 112650. [CrossRef]
36. Saeed, M.; Khan, I.; Adeel, M.; Akram, N.; Muneer, M. Synthesis of a CoO-ZnO photocatalyst for enhanced visible-light assisted photodegradation of methylene blue. *New J. Chem.* **2022**, *46*, 2224–2231. [CrossRef]
37. Wan, Q.; Liu, R.X.; Zhang, Z.; Wu, X.D.; Hou, Z.W.; Wang, L. Recent Advances in the Electrochemical Defluorinative Transformations of C—F Bonds. *Chin. J. Chem.* **2024**. [CrossRef]
38. Yang, L.; Li, Z.; Wang, X.; Li, L.; Chen, Z. Facile electrospinning synthesis of S-scheme heterojunction $CoTiO_3$/g-C_3N_4 nanofiber with enhanced visible light photocatalytic activity. *Chin. J. Catal.* **2024**, *59*, 237–249. [CrossRef]
39. Zheng, Y.; Liu, Y.; Guo, X.; Chen, Z.; Zhang, W.; Wang, Y.; Tang, X.; Zhang, Y.; Zhao, Y. Sulfur-doped g-C3N4/rGO porous nanosheets for highly efficient photocatalytic degradation of refractory contaminants. *J. Mater. Sci. Technol.* **2020**, *41*, 117–126. [CrossRef]
40. Pradhan, D.; Nayak, N.; Kanar, M.; Kumar Dash, S. Effortless Fabrication of Bi_2O_3-Co_3O_4 Nanocomposite Catalyst: Harnessing Photocatalytic Power for Efficient Cationic Dye Degradation. *ChemistrySelect* **2024**, *9*, e202400217. [CrossRef]
41. Liu, X.; Yang, Z.; Zhang, L. In-situ fabrication of 3D hierarchical flower-like β-Bi_2O_3@CoO Z-scheme heterojunction for visible-driven simultaneous degradation of multi-pollutants. *J. Hazard. Mater.* **2021**, *403*, 123566. [CrossRef]
42. Ma, H.; Yang, X.; Tang, X.; Cao, X.; Dai, R. Self-assembled Co-doped β-Bi_2O_3 flower-like structure for enhanced photocatalytic antibacterial effect under visible light. *Appl. Surf. Sci.* **2022**, *572*, 151348. [CrossRef]
43. Xu, X.; Feng, X.; Wang, W.; Song, K.; Ma, D.; Zhou, Y.; Shi, J.W. Construction of II-type and Z-scheme binding structure in P-doped graphitic carbon nitride loaded with ZnO and ZnTCPP boosting photocatalytic hydrogen evolution. *J. Colloid Interface Sci.* **2023**, *651*, 669–677. [CrossRef]

44. Majeed Khan, M.A.; Siwach, R.; Kumar, S.; Ahmed, M.; Ahmed, J. Investigations on microstructure, optical, magnetic, photocatalytic, and dielectric behaviours of pure and Co-doped ZnO NPs. *J. Mater. Sci. Mater. Electron.* **2020**, *31*, 6360–6371. [CrossRef]
45. Zhang, D.; Su, C.; Yao, S.; Li, H.; Pu, X.; Geng, Y. Facile in situ chemical transformation synthesis, boosted charge separation, and increased photocatalytic activity of $BiPO_4$/BiOCl p-n heterojunction photocatalysts under simulated sunlight irradiation. *J. Phys. Chem. Solids* **2020**, *147*, 109630. [CrossRef]
46. Tenri Ola, A.T.; Rahmat, R.; Fahri, A.N.; Heryanto, H.; Mutmainna, I.; Sesa, E.; Tahir, D. Synergistic Effect of Chitosan and Activated Carbon (AC) in Suppressing Recombination Charge of Composite $Ca_2Fe_2O_5$−AC/Chitosan for High Photodegradation of Fipronil Wastewater. *J. Polym. Environ.* **2022**, *30*, 3218–3229. [CrossRef]
47. Hirashima, S.; Amimoto, T.; Iwamoto, Y.; Takeda, K. Photodegradation of the insecticide fipronil in aquatic environments: Photo-dechlorination processes and products. *Environ. Sci. Pollut. Res.* **2023**, *30*, 89877–89888. [CrossRef]
48. Mianjy, J.G.; Niknafs, B.N. Photodegradation of fipronil in natural water by high intensity UV light under laboratory conditions. *Asian J. Chem.* **2013**, *25*, 2284–2288. [CrossRef]
49. Roulová, N.; Hrdá, K.; Kašpar, M.; Peroutková, P.; Josefová, D.; Palarčík, J. Removal of Chloroacetanilide Herbicides from Water Using Heterogeneous Photocatalysis with TiO2/UV-A. *Catalysts* **2022**, *12*, 597. [CrossRef]
50. Wang, Y.; Lin, C.; Liu, X.; Ren, W.; Huang, X.; He, M.; Ouyang, W. Efficient removal of acetochlor pesticide from water using magnetic activated carbon: Adsorption performance, mechanism, and regeneration exploration. *Sci. Total. Environ.* **2021**, *778*, 146353. [CrossRef]
51. Yi-Zhu, P.; Wan-Hong, M.; Man-Ke, J.; Xiao-Rong, Z.; Johnson, D.M.; Ying-Ping, H. Comparing the degradation of acetochlor to RhB using BiOBr under visible light: A significantly different rate-catalyst dose relationship. *Appl. Catal. B Environ.* **2016**, *181*, 517–523. [CrossRef]

catalysts

Review

A Comprehensive Review of Fine Chemical Production Using Metal-Modified and Acidic Microporous and Mesoporous Catalytic Materials

Joseph Lantos [1], Narendra Kumar [2,*] and Basudeb Saha [1,*]

[1] School of Engineering, Lancaster University, Lancaster LA1 4YW, UK; j.lantos@lancaster.ac.uk
[2] Laboratory of Industrial Chemistry and Reaction Engineering, Faculty of Science and Engineering, Johan Gadolin Process Chemistry Center, Åbo Akademi University, 20500 Turku, Finland
* Correspondence: narendra.kumar@abo.fi (N.K.); b.saha@lancaster.ac.uk (B.S.)

Abstract: Fine chemicals are produced in small annual volume batch processes (often <10,000 tonnes per year), with a high associated price (usually >USD 10/kg). As a result of their usage in the production of speciality chemicals, in areas including agrochemicals, fragrances, and pharmaceuticals, the need for them will remain high for the foreseeable future. This review article assesses current methods used to produce fine chemicals with heterogeneous catalysts, including both well-established and newer experimental methods. A wide range of methods, utilising microporous and mesoporous catalysts, has been explored, including their preparation and modification before use in industry. Their potential drawbacks and benefits have been analysed, with their feasibility compared to newer, recently emerging catalysts. The field of heterogeneous catalysis for fine chemical production is a dynamic and ever-changing area of research. This deeper insight into catalytic behaviour and material properties will produce more efficient, selective, and sustainable processes in the fine chemical industry. The findings from this article will provide an excellent foundation for further exploration and a critical review in the field of fine chemical production using micro- and mesoporous heterogeneous catalysts.

Keywords: fine chemicals; speciality chemicals; heterogeneous catalysis; microporous and mesoporous catalysts; metal organic frameworks (MOFs); zeolites; ion exchange; agrochemicals; fragrances; pharmaceuticals

Citation: Lantos, J.; Kumar, N.; Saha, B. A Comprehensive Review of Fine Chemical Production Using Metal-Modified and Acidic Microporous and Mesoporous Catalytic Materials. *Catalysts* **2024**, *14*, 317. https://doi.org/10.3390/catal14050317

Academic Editors: Antonio Eduardo Palomares and Gang Li

Received: 26 February 2024
Revised: 19 April 2024
Accepted: 7 May 2024
Published: 10 May 2024

1. Introduction

Fine chemicals are described as pure and complex substances, produced mainly for use in further processing to manufacture speciality, high-value chemicals. They are produced in small annual volume batch processes (often <10,000 tonnes per year) [1], with a high associated price (usually >USD 10/kg) [1]. As a result of their usage in the production of speciality chemicals, in areas including agrochemicals, fragrances, and pharmaceuticals, their necessity will remain high for the foreseeable future. As a result, the way in which fine chemicals are produced is of the utmost importance. At this moment in time, there is a scarcity of literature regarding the production of fine chemicals through the use of micro and mesoporous catalysts. Particularly in recent years, the focus has shifted away from reviewing existing catalytic methods, especially in the fine chemical industry.

The production of industrially significant complex molecules, such as fine and speciality chemicals, heavily relies on a range of organic synthetic techniques which employ several reagents and catalysts. As a result of this, significant advancements in the field of catalysis over the past two decades have been instrumental in improving precise and selective control of various reaction paths and products [2,3]. However, the main issue with soluble (typically precious metal) catalysts is often contamination of the reaction products, aside from their expense and toxicity. Additionally, these reaction promotors frequently experience moderate selectivity (towards the activation of the desired functional group), deactivation/degradation (reducing the number of turnovers or productivity), and complex product purification (from the high-added-value organic compounds) [4]. This emphasises the requirement for sophisticated catalysts to be used in various stages of organic synthesis that have the right design, reactivity, and long-term stability.

In an ideal scenario, heterogeneous catalysis provides a clearer means of recovering the reaction product (from the solid catalyst) in a liquid medium. This product can then be recovered using filtration or centrifugation, thereby reducing the likelihood of contaminating the intended synthetic product [5,6]. Furthermore, under more accommodating reaction conditions and longer reaction times, such a reactive solid could be used. It would work with a continuous flow of reactants to separate the product without requiring the reactor to be evacuated, and it would provide easy scale-up, step-economy, high yields, safety, and reproducibility [7,8]. Ordered porous solids are the most desirable among the various solids used as heterogeneous catalysts (such as polymers, crystalline metal oxides, amorphous carbon, etc.) in terms of the fundamental comprehension and control of the chemical transformation in pores (which have the same size, shape, and functionality throughout the crystal). This points towards a clear need to review existing catalytic porous solids, as well as experimentally promising materials in development.

The purpose of this review is to assess the current methods used to produce fine chemicals with heterogeneous catalysts, including both well-established methods, as well as newer experimental methods. A wide range of methods, with the utilisation of both microporous and mesoporous catalysts, will be explored, including their preparation and modification before use in industry. Their potential drawbacks, as well as benefits, will be analysed, with their feasibility compared to that of newer, recently emerging catalysts. The findings of this literature review will form a basis for a critical review of the same topic, providing a platform to further critique this field in greater detail.

The motivation for conducting this critical review on both microporous and mesoporous heterogeneous catalysts in the fine chemical industry is primarily driven by the lack of recent, comprehensive literature encapsulating this rapidly advancing field. As it stands, there is a significant amount of emerging literature surrounding individual catalysts and their applications; however, to the best of our knowledge, there are few to no concise and updated overall reviews. This paper aims to address this void by providing an updated and detailed examination of current advancements, methodologies, and applications of heterogeneous catalyst structures in the fine chemical industry.

2. Relevant Sections

2.1. What Are Fine Chemicals?

Fine chemicals are both complex and single pure chemical substances produced in relatively low annual volumes globally. These volumes are approximately <1000 MT per year, with a high price associated with them, (>USD 10 per kg) [1].

Chemicals that fall under the heading of "fine chemicals" can be found in a range of sectors, including the pharmaceutical, agrochemical, and life sciences sectors [1]. As stated by Ref. [9], fine chemicals are usually produced in batch processes, as opposed to bulk chemicals and commodities, which are produced in continuous processes. This is one of the reasons behind the smaller and more tightly controlled volume of these fine chemicals, increasing the value and thus, the market price, of these fine chemicals.

Additionally, fine chemicals are often required for very specific and complex processes. The fine chemical industry falls between the other two main sectors of chemicals: commodities and speciality chemicals, with each being the supplier and consumer of fine chemicals, respectively. As a result, fine chemicals possess uses in high-value production, as well as provide added value to cheaper chemical commodities. In 2021, fine and speciality chemicals represented approximately 17% of worldwide chemical exports, with an associated value of EUR 379 billion [10]. This is because often, there are further processing stages required for fine chemicals after their production, which leads to the final product being even greater in value. As stated by Ref. [11], fine chemicals are often produced to very exact specifications, defining what they are, as opposed to what they can do, as in the example of speciality chemicals. This is due to the fact that fine chemicals require further processing down the line for the production of other higher-value products like those speciality chemicals mentioned above.

According to Ref. [11], fine chemicals can be split into three major categories, "biocides, active pharmaceutical ingredients and speciality chemicals". These statements are supported by Ref. [12], in which fine chemicals are referred to in a similar manner to the description above as single pure substances based on exacting specifications for further processing. An additional explanation for their high value and small production volumes would be the strict regulations to which fine chemicals are often subjected since the majority end up facing human use or consumption.

Areas in which fine chemicals are often used, once being further developed into speciality chemicals, include adhesives, agrochemicals, biocides, catalysts, dyestuffs and pigments, enzymes, electronic chemicals, flavours and fragrances, food and feed additives, pharmaceuticals and speciality polymers.

2.2. *The Use of Heterogeneous Catalysts in the Production of Fine Chemicals*

Why Are Catalysts Used?

Catalysis, in any form, is essential for chemical reactions. Even in those few situations where reaction speed is already sufficient, reaction economy, yield, and selectivity can be improved by the utilisation of a catalyst. An ideal catalyst is one with both a high turnover frequency (TOF), as well as an infinite amount of product produced (TON) under room temperature and atmospheric pressure. This, however, is often not feasible in practice. There are high associated costs with maintaining a reactor at room temperature, particularly during highly exothermic reactions. As a result, a compromise between the operating conditions and the desired TOF/TON is often made. Alongside this, the reaction should feature no catalyst deactivation or poisoning.

2.3. *Catalyst Selection*

The catalysts used in the production of fine chemicals vary drastically, being heavily dependent on the reaction and the chemicals being produced. Moreover, the choice of which catalyst to use is reliant on a range of other factors. Several catalysts could be suitable for a reaction; however, others may be more optimal in terms of cost, yield, time taken for the reaction, selectivity, ability to resist catalyst poisoning or a form of deactivation, with minimal product damage or contamination. These factors are essential for catalyst choice in terms of its economy and lifespan consideration, as explained by Ref. [13]. Additionally, it is suggested that catalysts also be considered from a practical standpoint. This reference states that in an optimal process, a catalyst should be "wide in scope, easy to perform and insensitive to oxygen and water". It can be inferred from this that choosing the optimal catalyst is not easy. Moreover, the impact of the catalyst choice on a reaction illustrates the importance of selecting the right substance, from both an environmental and economic perspective.

Catalysts can be separated into two distinct categories, heterogeneous and homogenous. Heterogenous catalysis involves a reaction whereby the catalyst in use is in a different phase than the reactants. Often, this occurs when the catalyst is in a solid form, with the reactants being a liquid or gas. Conversely, homogenous catalysts are in the same phase as the reactants in question. Heterogenous catalysis is often favoured and used more often in the production of fine chemicals, mostly due to its ease of recovery in comparison to most homogenous catalysts.

The porosity of a catalyst is usually a concern regarding heterogeneous catalysts, unlike with homogenous catalysts. This is because fluids are unable to possess pores, whereas solids can. Since heterogeneous catalysis usually concerns a solid catalyst, it is evident why this is applicable here. In terms of porous catalytic structures, there are three major categories: microporous, mesoporous, and macroporous. Microporous catalysts feature much smaller pores than those found in mesoporous catalysts, with micropores considered to be 2 nm or smaller in diameter, whereas mesoporous materials are generally categorised as having pores with diameters between 2–50 nm, as evidenced in Ref. [14]. Often these porous materials are constructed from tetrahedral units, whereby they are commonly used in environmental remediation, as well as in treatment, purification, and separation.

2.4. Mesoporous Materials and Their Uses as Catalysts in the Production of Fine Chemicals

2.4.1. Background

Mesoporous materials have a porous structure commonly used in the field of fine chemical development. They are primarily constructed from a silica-based matrix and feature ordered, homogenous pores with a diameter range between 2–50 nm [14,15]. As a result of their extensive surface area and pore volume, accessibility to active sites within the structure is relatively uncomplicated. Moreover, any entering reactive species can undergo significantly more rapid diffusion into the structure [16].

One major advantage of mesoporous materials would be both their chemical and physical properties, which feature exceptionally high surface areas, large pore volumes, and notably, the ability to present adaptive pore sizes and shapes [17]. Having this property allows for the synthesis of catalysts that can be developed and tailored to specific reactions, improving both their overall efficiency and selectivity. In addition, their porous structure can demonstrate nanoscale effects within their internal mesochannels, which can significantly influence catalytic activity [17].

2.4.2. Types of Mesoporous Catalysts

Organo-Silica Based

Silica-based heterogeneous catalysts are typically mesoporous, with an amorphous pore structure. Their larger pore size often results in the easier embedding of functional groups and guest species, providing a high level of customisability in terms of their catalytic performance. Additionally, the use of (organo) silica-based catalysts offers a substantial reduction in the requirement of a solvent during its use in reactions, as explored during the analysis of the asymmetric aldol reaction [18]. Moreover, this study determined that the silica-based catalyst structure used can be recovered and reused several times with reactivation, providing a minimal loss of activity regarding traditionally used fresh organocatalysts. It further goes on to explain that the use of silicas as a base structure for the addition of an acid or base not only proves more affordable but shows more versatility than other solid-supported materials. This has been further supported in other literature exploring the effects of using a silica-based structure during the synthesis of fine organic chemicals [19,20].

There are two methods commonly used to bind a form of organic catalyst to a silica-based structure. The first method, "post-synthetic grafting", consists of a set of linear steps used to build up the catalyst on the silica surface [21–24]. The use of grafting as a covalent technique has been found to provide higher stability, as well as reduced leaching, under most mild conditions [25], both of which are ultimately desirable features during the selection of a catalyst. The second method, known as the "co-condensation method", involves the initial synthesis of any necessary precursors. Following this, the precursors can then be incorporated into the silica support structure [26–28]. According to Ref. [25], by using precursors, various types of active sites can be installed onto the structure in specific confined places. This is further supported and evidenced by research primarily focusing on organosilica structures and their contribution to the manufacture of value-added and fine chemicals [29–31]. The benefits of this method include easier modification of the catalyst, allowing its properties to be tailored to the desired outcome of the reaction. Such freedom to customise and alter reaction products is highly promising due to its usefulness in the fine chemicals industry. It is important to note that the development of silica-based catalyst structures can have a high associated production cost due to the often "costly surfactant structure directing agents", as stated by Ref. [25]. In many cases, high production costs pose a roadblock to more elaborate and in-depth research, resulting in discouragement of their use, particularly when more cost-efficient catalyst structures exist. It is also possible to develop a hybrid–synthetic approach, using both methods, as described by Ref. [18]. This paper determined the best catalyst for the use of a silica-based structure under solvent-free conditions, with the addition of an acid, a commonly used additive during the synthesis of catalysts.

The reasoning behind utilising an acid during the synthesis of a heterogeneous catalyst is complex, with its main advantages including easier recovery of the catalyst in comparison to that required when using a base, as supported by Ref. [32]. As a result of their editable pore dimensions, the use of silica structures enables catalytic reactions involving both bulky substrates and products. This widens the potential use for them across a greater range of reactions, providing an advantage over other mesoporous materials. However, in opposition to zeolites and MOFs, mesoporous silica consists of amorphous walls, which hinders their ability and thus, the activity of embedded active sites.

As a result, further research into creating crystalline materials with ordered mesopores and inherent microporosity is required, with the end goal of expanding catalytic design possibilities and generating a solution or potential use for amorphous walls in porous solids

Types of (Organo) Silica Structures

There is a wide range of existing mesoporous silica structures; however, not all provide utility in the context of fine chemical production. Table 1 illustrates several mesoporous silica nanoparticles (MSN) commonly used as fine chemical catalyst structures, along with their characterisation, properties, and an example of their use in the fine chemical industry.

Table 1. Some silica structures commonly used in the fine chemical industry.

MSN Family Group	MSN Structure Type	Pore Symmetry	Pore Volume (cm^3g^{-1})	Pore Size (nm)	References	Example of Use in Fine Chemical Production
M41S	MCM-41	2D hexagonal P6mm	>1	1.5–8	[33,34]	Continuous one-pot synthesis of citronellal to menthol [35]
	MCM-48	3D cubic Ia3d	>1	2–5	[33,34]	Oxidation of benzyl alcohol to benzaldehyde [36]
Santa Barbara Amorphous (SBA)	SBA-1	3D cubic Pm3n	-	1.5–3	[37,38]	Oxidation of veratryl alcohol to veratryldehyde [39]
	SBA-3	2D hexagonal p6mm	-	1.5–3.5	[40]	Hydrogenation of toluene to methylcyclohexane [41]
	SBA-12	3D hexagonal P63/mmc	0.83	3.1	[42–44]	Tandem reaction of aliphatic primary amine to amides [45]
	SBA-15	2D hexagonal p6mm	1.17	6–10	[34,46]	Hydrogenation of *p*-nitrophenol to *p*-aminophenol [47]
	SBA-16	3D cage-like cubic Im3m	0.91	5–15	[34]	Alcoholysis of 2-furan methanol into n-Butyl Levulinate [48]
KIT	KIT-5	3D cubic Fm3m	0.45	9.3	[49,50]	Acylation of 2-methoxynaphthalene to 2-acetyl-6-methoxynaphthalene [51]
	KIT-6	3D cubic Ia3d	-	4–12	[40,52]	The condensation of benzaldehyde and 1-heptanal to Amylcinnamaldehyde [53]
MSU	MSU.H	2D hexagonal p6mm	-	7.5–12	[54,55]	Selective hydrogenation of cinnamaldehyde to cinnamyl alcohol [56]

- MCM-41 (M41S) [57]

Since their discovery in 1992, mesoporous molecular sieves in the M41S family have become increasingly researched across several industries. This is mostly due to their large surface area, neatly arranged pore structure, and uniform pore size [58], all of which are appealing features from a catalytic perspective. The M41S family can be classified into three categories: hexagonal MCM-41 (a commonly used catalyst in the fine chemicals industry), with its nonintersecting channels in a honeycomb pattern; cubic MCM-48, known for its intricate three-dimensional channel system; and unstable lamellar MCM-50, which tends to collapse when the template structure is removed [59,60]. MCM-41 has gained recent industrial attention in fine chemical synthesis as a result of its advantageous properties over other silica structures. These include its particularly high surface area and thermal stability, as well as the fact that it inherently possesses mild acidic properties [61]. It is often used in reactions involving organic transformations, including acid/base catalysis and oxidative coupling. An example of MCM-41 being used in the production of a fine chemical would be its use during the manufacture of menthol, as investigated in Ref. [35]. The paper explores the continuous one-pot synthesis of menthol from citronellal, using MCM-41 as an alternative intermediary set of steps to the traditional synthetic route opted for in the industry, the Takasago five-step process. The benefits of using the Ru-modified MCM-41 structure over other multi-step traditional methods include an easier separation of the catalyst, as well as the re-use of the catalyst. Moreover, the inclination towards easier catalyst and product separation would indicate lower costs, as well as improved environmental friendliness, something the Takasago five-step process cannot easily deliver [62,63].

Similarly, the use of MCM-41, with a bi-functional powder catalyst, in batch processes (also to produce menthol) has been explored, with similar results. These studies have indicated the significant impact that an active metal modification (particularly on an aluminosilicate structure) can have on a reaction, with the various side reactions being significantly affected, depending on the metal used [64,65]. In addition to this, more reaction-specific catalysts can be derived from MCM-41 to utilise its specific properties. The use of such catalysts often results in an increased cost due to the greater complexity of catalyst production; however, the improved yield and selectivity often justify this expense. For example, a mesoporous Ce composite material has been derived from MCM-41 and has been used as a heterogeneous catalyst in the synthesis of monoterpenoid dioxinols. This study highlighted the advantages the MCM-41 structure can provide, with it producing the highest reaction selectivity in comparison to other catalyst structures (various zeolites and metal oxides) used in the procedure [66]. Potential reasons for this high selectivity include its higher nominal metal content (indicating catalytic activity from the Ce), the presence of both mild acidic and basic sites acting to minimise side reactions, and its proven stability to promote dioxinol over the backward reaction. The higher activity of the Ce-MCM-41 has been attributed to the mild acidity, high surface area, and the large mesopores of MCM-41 mesoporous materials. The synthesis route for this reaction followed a fairly unorthodox procedure in that the isopulegol underwent Prins cyclisation to first form tetrahydropyran as an intermediate, followed by its ring rearrangement to dioxinol. Minimal experimental literature on the transformation of tetrahydropyran to dioxinol currently exists. Thus, the research undertaken in this paper focuses on exploring this alternative route, which can be achieved through the utilisation of the Ce-composite MCM-41 structure. Compound (**3**) tetrahydropyran is an intermediate, which can be formed due to a rearrangement reaction. The associated reaction scheme is shown in Figure 1.

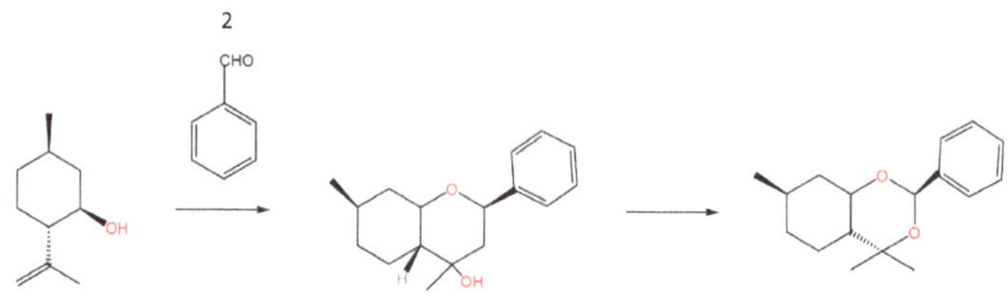

Figure 1. A diagram showing the reaction scheme for the two-step synthesis of monoterpenoid dioxinols (**4**) from isopulegol (**1**) and benzaldehyde (**2**), forming the intermediate tetrahydropyran (**3**), adapted from Ref. [66].

A similar study investigating the synthesis of compounds using tetrahydropyran moiety with different heterogeneous catalysts yielded similar results, with a Ce-MCM-41 structure producing both the highest yield and selectivity [67]. Another prominent modification made to MCM-41 involves the addition of ruthenium (Ru) to the sieve structure. Research in this field has shown that the modification method used can itself influence the resulting structure and behaviour of the catalyst, as explored in Ref. [61], with differing conversion and activity rates experienced when subsequently used in the selective hydrogenation of cinnamaldehyde to cinnamyl alcohol, a fine chemical substance used in the production of perfumes. Additionally, Ru-modified MCM-41 has been used experimentally as a heterogeneous catalyst for the ring opening of decalin, with differing preparation methods resulting in significantly varied results in terms of overall conversion, selectivity, and activity [68].

- SBA-15 [69]

The newly discovered Santa Barbara Amorphous (SBA) family has also gained popularity because of its stronger walls, making it more stable under high-temperature and high-pressure conditions, and its larger pore size in comparison to M41S. SBA-1, which possesses a cubic assembly of rounded micelles, and SBA-15, with its highly organized hexagonal structure and larger pores compared to MCM-41, are the two most common variants in the SBA family. The large pores that SBA-15 possesses are particularly useful in accommodating larger molecules, opening the range of possible reactions for which it can be used. The synthesis of all these mesoporous silicas generally occurs in acidic or basic environments and involves using surfactants or some type of amphiphilic triblock copolymers as structure-directing agents [58].

It is also common to produce hybrid catalytic systems by functionalizing SBA-15 with differing organic or inorganic moieties, such as ferrocene in the hydroxylation of benzene to produce phenol [70]. By doing so, the selectivity of the reaction, along with the catalytic activity, can be improved. Further examples of this include the bi-functionalisation of SBA-15 in preparation for its use as a catalyst during the synthesis of 5-Hydroxymethylfurfural [71], a commonly used starting material for fine chemical production.

Metal–Organic Frameworks

Metal–organic frameworks (MOFs) are a highly studied family of porous materials regularly used in the fine chemical industry. Often used as heterogeneous catalysts, they are a form of hybrid solid, capable of being structured in either a two- or three-dimensional manner. They are effectively formed by the self-construction of cationic systems, fulfilling a node role, with polytopic organic ligands acting as a form of linkers. MOFs themselves can be categorised in several different ways, i.e., based on their synthesis method and structure, or the integration of functional composites or active precursors [72]. Table 2 illustrates these categories, along with structure examples and their uses in the fine chemical industry.

Table 2. Classification for MOFs, along with their relevant properties and uses in the fine chemical industry.

Classification of MOF	Example MOF	Structure	Metal Nodes/Clusters	Ligands	Pore Volume [1] (cm^3 g^{-1})	Brunauer, Emmett, and Teller (BET) Surface Area (m^2 g^{-1})	Example of Use in Fine Chemical Production [2,3]
Isoreticular MOFs	IRMOF-3	Ocothedral crystalline	Zn_4O	2-amino-1,4-benzenedicarboxylate	1.07 [73]	996 [74]	Hydrogenation of levulinic acid to γ-valerolactone (GVL) [75]
Zeolitic Imidazolate Frameworks (ZIFs)	Co-ZIF-9	Zeolite topological structured (SOD)	ZnN_4, CoN_4, or CuN_4 tetrahedral clusters	Imidazolate	0.07–0.18 [76]	1428.37 [77]	Low-temperature liquid phase Knoevenagel reaction [78]
Porous Coordination Networks (PCNs)	PCN-57		$Zr_6O_4(OH)_4(RCO_2)_{12}$	Benzothiadiazole, tetramethyl-triphenylene dicarboxylate (TTDC)	1.36 [79]	3300 [80]	Epoxide Ring-Opening to trans-1,2-diols [81]
Materials Institute Lavoisier (MIL) MOFs	MIL-100(Cr)		Trimeric chromium(III) octahedral clusters	Benzene-1,3,5-tricarboxylate (BTC)	0.793 [82]	1720 [82]	The prins condensation of β-pinene and paraformaldehyde to form [2-(7,7-dimethyl-4-bicyclo[3.1.1]hept-3-enyl)ethanol] (nopol) [83]

[1] It is important to note that pore volume is highly dependent on synthesis and production methods, along with any precursors or post-synthesis composites. [2] The pore volume and BET surface area for each respective MOF structure vary slightly in relation to the provided fine chemical reaction example. This is because their individual properties are altered, depending on any precursor or post-synthesis composite. [3] Some of these reactions listed produced intermediates used in further reactions to produce fine chemicals, as opposed to producing fine chemicals themselves.

The catalytic abilities of MOFs are dependent on the type of MOF in question. There are typically five ways in which their catalytic activity can be employed, as follows:

- Solid acid-base catalysts;
- Inherent metal framework (centres) catalyst sites;
- Metal catalyst confinement (surface anchoring);
- Post-synthesis functional composites;
- Precursors/sacrificial templates.

Solid acid–base MOFs can occur in a number of ways. To begin with, the existing metal nodes within their structure can function as Lewis and Brønsted acid sites [84]. This is due to weakly coordinated moieties which can be removed through thermal activation (leaving behind unsaturated metal sites exhibiting Lewis acid activity) and internal species bonding to metal atoms in the structure, leading to the dissociation of protons (illustrating Brønsted acid activity). Additionally, functional linkers within the structure can act as acid sites. Due to their highly customisable shapes and dimensions, additional acid functional groups can be introduced, offering Brønsted acidity [85]. Finally, catalytically active sites can be added to MOFs through the introduction of an external guest species, using the MOF as a support. This is possible due to the high pore volumes that MOFs can possess, providing a large surface area for the inclusion of an acid guest species [86]. Similarly, guest metal species can be added to either the surface (utilising the MOF as a support) or integrated into the ligand structure.

Alternatively, MOFs can act as precursors for porous carbon catalytic structures. This is often achieved through pyrolysis, whereby the MOF is exposed to high temperatures under an inert atmosphere, producing carbons that can be impregnated with existing metal species within the MOF [87,88].

MOFs have gained popularity recently within the catalyst industry due to their customisable structures, allowing for more control over the catalyst's properties. As stated by Ref. [89], metal–organic frameworks hold several advantages over traditional zeolites. The number of possible zeolites for catalytic use is limited, whereas the number of potential metal–organic framework structures is almost infinite. Moreover, because of their high porosity and surface area, they can retain up to 50–150 wt% of occluded solvent [89], meaning that they can, in some cases, hold ten times as much solvent by weight as a zeolite. Additionally, their high surface area allows for greater exposure of active sites, facilitating improved catalytic efficiency and selectivity. This characteristic is particularly beneficial in fine chemical synthesis, where precise control over the reaction conditions and product quality is essential. Table 3, comparing relevant properties (within the context of fine chemical production) between MOFs and zeolites, is shown below.

Table 3. Comparison of relevant (generic) properties between MOFs and zeolites in the context of fine chemical production.

Properties	MOFs	Zeolites
(Typical) Pore volume (cm^3)	>1	0.1–0.5
Chemical stability	Limited chemical stability, especially towards water, in most cases.	Stable towards solvents, acids, and oxidizing/reducing agents.
Thermal stability	Unstable above 300 °C	Stable above 450 °C
BET surface area ($m^2\ g^{-1}$)	1000–10,000	200–500
Metal-site density	High	Low
Lewis acidity	Accessible framework metal ions	Accessible framework metal ions
Brønsted acidity	Through bridging Si(OH)/Al hydroxyl groups.	Possible through an organic linker.
Basicity	From oxygen atoms within the framework.	Possible through an organic linker.

In terms of their synthesis and development, MOFs can be tailored to incorporate various functional groups and metal centres, creating a platform for the design of catalysts with more specific activities and selectivities. This flexible aspect has widened the possibilities for catalytic development and provided stability in challenging catalytic processes. According to Ref. [90], newer self-assembly synthesis methods for MOFs have been studied in recent years, including hydro- or solvothermal processes, in a bid to develop more effective methods. With the benefits of utilising MOFs as heterogeneous catalysts, it is evident why research efforts have been pursued in this field.

However, the practical implementation of MOFs as heterogeneous catalysts also presents challenges related to their potential structural degradation over prolonged use, leading to a decrease in catalytic activity. Moreover, MOFs can express stability issues in particular situations, such as organic transformation under extreme conditions (acid/basic environments, high temperatures, etc.), as has been explained by Ref. [90]. Additionally, the potential diffusion limitations arising from their intricate pore structures can impact mass transport and reaction kinetics, influencing overall catalytic performance. Diffusion limitation often leads to further catalyst deactivation from pore blockages and poisoning. However, this becomes less of a problem in liquid-phase reactions with MOFs as a heterogeneous catalyst structure. Unlike traditional zeolites, which often experience diffusion limitations in liquid phase reactions and excel in gas phase reactions, MOFs traditionally possess a greater number of available pores and pore sizes, increasing their validity and uses in liquid phase reactions, which are used more regularly in fine chemical production [91].

Additionally, the advantages of MOFs (with regards to other micro/mesoporous materials) remain true for oxidation reactions, with the added benefit that MOFs contain a large amount of transition metals that are considered the conventional type used for oxidation sites. This is a major point of consideration since zeolites and other mesoporous aluminosilicates are null in terms of activity for oxidation reactions.

Some studies have explored the removal of metal from MOFs, leading to the development of carbon-based metal-free catalysts. This is achieved through carbonisation at high temperatures under inert atmospheric conditions, which can lead to an increased surface area and a larger pore volume. There is significant existing literature on such cases, focusing on the removal of metal from MOFs to form a carbon-based structure used for a range of reactions, such as Suzuki–Miyaura coupling reactions, including Refs. [92–96].

An additional field of research with high potential for MOFs includes the development of non-noble metal-based catalysts such as Ni and Fe. This is due to their intrinsic magnetic properties, which can promote easier and more successful recycling of the catalysts, as well as improve their life cycle, with predominant use in the oxidation of alcohols to esters [97–99].

Zeolites

Zeolites are crystalline aluminosilicates that are well known for their widespread uses as catalysts in the manufacturing of fine chemicals. Occurring naturally or resulting from chemical synthesis, the main reason behind their utility is their distinct molecular-scale structure of organised, linked channels and consistent pore size. Zeolites are generally represented by the following empirical formula [100]:

$$M_{2/n}O \cdot Al_2O_3 \cdot xSiO_2 \cdot yH_2O \quad (1)$$

Equation (1)—A general empirical formula representing a zeolite structure.

The x in this formula is typically greater than or equal to 2. This is because the AlO_4 tetrahedra is joined to the SiO_4 molecules within the structure. In this context, n represents the available cation valence (available cation ability to form chemical bonds). It is important to note that although zeolites are typically microporous, the recent emergence and development of mesoporous zeolites have significantly increased their utility, particularly in reactions involving big molecules. They are used in a vast range of reactions, with ion-exchanged zeolites being one of the most frequently used catalysts in history [101]. Depending on the form taken, their uses vary significantly. For example, in their proton exchange form, they fulfil a large role in the oil refining industry because of their strong acidic and shape-selective properties [102,103]. Further potential uses for zeolites include their application in various types of water purification, in large biomolecule separation, and the removal of some radioactive contaminants [104–106]. The use of zeolites for large molecule separation features across a vast range of reactions during fine chemical production, often capitalising upon the nature of their hierarchical structure. They act to increase the effective hydrogen-to-carbon ratio of chemical products through a range of intermediate dehydrogenation/hydrogenation, oligomerisation, and cracking reactions. Subsequent carbon bond-forming reactions, such as aldol condensation and Diels–Alder reactions, can then occur on the Zeolite structure to form larger molecules and aromatics [107–109]. An example of utilising the hierarchical nature of zeolites during processes involving large organic molecules includes such reactions as the pyrolysis of wood polymer to produce aromatics [110]. ZSM-5 is discussed as one of the more widely used zeolite catalysts in aromatics production due to its increased efficiency, with utilisation as fuel additives [111], solvents [112] or types of polymer synthesis [113]. Its microporous structure, however, is noted as a major limiting factor as a result of the reduced diffusivity for larger molecules. This further highlights the advantage of hierarchical zeolite structures, illustrating a strong case for the use of mesoporosity in zeolites.

Since the orientation of individual pores is usually random, and the sizes and shapes of the mesopores have no impact on the zeolite's crystal structure, the mesopore system in mesoporous zeolites can be considered a non-crystallographic pore system. Since this pore system is not atomically ordered, mesoporous zeolites are regarded as hierarchical porous materials, i.e., they possess more than two pore size distributions [114,115]. Hierarchical zeolite materials can be separated into three categories: hierarchical zeolite crystals, nanosized zeolite crystals, and supported zeolite crystals, with each type varying in regard to pore size and structure. Hierarchical zeolite crystals exhibit extra pores, either mesopores (under 50 nm) or macropores (over 50 nm), within each crystal and include an additional mesopore system alongside the standard micropores. Nanosized zeolite crystals are smaller, and their mesoporous system comes from their packing structure. Supported zeolite crystals are essentially dispersed in another material's pore system, leading to a mix of micro- and mesopores, (depending on the support structure). To alter a zeolite structure into a mesoporous form, the zeolite crystals themselves must be altered, with either a top-down or bottom-up approach [116]. The top-down method, known as post-treatment, involves the use of an acid or base component to remove any Si or Al species from the template. This method, known as demetalation, is commonly achieved via the use of dealumination and desilication, both of which have been studied in Refs. [117–119]. Conversely, the bottom-up approach, known as direct templating, entails the direct synthesis of mesoporous zeolites in the presence of mesoscale organic porogens and organic directing agents. Various bottom-up methods exist, with one common approach being the use of some type of solid templating. For example, templating can be completed through a range of carbon nanomaterials [120–123] to produce a variety of common zeolites such as ZSM-5, zeolite-β, zeolite-X, zeolite-A, and zeolite-Y. Carbon-based templates produced through carbonisation can also be built upon to produce different types of zeolites [124], as can the use of aerogel [125,126], polymer [127], resin [128,129], and solid biological templates [130].

Other templating approaches for producing hierarchical zeolites include delamination, the process of synthesizing layered precursors as lamellar precursors with an additionally intercalated surfactant. This surfactant can then be removed, leading to the collapse of the structure and the formation of an accessible zeolite material that exhibits mesoporosity [131,132].

Whilst the development of zeolites possessing mesoporosity exhibits advantages, the negative environmental drawbacks that their synthesis produces remain a sizeable issue. Significant emissions are produced during their manufacture due to the application of different multifunctional templates, their subsequent removal, or the release of acids and alkalis used in zeolite dealumination and desiccation [133]. Furthermore, even the processes currently in use to create microporous zeolites are not environmentally friendly because they all require the use of artificial chemicals containing silicon and aluminium, which are derived from natural silicate or aluminosilicate minerals. These derivations occur through labour-intensive procedures that result in significant waste production and energy consumption [134].

Types of Zeolites

The classification of a zeolite is governed by the silica:aluminium ratio within the structure [135]. For example, a zeolite labelled as high-silica would have a large Si:Al ratio, whereas low-silica or Al-rich would indicate that the structure has a low Si:Al ratio. In terms of the ratio number itself, its classification is dependent on the type of zeolite. For BEA zeolites, low-silica refers to a ratio less than 5, whereas high-silica refers to a ratio greater than 10 [136,137]. For X and Y zeolites, high silica indicates a ratio greater than 3, while low silica would correspond to ratios between 3 and 1.5 [138–140]. For ZSM-5 zeolites, high-silica corresponds to a ratio above 20, whilst low-silica indicates a ratio below 15 [136,141,142]. Table 4 shows several zeolites commonly used in the fine chemical industry, including their classification, relevant properties, and examples of their uses as catalytic materials in fine chemical production.

Table 4. Zeolite structures commonly utilised as heterogeneous catalysts in the fine chemical industry, along with their relevant classifications, properties, and examples of their uses.

Framework Type	Zeolite Name	Symmetry	Type of Channels (Dimensionality	Void Fraction [1]	Framework Density [2] (g/cc)	Framework Density (T/1000 Å^3)	Order	Common Reactions Used in the Fine Chemical Industry
Beta (BEA)	Zeolite-Beta	Tetraganol	3D	-	-	15.1	Partially disordered	Ortho-selective Phenol Hydroxylation forming catechol [143]
Faujasite (FAU)	Zeolite-x	Cubic	3D	0.5	1.31	13.3	Fully ordered	Knoevenagel condensation of benzaldehyde and ethyl cyanoacetate [144]
FAU	Zeolite-Y	Cubic	3D	0.48	1.25–1.29	13.3	Fully ordered	Acetalization of ethyl acetoacetate with ethylene glycol to 1,3-Dioxolane-2-acetic acid, 2-methyl-, ethyl ester (fructone) [145]
MFI	ZSM-5	Orthorhombic	3D	-	-	18.4	Fully ordered	Benzene methylation to xylene and toluene [146]
Mordenite (MOR)	Mordenite	Orthorhombic	2D	0.28	1.7	17	Fully ordered	Para-selective tert-Butylation of toluene to 4-tert-butyltoluene [147]
FER	Ferrierite	Orthorhombic	2D	0.28	1.76	17.6	Fully ordered	Isomerization of n-Butene to Isobutene (fine chemical intermediate) [148]
LTA	Zeolite-A	Cubic	3D	0.47	1.27	14.2	Fully ordered	Knoevenagel condensation of benzaldehyde and ethyl cyanoacetate [149]

[1] As stated in Ref. [100], the void fraction has been determined from the water content of the hydrated zeolite structure. [2] As stated in Ref. [100], the framework density is based on the dimensions of the unit cell of the hydrated zeolite and the framework contents only.

ZIFs

Zeolitic imidazolate frameworks (ZIFs), a popular porous hybrid structure developed in recent years, have become a focus in the heterogeneous catalytic industry for several reasons. Possessing a crystalline structure with the ability to express hierarchical porosity, they are a subclass of MOFs, combining some of the most desirable properties of both MOFs and zeolites. They differ in structure from traditional MOFs, since they are composed primarily of Zi (ii), Co (ii), and imidazolate linkers, as opposed to the wide range of metals and organic linkers of which MOFs can be composed. When compared to traditional MOFs, ZIFs typically exhibit much greater stability in terms of thermal, hydrothermal, and chemical properties [150,151]. This includes their ability to be boiled in various organic and alkaline solutions without the loss of crystallinity or a reduction in porosity [152].

In addition, the tetrahedral crystalline structure of a ZIF is like that of a zeolite, with the aluminium and/or silicon being replaced by either zinc or cobalt (transition metals). This mixed structure is one of the main advantages they hold over zeolites—they can be exposed to a greater range of surface modifications. Along with this, their tunable pores (due to their metal ions) and their high porosity make them extremely suitable as catalyst support structures for a range of reactions [153]. The main routes for the manufacture of ZIFs include the use of solvent-based or solvent-free synthesis [153]. The most commonly used solvents include water, methanol, or ethanol. It is not uncommon, however, to employ other solvents, such as dimethylformamide or diethyl formamide, with the solvent chosen to be dependent on the specific process. From here, the chemical route taken can vary, with a wide range of methods available, subject to the solvent chosen. Older methods include solvothermal synthesis, whereby organic solvents are used in the formation of a ZIF. As discussed above, the preliminary solvents used in experimental work include various alcohols, as explored in Ref. [152]. More recent works include the incorporation of bases to deprotonate specific linkers and provide a higher yield and greater rate of reaction. Such cases in the literature include pyridine [154] and triethylamine [155]. Findings from both of these papers provided a strong argument for the use of a basic solvent in conjunction with ZIF structures, with its main benefit being highlighted as the reusability of the ZIF structure (due to its retention of catalytic activity). However, both studies noted the impact that the reaction solvent can have on its catalytic performance, indicating some drawbacks in terms of the reactions for which they are suitable.

Whilst solvothermal synthesis can offer a wider range of choices regarding the type of ZIF developed and greater flexibility in terms of the solubility of the precursors, the associated drawbacks of organic solvents often reduce their industrial popularity. The negative environmental impacts, in addition to the high costs and in some cases, toxicity to humans, is driving research in ZIF development towards cleaner alternatives. Research shows that the use of aqueous mediums (hydrothermal synthesis) can produce higher yields than organic solvents in reduced timeframes, but they require additional linkers, such as in the case of the development of ZIF-8 nanocrystals [156], or even nano-sized ZIF-67 crystals [157]. Further, more recent research in this field has been conducted in a bid to develop this greener method, with various modifications to the process. These include the use of surfactants to regulate the size of the crystals in a ZIF structure. Successful examples include research conducted whereby the diameter of the crystals produced could be controlled [158], or whereby the impact of the surfactant used on the hierarchical structure of the ZIF was explored [159].

Newer solvent-based methods of ZIF production include microwave and ionothermal synthesis. The former involves the use of microwave-assisted heating technology and has been shown to drastically shorten the synthesis time, as well as to produce a higher yield whilst reducing the number of ligands present, without the use of deprotonating agents [159,160]. The latter, ionothermal synthesis, involves the use of ionic liquids (as solvents) in an open system. There has been minimal research in this field thus far, despite the possibilities it has been shown to possess. These include using the ionic solvent as a template to prevent competitive interaction, the wide range of novel structures that could be developed with further experimental work (as explored by Martins et Al in [161]), and its possibilities to be incorporated with other synthesis methods, such as microwave synthesis [162].

A small number of promising novel solvent-free synthesis routes exist. Despite their lower cost and their significantly eco-friendlier nature, very little research has been conducted in this area. Some studies include the synthesis of ZIFs using a dry-gel conversion, with the synthesised ZIF structures showing promising results. These include excellent reactivity and catalytic activity, as well as strong reusability [152,163].

The use of ZIFs as catalysts in the fine chemical industry is gaining traction, with catalysis roles in various scientific sectors, including pharmaceuticals [164,165], organic chemical production (hydrocarbon separation) [166], aromatics, and biomass conversion [167]. Notably, all the research conducted on ZIF structures as catalysts are remarkably novel, with their use appearing earlier in other fields, such as gas and energy storage [168–170], as well as drug delivery [171,172]. Whilst minimal, the positive findings from this early research are indicative of the potential uses of ZIFs as catalyst structures, and it can be anticipated that further investigation into their catalytic benefits will prove popular and essential as the fine chemical industry continues to evolve and seek cleaner and more efficient production methods.

Carbon-Based

Mesoporous carbon-based heterogeneous catalytic structures possess highly desirable physical and chemical features, particularly regarding the synthesis of products in the fine chemical industry. Notable properties include a high surface area (contributing to higher catalytic activity), significant pore volume, good thermostability due to their composition of carbon atoms linked in a framework, enhanced mass transfer, and relatively uncomplicated diffusion [173]. Due to their excellent conductive abilities, carbon-based catalysts are often used in electrochemical-based reactions [174,175]. The processes for the synthesis of carbon-based catalysts have changed over time due to the development of novel methods, with greater simplicity and fewer environmental drawbacks. Typically, carbon-based structures are microporous, so methods to introduce mesoporosity have been created. Initially, mesopores in carbon structures were created in the spaces between carbon particles, such as in carbon aerogels, or by enlarging micropores via oxidation during the activation process, such as in activated carbons [176]. These methods, however, resulted in difficulties such as the struggle to regulate the structure's shape or a loss in carbon yield.

There are three major categories of methods for the synthesis of carbon-based structures, including:

- Activation methods;
- Catalytic activation methods (using metal ions);
- Template methods.

Activation methods are regarded to be the most widely used techniques [173], with two subcategories: chemical methods and physical methods. In the physical methods, materials are heated to between 400–900 °C in an oxygen-free environment (using gases like nitrogen or helium) to create a carbon-rich "char". This char is then exposed to higher temperatures (800–1000 °C) and oxidising agents (such as steam or carbon dioxide) to develop a mesoporous structure [177,178]. Chemical activation, on the other hand, streamlines this process by integrating the heating and exposure to oxidising chemicals into a single phase. Usually, chemical activation entails impregnating a carbon source with specific compounds and heating it to 400–700 °C [179] to cause it to thermally breakdown. Zinc chloride ($ZnCl_2$), aluminium chloride ($AlCl_3$), magnesium chloride ($MgCl_2$), potassium hydroxide (KOH), sodium carbonate (Na_2CO_3), phosphoric acid (H_3PO_4), and sodium hydroxide (NaOH) are common chemicals utilised in this procedure [179–181]. The choice of chemical is very important because it has a big impact on the activation process's mechanism and necessary temperature.

Chemical activation offers several advantages over physical activation, including lower energy consumption and activation temperatures, higher carbon yields, and faster processing times. However, in some cases, physical activation is preferred due to its relatively smaller impact on the environment, its readily available activating agents, and its simple technology [182–184]. In some cases, combining both types of activation methods produces better activation; thus, it is employed regularly.

Catalytic activation utilising metal ions can also be employed to create mesoporous carbon-based catalysts. By acting as catalysts to activate carbon precursors, metal ions such as iron (Fe), nickel (Ni), and cobalt (Co) can speed up the activation process. They can help provide stability and control the development of the mesopores. These metal ions are particularly efficient at accelerating the char activation process, which leads to a greater volume of mesopores [185]. In some cases, where any water-based solutions are used, the metal ions can leach out of the structure, reducing its lifespan and in turn, its activity.

The third method, templating, has two distinct sub-categories; hard and soft templating. Hard templating uses pre-made inorganic or organic templates such as colloidal silica or mesoporous silica. These templates are filled with carbon precursors and then carbonised. The template is then removed by chemical etching, using an acid or base, leaving behind the carbon-based structure [176].

The soft-templating method uses self-assembling molecules, such as metal–organic frameworks and surfactants, as templates. These materials are then bonded (often through hydrogen bonding) with carbon precursors. A calcination process is then used at temperatures up to 900 °C [186,187] to remove the templates, leaving behind the mesoporous structure [188]. One reason behind the development of mesoporous molecular sieves over zeolites is their ability to process heavy oil fractions, something other mesoporous structures cannot deliver [189–191]. Moreover, mesoporous carbon structures can exhibit high selectivity, and research has proved their ability to increase the reaction rate, as evidenced by literature investigating the dehydrogenation of ethanol to acetaldehyde [192]. Their excellent selectivity was determined to be significantly greater than that produced using SBA-15 as a catalyst support. These findings result from the inert nature of the carbon structure, inhibiting any additional secondary or side reactions. Whilst acetaldehyde itself is not strictly a fine chemical, such findings prove particularly beneficial in the fine chemical industry, where high selectivity of desired products is paramount to increase the efficiency of valuable and expensive processes. Additionally, the natural hydrophobic properties of the carbon structure provide an enrichment potential for organic compounds, expanding its uses as a highly selective catalyst support for organic reactions. An additional study utilising mesoporous carbon structures incorporated with the catalyst (as opposed to supporting it, as traditional research investigates) highlighted their reusability in the production of γ-valerolactone [193]. In addition, the study illustrates their stability in an acidic aqueous medium. It is this high stability that proves that the development of mesoporous carbon structures directly incorporating catalysts carries excellent potential, drawing further attention to carbon-based structures and the increased benefits that can be evoked through additional research.

2.5. *Microporous Materials and Their Uses as Catalysts in the Production of Fine Chemicals*

2.5.1. Background

Microporous materials contain crystalline structures with interconnected cages or channels. Their classification as microporous is determined by their smaller pore diameter of less than 2 nm [14].

The two main types of microporous structures used in industry are crystalline zeolites (aluminosilicates) and activated carbons [194]. Both types of structures, along with other less common microporous structures such as metal–organic frameworks (MOFs), can be modified to be mesoporous. Further types of microporous structures such as metal oxides and aluminophosphate will be reviewed below. Typically, microporous materials are synthesised under solvothermal conditions, with the individual parameters being altered, depending on the reaction. The most altered parameters when developing microporous structures include the starting compound, any solvent used, the temperature, the pressure, and any additional charge-compensating ions [195].

2.5.2. Types of Microporous Catalysts

Zeolites

Microporous zeolites offer similar advantages to those of mesoporous/hierarchal zeolites, however, with less methodology and fewer steps in terms of their development. As described previously, zeolites can occur naturally or can be synthetically developed.

Worldwide, both academic and industrial facilities have conducted extensive research on differing types of microporous zeolites. The reasoning behind this includes their desirable abilities and properties, such as consistent channel systems, tunable Brønsted and Lewis acidic sites, coke resistance, ion-exchange, and thermal stability features. The procedure for developing synthetic zeolites varies slightly from the method mentioned above to produce mesoporous zeolites. Since zeolites are generally microporous beforehand, with further processing required to make them mesoporous, there are usually fewer steps involved in this process.

There are a range of different methods available for the synthesis of zeolites, with the most common ones listed below, as taken from Ref. [196]:

- Hydrothermal synthesis;
- Solvothermal synthesis;
- Ionothermal synthesis;
- F-synthesis;
- Microwave-assisted hydrothermal synthesis;
- Microemulsion-based hydrothermal synthesis;
- Dry-gel conversion synthesis;
- Combinational synthesis.

Despite the desired properties of microporous zeolites, there is a current drive for extensive research around modifying and developing them further. The reasoning behind this is to enhance their uses and resolve their inability to catalyse reactions with larger particles. One potential route for doing so would be their modification to a hierarchal zeolite, as described above in the mesoporous zeolite section. There are several alternative routes, however. In the editorial section prepared for microporous zeolites and their applications [197], Kumar discusses the approach of embedding microporous zeolites in amorphous silica, referring to a study conducted on selective catalysts for the methanol to olefin process [198]. Whilst olefins themselves are not strictly fine chemicals, they are often used as building blocks in further reactions to produce various aromatics and detergents. The cause for this methodology lies in the potential for customising and specifically tailoring available acid sites in the catalyst structure. The paper revealed the higher catalytic activity of the synthesised zeolite, as well as the much higher selectivity ratio. These findings support explanations in wider research in the literature proposing that the strength, availability, and type of acid sites available in a catalyst can alter the propylene selectivity and rate of formation of intermediary products.

The impact that modifications of a zeolite structure can have on a reaction is analysed in Ref. [199], with its influence and impacts explored. The paper explains how preparation, precursors, or pre-treatment can impact and alter the physicochemical and catalytic properties of a material. It also elects to take an environmentally friendly approach during the preparation of the catalyst by choosing to leave out aqueous or organic solvents and instead using a solid-state ion exchange method. One notable finding from the paper regarding zeolite performance, however, concerned the deactivation of the zeolite structures with extended reaction times caused by carbonaceous deposits, resulting in pore blockages in the zeolite. These findings suggest that an avenue for future research could incorporate exploration into the reduction or prevention of deactivation to extend the lifespan of zeolite structures and reap their full selectivity benefits.

Similarly, Kumar has assessed the impact of the chosen preparation method on the catalytic behaviour of modified zeolite extrudates [200]. The study revealed that the different methods of platinum deposition onto the structure impact the metal-to-acid site ratio, subsequently altering the acidity and strength of the catalyst structure. It also concluded that the choice of the synthesis method impacts the overall conversion, with zeolite extrudates prepared via in situ synthesis producing the highest conversion. Alternatively, zeolite catalysts can be prepared via the use of various impregnation methods. In Ref. [201], H-Y5.1 zeolite catalyst, prepared via wetness impregnation, was used across a range of reactions in the production of upgraded fuels and aromatics, with the ratio between Fe and Ni in the catalyst being varied. This involved the co-processing of hexadecane with isoeugenol (2-methoxy-4-propenylphenol), with the isoeugenol undergoing hydrodeoxygenation and the hexadecane undergoing hydroisomerisation–hydrocracking. The reaction schemes for this process can be seen in Figures 2 and 3 below:

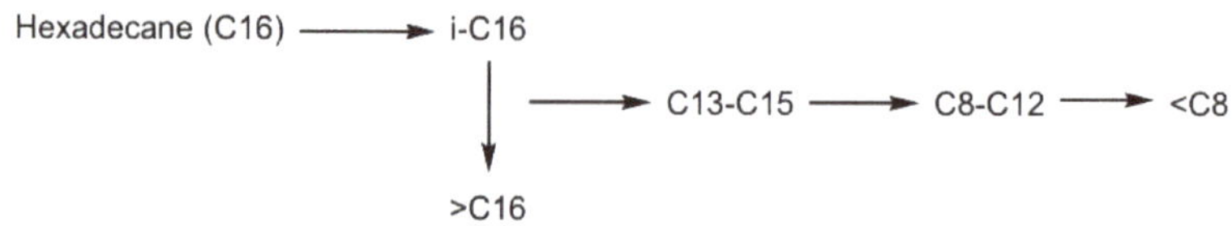

Figure 2. The reaction scheme for n-hexadecane hydroisomerisation–hydrocracking, adapted from Ref. [201].

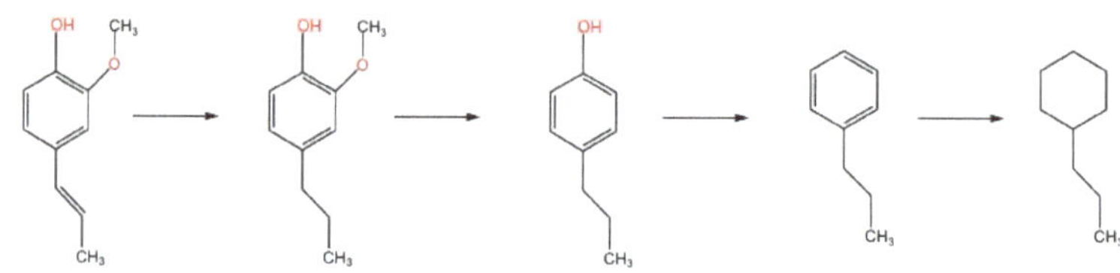

Figure 3. The reaction scheme for the hydrodeoxygenation of isoeugenol, adapted from Ref. [201].

Findings from the study showed a reduction in specific surface area and a decrease in pore volume due to a blockage in the larger pores in the material. The study also showed a change in the acid sites, dependent on the catalyst structure used, with Fe leading to a decrease in strong acid sites and an increase in medium Brønsted sites, whilst Ni increased the weak and medium Lewis sites. This change in acid site presence can be attributed to interactions between the metal ions with the zeolite's acidic surface, leading to coordination with any present oxygen/hydroxyl groups and the formation of metal clusters. This in turn can block or inhibit the active sites. The balance between both Lewis and Brønsted acid sites can heavily impact the selectivity of a reaction. For example, a greater number of Lewis acid sites (because of the nickel impregnation) contributes towards the hydrodeoxygenation (HDO) of isoeugenol. This is a result of their ability to promote the adsorption of oxygenated compounds by forming a complex with present oxygen atoms. This in turn makes the C-O bond more susceptible to cleavage in the HDO process. Additionally, these papers illustrate the effect of the preparation method and support design on a reaction and highlight the drawbacks of zeolites in terms of pore blockage tendencies. A clear link between the acidity and ending selectivity of a reaction can be seen, indicating the importance of correct catalyst preparation. This is especially important in the fine chemical industry, where the economy of reactions and selectivity to desired products can dictate the financial outcome and viability of a particular product. It could be argued that zeolites are the most commonly used porous catalyst in both the fine chemical industry, as well as in wider chemical production. Extensive literature detailing their vast applications exists, with zeolite-β featuring heavily in small-scale fine chemical production. Further examples of the use of zeolite-β include:

- The use of β zeolites in the cyclization of citronellal [202,203];
- The use of H-Beta-25 in the transformation of glucose to methyl levulinate [204];
- The use of Sn-Beta zeolite in the production of methyl lactate from glucose [205];
- The use of H-beta zeolite catalysts in the prins cyclization of (−)-isopulegol [206];
- The use of H and Fe-modified beta zeolites in the production of trans-carveol from α-pinene oxide [207].

Activated Carbons

Carbon-based structures, which are mesoporous, can also possess micropores. Whilst their ability to adsorb large organic materials is reduced in comparison to mesoporous carbon-based structures, they feature an enlarged specific surface area. This aids their adsorption of volatile molecules, as well as their hydrogen storage and CO_2 capture abilities [208,209].

One common approach to developing activated carbon-based catalysts involves the use of developing biomass as a precursor [210–212]. Once the carbon has been obtained, it can be activated using pyrolysis in an inert atmosphere, just as mesoporous carbon is activated. It is then developed through physical or chemical activation to improve its porosity [213]. Several studies have found that the use of biomass as a precursor provides a high oxygen content, leading to greater porosity and surface area [214].

The surface reactions on carbon can be complex due to the interaction of numerous oxygen groups. These can be divided into two main groups, the basic and acidic oxygen groups [215,216]. The surface activity and catalytic behaviour of the structure are usually dependent on these surface groups, with studies conducted regarding the impact these groups have [216,217].

Like mesoporous carbon-based catalysts, they are often modified, or "doped" with additional materials, to alter the available specific surface area and increase micropore development, i.e., through nitrogen doping [218–221]. There are countless other examples of such experimental work, including the modification of carbon nanospheres with phosphorus [222], the synthesis of a dual-modified phosphomolybdic acid/silver carbon composite [223], the production of jet fuel through the hydrodeoxygenation of isoeugenol over a carbon-supported platinum catalyst [224], and many more [225,226]. One common drawback that occurs when a microporous carbon structure is utilised is the frequency of pore blockage. This is a common occurrence in microporous structures due to their smaller pore size, resulting in a subsequent loss of activity and catalytic deactivation. Whilst microporous carbon structures possess several advantages over their mesoporous counterparts, it is clear that the pore blockage tendencies are much higher, illustrating a major hindrance to their wider use.

A study conducted on the kinetics and deactivation of catalysts in the hydrogenation of β-sitosterol to β-sitostanol, utilising both micro and mesoporous carbon supports [227], reiterated catalyst deactivation as a prominent issue when using carbon-based catalyst structures, suggesting the utilisation of a larger amount of the catalyst to combat this problem. The reasoning behind this is to counter the catalysts' adsorbent properties so that the activity can remain higher for prolonged periods. Additionally, the differing conversions obtained between the micro- and mesoporous carbon were analysed, whereby it was found that the mesoporous carbon structure provided higher overall conversions than its counterpart. In addition, deactivation occurred at a slower rate for the mesoporous carbon catalyst than for the microporous example. This difference can be credited to the fact that the mesoporous structure contained larger pores, meaning that the effects of coking are less significant and require a longer time span to build up.

In the fine chemical industry, where activity and selectivity are paramount, the use of microporous carbon catalysts carries several advantages but, in most cases, these are outweighed by its disadvantages. Their large specific surface area is particularly beneficial for the adsorption of smaller molecules, which can prove useful in purification or pollutant particle-capturing processes. Despite these advantages, however, the challenges posed by pore blockage are significant and cannot be overlooked, particularly when blockages can lead to a loss of activity and catalytic deactivation. The comparison with mesoporous carbon structures further underlines this issue, as it becomes clear that these larger pores offer higher overall conversions and slower deactivation rates. Considering these factors, it can be concluded that mesoporous carbon structures are generally more beneficial for applications in the fine chemical industry in processes in which longer catalyst life and higher resistance to catalyst fouling and deactivation are required.

Aluminophosphate

There are several other microporous structures used as heterogeneous catalysts for both the production of fine chemicals and the chemical industry as a whole.

One of these commonly used structures is aluminophosphate. Aluminophosphate is usually expressed in a crystalline structure formed by connecting oxygen atoms between alternating phosphorus tetrahedra and aluminium polyhedra. Aluminophosphates come in a variety of polymorphs that can create pore frameworks with varying dimensions. These crystalline aluminophosphates often feature uniform pores with diameters above 1 nm, placing them in the category of extra-large microporous crystalline materials [228]. The structure of these frameworks provides a number of sites for chemical reactions to occur, which is essential for their activity and effectiveness as acid catalysts [229]. Often, aluminophosphates are doped with specific metals, whereby the metal ions replace the aluminium ions within the framework. This process is frequently used in the manufacture of single-site catalysts [230], allowing for its use in particular reactions, such as in the oxidation of hydrocarbons [231] or the esterification of acetic acid [232]. One notable example of the use of the metal aluminophosphate being used in the fine chemical industry features in a research paper on the production of biphenyl urea [233]. Findings from the paper indicated cobalt-aluminophosphate as the catalyst providing the highest yield of biphenyl urea. The paper also concluded that under mild conditions, metal-aluminophosphate catalysts are a viable alternative to traditionally used environmentally hazardous and costly alternatives.

When compared to their crystalline counterparts, amorphous aluminophosphates are a more affordable and often more thermally stable catalyst [234]. They only require a straightforward co-precipitation step in their manufacture, whereas the synthesis of crystalline aluminophosphates requires the use of organic agents that direct structure, like trialkyl amines or copolymers, which are then burned off after the synthesis is complete.

MOFs

Microporous MOFS, like mesoporous MOFs, consist of self-constructing cationic systems, with polytopic organic ligands acting as linkers. In comparison to mesoporous MOFs, they possess a much larger specific surface area, making them more suitable for interactions requiring high surface contact [235]. Along with this, their smaller pore size becomes useful in situations where their ability to process and discriminate between smaller molecules proves advantageous. A notable feature of microporous MOFs is their ability to adsorb carbon dioxide [236]. An article produced by Pal et al. investigated a co-MOF and assessed its selectivity in terms of CO_2 sorption through physisorption, as opposed to a chemical reaction involving Aryl C-H$\cdots$O=C=O Interactions. The study presents the use of microporous MOFs as an effective carbon capture solution, illustrating the co-MOFs' ability to selectively predominantly adsorb carbon dioxide over other gases in the mixture. This can be explained by the use of grand canonical Monte Carlo (GCMC) computer simulations performed on the Co-MOF structure, which showed how the CO_2 molecules had been electrostatically trapped as a result of interactions between the oxygen atoms in the CO_2 molecules and the hydrogen atoms attached to the pyridine rings in the framework's spacers. These results are supported by additional literature in this field of research, such as in this study similarly investigating the selective adsorption properties of microporous MOFs [237]. These findings prove the value of microporous MOFs within the field of green chemistry, laying a foundation for further investigation into their CO_2 adsorption abilities, given their potential to actively reduce atmospheric greenhouse gas emissions.

In terms of their catalytic applications, particularly in the fine chemical industry, they can be employed in the Hantzsch reaction to produce polyhydroquinolines, whereby their derivates are utilised in the pharmaceutical industry due to their medicinal functions, as discussed in Ref. [238]. The reaction scheme for this process is shown in Figure 4.

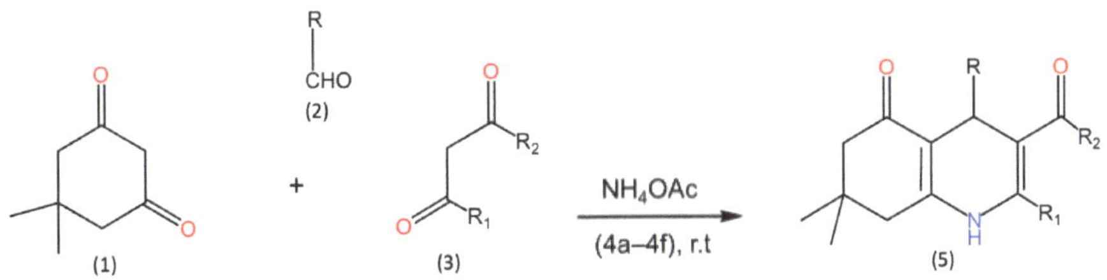

Figure 4. A reaction scheme illustrating the organocatalysed (**4a–4f**) unsymmetric Hantzsch reaction using dimedone (**1**), substituted aldehyde (**2**), and acetoacetate ester or acetylacetone (**3**) to produce polyhydroquinoline derivatives (**5**), adapted from Ref. [238].

In this paper [239], the Zn-MOF conveyed high efficiency as a result of its hierarchical microporous structure, which provided a large surface area for reactions to occur on the structure. Along with this, the paper also discussed how the Zn-MOF catalyst could be reused multiple times, with only a moderate decrease in performance, demonstrating its stability and potential long-term financial viability. Notably, however, the intricacies of the potential deactivation mechanisms (such as coking) were not extensively discussed. The potential for deactivation of the catalyst is a pivotal factor in terms of its suitability over traditional catalysts for such a reaction, where such drawbacks would be crucial on an industrial scale. Similar experimental work conducted on the use of microporous Zn-MOFs as heterogeneous catalysts has been conducted by Roy et al., whereby its impact on various organic transformations has been explored [240]. The Zn-MOF studied in this paper exhibited high catalytic efficiency during reactions such as the cycloaddition of CO_2 with epoxides to form cyclic carbonates (a valuable intermediate in the fine chemical industry). One advantage of using the MOF structure was highlighted as its ability to catalyse reactions under mild conditions. This could prove particularly beneficial for large-scale industrial reactions as the sector shifts towards more sustainable and eco-friendly chemical processes. Moreover, the paper addresses the low price associated with zinc metal, linking this to its large surface area-to-volume ratio and divalent oxidation state. This presents the Zn-MOF catalyst as an overall suitable alternative for several organic processes, including high-temperature reactions and the involvement of hazardous materials.

Metal Oxides

Microporous metal oxide structures feature metal centres of either single ions, clusters, or mixed-metal clusters, linked together through oxygen atoms. These linkages, which contribute towards its stability, can then form a variety of structures such as rings, chains, and layers, which lead to the formation of its porous structure. Similar to other microporous structures, they can be organised in either a crystalline or amorphous structure. Zeolites are a type of metal oxide; however, other types of metal oxides do exist, yet these are used much less frequently in the fine chemical industry. Other types of metal oxides used as microporous catalysts include MoVNbTeOx [241] and surface-phosphate nickel oxide [242], which can be used in the dehydrogenation of ethane to ethylene.

An alternative avenue for future exploration would be the use of heterogeneous transition metal oxide catalysts, which have applications in several areas, including the conversion of hydrocarbons to value-added chemicals and geochemical redox processes [243]. Their production has been explored in Ref. [244], where MoO_3 and WO_3 catalysts were synthesised. The paper draws comparisons between them and traditional zeolites, explaining how these transition metal oxides combine the molecular-sieve attributes of Si-zeolites with the enhanced acidic strength and redox properties of metal cations. Combining these properties into one structure provides versatility in terms of catalytic performance, which is highly sought after in the fine chemical industry. The drawbacks of transition metal oxide catalysts were discussed, with the most significant being structural stability issues and synthesis complexity. These issues provide a hindrance to the practical implementation of transition metal oxides, indicating that significant further validation, both experimentally and practically, would be required to fully leverage their advantages and to result in widespread industrial implementation.

Polymer-Supported Catalysts

Porous polymer structures are a class of amorphous, crosslinked structures capable of featuring a range of pore diameters, with the ability to possess hierarchical porosity in the majority of cases [245]. This cross-linked porous structure provides an extensive surface area, with the ability to incorporate additional functional groups and embed a range of active catalytic sites. Additionally, their extensive crosslinking imparts high thermal and chemical stability, giving rise to a range of potential uses as catalytic support structures. An additional feature of the majority of these polymer structures is their recyclability and the opportunities to recover and reuse these structures. In relation to fine chemical production, high recyclability offers a major advantage in terms of reducing production costs [246].

Alkene epoxidation is a vital chemical reaction utilised in the fine chemical industry to produce various compounds, including epoxides, a valuable intermediate in the synthesis of pharmaceutical and agrochemical products [247]. The potential use of polymer-based structures as catalyst supports for these reactions has been the subject of focus in recent years, as they have exhibited greater potential in terms of stability, as well as in regard to providing a potentially greener production route. Subsequently, new, modern methods have been developed and patented, including several patents regarding liquid phase epoxidation and incorporated heterogeneous catalysts [248,249]. These polymer structures provide a support matrix that facilitates and subsequently immobilises the catalyst.

A study conducted on the epoxidation of alkenes using a polymer-supported Mo(VI) catalyst [250] evaluated the stability and activity, as well as the catalytic leaching, under a range of conditions. It highlighted current industrial concerns regarding the potential leaching during the use of long-term heterogeneous catalysts, mentioning the recent research conducted concerning the use of alternative inorganic catalyst structures and their lack of suitability [251–253], underlining the need to develop and scale-up a stable method utilising a polymer-based structure instead. Whilst this research employed a batch process, the focus was to investigate the catalyst structure in regard to its future uses in continuous processes. The findings from the paper include enhanced reaction stability and efficiency, as well as the potential for the reuse of the catalyst. One drawback to note included the importance of utilising optimal conditions in the process to ensure that efficiency was optimal. However, such a drawback exists for almost all chemical reactions regardless of sector, outlining the importance of weighing the benefits of using a polymer structure as a heterogeneous catalyst over any negative aspects. Similar studies investigating the preparation of epoxides using a polybenzimidazole-supported Mo(VI) catalyst [254,255] produced similar results. Whilst the alkene being epoxidised varied between the studies, the findings illustrated that a polymer-based catalyst structure (in this case, polybenzimidazole) is active in these reactions, and it exhibits particularly high activity under reaction-specific optimised conditions. Both studies showed greater reaction efficiency, with a notable catalyst leaching recorded in Ref. [255]. However, these studies were conducted under batch conditions, and it has been suggested that when transferred to a continuous process, leaching and loss in catalytic activity would no longer be an issue. This is of great importance in terms of the scale-up and adoption of these processes on a larger scale, particularly in the fine chemical industry, wherein both prolonged catalyst activity and lifespan are highly desirable.

Various studies in the literature portray polymers as suitable catalyst supports as a result of them being inert, insoluble, and non-toxic [256,257]. In particular, being inert is an essential characteristic to maximise the efficiency of catalytic reactions. This reduces the risk of catalyst poisoning and facilitates the reuse and recovery of the catalyst, providing a potential economic advantage in the long run in terms of cost savings in regard to process scale-up. Their non-toxic properties are essential in the fine chemical industry in terms of safety assurance and regulatory compliance, particularly in industries where products may subsequently face human consumption. These findings are further supported in Ref. [258], wherein a Ps·AMP·Mo catalyst was used in the epoxidation of 1-hexene and 4-vinyl-1-cyclohexene. Continuous and batch operations were analysed, with the catalyst exhibiting high activity and selectivity in both circumstances. In addition to this, epoxidation in a FlowSyn continuous flow reactor illustrated noticeable time savings, highlighting the benefits and potential streamlining of operations provided by the use of a polymer-based catalyst structure.

The importance of optimising these processes to fully reap their benefits has been a consistent theme throughout most of the literature regarding the epoxidation of alkenes using polymer-based catalyst structures. The increased catalytic activity and efficiency are diminished under non-ideal reaction conditions, regardless of the polymer structure used, as evidenced in Ref. [259]. In this research paper, the catalyst used was a polystyrene 2-(aminomethyl) pyridine-supported Mo(VI) catalyst, yet the importance of optimising the reaction conditions was further stressed, regardless of whether it was a batch or continuous process. Additionally, the optimisation of a process reduces its environmental impact, as well as its operating and energy consumption costs. To further verify the results produced during the physical optimisation of a polymer-based catalytic epoxidation reaction, an artificial neural network model has been employed, and the results of this model were compared to those produced experimentally [260]. This has highlighted not only the importance of optimising a process in terms of cost savings and efficiency but also regarding the benefits it supplies in terms of continuous epoxidation experiments as an elaboration of those carried out in batch reactors. The importance of greener epoxidation processes has been explored in Ref. [261], with the main benefits of developing greener methods including reduced waste, further potential savings in regard to energy and material costs, as well as exhibiting progressive movements towards more environmentally conscious processes and tighter emissions regulations.

2.6. Comparison between Mesoporous and Microporous Structures and Their Uses as Catalysts in the Production of Fine Chemicals

After reviewing the existing literature concerning various types of mesoporous and microporous catalysts, one key takeaway would be that many of these structures can alternate between mesoporous, microporous, or can sometimes exhibit both (hierarchical). For example, both micro- and mesoporous zeolites exist, yet their uses differ, depending on the reaction or process in question. A similar case exists for MOFs and carbon-based structures, along with many others. However, in the context of fine chemicals, some pore structures, along with specific materials, are significantly better suited to this application. A clear example of this would be the carbon-based structures. According to the definition provided by the International Federation of Pure and Applied Chemistry (IUPAC), porous carbon-based materials can be split into three categories, based on their pore diameter size: microporous carbon (<2 nm), mesoporous carbon (between 2 and 50 nm), and macroporous carbon (>50 nm) [262]. Much of the literature on microporous carbon-based structures falls into the field of electrochemistry, as opposed to the industry of fine chemicals. One explanation for this may be due to its strong conductive abilities over other heterogeneous catalysts. Apart from this, many of the other advantages that it provides are shared with other catalytic structures. Despite this, its properties are not often harnessed in fine chemical production, yet in its mesoporous form, its use is significantly more common.

Therefore, several materials explored in this review, such as microporous MOFs, can possess either a crystalline or amorphous structure. However, the crystalline structure often features more prominently in catalytic research, whereas in industry, amorphous structures are typically used more often due to their lower cost and greater flexibility [263–265].

In terms of determining the most efficient or suitable catalyst, it could be argued that this is nothing but a trivial matter. Despite some structures exhibiting more advantages than others, it is solely a matter of the process in question. For example, compared with the most used catalyst support structures, such as silica and zeolites, carbon-based materials possess several advantages, such as stability in both basic and acidic mediums. Along with this, activated carbon obtained from biochar can yield a highly porous structure, with surface areas reaching as high as 2000 m^2/g (depending on the source and synthesis/activation method). This is significantly larger in comparison to other common material-specific surface areas; for example, silica can only reach 750 m^2/g and zeolites up to 800 m^2/g [211]. However, in the case of an acid catalysis reaction, the use of a zeolite structure would be strongly preferred due to its strong Brønsted and Lewis acid sites, something that a carbon-based catalyst could not offer on its own.

2.7. Synthesis and Modification

Often, during the production of fine chemicals, a large number of preparation steps are required. This situation has been reviewed above for specific processes with certain materials. This section, however, will analyse these modification and preparation methods more generically. Often, the reactants being used must be altered or treated before use. Additionally, the catalyst being used will often require some prior steps, be it modification with alternative compounds, thermal treatment, or the application of support [266]. According to Ref. [61], the approach taken to synthesise and prepare the catalyst structure will heavily influence the formed structure in terms of the dispersion of the catalyst, as well as its surface area and concentration. These findings are also supported by Refs. [267–269], indicating that the method used in catalyst development and preparation is crucial in dictating the reaction's performance and outcome. The additional costs incurred during any preparation, in terms of additional materials and energy requirements, further add to the value of the final product being manufactured; hence, the choice of a catalyst is not to be taken lightly.

The majority of the literature in this field often focuses on the effectiveness of the catalyst used for a specific process, with the type of processes being analysed varying heavily. Often, the processes being investigated consist of either hydrogenation, dehydrogenation, hydrocracking, and decarbonylation (which occur on metal sites); or dehydration, isomerization, and hydrogenolysis (which occur on acid sites) [270]. A significant amount of literature has been produced in recent years with a focus on modifying and enhancing existing micro and mesoporous catalysts to improve performance. The primary focus of research in this field has been the synthesis, characterisation, and applications of acidic, noble, and transition metal modification.

In the context of heterogeneous catalysts, modification occurs when the base catalyst substance is combined or added to an alternative substance, often a liquid or solid, to alter and extrude-specific properties that may be desired for a reaction. Unlike the direct synthesis of a catalyst, it is not being generated solely from reactants, but already exists and is being both chemically and physically altered. For example, in the case of the synthesis of florol, as explored by Ref. [271], the H-USY-30 catalyst was modified via treatment with an aqueous NaOH alkaline solution. This modification consisted of combining the two in an Easymax 106 reactor for a specific period, followed by ion exchange to convert it to its protonic form. The NaOH treatment is vital for improving the overall catalytic ability of the H-USY-30. The NaOH acts as a baseleaching agent, leading to the dissolution of the amorphous silica and aluminium within the structure. This increases porosity and leads to an alteration in the distribution and strength of the acid sites. This then minimizes side reactions and increases overall reaction selectivity. The associated reaction scheme of the desilication alkaline treatment can be seen below in Figure 5.

There are countless other examples of catalyst modification across all aspects of the chemical industry; for example, in Ref. [68], Kumar investigates the ring opening of decalin using MCM-41, modified with Ru. Subsequently, according to Ref. [61], the most common methods of mesoporous catalyst preparation or modification (zeolites, mesoporous silicas, etc.) are ion exchange and impregnation, with in situ synthesis being scarcely used, owing to it being a relatively new method.

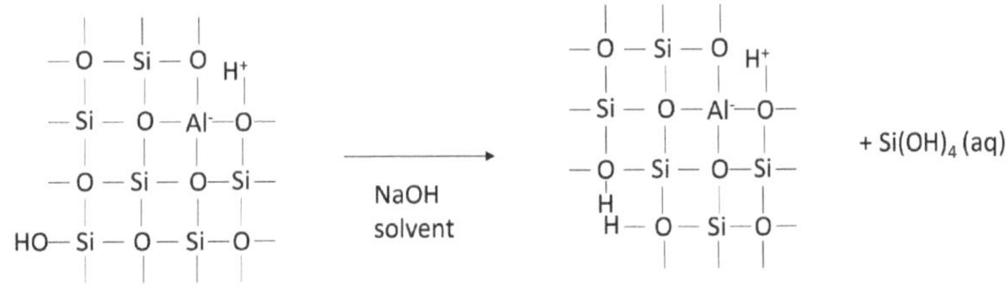

Figure 5. A diagram illustrating the role that an alkaline treatment can play in the modification and preparation of an aluminosilicate structure, adapted from Ref. [271].

There are a range of ways in which catalysts can be modified, depending on their existing structure and the desired outcome. These include ion exchange, impregnation of a catalyst, as well as in situ and composite methods. These will be explored in further detail below, with their relevance in the production of fine chemicals assessed and reviewed across the existing literature and experimental papers/journals.

2.7.1. Ion-Exchange

The ion exchange method is a widely used technique in catalysis, particularly for materials with micro- and mesoporous structures. It involves the replacement of ions within the material's framework with other ions from a solution. This method is often employed for catalysts with low metal loadings, as detailed in Ref. [68], wherein MCM-41 was utilised for the ring-opening decalin. This method then allows for the controlled introduction of specific ions or molecules to modify the catalytic properties of the material. In the context of catalyst preparation, the ion exchange method helps put certain metal ions into the molecular sieve's structure. This exchange process, in most cases, is crucial for tailoring the surface properties, as well as for enhancing the catalytic activity of the material. The reasoning behind using ion exchange in this way is to control the types and concentrations of ions introduced so that the acidity and selectivity of the catalyst can be managed, enabling it to be used more efficiently in specific chemical reactions. As mentioned previously, preparing a catalyst in this way often requires many steps to ensure that the properties of the catalyst are controlled. Even slight deteriorations from the correct process can lead to a change in the catalyst, rendering it inefficient or perhaps no longer viable. Moreover, ion exchange can be an energy-intensive process, leading to the conclusion that an ion exchange method possesses both pros and cons. This explains its popularity in the use of heterogeneous catalysts but points towards the use of other modifications and preparation steps to obtain a more ideal catalyst.

2.7.2. Evaporation Impregnation

Metal precursors are dissolved in distilled water and added to the microporous zeolites or mesoporous materials in a flask. Metal modifications are carried out for 24 h in a rotatory evaporator equipped with a water bath. After completion of the synthesis, the aqueous phase is evaporated, and the catalyst is removed from the flask, followed by drying and calcination in a muffle oven [272].

2.7.3. Deposition Precipitations

In the deposition precipitation method, the pH of the dissolved metal precursor and the microporous zeolites or mesoporous materials is enhanced to 9–10, and synthesis is carried out for 24 h in alkaline media. After completion of the synthesis, the catalyst is filtered, washed with distilled water, dried, and calcined in a muffle oven.

2.7.4. Physicochemical Characterisation

The physicochemical characterisation of catalytic materials applied in the production of fine chemicals is of immense importance for understanding the reaction mechanism.

Furthermore, physicochemical characterisation results are utilized in the development of new efficient and stable catalysts with high selectivity to desired fine chemicals. Instruments and techniques applied in the physicochemical characterisations are as follows: X-ray powder diffraction (XRD), used for structure and phase purity determination; scanning electron microscopy (SEM), for determination of the morphological features, such as crystal shape, size, and distributions; energy dispersive X-ray analyses (EDXA), for measurement of the chemical compositions. Textural properties such as surface area, pore volume, and pore size distributions, of the pristine solid acid, as well as the transition and noble metal modified heterogeneous catalysts, are determined using N_2-physisorption. The number and strength of the Brønsted and Lewis acid sites in the catalytic materials are measured using FTIR-Pyridine. The transition and noble metal nanoparticle size, shape, and distributions are measured using transmission electron microscopy (TEM) and high-resolution transmission electron microscopy (HRTEM). Furthermore, HRTEM and TEM are also applied for the measurement of pore dimensions, the periodicity of pores, and the channel systems in the structured microporous and mesoporous materials utilised for fine chemical synthesis. Framework Al and extra-framework Al, important for the creation of Brønsted and Lewis acid sites, are measured using ^{27}Al-NMR and ^{29}Si MAS NMR spectroscopies. The oxidation states of the transition and noble metals are important to produce fine chemicals using heterogeneous catalysts. X-ray photoelectron spectroscopy (XPS) is used for the determination of the oxidation states of the metal-modified catalysts.

2.7.5. Acid Preparation

Acids can play a crucial role in the development, modification, or synthesis of catalysts. They can perform several roles, with the main objective being to chemically alter the catalyst to increase conversion and promote greater selectivity towards the desired products of a reaction. Common ways in which acids are used in this process include template removal, pore diameter adjustment, and forms of surface modification.

Template removal incorporates the use of an acid to remove organic templates or surfactants from the synthesized catalysts. The acids are used to dissolve or remove the organic components from the synthesized material, leaving behind the desired porous structure. It must be noted, however, that paying close attention to the removal process is crucial to maintaining the material's structure and desired qualities. Incomplete or improper removal could lead to structural flaws, blocked pores, or unwanted contaminants, ultimately impacting the catalyst's performance.

Pore size adjustment allows for the diameter of the pores within the material to be altered. By doing so, the ability of reactant molecules to enter the structure and contact active sites can be modified, providing the ability to change the catalyst's conversion and selectivity.

Altering the surface of the material using an acid can not only change its structure and the number of active sites available for a reaction, but it can also enhance the catalyst's stability and reduce the risk of catalyst contamination or deactivation. This is vital in extending the life of the catalyst, meaning it can be recovered and reused more often, reducing catalytic costs in the long term.

2.7.6. Base Preparation

Utilising a base, either in the preparation or synthesis of a catalyst, generally achieves similar objectives as using an acid; however, the preparation progresses along a different chemical route. Bases can be used to control the pH of the catalyst, as well as to perform surface adjustments on the catalytic material. This includes adding or removing specific functional groups from the catalyst surface, which can in turn lead to a different chemical structure and to altered reaction parameters. Bases effectively allow for the customization of the structural characteristics of the catalyst, providing an enhancement to its efficiency, selectivity, and conversion. In comparison to the route followed by an acid, the alkaline route features both hydrolysis and condensation, leading to a significantly more compact and refined structure [273].

The use of either a base or acid as a heterogeneous catalyst has recently been featured in extensive research, with attempts made to generate a hybrid catalyst featuring both. Such a catalyst has been growing in popularity due to its faster reaction times and lower corrosion or deactivation potential, as evidenced in Refs. [274,275].

2.7.7. Comparison

A comparison between using an acid, a base, and an ion exchange resin as a heterogeneous catalyst during the production of biodiesel has been explored in Ref. [32]. The advantages of using an acid heterogeneous catalyst have been listed as the recovery and reuse aspect of the catalyst, with the main drawback being a loss of the catalyst due to leaching, as well as the fact that a high temperature is required, along with a prolonged reaction time. This has also been mentioned in a similar context in Refs. [276–278]. Conversely, the paper suggests that using a base as a heterogeneous catalyst produces a higher reaction rate. However, a loss of the catalyst also occurred due to leaching, as well as to the fact that it was sensitive to the materials used during the reaction. Finally, the paper compares these catalysts to the use of an ion-exchange resin. The ion-exchange resin featured the complete removal of by-products, as well as easy product separation from the reactants and catalyst, in addition to the ability to catalyse the reaction as either a base or an acid. Its drawbacks, however, included the fact that its resin active sites suffered from deactivation, and it exhibited a slow reaction rate.

2.7.8. Impregnation

The technique of impregnation features the absorption and/or penetration of an active catalyst species into the pores of a solid support material, such as activated carbon, alumina, or silica. The impregnated catalyst can be in the form of a solution, dispersion, or suspension, which is deposited onto the supporting structure through several physicochemical interactions.

This method of preparation allows for a relatively high concentration of active species on the support material, providing greater catalytic activity and selectivity. This is achieved through careful control of the active species and the impregnation time.

A key advantage of the impregnation method is its flexibility in accommodating a range of active species, including metals, metal oxides, and metal complexes.

Potential drawbacks regarding the use of impregnation as a catalyst synthesis method include the possible uneven distribution of the active catalyst within the porous structure of the support material, which can affect the overall catalytic performance and stability. One proposal to solve this issue would be to employ careful control of the impregnation parameters and post-impregnation treatments, e.g., drying and calcination. The use of some form of post-treatment would ensure uniform distribution, as well as strong adhesion of the active species within the structure of the support material.

The porous structure undergoing impregnation can vary, depending on the specific reaction. Often, the structure must undergo a form of preparation or treatment, usually calcination, before the impregnation step [279,280]. Following this, the impregnation occurs either through wet impregnation or incipient wetness impregnation. In some specific cases, for example, Ref. [279], impregnation of a catalyst occurs in a solventless environment, leading to what is known as "precursor dry impregnation". Alternatively, evaporation-impregnation can be used, such as in the preparation of ceria-supported catalysts [281]. Here, evaporation–impregnation was achieved by stirring in a rotary evaporator, followed by water evaporation under vacuum.

One important aspect of the impregnation method to note is the preparation of a precursor solution and the following drying and/or calcination that occurs. This pre-step is similar to the methodology used in the ion-exchange method. However, the impregnation method is regarded to be more straightforward and is claimed to be one of the most popular methods, as per Ref. [282].

2.7.9. Oxygen Doping

Oxygen doping is an approach used on mesoporous carbon bases to improve the hydrophilicity of the surface. The surface can then provide active sites for catalytic reactions or the selective adsorption of cationic compounds. Additionally, it can aid in the dispersion of metals within the structure [283,284]. In general, oxygen doping can introduce additional active sites to the structure, increasing activity, as well as thermal stability, which in some cases, can prevent sintering, extending the life of the catalyst.

3. Conclusions

Overall, this literature review explored the complex and wide-ranging field of microporous and mesoporous catalysts within the fine chemical industry. The efficiency, selectivity, and sustainability of the chemical reactions involved are strongly influenced by the catalyst of choice, regardless of its mesoporous or microporous nature. Because their pores are smaller (<2 nm), microporous materials—like zeolites, aluminophosphate, and some metal–organic frameworks (MOFs)—offer excellent selectivity and are especially well-suited for reactions involving smaller molecules. They often require significantly fewer steps during their manufacture and synthesis, resulting in fewer harmful emissions and a lower cost of production. The most-used microporous structure, as a heterogeneous catalyst, is zeolites. However, the rise of MOFs and other alternatives that provide a wider range of customisability in terms of pore size and volume are under development every day, with their significance rapidly growing as a result of their potential promise as better heterogeneous catalysts.

Mesoporous materials, including various silica-based structures, zeolites, ZIFs and some MOFs, provide the advantage of accommodating larger molecules and demonstrating enhanced diffusion. This makes them ideal for a broader range of reactions. Moreover, they possess the ability to express adaptive pore shapes and sizes, meaning that the catalytic structure can be tailored to the needs of individual reactions, massively increasing its effectiveness due to its high selectivity and activity. The most popular mesoporous catalysts in the industry include organo-silica-based MCM-41, as well as mesoporous zeolites. In a similar fashion to microporous materials, the use of MOFs in place of traditional mesoporous catalysts is growing. The reason for this trend includes the almost infinite number of possible structures that MOFs can exhibit, meaning that they can be tailored for almost every reaction, and there is always the possibility of developing a catalytically improved structure. Moreover, the industrial catalytic use of ZIF structures as heterogeneous catalysts is beginning to rise, not just in the fine chemical industry but across a range of chemical sectors. This is a result of their high specific surface areas, coordination topology features, and in some cases, nitrogen-rich structures, leading to a high density, as well as a uniform distribution, of active sites.

This review has also demonstrated the importance of catalyst modification and preparation, showing how techniques like ion exchange, impregnation, and oxygen doping can significantly alter the properties of these catalysts, making them suitable for certain uses in the fine chemical industry. These modifications can enhance properties such as surface area, pore volume, and chemical functionality, therefore improving the catalytic activity and selectivity of a reaction.

In conclusion, the field of heterogeneous catalysis for fine chemical production is a dynamic and ever-changing area of research. Continued research regarding micro- and mesoporous catalyst design, combined with a clearer understanding of structural impact on activity, will significantly advance the field. This deeper insight into catalytic behaviour and material properties will lead to the production of more efficient, selective, and sustainable processes in the fine chemical industry. It is noteworthy to mention that solid acid, transition, and noble metal-modified heterogeneous catalysts undergo catalyst deactivation during the production of fine and speciality chemicals. However, most of the spent catalytic materials after the reaction could be regenerated and reused.

The findings from this comprehensive literature review provide a solid foundation for further exploration and a critical review in the field of fine chemical production using micro- and mesoporous heterogeneous catalysts.

4. Future Directions

There is a rapidly growing interest in the use of both microporous and mesoporous heterogeneous catalysts in the fine chemical industry. As it stands, the current work in this field is predominantly focused on small-scale investigation of these structures and their individual applications. A key area for future research in this field would involve investigations into bridging the gap between lab-scale experiments and industrial-level applications of meso- and microporous heterogeneous catalysts. It is evident through a review of the existing literature that a case for the use of various porous catalyst structures, including MOFs and ZIFs, exists. However, a lack of experimental work focused on increasing the utilisation scale and production volumes remains, indicating that further work in this field is required before the widespread adoption of porous catalysts in the fine chemical industry becomes a possibility.

Additionally, a future exploration into innovative synthesis methods to enhance selectivity, stability, and efficiency, as well as additional work on the alteration of catalyst structures, both physically and chemically, to modify specific properties is highly recommended. Doing so would open a gateway into not only a wider adoption of the catalyst structures reviewed but also into the use of more environmentally and economically beneficial processes, which is something that would be highly valuable in a sector such as the fine chemicals industry, wherein even marginal gains and advancements would prove beneficial.

Several solid acids, transition, and noble metal-modified microporous and mesoporous heterogeneous catalysts are currently applied in the production of fine chemicals, as described in this paper. Plausible explanations for the advantages of using these catalytic materials include the possibility of easy separation from the reaction media, as well as its regeneration and reuse in the production of fine chemicals. However, taking into consideration the sustainable development of human society and the efficient use of limited natural resources, new environmentally friendly catalytic materials need to be discovered and utilized for industrial processes. Furthermore, new green process technology should be developed for the rational utilisation of natural resources and the minimal generation of industrial residuals.

Finally, the incorporation of computational modelling and machine learning into catalyst design would provide additional optimisation of reactions and chemical processes, providing the opportunity to enhance the efficiency and sustainability of fine chemical production. Such optimisation could lead to reduced chemical waste, experimentation costs, and development times, all of which are critical for maintaining competitiveness and meeting regulatory standards in the fine chemical industry.

Author Contributions: J.L.: writing—original draft preparation, formal analysis, reference collections, and revision; N.K.: conceptualization, formal analysis, writing—review and editing, consultations, and comments; B.S.: conceptualization, supervision, formal analysis, consultations, writing—review and editing, resources, and project administration. All authors have read and agreed to the published version of the manuscript.

Funding: This research received no external funding.

Acknowledgments: This work is a result of the cooperative activities between the Laboratory of Industrial Chemistry and Reaction Engineering, the Faculty of Science and Engineering, Johan Gadolin Process Chemistry Center, Åbo Akademi University, Turku, Finland, and the School of Engineering, Lancaster University, United Kingdom.

Conflicts of Interest: The authors declare no conflict of interest.

Abbreviations

MT—megatonnes; TOF—turnover frequency; TON—turnover number; MOF—metal-organic framework; ZIF—zeolitic imidazolate framework; GCMC—grand canonical Monte Carlo; IUPAC—International Federation of Pure and Applied Chemistry.

References

1. Panizza, M. Chapter 13–Fine chemical industry, pulp and paper industry, petrochemical industry and pharmaceutical industry. In *Electrochemical Water and Wastewater Treatment*; Martínez-Huitle, C.A., Rodrigo, M.A., Scialdone, O., Eds.; Elsevier: Amsterdam, The Netherlands, 2018. [CrossRef]
2. Blaser, H.-U.; Studer, M. The role of catalysis for the clean production of fine chemicals. *Appl. Catal. A Gen.* **1999**, *189*, 191–204. [CrossRef]
3. Busacca, C.A.; Fandrick, D.R.; Song, J.J.; Senanayake, C.H. The Growing Impact of Catalysis in the Pharmaceutical Industry. *Adv. Synth. Catal.* **2011**, *353*, 1825–1864. [CrossRef]
4. Hayler, J.D.; Leahy, D.K.; Simmons, E.M. A Pharmaceutical Industry Perspective on Sustainable Metal Catalysis. *Organometallics* **2019**, *38*, 36–46. [CrossRef]
5. Shokouhimehr, M. Magnetically Separable and Sustainable Nanostructured Catalysts for Heterogeneous Reduction of Nitroaromatics. *Catalysts* **2015**, *5*, 534–560. [CrossRef]
6. de Sousa, L.B.; Speranza, J.T.; Condotta, R.; Cella, R. Heterogeneous catalyzed isomerization of turpentine oil by ordered mesoporous materials like M41S structures. *Can. J. Chem. Eng.* **2023**, *101*, 4106–4117. [CrossRef]
7. Malet-Sanz, L.; Susanne, F. Continuous Flow Synthesis. A Pharma Perspective. *J. Med. Chem.* **2012**, *55*, 4062–4098. [CrossRef] [PubMed]
8. Porta, R.; Benaglia, M.; Puglisi, A. Flow Chemistry: Recent Developments in the Synthesis of Pharmaceutical Products. *Org. Process. Res. Dev.* **2016**, *20*, 2–25. [CrossRef]
9. Bulk Chemicals vs. Fine Chemicals: The Difference. Available online: https://capitalresin.com/bulk-chemicals-vs-fine-chemicals-the-difference/#:~:text=Fine%20chemicals%20are%20more%20complex,surface%2Dlevel%20differences%2C%20however (accessed on 11 May 2023).
10. Anonymous. Global Chemicals Export Value by Segment. Available online: https://www.statista.com/statistics/1380198/global-chemicals-export-value-by-segment/ (accessed on 11 May 2023).
11. Things You Should Know about Fine Chemicals. Available online: https://www.qinmuchem.com/news/things-you-should-know-about-fine-chemicals.html (accessed on 11 May 2023).
12. Pollak, P. *Fine Chemicals the Industry and the Business*, 2nd ed.; John Wiley & Sons: Hoboken, NJ, USA, 2011.
13. Ballini, R. *Eco-Friendly Synthesis of Fine Chemicals*; Royal Society of Chemistry: London, UK, 2009.
14. Schwanke, A.J.; Balzer, R.; Pergher, S. Microporous and mesoporous materials from natural and inexpensive sources. In *Handbook of Ecomaterials*; Martínez, L.M.T., Kharissova, O.V., Kharisov, B.I., Eds.; Springer: Berlin/Heidelberg, Germany, 2017.
15. Mallick, K.K.; Winnett, J. 6-3D bioceramic foams for bone tissue engineering. In *Bone Substitute Biomaterials*; Mallick, K., Ed.; Elsevier: Amsterdam, The Netherlands, 2014. [CrossRef]
16. Singh, B.K.; Kim, Y.; Kwon, S.; Na, K. Synthesis of Mesoporous Zeolites and Their Opportunities in Heterogeneous Catalysis. *Catalysts* **2021**, *11*, 1541. [CrossRef]
17. Zu, L.; Zhang, W.; Qu, L.; Liu, L.; Li, W.; Yu, A.; Zhao, D. Mesoporous Materials for Electrochemical Energy Storage and Conversion. *Adv. Energy Mater.* **2020**, *10*, 2152. [CrossRef]
18. Sánchez-Antonio, O.; Romero-Sedglach, K.A.; Vázquez-Orta, E.C.; Juaristi, E. New Mesoporous Silica-Supported Organocatalysts Based on (2*S*)-(1,2,4-Triazol-3-yl)-Proline: Efficient, Reusable, and Heterogeneous Catalysts for the Asymmetric Aldol Reaction. *Molecules* **2020**, *25*, 4532. [CrossRef]
19. Xie, Y.; Sharma, K.K.; Anan, A.; Wang, G.; Biradar, A.V.; Asefa, T. Efficient solid-base catalysts for aldol reaction by optimizing the density and type of organoamine groups on nanoporous silica. *J. Catal.* **2009**, *265*, 131–140. [CrossRef]
20. Corma, A.; Garcia, H. Silica-Bound Homogenous Catalysts as Recoverable and Reusable Catalysts in Organic Synthesis. *Adv. Synth. Catal.* **2006**, *348*, 1391–1412. [CrossRef]
21. Scatena, G.S.; de la Torre, A.F.; Cass, Q.B.; Rivera, D.G.; Paixão, M.W. Multicomponent Approach to Silica-Grafted Peptide Catalysts: A 3 D Continuous-Flow Organocatalytic System with On-line Monitoring of Conversion and Stereoselectivity. *ChemCatChem* **2014**, *6*, 3208–3214. [CrossRef]
22. Puglisi, A.; Benaglia, M.; Annunziata, R.; Chiroli, V.; Porta, R.; Gervasini, A. Chiral Hybrid Inorganic–Organic Materials: Synthesis, Characterization, and Application in Stereoselective Organocatalytic Cycloadditions. *J. Org. Chem.* **2013**, *78*, 11326–11334. [CrossRef] [PubMed]
23. Brühwiler, D. Postsynthetic functionalization of mesoporous silica. *Nanoscale* **2010**, *2*, 887–892. [CrossRef] [PubMed]
24. Jiang, D.M.; Gao, J.S.; Yang, Q.H.; Li, C. Large-pore mesoporous ethane-silicas as efficient heterogeneous asymmetric catalysts. In *Studies in Surface Science and Catalysis*; Elsevier: Amsterdam, The Netherlands, 2007; Volume 170, pp. 1252–1259. [CrossRef]

25. Martín, N.; Cirujano, F.G. Organic synthesis of high added value molecules with MOF catalysts. *Org. Biomol. Chem.* **2020**, *18*, 8058–8073. [CrossRef] [PubMed]
26. Han, Y.R.; Park, J.-W.; Kim, H.; Ji, H.; Lim, S.H.; Jun, C.-H. A one-step co-condensation method for the synthesis of well-defined functionalized mesoporous SBA-15 using trimethallylsilanes as organosilane sources. *Chem. Commun.* **2015**, *51*, 17084–17087. [CrossRef] [PubMed]
27. Ferré, M.; Pleixats, R.; Man, M.W.C.; Cattoën, X. Recyclable organocatalysts based on hybrid silicas. *Green Chem.* **2016**, *18*, 881–922. [CrossRef]
28. Putz, A.-M.; Almásy, L.; Len, A.; Ianăşi, C. Functionalized silica materials synthesized via co-condensation and post-grafting methods. *Full-Nanotub. Carbon Nanostruct.* **2019**, *27*, 323–332. [CrossRef]
29. Liang, J.; Liang, Z.; Zou, R.; Zhao, Y. Heterogeneous Catalysis in Zeolites, Mesoporous Silica, and Metal–Organic Frameworks. *Adv. Mater.* **2017**, *29*, 1139. [CrossRef]
30. Taguchi, A.; Schüth, F. Ordered mesoporous materials in catalysis. *Microporous Mesoporous Mater.* **2005**, *77*, 1–45. [CrossRef]
31. Brunel, D.; Blanc, A.C.; Galarneau, A.; Fajula, F. New trends in the design of supported catalysts on mesoporous silicas and their applications in fine chemicals. *Catal. Today* **2002**, *73*, 139–152. [CrossRef]
32. Jayakumar, M.; Karmegam, N.; Gundupalli, M.P.; Gebeyehu, K.B.; Asfaw, B.T.; Chang, S.W.; Balasubramani, R.; Awasthi, M.K. Heterogeneous base catalysts: Synthesis and application for biodiesel production—A review. *Bioresour. Technol.* **2021**, *331*, 125054. [CrossRef] [PubMed]
33. Kumar, D.; Schumacher, K.; von Hohenesche, C.D.F.; Grün, M.; Unger, K. MCM-41, MCM-48 and related mesoporous adsorbents: Their synthesis and characterisation. *Colloids Surf. A Physicochem. Eng. Asp.* **2001**, *187–188*, 109–116. [CrossRef]
34. Wang, S. Ordered mesoporous materials for drug delivery. *Microporous Mesoporous Mater.* **2009**, *117*, 1–9. [CrossRef]
35. Vajglová, Z.; Kumar, N.; Peurla, M.; Eränen, K.; Mäki-Arvela, P.; Murzin, D.Y. Cascade transformations of (±)-citronellal to menthol over extruded Ru-MCM-41 catalysts in a continuous reactor. *Catal. Sci. Technol.* **2020**, *10*, 8108–8119. [CrossRef]
36. Endud, S.; Wong, K.-L. Mesoporous silica MCM-48 molecular sieve modified with $SnCl_2$ in alkaline medium for selective oxidation of alcohol. *Microporous Mesoporous Mater.* **2007**, *101*, 256–263. [CrossRef]
37. Huo, Q.; Margolese, D.I.; Ciesla, U.; Feng, P.; Gier, T.E.; Sieger, P.; Leon, R.; Petroff, P.M.; Schüth, F.; Stucky, G.D. Generalized synthesis of periodic surfactant/inorganic composite materials. *Nature* **1994**, *368*, 317–321. [CrossRef]
38. Huo, Q.; Margolese, D.I.; Ciesla, U.; Demuth, D.G.; Feng, P.; Gier, T.E.; Sieger, P.; Firouzi, A.; Chmelka, B.F. Organization of Organic Molecules with Inorganic Molecular Species into Nanocomposite Biphase Arrays. *Chem. Mater.* **1994**, *6*, 1176–1191. [CrossRef]
39. Parida, N.; Badamali, S.K. Facile synthesis and catalytic activity of nanoporous SBA-1. *J. Porous Mater.* **2022**, *29*, 161–167. [CrossRef]
40. Huo, Q.; Leon, R.; Petroff, P.M.; Stucky, G.D. Mesostructure Design with Gemini Surfactants: Supercage Formation in a Three-Dimensional Hexagonal Array. *Science* **1995**, *268*, 1324–1327. [CrossRef] [PubMed]
41. Kot, M.; Kiderys, A.; Janiszewska, E.; Pietrowski, M.; Yang, C.-M.; Zieliński, M. Hydrogenation of toluene over nickel nanoparticles supported on SBA-3 and AlSBA-3 materials. *Catal. Today* **2020**, *356*, 64–72. [CrossRef]
42. Lu, G.Q.; Zhao, X.S. *Nanoporous Materials: Science and Engineering*; World Scientific Publishing: Singapore, 2004.
43. Ge, S.; Geng, W.; He, X.; Zhao, J.; Zhou, B.; Duan, L.; Wu, Y.; Zhang, Q. Effect of framework structure, pore size and surface modification on the adsorption performance of methylene blue and Cu2+ in mesoporous silica. *Colloids Surf. A Physicochem. Eng. Asp.* **2018**, *539*, 154–162. [CrossRef]
44. Mayoral, A.; Blanco, R.M.; Diaz, I. Location of enzyme in lipase-SBA-12 hybrid biocatalyst. *J. Mol. Catal. B Enzym.* **2013**, *90*, 23–25. [CrossRef]
45. Kumar, A.; Nepak, D.; Srinivas, D. Direct synthesis of amides from amines using mesoporous Mn-SBA-12 and Mn-SBA-16 catalysts. *Catal. Commun.* **2013**, *37*, 36–40. [CrossRef]
46. Galarneau, A.; Cambon, H.; Di Renzo, F.; Ryoo, R.; Choi, M.; Fajula, F. Microporosity and connections between pores in SBA-15 mesostructured silicas as a function of the temperature of synthesis. *New J. Chem.* **2003**, *27*, 73–79. [CrossRef]
47. Gao, D.; Zhang, X.; Dai, X.; Qin, Y.; Duan, A.; Yu, Y.; Zhuo, H.; Zhao, H.; Zhang, P.; Jiang, Y.; et al. Morphology-selective synthesis of active and durable gold catalysts with high catalytic performance in the reduction of 4-nitrophenol. *Nano Res.* **2016**, *9*, 3099–3115. [CrossRef]
48. Sankar, E.S.; Reddy, K.S.; Jyothi, Y.; Raju, B.D.; Rao, K.S.R. Alcoholysis of Furfuryl Alcohol into n-Butyl Levulinate Over SBA-16 Supported Heteropoly Acid Catalyst. *Catal. Lett.* **2017**, *147*, 2807–2816. [CrossRef]
49. Kleitz, F.; Liu, D.; Anilkumar, G.M.; Park, I.-S.; Solovyov, L.A.; Shmakov, A.N.; Ryoo, R. Large Cage Face-Centered-Cubic *Fm3m* Mesoporous Silica: Synthesis and Structure. *J. Phys. Chem. B* **2003**, *107*, 14296–14300. [CrossRef]
50. Kalbasi, R.J.; Zirakbash, A. Synthesis, characterization and drug release studies of poly(2-hydroxyethyl methacrylate)/KIT-5 nanocomposite as an innovative organic–inorganic hybrid carrier system. *RSC Adv.* **2015**, *5*, 12463–12471. [CrossRef]
51. Bejblová, M.; Procházková, D.; Čejka, J. Acylation Reactions over Zeolites and Mesoporous Catalysts. *ChemSusChem* **2009**, *2*, 486–499. [CrossRef] [PubMed]
52. Kleitz, F.; Choi, S.H.; Ryoo, R. Cubic Ia3d large mesoporous silica: Synthesis and replication to platinum nanowires, carbon nanorods and carbon nanotubes. *Chem. Commun.* **2003**, *17*, 2136–2137. [CrossRef] [PubMed]

53. Prabhu, A.; Palanichamy, M. Mesoporous cubic Ia3d materials for the preparation of fine chemicals: Synthesis of jasminaldehyde. *Microporous Mesoporous Mater.* **2013**, *168*, 126–131. [CrossRef]
54. Kim, S.-S.; Pauly, T.R.; Pinnavaia, T.J. Non-ionic surfactant assembly of ordered, very large pore molecular sieve silicas from water soluble silicates. *Chem. Commun.* **2000**, *17*, 1661–1662. [CrossRef]
55. Kleitz, F. Ordered mesoporous materials. In *Handbook of Heterogeneous Catalysis Anonymous;* John Wiley & Sons: Hoboken, NJ, USA, 2008. [CrossRef]
56. Shimizu, T.; Ota, M.; Sato, Y.; Inomata, H. Effect of pore structure on catalytic properties of mesoporous silica supported rhodium catalysts for the hydrogenation of cinnamaldehyde. *Chem. Eng. Res. Des.* **2015**, *104*, 174–179. [CrossRef]
57. Kresge, C.T.; Leonowicz, M.E.; Roth, W.J.; Vartuli, J.C.; Beck, J.S. Ordered mesoporous molecular sieves synthesized by a liquid-crystal template mechanism. *Nature* **1992**, *359*, 710–712. [CrossRef]
58. Shao, Y.; Wang, L.; Zhang, J.; Anpo, M. Novel synthesis of high hydrothermal stability and long-range order MCM-48 with a convenient method. *Microporous Mesoporous Mater.* **2005**, *86*, 314–322. [CrossRef]
59. Sayari, A. Catalysis by Crystalline Mesoporous Molecular Sieves. *Chem. Mater.* **1996**, *8*, 1840–1852. [CrossRef]
60. Vinu, A.; Murugesan, V.; Hartmann, M. Pore Size Engineering and Mechanical Stability of the Cubic Mesoporous Molecular Sieve SBA-1. *Chem. Mater.* **2003**, *15*, 1385–1393. [CrossRef]
61. Kumar, N.; Mäki-Arvela, P.; Hajek, J.; Salmi, T.; Murzin, D.; Heikkilä, T.; Laine, E.; Laukkanen, P.; Väyrynen, J. Physico-chemical and catalytic properties of Ru–MCM-41 mesoporous molecular sieve catalyst: Influence of Ru modification methods. *Microporous Mesoporous Mater.* **2004**, *69*, 173–179. [CrossRef]
62. da Silva, K.A.; Robles-Dutenhefner, P.A.; Sousa, E.M.; Kozhevnikova, E.F.; Kozhevnikov, I.V.; Gusevskaya, E.V. Cyclization of (+)-citronellal to (−)-isopulegol catalyzed by H3PW12O40/SiO_2. *Catal. Commun.* **2004**, *5*, 425–429. [CrossRef]
63. Makiarvela, P.; Mäki-Arvela, P.; Kumar, N.; Nieminen, V.; Sjöholm, R.; Salmi, T.; Murzin, D.Y. Cyclization of citronellal over zeolites and mesoporous materials for production of isopulegol. *J. Catal.* **2004**, *225*, 155–169. [CrossRef]
64. Nie, Y.; Niah, W.; Jaenicke, S.; Chuah, G.-K. A tandem cyclization and hydrogenation of (±)-citronellal to menthol over bifunctional Ni/Zr-beta and mixed Zr-beta and Ni/MCM-41. *J. Catal.* **2007**, *248*, 1–10. [CrossRef]
65. Balu, A.M.; Campelo, J.M.; Luque, R.; Romero, A.A. One-step microwave-assisted asymmetric cyclisation/hydrogenation of citronellal to menthols using supported nanoparticles on mesoporous materials. *Org. Biomol. Chem.* **2010**, *8*, 2845–2849. [CrossRef] [PubMed]
66. Stekrova, M.; Mäki-Arvela, P.; Leino, E.; Valkaj, K.M.; Eränen, K.; Aho, A.; Smeds, A.; Kumar, N.; Volcho, K.P.; Salakhutdinov, N.F.; et al. Two-step synthesis of monoterpenoid dioxinols exhibiting analgesic activity from isopulegol and benzaldehyde over heterogeneous catalysts. *Catal. Today* **2017**, *279*, 56–62. [CrossRef]
67. Stekrova, M.; Mäki-Arvela, P.; Kumar, N.; Behravesh, E.; Aho, A.; Balme, Q.; Volcho, K.P.; Salakhutdinov, N.F.; Murzin, D.Y. Prins cyclization: Synthesis of compounds with tetrahydropyran moiety over heterogeneous catalysts. *J. Mol. Catal. A Chem.* **2015**, *410*, 260–270. [CrossRef]
68. Kumar, N.; Kubicka, D.; Garay, A.L.; Mäki-Arvela, P.; Heikkilä, T.; Salmi, T.; Murzin, D.Y. Synthesis of Ru-modified MCM-41 Mesoporous Material, Y and Beta Zeolite Catalysts for Ring Opening of Decalin. *Top. Catal.* **2009**, *52*, 380–386. [CrossRef]
69. Zhao, D.; Feng, J.; Huo, Q.; Melosh, N.; Fredrickson, G.H.; Chmelka, B.F.; Stucky, G.D. Triblock Copolymer Syntheses of Mesoporous Silica with Periodic 50 to 300 Angstrom Pores. *Science* **1998**, *279*, 548–552. [CrossRef] [PubMed]
70. Li, L.; Shi, J.-L.; Yan, J.-N.; Zhao, X.-G.; Chen, H.-G. Mesoporous SBA-15 material functionalized with ferrocene group and its use as heterogeneous catalyst for benzene hydroxylation. *Appl. Catal. A Gen.* **2004**, *263*, 213–217. [CrossRef]
71. Bhanja, P.; Modak, A.; Chatterjee, S.; Bhaumik, A. Bifunctionalized Mesoporous SBA-15: A New Heterogeneous Catalyst for the Facile Synthesis of 5-Hydroxymethylfurfural. *ACS Sustain. Chem. Eng.* **2017**, *5*, 2763–2773. [CrossRef]
72. Tong, P.; Liang, J.; Jiang, X.; Li, J. Research Progress on Metal-Organic Framework Composites in Chemical Sensors. *Crit. Rev. Anal. Chem.* **2020**, *50*, 376–392. [CrossRef] [PubMed]
73. Britt, D.; Tranchemontagne, D.; Yaghi, O.M. Metal-organic frameworks with high capacity and selectivity for harmful gases. *Proc. Natl. Acad. Sci. USA* **2008**, *105*, 11623–11627. [CrossRef] [PubMed]
74. Nuri, A.; Vucetic, N.; Smått, J.-H.; Mansoori, Y.; Mikkola, J.-P.; Murzin, D.Y. Pd Supported IRMOF-3: Heterogeneous, Efficient and Reusable Catalyst for Heck Reaction. *Catal. Lett.* **2019**, *149*, 1941–1951. [CrossRef]
75. Herbst, A.; Janiak, C. MOF catalysts in biomass upgrading towards value-added fine chemicals. *CrystEngComm* **2017**, *19*, 4092–4117. [CrossRef]
76. Pendashteh, A.; Vilela, S.M.; Krivtsov, I.; Ávila-Brande, D.; Palma, J.; Horcajada, P.; Marcilla, R. Bimetal zeolitic imidazolate framework (ZIF-9) derived nitrogen-doped porous carbon as efficient oxygen electrocatalysts for rechargeable Zn-air batteries. *J. Power Sources* **2019**, *427*, 299–308. [CrossRef]
77. Li, Q.; Kim, H. Hydrogen production from NaBH4 hydrolysis via Co-ZIF-9 catalyst. *Fuel Process. Technol.* **2012**, *100*, 43–48. [CrossRef]
78. Nguyen, L.T.L.; Le, K.K.A.; Truong, H.X.; Phan, N.T.S. Metal–organic frameworks for catalysis: The Knoevenagel reaction using zeolite imidazolate framework ZIF-9 as an efficient heterogeneous catalyst. *Catal. Sci. Technol.* **2012**, *2*, 521–528. [CrossRef]
79. Babarao, R.; Jiang, Y.; Medhekar, N.V. Postcombustion CO_2 Capture in Functionalized Porous Coordination Networks. *J. Phys. Chem. C* **2013**, *117*, 26976–26987. [CrossRef]

80. Goswami, S.; Miller, C.E.; Logsdon, J.L.; Buru, C.T.; Wu, Y.-L.; Bowman, D.N.; Islamoglu, T.; Asiri, A.M.; Cramer, C.J.; Wasielewski, M.R.; et al. Atomistic Approach toward Selective Photocatalytic Oxidation of a Mustard-Gas Simulant: A Case Study with Heavy-Chalcogen-Containing PCN-57 Analogues. *ACS Appl. Mater. Interfaces* **2017**, *9*, 19535–19540. [CrossRef]
81. García-García, P.; Corma, A. Hf-based Metal-Organic Frameworks in Heterogeneous Catalysis. *Isr. J. Chem.* **2018**, *58*, 1062–1074. [CrossRef]
82. Hamon, L.; Heymans, N.; Llewellyn, P.L.; Guillerm, V.; Ghoufi, A.; Vaesen, S.; Maurin, G.; Serre, C.; De Weireld, G.; Pirngruber, G.D. Separation of CO_2–CH_4 mixtures in the mesoporous MIL-100(Cr) MOF: Experimental and modelling approaches. *Dalton Trans.* **2012**, *41*, 4052–4059. [CrossRef] [PubMed]
83. Opanasenko, M.; Dhakshinamoorthy, A.; Hwang, Y.K.; Chang, J.; Garcia, H.; Čejka, J. Superior Performance of Metal–Organic Frameworks over Zeolites as Solid Acid Catalysts in the Prins Reaction: Green Synthesis of Nopol. *ChemSusChem* **2013**, *6*, 865–871. [CrossRef] [PubMed]
84. Dissegna, S.; Epp, K.; Heinz, W.R.; Kieslich, G.; Fischer, R.A. Defective metal-organic frameworks. *Adv. Mater.* **2018**, *30*, 1704501. [CrossRef] [PubMed]
85. Zhang, Z.; Liu, Y.-W.; Tian, H.-R.; Li, X.-H.; Liu, S.-M.; Lu, Y.; Sun, Z.-X.; Liu, T. Polyoxometalate-Based Metal-Organic Framework Fractal Crystals. *Matter* **2020**, *2*, 250–260. [CrossRef]
86. Chen, L.; Xu, Q. Metal-Organic Framework Composites for Catalysis. *Matter* **2019**, *1*, 57–89. [CrossRef]
87. Hao, M.; Qiu, M.; Yang, H.; Hu, B.; Wang, X. Recent advances on preparation and environmental applications of MOF-derived carbons in catalysis. *Sci. Total. Environ.* **2021**, *760*, 143333. [CrossRef] [PubMed]
88. Pérez-Mayoral, E.; Godino-Ojer, M.; Matos, I.; Bernardo, M. Opportunities from Metal Organic Frameworks to Develop Porous Carbons Catalysts Involved in Fine Chemical Synthesis. *Catalysts* **2023**, *13*, 541. [CrossRef]
89. Wilson, K.; Lee, A.F. *Heterogeneous Catalysts for Clean Technology: Spectroscopy, Design, and Monitoring*; John Wiley & Sons: Hoboken, NJ, USA, 2014.
90. Konnerth, H.; Matsagar, B.M.; Chen, S.S.; Prechtl, M.H.; Shieh, F.-K.; Wu, K.C.-W. Metal-organic framework (MOF)-derived catalysts for fine chemical production. *Coord. Chem. Rev.* **2020**, *416*, 213319. [CrossRef]
91. Dhakshinamoorthy, A.; Opanasenko, M.; Čejka, J.; Garcia, H. Metal organic frameworks as heterogeneous catalysts for the production of fine chemicals. *Catal. Sci. Technol.* **2013**, *3*, 2509–2540. [CrossRef]
92. Chaikittisilp, W.; Ariga, K.; Yamauchi, Y. A new family of carbon materials: Synthesis of MOF-derived nanoporous carbons and their promising applications. *J. Mater. Chem. A* **2013**, *1*, 14–19. [CrossRef]
93. Wang, X.; Li, Y. Nanoporous carbons derived from MOFs as metal-free catalysts for selective aerobic oxidations. *J. Mater. Chem. A* **2016**, *4*, 5247–5257. [CrossRef]
94. Chen, Y.-Z.; Cai, G.; Wang, Y.; Xu, Q.; Yu, S.-H.; Jiang, H.-L. Palladium nanoparticles stabilized with N-doped porous carbons derived from metal–organic frameworks for selective catalysis in biofuel upgrade: The role of catalyst wettability. *Green Chem.* **2016**, *18*, 1212–1217. [CrossRef]
95. Dong, W.; Zhang, L.; Wang, C.; Feng, C.; Shang, N.; Gao, S.; Wang, C. Palladium nanoparticles embedded in metal–organic framework derived porous carbon: Synthesis and application for efficient Suzuki–Miyaura coupling reactions. *RSC Adv.* **2016**, *6*, 37118–37123. [CrossRef]
96. Van Nguyen, C.; Boo, J.R.; Liu, C.-H.; Ahamad, T.; Alshehri, S.M.; Matsagar, B.M.; Wu, K.C.-W. Oxidation of biomass-derived furans to maleic acid over nitrogen-doped carbon catalysts under acid-free conditions. *Catal. Sci. Technol.* **2020**, *10*, 1498–1506. [CrossRef]
97. Zhong, W.; Liu, H.; Bai, C.; Liao, S.; Li, Y. Base-Free Oxidation of Alcohols to Esters at Room Temperature and Atmospheric Conditions using Nanoscale Co-Based Catalysts. *ACS Catal.* **2015**, *5*, 1850–1856. [CrossRef]
98. Yao, X.; Bai, C.; Chen, J.; Li, Y. Efficient and selective green oxidation of alcohols by MOF-derived magnetic nanoparticles as a recoverable catalyst. *RSC Adv.* **2016**, *6*, 26921–26928. [CrossRef]
99. Tang, B.; Song, W.-C.; Yang, E.-C.; Zhao, X.-J. MOF-derived Ni-based nanocomposites as robust catalysts for chemoselective hydrogenation of functionalized nitro compounds. *RSC Adv.* **2017**, *7*, 1531–1539. [CrossRef]
100. Breck, D.W. Zeolite Molecular Sieves: Structure, Chemistry, and Use. *J. Chromatogr. Sci.* **1974**, *13*, 18A.
101. Rosas-Arbelaez, W.; Fijneman, A.J.; Friedrich, H.; Palmqvist, A.E.C. Hierarchical micro-/mesoporous zeolite microspheres prepared by colloidal assembly of zeolite nanoparticles. *RSC Adv.* **2020**, *10*, 36459–36466. [CrossRef]
102. Corma, A. State of the art and future challenges of zeolites as catalysts. *J. Catal.* **2003**, *216*, 298–312. [CrossRef]
103. Primo, A.; Garcia, H. Zeolites as catalysts in oil refining. *Chem. Soc. Rev.* **2014**, *43*, 7548–7561. [CrossRef]
104. Xu, F.; Wang, Y.; Wang, X.; Zhang, Y.; Tang, Y.; Yang, P.; Xu, F.; Wang, Y.; Wang, X.; Zhang, Y.; et al. A Novel Hierarchical Nanozeolite Composite as Sorbent for Protein Separation in Immobilized Metal-Ion Affinity Chromatography. *Adv. Mater.* **2003**, *15*, 1751–1753. [CrossRef]
105. Bacakova, L.; Vandrovcova, M.; Kopova, I.; Jirka, I. Applications of zeolites in biotechnology and medicine—A review. *Biomater. Sci.* **2018**, *6*, 974–989. [CrossRef] [PubMed]
106. Misaelides, P. Application of natural zeolites in environmental remediation: A short review. *Microporous Mesoporous Mater.* **2011**, *144*, 15–18. [CrossRef]
107. García-Martínez, J.; Li, K.; Davis, M.E. *Mesoporous Zeolites–Preparation, Characterization and Applications*; John Wiley & Sons: Hoboken, NJ, USA, 2015.

108. Huber, G.W.; Corma, A.; Huber, G.W.; Corma, A. Synergies between Bio- and Oil Refineries for the Production of Fuels from Biomass. *Angew. Chem. Int. Ed.* **2007**, *46*, 7184–7201. [CrossRef] [PubMed]
109. Corma, A.; Huber, G.W.; Sauvanaud, L.; Oconnor, P. Processing biomass-derived oxygenates in the oil refinery: Catalytic cracking (FCC) reaction pathways and role of catalyst. *J. Catal.* **2007**, *247*, 307–327. [CrossRef]
110. Kim, Y.-M.; Jeong, J.; Ryu, S.; Lee, H.W.; Jung, J.S.; Siddiqui, M.Z.; Jung, S.-C.; Jeon, J.-K.; Jae, J.; Park, Y.-K. Catalytic pyrolysis of wood polymer composites over hierarchical mesoporous zeolites. *Energy Convers. Manag.* **2019**, *195*, 727–737. [CrossRef]
111. Carlson, T.R.; Tompsett, G.A.; Conner, W.C.; Huber, G.W. Aromatic Production from Catalytic Fast Pyrolysis of Biomass-Derived Feedstocks. *Top. Catal.* **2009**, *52*, 241–252. [CrossRef]
112. Clark, J.H.; Deswarte, F.E.I.; Farmer, T.J. The integration of green chemistry into future biorefineries. *Biofuels, Bioprod. Biorefining* **2008**, *3*, 72–90. [CrossRef]
113. Erdmenger, T.; Guerrero-Sanchez, C.; Vitz, J.; Hoogenboom, R.; Schubert, U.S. Recent developments in the utilization of green solvents in polymer chemistry. *Chem. Soc. Rev.* **2010**, *39*, 3317–3333. [CrossRef]
114. Zhang, J.; Wang, L.; Ji, Y.; Chen, F.; Xiao, F.-S. Mesoporous zeolites for biofuel upgrading and glycerol conversion. *Front. Chem. Sci. Eng.* **2018**, *12*, 132–144. [CrossRef]
115. Egeblad, K.; Christensen, C.H.; Kustova, M.; Christensen, C.H. Templating Mesoporous Zeolites. *Chem. Mater.* **2008**, *20*, 946–960. [CrossRef]
116. Yue, Y.; Liu, H.; Yuan, P.; Li, T.; Yu, C.; Bi, H.; Bao, X. From natural aluminosilicate minerals to hierarchical ZSM-5 zeolites: A nanoscale depolymerization–reorganization approach. *J. Catal.* **2014**, *319*, 200–210. [CrossRef]
117. Groen, J.C.; Moulijn, J.A.; Pérez-Ramírez, J. Desilication: On the controlled generation of mesoporosity in MFI zeolites. *J. Mater. Chem.* **2006**, *16*, 2121–2131. [CrossRef]
118. Tao, Y.; Kanoh, H.; Abrams, L.; Kaneko, K. Mesopore-Modified Zeolites: Preparation, Characterization, and Applications. *Chem. Rev.* **2006**, *106*, 896–910. [CrossRef] [PubMed]
119. van Donk, S.; Janssen, A.H.; Bitter, J.H.; de Jong, K.P. Generation, Characterization, and Impact of Mesopores in Zeolite Catalysts. *Catal. Rev.* **2003**, *45*, 297–319. [CrossRef]
120. Madsen, C.; Jacobsen, C.J.H. Nanosized zeolite crystals—Convenient control of crystal size distribution by confined space synthesis. *Chem. Commun.* **1999**, *8*, 673–674. [CrossRef]
121. Schmidt, I.; Madsen, C.; Jacobsen, C.J.H. Confined Space Synthesis. A Novel Route to Nanosized Zeolites. *Inorg. Chem.* **2000**, *39*, 2279–2283. [CrossRef]
122. Jacobsen, C.J.; Madsen, C.; Janssens, T.V.; Jakobsen, H.J.; Skibsted, J. Zeolites by confined space synthesis—Characterization of the acid sites in nanosized ZSM-5 by ammonia desorption and 27Al/29Si-MAS NMR spectroscopy. *Microporous Mesoporous Mater.* **2000**, *39*, 393–401. [CrossRef]
123. Tao, Y.; Kanoh, H.; Kaneko, K. Uniform Mesopore-Donated Zeolite Y Using Carbon Aerogel Templating. *J. Phys. Chem. B* **2003**, *107*, 10974–10976. [CrossRef]
124. Kim, S.-S.; Shah, J.; Pinnavaia, T.J. Colloid-Imprinted Carbons as Templates for the Nanocasting Synthesis of Mesoporous ZSM-5 Zeolite. *Chem. Mater.* **2003**, *15*, 1664–1668. [CrossRef]
125. Tao, Y.; Kanoh, H.; Kaneko, K. ZSM-5 Monolith of Uniform Mesoporous Channels. *J. Am. Chem. Soc.* **2003**, *125*, 6044–6045. [CrossRef]
126. Tao, Y.; Kanoh, H.; Hanzawa, Y.; Kaneko, K. Template synthesis and characterization of mesoporous zeolites. *Colloids Surf. A Physicochem. Eng. Asp.* **2004**, *241*, 75–80. [CrossRef]
127. Xiao, F.S.; Wang, L.; Yin, C.; Lin, K.; Di, Y.; Li, J.; Xu, R.; Su, D.S.; Schlögl, R.; Yokoi, T.; et al. Catalytic properties of hierarchical mesoporous zeolites templated with a mixture of small organic ammonium salts and mesoscale cationic polymers. *Angew. Chem.-Int. Ed.* **2006**, *45*, 3090–3093. [CrossRef]
128. Wang, H.; Pinnavaia, T.J. MFI Zeolite with Small and Uniform Intracrystal Mesopores. *Angew. Chem.* **2006**, *118*, 7765–7768. [CrossRef]
129. Naydenov, V.; Tosheva, L.; Sterte, J. Vanadium modified AlPO-5 spheres through resin macrotemplating. *Microporous Mesoporous Mater.* **2003**, *66*, 321–329. [CrossRef]
130. Valtchev, V.; Smaihi, M.; Faust, A.C.; Vidal, L. Dual templating function of Equisetum arvense in the preparation of zeolite macrostructures. *Stud. Surf. Sci. Catal.* **2004**, *154*, 588–592. [CrossRef]
131. Corma, A.; Fornes, V.; Rey, F. Delaminated Zeolites: An Efficient Support for Enzymes. *Adv. Mater.* **2002**, *14*, 71–74. [CrossRef]
132. Corma, A.; Fornes, V.; Pergher, S.B.; Maesen, T.L.M.; Buglass, J.G. Delaminated zeolite precursors as selective acidic catalysts. *Nature* **1998**, *396*, 353–356. [CrossRef]
133. Nielsen, M.; Brogaard, R.Y.; Falsig, H.; Beato, P.; Swang, O.; Svelle, S. Kinetics of Zeolite Dealumination: Insights from H-SSZ-13. *ACS Catal.* **2015**, *5*, 7131–7139. [CrossRef]
134. Poliakoff, M.; Licence, P. Green chemistry. *Nature* **2007**, *450*, 810–812. [CrossRef] [PubMed]
135. Li, J.; Gao, M.; Yan, W.; Yu, J. Regulation of the Si/Al ratios and Al distributions of zeolites and their impact on properties. *Chem. Sci.* **2023**, *14*, 1935–1959. [CrossRef] [PubMed]
136. Luo, P.; Xu, H.; Xue, T.; Jiang, J.-G.; Wu, H.-H.; He, M.; Wu, P. Postsynthesis of high silica beta by cannibalistic dealumination of OSDA-free beta and its catalytic applications. *Inorg. Chem. Front.* **2021**, *8*, 1574–1587. [CrossRef]

137. Pilar, R.; Moravkova, J.; Sadovska, G.; Sklenak, S.; Brabec, L.; Pastvova, J.; Sazama, P. Controlling the competitive growth of zeolite phases without using an organic structure-directing agent. Syn-thesis of Al-rich*BEA. *Microporous Mesoporous Mater.* **2022**, *333*, 111726. [CrossRef]
138. Agostini, G.; Lamberti, C.; Palin, L.; Milanesio, M.; Danilina, N.; Xu, B.; Janousch, M.; van Bokhoven, J.A. In Situ XAS and XRPD Parametric Rietveld Refinement To Understand Dealumination of Y Zeolite Catalyst. *J. Am. Chem. Soc.* **2010**, *132*, 667–678. [CrossRef] [PubMed]
139. Oleksiak, M.D.; Muraoka, K.; Hsieh, M.F.; Conato, M.T.; Shimojima, A.; Okubo, T.; Chaikittisilp, W.; Rimer, J.D. Back Cover: Organic-Free Synthesis of a Highly Siliceous Faujasite Zeolite with Spatially Biased Q4 (nAl) Si Speciation. *Angew. Chem. Int. Ed.* **2017**, *56*, 13532. [CrossRef]
140. Meng, B.; Ren, S.; Li, Z.; Nie, S.; Zhang, X.; Song, W.; Guo, Q.; Shen, B. A facile organic-free synthesis of high silica zeolite Y with small crystal in the presence of Co^{2+}. *Microporous Mesoporous Mater.* **2021**, *323*, 111248. [CrossRef]
141. Zhang, H.; Wu, C.; Song, M.; Lu, T.; Wang, W.; Wang, Z.; Yan, W.; Cheng, P.; Zhao, Z. Accelerated synthesis of Al-rich zeolite beta via different radicalized seeds in the absence of organic templates. *Microporous Mesoporous Mater.* **2021**, *310*, 110633. [CrossRef]
142. Loewenstein, W. The distribution of aluminum in the tetrahedra of silicates and aluminates. *Am. Mineral. J. Earth Planet. Mater.* **1954**, *39*, 92–96.
143. He, Z.; Wu, J.; Gao, B.; He, H. Hydrothermal Synthesis and Characterization of Aluminum-Free Mn-β Zeolite: A Catalyst for Phenol Hydroxylation. *ACS Appl. Mater. Interfaces* **2015**, *7*, 2424–2432. [CrossRef]
144. Grass, J.-P.; Klühspies, K.; Reiprich, B.; Schwieger, W.; Inayat, A. Layer-Like Zeolite X as Catalyst in a Knoevenagel Condensation: The Effect of Different Preparation Pathways and Cation Exchange. *Catalysts* **2021**, *11*, 474. [CrossRef]
145. Zhang, F.; Yuan, C.; Wang, J.; Kong, Y.; Zhu, H.; Wang, C. Synthesis of fructone over dealmuinated USY supported heteropoly acid and its salt catalysts. *J. Mol. Catal. A Chem.* **2006**, *247*, 130–137. [CrossRef]
146. Wang, Q.; Han, W.; Lyu, J.; Zhang, Q.; Guo, L.; Li, X. *In situ* encapsulation of platinum clusters within H-ZSM-5 zeolite for highly stable benzene methylation catalysis. *Catal. Sci. Technol.* **2017**, *7*, 6140–6150. [CrossRef]
147. Kostrab, G.; Mravec, D.; Bajus, M.; Janotka, I.; Sugi, Y.; Cho, S.; Kim, J. tert-Butylation of toluene over mordenite and cerium-modified mordenite catalysts. *Appl. Catal. A Gen.* **2006**, *299*, 122–130. [CrossRef]
148. Pellet, R.; Casey, D.; Huang, H.; Kessler, R.; Kuhlman, E.; Oyoung, C.; Sawicki, R.; Ugolini, J. Isomerization of n-Butene to Isobutene by Ferrierite and Modified Ferrierite Catalysts. *J. Catal.* **1995**, *157*, 423–435. [CrossRef]
149. da Silva, J.F.; Ferracine, E.D.d.S.; Cardoso, D. Improved accessibility of Na-LTA zeolite catalytic sites for the Knoevenagel condensation reaction. *Microporous Mesoporous Mater.* **2022**, *331*, 111640. [CrossRef]
150. Yaghi, O.M.; O'Keeffe, M.; Ockwig, N.W.; Chae, H.K.; Eddaoudi, M.; Kim, J. Reticular synthesis and the design of new materials. *Nature* **2003**, *423*, 705–714. [CrossRef]
151. Phan, A.; Doonan, C.J.; Uribe-Romo, F.J.; Knobler, C.B.; O'Keeffe, M.; Yaghi, O.M. Synthesis, Structure, and Carbon Dioxide Capture Properties of Zeolitic Imidazolate Frameworks. *Accounts Chem. Res.* **2009**, *43*, 58–67. [CrossRef] [PubMed]
152. Park, K.S.; Ni, Z.; Côté, A.P.; Choi, J.Y.; Huang, R.; Uribe-Romo, F.J.; Chae, H.K.; O'Keeffe, M.; Yaghi, O.M. Exceptional chemical and thermal stability of zeolitic imidazolate frameworks. *Proc. Natl. Acad. Sci. USA* **2006**, *103*, 10186–10191. [CrossRef]
153. Kouser, S.; Hezam, A.; Khadri, M.J.N.; Khanum, S.A. A review on zeolite imidazole frameworks: Synthesis, properties, and applications. *J. Porous Mater.* **2022**, *29*, 663–681. [CrossRef]
154. Yang, T.; Chung, T.-S. Room-temperature synthesis of ZIF-90 nanocrystals and the derived nano-composite membranes for hydrogen separation. *J. Mater. Chem. A* **2013**, *1*, 6081–6090. [CrossRef]
155. Gross, A.F.; Sherman, E.; Vajo, J.J. Aqueous room temperature synthesis of cobalt and zinc sodalite zeolitic imidizolate frameworks. *Dalton Trans.* **2012**, *41*, 5458–5460. [CrossRef] [PubMed]
156. Pan, Y.; Liu, Y.; Zeng, G.; Zhao, L.; Lai, Z. Rapid synthesis of zeolitic imidazolate framework-8 (ZIF-8) nanocrystals in an aqueous system. *Chem. Commun.* **2011**, *47*, 2071–2073. [CrossRef] [PubMed]
157. Qian, J.; Sun, F.; Qin, L. Hydrothermal synthesis of zeolitic imidazolate framework-67 (ZIF-67) nanocrystals. *Mater. Lett.* **2012**, *82*, 220–223. [CrossRef]
158. Pan, Y.; Heryadi, D.; Zhou, F.; Zhao, L.; Lestari, G.; Su, H.; Lai, Z. Tuning the crystal morphology and size of zeolitic imidazolate framework-8 in aqueous solution by surfactants. *CrystEngComm* **2011**, *13*, 6937–6940. [CrossRef]
159. Xiong, W.; Zhang, Q. Surfactants as Promising Media for the Preparation of Crystalline Inorganic Materials. *Angew. Chem. Int. Ed.* **2015**, *54*, 11616–11623. [CrossRef] [PubMed]
160. Ni, Z.; Masel, R.I. Rapid Production of Metal–Organic Frameworks via Microwave-Assisted Solvothermal Synthesis. *J. Am. Chem. Soc.* **2006**, *128*, 12394–12395. [CrossRef]
161. Martins, G.A.V.; Byrne, P.J.; Allan, P.; Teat, S.J.; Slawin, A.M.Z.; Li, Y.; Morris, R.E. The use of ionic liquids in the synthesis of zinc imidazolate frameworks. *Dalton Trans.* **2010**, *39*, 1758–1762. [CrossRef]
162. Yang, L.; Lu, H. Microwave-assisted Ionothermal Synthesis and Characterization of Zeolitic Imidazolate Framework-8. *Chin. J. Chem.* **2012**, *30*, 1040–1044. [CrossRef]
163. Shi, Q.; Chen, Z.; Song, Z.; Li, J.; Dong, J. Synthesis of ZIF-8 and ZIF-67 by Steam-Assisted Conversion and an Investigation of Their Tribological Behaviors. *Angew. Chem.* **2011**, *50*, 672–675. [CrossRef]
164. Jiang, H.; Liu, M.; Zhou, M.; Du, Y.; Chen, R. Hierarchical Pd@ZIFs as Efficient Catalysts for *p*-Nitrophenol Reduction. *Ind. Eng. Chem. Res.* **2021**, *60*, 15045–15055. [CrossRef]

165. Luo, Z.; Chaemchuen, S.; Zhou, K.; Verpoort, F. Ring-Opening Polymerization of l-Lactide to Cyclic Poly(Lactide) by Zeolitic Imidazole Framework ZIF-8 Catalyst. *ChemSusChem* **2017**, *10*, 4135–4139. [CrossRef]
166. Peralta, D.; Chaplais, G.; Simon-Masseron, A.; Barthelet, K.; Chizallet, C.; Quoineaud, A.-A.; Pirngruber, G.D. Comparison of the Behavior of Metal–Organic Frameworks and Zeolites for Hydrocarbon Separations. *J. Am. Chem. Soc.* **2012**, *134*, 8115–8126. [CrossRef]
167. Chizallet, C.; Lazare, S.; Bazer-Bachi, D.; Bonnier, F.; Lecocq, V.; Soyer, E.; Quoineaud, A.-A.; Bats, N. Catalysis of Transesterification by a Nonfunctionalized Metal−Organic Framework: Acido-Basicity at the External Surface of ZIF-8 Probed by FTIR and *ab Initio* Calculations. *J. Am. Chem. Soc.* **2010**, *132*, 12365–12377. [CrossRef]
168. Thomas, A.; Prakash, M. The Role of Binary Mixtures of Ionic Liquids in ZIF-8 for Selective Gas Storage and Separation: A Perspective from Computational Approaches. *J. Phys. Chem. C* **2020**, *124*, 26203–26213. [CrossRef]
169. Denning, S.; Majid, A.A.A.; Lucero, J.M.; Crawford, J.M.; Carreon, M.A.; Koh, C.A. Methane Hydrate Growth Promoted by Microporous Zeolitic Imidazolate Frameworks ZIF-8 and ZIF-67 for Enhanced Methane Storage. *ACS Sustain. Chem. Eng.* **2021**, *9*, 9001–9010. [CrossRef]
170. Wang, Q.; Liu, T.; Chen, Y.; Wang, Q. The Ni/Ni3S2 nanocomposite derived from Ni-ZIF with superior energy storage performance as cathodes for asymmetric supercapacitor and rechargeable aqueous zinc ion battery. *J. Alloys Compd.* **2022**, *891*, 161935. [CrossRef]
171. Dou, J.; Bian, W.; Zheng, X.; Yue, Q.; Song, Q.; Deng, S.; Wang, L.; Tan, W.; Li, W.; Zhou, B. A ZIF-based drug delivery system as three-in-one platform for joint cancer therapy. *Mater. Chem. Phys.* **2023**, *297*, 127345. [CrossRef]
172. Shi, L.; Wu, J.; Qiao, X.; Ha, Y.; Li, Y.; Peng, C.; Wu, R. In Situ Biomimetic Mineralization on ZIF-8 for Smart Drug Delivery. *ACS Biomater. Sci. Eng.* **2020**, *6*, 4595–4603. [CrossRef]
173. Mehdipour-Ataei, S.; Aram, E. Mesoporous Carbon-Based Materials: A Review of Synthesis, Modification, and Applications. *Catalysts* **2023**, *13*, 2. [CrossRef]
174. Wang, J.; Chen, Y.; Zhang, S.; Yang, C.; Zhang, J.Y.; Su, Y.; Zheng, G.; Fang, X. Controllable States and Porosity of Cu-Carbon for CO_2 Electroreduction to Hydrocarbons. *Small* **2022**, *18*, 2238. [CrossRef] [PubMed]
175. Wu, J.; Sharifi, T.; Gao, Y.; Zhang, T.; Ajayan, P.M. Emerging Carbon-Based Heterogeneous Catalysts for Electrochemical Reduction of Carbon Dioxide into Value-Added Chemicals. *Adv. Mater.* **2019**, *31*, e1804257. [CrossRef] [PubMed]
176. Inagaki, M.; Toyoda, M.; Soneda, Y.; Tsujimura, S.; Morishita, T. Templated mesoporous carbons: Synthesis and applications. *Carbon* **2016**, *107*, 448–473. [CrossRef]
177. Mui, E.L.; Ko, D.C.; McKay, G. Production of active carbons from waste tyres—A review. *Carbon* **2004**, *42*, 2789–2805. [CrossRef]
178. Yang, T.; Lua, A.C. Characteristics of activated carbons prepared from pistachio-nut shells by physical activation. *J. Colloid Interface Sci.* **2003**, *267*, 408–417. [CrossRef] [PubMed]
179. Bergna, D.; Hu, T.; Prokkola, H.; Romar, H.; Lassi, U. Effect of Some Process Parameters on the Main Properties of Activated Carbon Produced from Peat in a Lab-Scale Process. *Waste Biomass-Valorization* **2020**, *11*, 2837–2848. [CrossRef]
180. Marsh, H.; Rodríguez-Reinoso, F. *Activated Carbon*, 1st ed.; Elsevier: Amsterdam, The Netherlands, 2006.
181. Xu, B.; Chen, Y.; Wei, G.; Cao, G.; Zhang, H.; Yang, Y. Activated carbon with high capacitance prepared by NaOH activation for supercapacitors. *Mater. Chem. Phys.* **2010**, *124*, 504–509. [CrossRef]
182. Guo, Y.; Rockstraw, D.A. Physical and chemical properties of carbons synthesized from xylan, cellulose, and Kraft lignin by H3PO4 activation. *Carbon* **2006**, *44*, 1464–1475. [CrossRef]
183. Xing, Z.; Qi, Y.; Tian, Z.; Xu, J.; Yuan, Y.; Bommier, C.; Lu, J.; Tong, W.; Jiang, D.-E.; Ji, X. Identify the Removable Substructure in Carbon Activation. *Chem. Mater.* **2017**, *29*, 7288–7295. [CrossRef]
184. Smith, M.R.; Bittner, E.W.; Shi, W.; Johnson, J.K.; Bockrath, B.C. Chemical Activation of Single-Walled Carbon Nanotubes for Hydrogen Adsorption. *J. Phys. Chem. B* **2003**, *107*, 3752–3760. [CrossRef]
185. Kyotani, T. Control of pore structure in carbon. *Carbon* **2000**, *38*, 269–286. [CrossRef]
186. Tanaka, S.; Nishiyama, N.; Egashira, Y.; Ueyama, K. Synthesis of ordered mesoporous carbons with channel structure from an organic–organic nanocomposite. *Chem. Commun.* **2005**, *16*, 2125–2127. [CrossRef] [PubMed]
187. Tanaka, S.; Katayama, Y.; Tate, M.P.; Hillhouse, H.W.; Miyake, Y. Fabrication of continuous mesoporous carbon films with face-centered orthorhombic symmetry through a soft templating pathway. *J. Mater. Chem.* **2007**, *17*, 3639–3645. [CrossRef]
188. Javed, H.; Pani, S.; Antony, J.; Sakthivel, M.; Drillet, J.-F. Synthesis of mesoporous carbon spheres *via* a soft-template route for catalyst supports in PEMFC cathodes. *Soft Matter* **2021**, *17*, 7743–7754. [CrossRef] [PubMed]
189. Bonneviot, L.; Béland, F.; Danumah, C.; Giasson, S.; Kaliaguine, S. *Mesoporous Molecular Sieves;* Elsevier: Amsterdam, The Netherlands, 1998.
190. Corma, A.; Iborra, S.; Miquel, S.; Primo, J. Catalysts for the Production of Fine Chemicals: Production of Food Emulsifiers, Monoglycerides, by Glycerolysis of Fats with Solid Base Catalysts. *J. Catal.* **1998**, *173*, 315–321. [CrossRef]
191. Jaroniec, C.P.; Kruk, M.; Jaroniec, M.; Sayari, A. Tailoring Surface and Structural Properties of MCM-41 Silicas by Bonding Organosilanes. *J. Phys. Chem. B* **1998**, *102*, 5503–5510. [CrossRef]
192. Wang, Q.; Shi, L.; Lu, A. Highly Selective Copper Catalyst Supported on Mesoporous Carbon for the Dehydrogenation of Ethanol to Acetaldehyde. *ChemCatChem* **2015**, *7*, 2846–2852. [CrossRef]
193. Liu, X.; Lan, G.; Su, P.; Qian, L.; Reina, T.R.; Wang, L.; Li, Y.; Liu, J. Highly stable Ru nanoparticles incorporated in mesoporous carbon catalysts for production of γ-valerolactone. *Catal. Today* **2020**, *351*, 75–82. [CrossRef]

194. Budd, P.M.; Makhseed, S.M.; Ghanem, B.S.; Msayib, K.J.; Tattershall, C.; McKeown, N.B. Microporous polymeric materials. *Mater. Today* **2004**, *7*, 40–46. [CrossRef]
195. Attfield, M.P. Microporous Materials. *Sci. Prog.* **2002**, *85*, 319–345. [CrossRef]
196. CejkaM, J. *Introduction to Zeolite Science and Practice*, 3rd ed.; Elsevier: Amsterdam, The Netherlands, 1998.
197. Kumar, N. Microporous Zeolites and Related Nanoporous Materials: Synthesis, Characterization and Application in Catalysis. *Catalysts* **2021**, *11*, 382. [CrossRef]
198. Kamaluddin, H.S.; Basahel, S.N.; Narasimharao, K.; Mokhtar, M. H-ZSM-5 Materials Embedded in an Amorphous Silica Matrix: Highly Selective Catalysts for Propylene in Methanol-to-Olefin Process. *Catalysts* **2019**, *9*, 364. [CrossRef]
199. Kumar, N.; Mäki-Arvela, P.; Diáz, S.F.; Aho, A.; Demidova, Y.; Linden, J.; Shepidchenko, A.; Tenhu, M.; Salonen, J.; Laukkanen, P.; et al. Isomerization of α-Pinene Oxide Over Iron-Modified Zeolites. *Top. Catal.* **2013**, *56*, 696–713. [CrossRef]
200. Vajglová, Z.; Kumar, N.; Peurla, M.; Hupa, L.; Semikin, K.; Sladkovskiy, D.A.; Murzin, D.Y. Effect of the Preparation of Pt-Modified Zeolite Beta-Bentonite Extrudates on Their Catalytic Behavior in *n*-Hexane Hydroisomerization. *Ind. Eng. Chem. Res.* **2019**, *58*, 10875–10885. [CrossRef]
201. Vajglová, Z.; Gauli, B.; Mäki-Arvela, P.; Kumar, N.; Eränen, K.; Wärnå, J.; Lassfolk, R.; Simakova, I.L.; Prosvirin, I.P.; Peurla, M.; et al. Interactions between Iron and Nickel in Fe–Ni Nanoparticles on Y Zeolite for Co-Processing of Fossil Feedstock with Lignin-Derived Isoeugenol. *ACS Appl. Nano Mater.* **2023**, *6*, 10064–10077. [CrossRef]
202. Vajglová, Z.; Kumar, N.; Mäki-Arvela, P.; Eränen, K.; Peurla, M.; Hupa, L.; Nurmi, M.; Toivakka, M.; Murzin, D.Y. Synthesis and Physicochemical Characterization of Shaped Catalysts of β and Y Zeolites for Cyclization of Citronellal. *Ind. Eng. Chem. Res.* **2019**, *58*, 18084–18096. [CrossRef]
203. Vajglová, Z.; Kumar, N.; Mäki-Arvela, P.; Eränen, K.; Peurla, M.; Hupa, L.; Murzin, D.Y. Effect of Binders on the Physicochemical and Catalytic Properties of Extrudate-Shaped Beta Zeolite Catalysts for Cyclization of Citronellal. *Org. Process. Res. Dev.* **2019**, *23*, 2456–2463. [CrossRef]
204. Aho, A.; Kumar, N.; Eränen, K.; Mäki-Arvela, P.; Salmi, T.; Peurla, M.; Angervo, I.; Hietala, J.; Murzin, D.Y. Catalytic conversion of glucose to methyl levulinate over metal-modified Beta zeolites. *React. Kinet. Catal. Lett.* **2022**, *135*, 1971–1986. [CrossRef]
205. Aho, A.; Kumar, N.; Eränen, K.; Lassfolk, R.; Mäki-Arvela, P.; Salmi, T.; Peurla, M.; Angervo, I.; Hietala, J.; Murzin, D.Y. Improving the methyl lactate yield from glucose over Sn–Al-Beta zeolite by catalyst promoters. *Microporous Mesoporous Mater.* **2023**, *351*, 112483. [CrossRef]
206. Zaykovskaya, A.O.; Kumar, N.; Kholkina, E.A.; Li-Zhulanov, N.S.; Mäki-Arvela, P.; Aho, A.; Peltonen, J.; Peurla, M.; Heinmaa, I.; Kusema, B.T.; et al. Synthesis and physico-chemical characterization of Beta zeolite catalysts: Evaluation of catalytic properties in Prins cyclization of (−)-isopulegol. *Microporous Mesoporous Mater.* **2020**, *302*, 110236. [CrossRef]
207. Stekrova, M.; Kumar, N.; Díaz, S.; Mäki-Arvela, P.; Murzin, D.Y. H- and Fe-modified zeolite beta catalysts for preparation of trans-carveol from α-pinene oxide. *Catal. Today* **2015**, *241*, 237–245. [CrossRef]
208. Ma, H.; Chen, J.-J.; Tan, L.; Bu, J.-H.; Zhu, Y.; Tan, B.; Zhang, C. Nitrogen-Rich Triptycene-Based Porous Polymer for Gas Storage and Iodine Enrichment. *ACS Macro Lett.* **2016**, *5*, 1039–1043. [CrossRef]
209. Hu, X.; Radosz, M.; Cychosz, K.A.; Thommes, M. CO_2-Filling Capacity and Selectivity of Carbon Nanopores: Synthesis, Texture, and Pore-Size Distribution from Quenched-Solid Density Functional Theory (QSDFT). *Environ. Sci. Technol.* **2011**, *45*, 7068–7074. [CrossRef]
210. Yahya, M.A.; Mansor, M.H.; Zolkarnaini, W.A.A.W.; Rusli, N.S.; Aminuddin, A.; Mohamad, K.; Sabhan, F.A.M.; Atik, A.A.A.; Ozair, L.N. A brief review on activated carbon derived from agriculture by-product. In Proceedings of the International Conference on Recent Advancements in Science and Technology 2017 (ICoRAST2017), Melaka, Malaysia, 7–8 November 2017; p. 030023.
211. Adeleye, A.T.; Akande, A.A.; Odoh, C.K.; Philip, M.; Fidelis, T.T.; Amos, P.I.; Banjoko, O.O. Efficient synthesis of bio-based activated carbon (AC) for catalytic systems: A green and sustainable approach. *J. Ind. Eng. Chem.* **2021**, *96*, 59–75. [CrossRef]
212. Mohammed, M.H.; Gheni, S.A.; Hamad, K.I.; Mohammed, A.E.; Hmood, H.M.; Mahomood, M.A.; Mohammed, H.R.; Abdulwahab, Z.T.; Ahmed, S.M.; Hassan, A.A. Microporous activated carbon catalyst for an efficient and deactivation resistive supercritical water upgrading process of sour crude oil. *Diam. Relat. Mater.* **2023**, *135*, 109887. [CrossRef]
213. Bedia, J.; Peñas-Garzón, M.; Gómez-Avilés, A.; Rodriguez, J.J.; Belver, C. A Review on the Synthesis and Characterization of Biomass-Derived Carbons for Adsorption of Emerging Contaminants from Water. *J. Carbon Res.* **2018**, *4*, 63. [CrossRef]
214. Al-Qayim, K.; Nimmo, W.; Hughes, K.; Pourkashanian, M. Kinetic parameters of the intrinsic reactivity of woody biomass and coal chars via thermogravimetric analysis. *Fuel* **2017**, *210*, 811–825. [CrossRef]
215. Zhou, J.-H.; Sui, Z.-J.; Li, P.; Chen, D.; Dai, Y.-C.; Yuan, W.-K. Structural characterization of carbon nanofibers formed from different carbon-containing gases. *Carbon* **2006**, *44*, 3255–3262. [CrossRef]
216. Kvande, I.; Zhu, J.; Zhao, T.-J.; Hammer, N.; Rønning, M.; Raaen, S.; Walmsley, J.C.; Chen, D. Importance of Oxygen-Free Edge and Defect Sites for the Immobilization of Colloidal Pt Oxide Particles with Implications for the Preparation of CNF-Supported Catalysts. *J. Phys. Chem. C* **2010**, *114*, 1752–1762. [CrossRef]
217. Kvande, I.; Chen, D.; Zhao, T.-J.; Skoe, I.M.; Walmsley, J.C.; Rønning, M. Hydrogen Oxidation Catalyzed by Pt Supported on Carbon Nanofibers with Different Graphite Sheet Orientations. *Top. Catal.* **2009**, *52*, 664–674. [CrossRef]
218. Zhang, L.-Y.; Wang, M.-R.; Lai, Y.-Q.; Li, X.-Y. Nitrogen-doped microporous carbon: An efficient oxygen reduction catalyst for Zn-air batteries. *J. Power Sources* **2017**, *359*, 71–79. [CrossRef]

219. Abushawish, A.; Almanassra, I.W.; Backer, S.N.; Jaber, L.; Khalil, A.K.; Abdelkareem, M.A.; Sayed, E.T.; Alawadhi, H.; Shanableh, A.; Atieh, M.A. High-efficiency removal of hexavalent chromium from contaminated water using nitrogen-doped activated carbon: Kinetics and isotherm study. *Mater. Chem. Phys.* **2022**, *291*, 126758. [CrossRef]
220. Quan, C.; Jia, X.; Gao, N. Nitrogen-doping activated biomass carbon from tea seed shell for CO_2 capture and supercapacitor. *Int. J. Energy Res.* **2020**, *44*, 1218–1232. [CrossRef]
221. Li, X.; Wang, Q.; Ding, C.; Meng, Y.; Wang, J.; Zhang, K.; Liu, P. Nickel–copper catalysts supported by boron and nitrogen co-doped activated carbon for gas phase carbonylation of ethanol. *J. Porous Mater.* **2023**, *30*, 1575–1585. [CrossRef]
222. Pan, S.-F.; Yin, J.-L.; Zhu, X.-L.; Guo, X.-J.; Hu, P.; Yan, X.; Lang, W.-Z.; Guo, Y.-J. P-modified microporous carbon nanospheres for direct propane dehydrogenation reactions. *Carbon* **2019**, *152*, 855–864. [CrossRef]
223. Jiao, C.; Xu, J.L.; Chen, X.Y.; Zhang, Z.J. Design and synthesis of phosphomolybdic acid/silver dual-modified microporous carbon composite for high performance supercapacitors. *J. Alloys Compd.* **2019**, *791*, 1005–1014. [CrossRef]
224. Martínez-Klimov, M.E.; Mäki-Arvela, P.; Vajglova, Z.; Alda-Onggar, M.; Angervo, I.; Kumar, N.; Eränen, K.; Peurla, M.; Calimli, M.H.; Muller, J.; et al. Hydrodeoxygenation of Isoeugenol over Carbon-Supported Pt and Pt–Re Catalysts for Production of Renewable Jet Fuel. *Energy Fuels* **2021**, *35*, 17755–17768. [CrossRef]
225. Wang, D.; Huang, B.; Shi, Z.; Long, H.; Li, L.; Yang, Z.; Dai, M. Influence of cerium doping on Cu–Ni/activated carbon low-temperature CO-SCR denitration catalysts. *RSC Adv.* **2021**, *11*, 18458–18467. [CrossRef]
226. Abdulkareem-Alsultan, G.; Asikin-Mijan, N.; Mustafa-Alsultan, G.; Lee, H.V.; Wilson, K.; Taufiq-Yap, Y.H. Efficient deoxygenation of waste cooking oil over Co_3O_4–La_2O_3-doped activated carbon for the production of diesel-like fuel. *RSC Adv.* **2020**, *10*, 4996–5009. [CrossRef]
227. Mäki-Arvela, P.; Martin, G.; Simakova, I.; Tokarev, A.; Wärnå, J.; Hemming, J.; Holmbom, B.; Salmi, T.; Murzin, D. Kinetics, catalyst deactivation and modeling in the hydrogenation of β-sitosterol to β-sitostanol over microporous and mesoporous carbon supported Pd catalysts. *Chem. Eng. J.* **2009**, *154*, 45–51. [CrossRef]
228. Davis, M.E. Ordered porous materials for emerging applications. *Nature* **2002**, *417*, 813–821. [CrossRef] [PubMed]
229. Pastore, H.; Coluccia, S.; Marchese, L. Porous aluminophosphates: From Molecular Sieves to Designed Acid Catalysts. *Annu. Rev. Mater. Res.* **2005**, *35*, 351–395. [CrossRef]
230. Corà, F.; Gómez-Hortigüela, L.; Catlow, C.R.A. Aerobic oxidation of hydrocarbons in Mn-doped aluminophosphates: A computational perspective to understand mechanism and selectivity. *Proc. R. Soc. A Math. Phys. Eng. Sci.* **2012**, *468*, 2053–2069. [CrossRef]
231. Gómez-Hortigüela, L.; Corà, F.; Catlow, C.R.A. Complementary mechanistic properties of Fe- and Mn-doped aluminophosphates in the catalytic aerobic oxidation of hydrocarbons. *Phys. Chem. Chem. Phys.* **2013**, *15*, 6870–6874. [CrossRef] [PubMed]
232. Dawaymeh, F.; Elmutasim, O.; Gaber, D.; Gaber, S.; Reddy, K.S.K.; Basina, G.; Polychronopoulou, K.; Al Wahedi, Y.; Karanikolos, G.N. Metal substitution effects of aluminophosphate AlPO4-5 as solid acid catalyst for esterification of acetic acid with ethanol. *Mol. Catal.* **2021**, *501*, 111371. [CrossRef]
233. Nagaraju, N.; Kuriakose, G. A new catalyst for the synthesis of N,N-biphenylurea from aniline and dimethyl carbonate. *Green Chem.* **2002**, *4*, 269–271. [CrossRef]
234. Vijayasankar, A.V.; Mahadevaiah, N.; Bhat, Y.S.; Nagaraju, N. Mesoporous aluminophosphate materials: Influence of method of preparation and iron loading on textural properties and catalytic activity. *J. Porous Mater.* **2011**, *18*, 369–378. [CrossRef]
235. Ahmadi, M.; Ebrahimnia, M.; Shahbazi, M.-A.; Keçili, R.; Ghorbani-Bidkorbeh, F. Microporous metal–organic frameworks: Synthesis and applications. *J. Ind. Eng. Chem.* **2022**, *115*, 1–11. [CrossRef]
236. Pal, A.; Chand, S.; Madden, D.G.; Franz, D.; Ritter, L.; Johnson, A.; Space, B.; Curtin, T.; Das, M.C. A Microporous Co-MOF for Highly Selective CO_2 Sorption in High Loadings Involving Aryl C–H···O=C=O Interactions: Combined Simulation and Breakthrough Studies. *Inorg. Chem.* **2019**, *58*, 11553–11560. [CrossRef]
237. Pal, A.; Chand, S.; Elahi, S.M.; Das, M.C. A microporous MOF with a polar pore surface exhibiting excellent selective adsorption of $CO_2$from CO_2–$N_2$and CO_2–CH_4gas mixtures with high CO_2loading. *Dalton Trans.* **2017**, *46*, 15280–15286. [CrossRef]
238. Kumar, A.; Maurya, R.A. Synthesis of polyhydroquinoline derivatives through unsymmetric Hantzsch reaction using organocatalysts. *Tetrahedron* **2007**, *63*, 1946–1952. [CrossRef]
239. Ramish, S.M.; Ghorbani-Choghamarani, A.; Mohammadi, M. Microporous hierarchically Zn-MOF as an efficient catalyst for the Hantzsch synthesis of polyhydroquinolines. *Sci. Rep.* **2022**, *12*, 1479. [CrossRef]
240. Roy, D.; Kumar, P.; Soni, A.; Nemiwal, M. A versatile and microporous Zn-based MOFs as a recyclable and sustainable heterogeneous catalyst for various organic transformations: A review (2015-present). *Tetrahedron* **2023**, *138*, 133408. [CrossRef]
241. Chen, Y.; Yan, B.; Cheng, Y. Microporous exposure on catalytic performance of MoVNbTeOx mixed metal oxides in the oxidative dehydrogenation of ethane. *J. Catal.* **2023**, *426*, 308–318. [CrossRef]
242. Ivan, B.; Popescu, I.; Fechete, I.; Garin, F.; Pârvulescu, V.I.; Marcu, I.-C. The effect of phosphorus on the catalytic performance of nickel oxide in ethane oxidative dehydrogenation. *Catal. Sci. Technol.* **2016**, *6*, 6953–6964. [CrossRef]
243. Kung, H.H. *Transition Metal Oxides: Surface Chemistry and Catalysis*, 1st ed.; Elsevier: Amsterdam, The Netherlands, 1989; Volume 45. [CrossRef]
244. Corà, F.; Catlow, C.; Lewis, D. Design of microporous transition metal oxide catalysts and investigation of their synthesis conditions. *J. Mol. Catal. A Chem.* **2001**, *166*, 123–134. [CrossRef]

245. Sun, Q.; Dai, Z.; Meng, X.; Xiao, F.-S. Porous polymer catalysts with hierarchical structures. *Chem. Soc. Rev.* **2015**, *44*, 6018–6034. [CrossRef]
246. Zhang, Y.; Riduan, S.N. Functional porous organic polymers for heterogeneous catalysis. *Chem. Soc. Rev.* **2012**, *41*, 2083–2094. [CrossRef]
247. Widersten, M.; Gurell, A.; Lindberg, D. Structure–function relationships of epoxide hydrolases and their potential use in biocatalysis. *Biochim. Biophys. Acta* **2010**, *1800*, 316–326. [CrossRef]
248. Saha, B.; Ambroziak, K.; Sherrington, D.C.; Mbeleck, R.A. *A Continuous Process for the Liquid Phase Epoxidation of an Olefinic*; London South Bank University: London, UK, 2018.
249. Saha, B.; Ambroziak, K.; Sherrington, D.C.; Mbeleck, R. Liquid Phase Epoxidation Process. European Patent Number EP2459545, 28 February 2019.
250. Ambroziak, K.; Mbeleck, R.; He, Y.; Saha, B.; Sherrington, D.C. Investigation of Batch Alkene Epoxidations Catalyzed by Polymer-Supported Mo(VI) Complexes. *Ind. Eng. Chem. Res.* **2009**, *48*, 3293–3302. [CrossRef]
251. Bakala, P.C.; Briot, E.; Salles, L.; Brégeault, J.-M. Comparison of liquid-phase olefin epoxidation over MoOx inserted within mesoporous silica (MCM-41, SBA-15) and grafted onto silica. *Appl. Catal. A Gen.* **2006**, *300*, 91–99. [CrossRef]
252. Bruno, S.M.; Fernandes, J.A.; Martins, L.S.; Gonçalves, I.S.; Pillinger, M.; Ribeiro-Claro, P.; Rocha, J.; Valente, A.A. Dioxomolybdenum(VI) modified mesoporous materials for the catalytic epoxidation of olefins. *Catal. Today* **2006**, *114*, 263–271. [CrossRef]
253. Sakthivel, A.; Zhao, J.; Kühn, F.E. Cyclopentadienyl molybdenum complexes grafted on zeolites—Synthesis and catalytic application. *Catal. Lett.* **2005**, *102*, 115–119. [CrossRef]
254. Bhuiyan, M.R.; Mohammed, M.L.; Saha, B. Greener and Efficient Epoxidation of 1,5-Hexadiene with *tert*-Butyl Hydroperoxide (TBHP) as an Oxidising Reagent in the Presence of Polybenzimidazole Supported Mo(VI) Catalyst. *Reactions* **2022**, *3*, 537–552. [CrossRef]
255. Mbeleck, R.; Mohammed, M.L.; Ambroziak, K.; Sherrington, D.C.; Saha, B. Efficient epoxidation of cyclododecene and dodecene catalysed by polybenzimidazole supported Mo(VI) complex. *Catal. Today* **2015**, *256*, 287–293. [CrossRef]
256. Saha, B. *Catalytic Reactors*; De Gruyter: Berlin, Germany, 2016; pp. 1–338, ISBN 978-3-11-033296-4.
257. Mbeleck, R.; Ambroziak, K.; Saha, B.; Sherrington, D.C. Stability and recycling of polymer-supported Mo(VI) alkene epoxidation catalysts. *React. Funct. Polym.* **2007**, *67*, 1448–1457. [CrossRef]
258. Mohammed, M.L.; Mbeleck, R.; Saha, B. Efficient and selective molybdenum based heterogeneous catalyst for alkene epoxidation using batch and continuous reactors. *Polym. Chem.* **2015**, *6*, 7308–7319. [CrossRef]
259. Mohammed, M.L.; Mbeleck, R.; Patel, D.; Niyogi, D.; Sherrington, D.C.; Saha, B. Greener and efficient epoxidation of 4-vinyl-1-cyclohexene with polystyrene 2-(aminomethyl)pyridine supported Mo(VI) catalyst in batch and continuous reactors. *Chem. Eng. Res. Des.* **2015**, *94*, 194–203. [CrossRef]
260. Mohammed, M.L.; Patel, D.; Mbeleck, R.; Niyogi, D.; Sherrington, D.C.; Saha, B. Optimisation of alkene epoxidation catalysed by polymer supported Mo(VI) complexes and application of artificial neural network for the prediction of catalytic performances. *Appl. Catal. A Gen.* **2013**, *466*, 142–152. [CrossRef]
261. Mohammed, M.L.; Saha, B. Recent Advances in Greener and Energy Efficient Alkene Epoxidation Processes. *Energies* **2022**, *15*, 2858. [CrossRef]
262. Li, B.; Xiong, H.; Xiao, Y. Progress on Synthesis and Applications of Porous Carbon Materials. *Int. J. Electrochem. Sci.* **2020**, *15*, 1363–1377. [CrossRef]
263. Annath, H.; Manayil, J.C.; Thompson, J.; Marr, A.C.; Raja, R. Contrasting structure-property relationships in amorphous, hierarchical and microporous aluminophosphate catalysts for Claisen-Schmidt condensation reactions. *Appl. Catal. A Gen.* **2021**, *627*, 118376. [CrossRef]
264. Zallen, R. *The Physics of Amorphous Solids*; John Wiley & Sons: Hoboken, NJ, USA, 2008.
265. Goldsmith, B.R.; Peters, B.; Johnson, J.K.; Gates, B.C.; Scott, S.L. Beyond Ordered Materials: Understanding Catalytic Sites on Amorphous Solids. *ACS Catal.* **2017**, *7*, 7543–7557. [CrossRef]
266. Sá, J.; Medlin, J.W. *On-the-fly* Catalyst Modification: Strategy to Improve Catalytic Processes Selectivity and Understanding. *ChemCatChem* **2019**, *11*, 3355–3365. [CrossRef]
267. Hartmann, M.; Bischof, C.; Luan, Z.; Kevan, L. Preparation and characterization of ruthenium clusters on mesoporous supports. *Microporous Mesoporous Mater.* **2001**, *44–45*, 385–394. [CrossRef]
268. Mercadante, L.; Neri, G.; Milone, C.; Donato, A.; Galvagno, S. Hydrogenation of α,β-unsaturated aldehydes over Ru/Al2O3 catalysts. *J. Mol. Catal. A Chem.* **1996**, *105*, 93–101. [CrossRef]
269. Consonni, M.; Jokic, D.; Murzin, D.Y.; Touroude, R. High Performances of Pt/ZnO Catalysts in Selective Hydrogenation of Crotonaldehyde. *J. Catal.* **1999**, *188*, 165–175. [CrossRef]
270. Mäki-Arvela, P.; Murzin, D.Y. Hydrodeoxygenation of Lignin-Derived Phenols: From Fundamental Studies towards Industrial Applications. *Catalysts* **2017**, *7*, 265. [CrossRef]
271. Lasne, B.; Mäki-Arvela, P.; Aho, A.; Vajglova, Z.; Eränen, K.; Kumar, N.; Sánchez-Velandia, J.E.; Peurla, M.; Mondelli, C.; Pérez-Ramírez, J.; et al. Synthesis of Florol via Prins cyclization over heterogeneous catalysts. *J. Catal.* **2022**, *405*, 288–302. [CrossRef]

272. Saeid, S.; Kråkström, M.; Tolvanen, P.; Kumar, N.; Eränen, K.; Peurla, M.; Mikkola, J.-P.; Maël, L.; Kronberg, L.; Eklund, P.; et al. Synthesis and Characterization of Metal Modified Catalysts for Decomposition of Ibuprofen from Aqueous Solutions. *Catalysts* **2020**, *10*, 786. [CrossRef]
273. Wen, J.; Wilkes, G.L. Organic/Inorganic Hybrid Network Materials by the Sol−Gel Approach. *Chem. Mater.* **1996**, *8*, 1667–1681. [CrossRef]
274. Ramadhas, A.; Jayaraj, S.; Muraleedharan, C. Biodiesel production from high FFA rubber seed oil. *Fuel* **2005**, *84*, 335–340. [CrossRef]
275. Berchmans, H.J.; Hirata, S. Biodiesel production from crude *Jatropha curcas* L. seed oil with a high content of free fatty acids. *Bioresour. Technol.* **2008**, *99*, 1716–1721. [CrossRef] [PubMed]
276. Behera, B.; Dey, B.; Balasubramanian, P. Algal biodiesel production with engineered biochar as a heterogeneous solid acid catalyst. *Bioresour. Technol.* **2020**, *310*, 123392. [CrossRef] [PubMed]
277. Lam, M.K.; Lee, K.T.; Mohamed, A.R. Homogeneous, heterogeneous and enzymatic catalysis for transesterification of high free fatty acid oil (waste cooking oil) to biodiesel: A review. *Biotechnol. Adv.* **2010**, *28*, 500–518. [CrossRef] [PubMed]
278. Kulkarni, M.G.; Dalai, A.K. Waste Cooking OilAn Economical Source for Biodiesel: A Review. *Ind. Eng. Chem. Res.* **2006**, *45*, 2901–2913. [CrossRef]
279. Tiozzo, C.; Bisio, C.; Carniato, F.; Guidotti, M. Grafted non-ordered niobium-silica materials: Versatile catalysts for the selective epoxidation of various unsaturated fine chemicals. *Catal. Today* **2014**, *235*, 49–57. [CrossRef]
280. Vilé, G.; Liu, J.; Zhang, Z. Surface engineering of a Cu-based heterogeneous catalyst for efficient azide–alkyne click cycloaddition. *React. Chem. Eng.* **2021**, *6*, 1878–1883. [CrossRef]
281. Leino, E.; Kumar, N.; Mäki-Arvela, P.; Rautio, A.-R.; Dahl, J.; Roine, J.; Mikkola, J.P. Synthesis and characterization of ceria-supported catalysts for carbon dioxide transformation to diethyl carbonate. *Catal. Today* **2018**, *306*, 128–137. [CrossRef]
282. Yamada, Y.; Akita, T.; Ueda, A.; Shioyama, H.; Kobayashi, T. Instruments for preparation of heterogeneous catalysts by an impregnation method. *Rev. Sci. Instrum.* **2005**, *76*, 062226. [CrossRef]
283. Stein, A.; Wang, Z.; Fierke, M.A. Functionalization of Porous Carbon Materials with Designed Pore Architecture. *Adv. Mater.* **2009**, *21*, 265–293. [CrossRef]
284. Wang, J.; Han, W. A Review of Heteroatom Doped Materials for Advanced Lithium–Sulfur Batteries. *Adv. Funct. Mater.* **2022**, *32*, 2107166. [CrossRef]

catalysts

MDPI

Article

Hydrodesulfurization of Thiophene in *n*-Heptane Stream Using CoMo/SBA-15 and CoMo/AlSBA-15 Mesoporous Catalysts

Ana Carla S. L. S. Coutinho [1], Joana M. F. Barros [2], Marcio D. S. Araujo [3], Jilliano B. Silva [4], Marcelo J. B. Souza [5], Regina C. O. B. Delgado [6], Valter J. Fernandes Jr. [7] and Antonio S. Araujo [8,*]

1 School of Engineering, Fluminense Federal University, Niteroi 24210-253, RJ, Brazil; acoutinho@id.uff.br
2 Academic Unit of Biology and Chemistry, Center of Education and Health, Federal University of Campina Grande, Cuite 58175-000, PB, Brazil; joana.maria@professor.ufcg.edu.br
3 Institute of Chemistry, Federal University of Rio Grande do Norte, Natal 59078-970, RN, Brazil; marcio.ifrn@gmail.com
4 Post-Graduate Program in Petroleum and Energy, Federal University of Rio Grande do Norte, Natal 59078-970, RN, Brazil; jilliano.lcl@gmail.com
5 Department of Chemical Engineering, Federal University of Sergipe, Sao Cristovao 59078-970, SE, Brazil; marcelojbs@ufs.br
6 Department of Engineering and Technology, Federal Rural University of Semi-Arid, Mossoro 59625-900, RN, Brazil; regina.brasil@ufersa.edu.br
7 Laboratory of Fuels and Lubricants, Institute of Chemistry, Federal University of Rio Grande do Norte, Natal 59078-970, RN, Brazil; valter.fernandes@ufrn.br
8 Laboratory of Catalysis and Petrochemistry, Institute of Chemistry, Federal University of Rio Grande do Norte, Natal 59078-970, RN, Brazil
* Correspondence: antonio.araujo@ufrn.br

Abstract: Heterogeneous catalysts containing cobalt and molybdenum supported on mesoporous materials types SBA-15 and AlSBA-15 were synthesized for application in the HDS reactions of thiophene in the n-heptane stream. The materials were synthesized by the hydrothermal method using Pluronic P123 as a template. The calcined SBA-15 and AlSBA-15 supports were submitted to co-impregnation with solutions of cobalt nitrate and ammonium heptamolybdate, aiming for the production of 15% in mass of metal loading with an atomic ratio of [Co/(Co + Mo)] = 0.45. The obtained materials were dried and calcined to obtain the mesoporous catalysts in the forms of CoMo/SBA-15 and CoMo/AlSBA-15. The catalysts were characterized by XRD, TG/DTG, SEM, and nitrogen adsorption. From XRD analysis, it was verified that after the decomposition of the cobalt and molybdenum salts, MoO_3, Co_3O_4, and $CoMoO_4$ oxides were formed on the supports, being attributed to these chemical species, the activity for the HDS reactions. The catalytic activity of the obtained catalysts was evaluated in a continuously flowing tubular fixed-bed microreactor coupled on-line to a gas chromatograph, using an n-heptane stream containing 12,070 ppm of thiophene (ca. 5100 ppm of sulfur) as a model compound. The synthesized catalysts presented suitable activity for the HDS reaction, and the main obtained products were cis- and trans-2-butene, 1-butene, n-butane, and low amounts of isobutane. The presence of 1,3-butadiene and tetrahydrothiophene (THT) was not detected. A mechanism of the primary and secondary reactions and subsequent formation of the olefins and paraffins in the CoMo/SBA-15 and CoMo/AlSBA-15 mesoporous catalysts was proposed, considering steps of desulfurization, hydrogenation, dehydrogenation, THT decyclization, and isomerization.

Keywords: SBA-15; nanostructured materials; cobalt; molybdenum; hydrodesulfurization; environmental catalysis; petroleum refining

Citation: Coutinho, A.C.S.L.S.; Barros, J.M.F.; Araujo, M.D.S.; Silva, J.B.; Souza, M.J.B.; Delgado, R.C.O.B.; Fernandes Jr., V.J.; Araujo, A.S. Hydrodesulfurization of Thiophene in *n*-Heptane Stream Using CoMo/SBA-15 and CoMo/AlSBA-15 Mesoporous Catalysts. *Catalysts* **2024**, *14*, 198. https://doi.org/10.3390/catal14030198

Academic Editor: Narendra Kumar

Received: 20 January 2024
Revised: 25 February 2024
Accepted: 28 February 2024
Published: 18 March 2024

1. Introduction

The removal of sulfur from fossil fuels is one of the most important processes in modern refineries. It is accomplished via hydrodesulfurization (HDS) of the sulfur organic compounds in the presence of heterogeneous catalysts. Currently, to comply with environmental legislation in many countries, there is great interest in developing processes to remove atmospheric contaminants, mainly sulfur derivatives, in addition to other hydrotreatment reactions. Classical catalysts used for such reactions in refining crude oil or liquid fuels generally consist of molybdenum supported on high surface area aluminas containing cobalt or nickel as promoters.

Hydrotreating processes (HDT) were developed from hydrogenation and cracking processes, and the most important HDT reaction was the removal of sulfur from the various fractions of petroleum and liquid coal derivatives, a process called hydrodesulfurization (HDS). Catalytic hydrotreatment consists of a variety of hydrogenation processes where oil and its various fractions react catalytically with hydrogen to remove S, N, O, and metals. Nowadays, hydrotreating is widely used to convert heavy oil loads and to improve the quality of various products. It is used in the pretreatment of charges for other refining processes such as catalytic reforming, catalytic cracking (FCC), and hydrocracking catalyst (HCC). Such pretreatment aims, among others, to protect the catalsts used in many consecutive stages in refining processes; reduce NO_X and SO_X emissions that may appear in the combustion of organic molecules, thus preventing premature corrosion of equipment; promote the improvement of the final properties of products from refineries (color, smell, stability, etc.); and add value to heavy distillates [1–3].

The use of increasingly heavier loads requires, in addition to hydrodesulfurization, conversion of larger molecules into smaller ones (hydrocracking—HCC) and removal of metals (hydrodemetallization—HDM), nitrogen (hydrodenitrogenation—HDN), and, in some cases, oxygen (hydrodeoxygenation—HDO). With the development of the petrochemical industry, other processes also gained prominence, such as aromatics degradation: hydrodearomatization (HDA); interconversion of organic molecules: isomerization (ISM); breaking of C-C bonds: hydrocracking (HCC); and olefin saturation: hydrogenation (HYD). Purification processes using hydrogen are applied to practically all distillate fractions. The complexity of the load and the lack of detailed knowledge about the nature of the compounds present in crude oil are two of the difficulties of hydrotreatment. It can be said that petroleum contains mainly hydrocarbons and, depending on its origin, may also contain large concentrations of heteroatoms [4–6].

The most industrially used HDS catalysts are based on $Co(Ni)Mo/Al_2O_3$, which have high mechanical strength and a large specific surface area [7–9]. The strong interactions between metals and alumina promote the formation of Mo-O-Al phases, resulting in an active phase for the process [10,11]. However, the phase formation between Co/Ni and Al_2O_3 is unfavorable for regulating the catalytic performance of MoS_2 sites. Active metal loading and unfavorable interactions have limited supported HDS catalysts to produce ultra-low sulfur fuels. Thus, several studies have been carried out aiming to improve the performance of HDS in relation to the following aspects: modification of zeolite supports, such as ZSM-5 [12–14] and Y zeolite [15], materials based on mesoporous silica, such as MCM-41, SBA-15, and SBA-16 [16–19], oxides single and mixed metallic [20–26], and carbon-based materials [27–33]. However, the development of new catalysts containing well-defined micro- and mesopores requires further research and additional modifications for evaluation as new supports for ultra-deep desulfurization.

The advantage of HDS over oxidative or adsorptive desulfurization methods lies in its effectiveness in removing a wide range of sulfur compounds from hydrocarbon streams at relatively mild conditions. HDS operates at high temperatures and pressures, typically using metal catalysts such as molybdenum or cobalt, which enables the efficient removal of sulfur compounds, including thiols, sulfides, and thiophenes. This method is effective in reducing sulfur content to meet stringent environmental regulations, and it does not produce harmful waste products as in some oxidative methods. Additionally, HDS is well suited for treating a wide variety of feedstocks, making it a versatile and widely used desulfurization method in the petroleum industry. Specifically, some of the key advantages of HDS include selectivity, efficiency, and compatibility with existing infrastructure. HDS is highly selective for sulfur removal, allowing for the removal of sulfur compounds without significantly affecting the hydrocarbon components. This selectivity is crucial for refining high-quality fuels. HDS processes can achieve high desulfurization efficiency, even at relatively mild operating conditions, which can result in lower energy consumption compared to other methods. These advantages make HDS a popular choice for desulfurization in the petroleum refining industry.

The composition of the hydrotreatment charge varies widely depending on the origin of the oil. Sulfur is the most abundant of the heteroatoms present in crude oil, in general, ranging from 0.1 to 5% w/w. The content of nitrogen compounds present in petroleum varies from 0.1 to 1% w/w and is generally concentrated in the heaviest fractions, mainly containing pyridine nuclei. Oxygen compounds are generally present in smaller quantities, values below 0.1% w/w, and are found in the forms of carboxylic acids and phenols. In Figure 1, some of the different types of aromatic compounds and sulfur-containing compounds commonly found in crude oil fractions are presented. The need to use increasingly heavier loads has led several countries in America and the European Union to increase efforts to control and prevent the emission of these pollutants. Thus, policies regulating the amounts of toxic compounds in fuels from refineries are being established. In Brazil, the National Agency of Petroleum, Natural Gas and Biofuels controls the concentration of sulfur in oil, gas, and fuels.

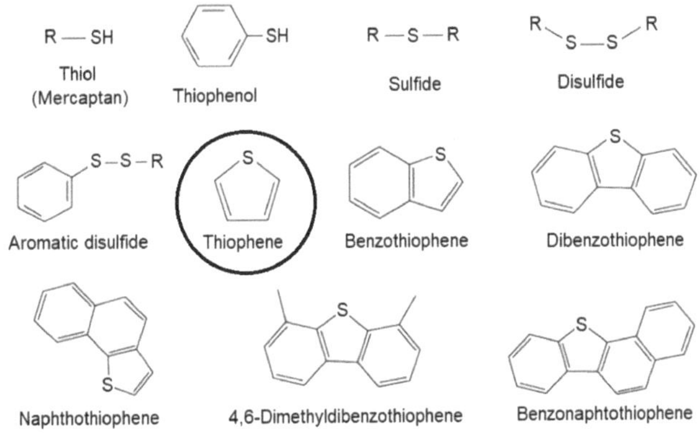

Figure 1. Some typical aromatic compounds and sulfur-containing compounds found in petroleum fractions, highlighting the molecular structure of thiophene.

Thiophene is a highly reactive molecule that contains a five-membered ring consisting of four carbon atoms and one sulfur atom. Thiophene readily undergoes various reactions, including nucleophilic and electrophilic substitutions, cyclization, and oxidation. For this reason, it was chosen as a probe molecule for this research.

Thiophene is commonly used in hydrodesulfurization (HDS) research for several specific/scientific reasons: (i) Thiophene is a heterocyclic compound containing sulfur, which is structurally similar to many sulfur-containing compounds found in crude oil; therefore, studying thiophene allows researchers to understand the behavior of more complex sulfur-containing compounds found in petroleum. (ii) Thiophene serves as a model compound for studying the mechanisms and kinetics of hydrodesulfurization reactions. Its relatively simple structure makes it suitable for detailed experimental and theoretical investigations, providing insights into the fundamental processes involved in HDS reactions. (iii) Researchers use thiophene as a probe molecule to evaluate the performance of various catalysts in hydrodesulfurization processes. Thus, by studying the effect of different catalysts on the conversion of thiophene to its desulfurized products, it is possible to identify catalysts with improved activity, selectivity, and stability for industrial HDS applications. Thiophene is also employed in mechanistic studies aimed at understanding the detailed reaction pathways involved in hydrodesulfurization. The use of thiophene in hydrodesulfurization research allows us to elucidate fundamental principles concerning sulfur removal from fossil fuels, leading to the development of more efficient and environmentally friendly HDS processes.

This work aimed to propose the synthesis of hydrodesulfurization catalysts (HDS) based on mesoporous molecular sieves of the SBA-15 and AlSBA-15 containing cobalt and molybdenum oxides deposited on their surface. Typically, SBA-15-type mesoporous materials have a high specific surface area and large pore diameter, perfectly adaptable to the kinetic diameters of the largest sulfur compound molecules. These structural characteristics of the support are fundamental to maximizing the metal dispersion of the active phases as well as improving the accessibility of the largest sulfur compounds, like thiophene, to metal sites, improving the efficiency of HDS processes.

2. Results and Discussion

2.1. Thermal Analysis of the Supports and Catalysts

Heat treatment or calcination is a very important step in obtaining high-quality SBA-15 and AlSBA-15 mesoporous materials. In this step, all the P123 triblock copolymers used as a structure template are removed. Thermogravimetry is a technique used to determine the best calcination conditions, aiming to remove all organic material and preserve the well-ordered hexagonal structure. Figure 2 shows the TG and DTG curves for the SBA-15 and AlSBA-15 samples in the non-calcined form at three different heating rates (β = 5, 10, and 20 °C/min).

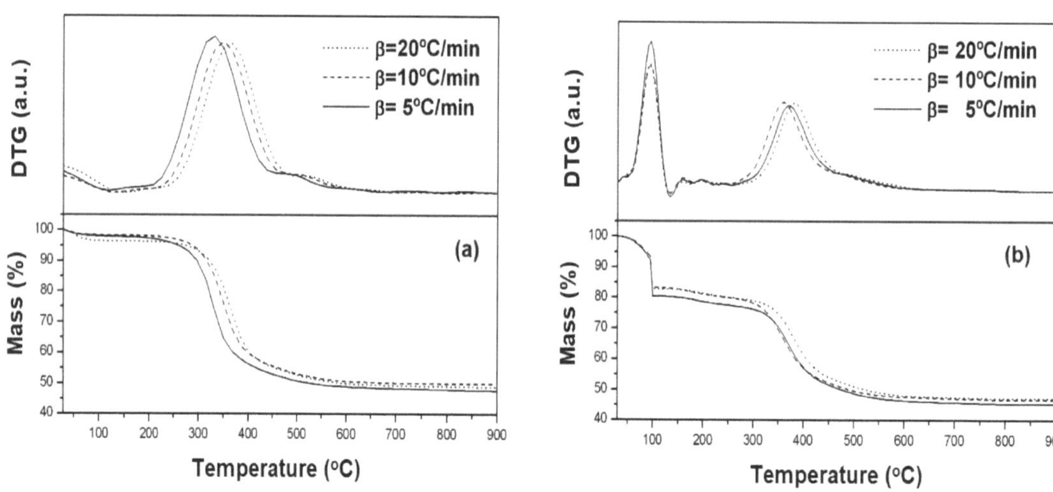

Figure 2. Thermogravimetry (TG) and derivative thermogravimetry (DTG) curves for the mesoporous supports at different heating rates: (**a**) SBA-15 and (**b**) AlSBA-15.

As can be seen from the TG/DTG curves, three mass loss events were typically obtained (Table 1). These events are attributed to (i) range of 30–130 °C, desorption of physisorbed water in the pores of the material; (ii) range of 130–450 °C, removal of the directing molecules (P123); and (iii) range of 450–650 °C, residual removal of template and release of interstitial water resulting from the silanol condensation process. This behavior is typical for mesoporous materials, such as MCM-41 and SBA-15 [34–39].

Table 1. Percentage of mass losses and respective temperature ranges for the samples SBA-15 and AlSBA-15 (Si/Al = 50) at a heating rate of 10 °C/min.

Mesoporous Material	Temperature Range (°C) and Mass Loss (%)				
	30–130	130–450	450–650	30–650	30–900
SBA-15	2.1	45.2	3.9	51.1	52.2
AlSBA-15	12.9	32.8	5.6	51.3	52.7

According to the data in Table 1, it is observed that comparing the total mass loss from 30 to 900 °C for the samples, there is no significant variation (from 52.3 to 52.7%), around 0.5% in mass. The difference in the percentage of mass loss among the materials relative to the first event, the removal of physisorbed water in the pores of the materials, can be attributed to the humidity to which each sample was exposed before thermogravimetric analysis. Thus, the highest percentage of mass loss in the AlSBA-15 sample (Si/Al = 50) may be related to physically adsorbed water due to the presence of aluminum and the template removal, which would be explained by the minor mass loss in the second event. The variation in the percentage of mass loss presented in the third event associated with the interstitial water may be an indication that the aluminum incorporated into the synthesis gel of the SBA-15 material interferes with the condensation of the silanol groups. Aluminum can be incorporated into the SBA-15 network both inside the network and on the surface of the material. The greater mass loss due to silanol condensation is evidence that there is more aluminum on the surface of the materials.

It was also observed that by increasing the heating rate from 5 to 10 and 20 °C/min, the temperature shifts to higher values, suggesting a variation in energy in the process for remotion of the P123 template from the pores of the materials. Thus, by applying multiple heating rate kinetic models [40–43], the values of the apparent activation energy (Ea) for this heat treatment process were in the range of Ea = 158–162 kJ/mol for SBA-15, and from Ea = 175–243 kJ/mol for AlSBA-15. The introduction of Al onto the SBA-15 structure suggests that Al^{3+} ions are present in the hexagonal structure, forming (-Si-O-Al-) interactions typical of aluminosilicate, consequently generating surface acid sites [44–49].

From the TG/DTG data, the temperature of 550 °C was selected for calcination of both materials. After the support calcination process, the active phases of cobalt and molybdenum were deposited on the mesoporous materials SBA-15 and AlSBA-15 using the excess solvent co-impregnation method. Thermogravimetry was also used to analyze the decomposition profiles of cobalt nitrate and ammonium heptamolybdate and thus determine the best conditions for calcining the catalysts. Figure 3 shows the TG/DTG curves of the decomposition of the precursor salts of Mo and Co supported on the SBA-15 and AlSBA-15 materials. The materials were calcined again in an air atmosphere to decompose the cobalt and molybdenum precursor salts into the respective oxides on the surface of the mesoporous supports and thus obtain the CoMo/SBA-15- and CoMo/AlSBA-15-supported catalysts.

From TG/DTG curves (Figure 3), four mass loss events were observed in the following temperature ranges: (i) 30–130 °C, (ii) 130–240 °C, (iii) 240–310 °C, and (iv) 310–410 °C, corresponding to steps for decomposition of the precursor salts inside the mesoporous supports SBA-15 and AlSBA-15. The values of mass loss relative to each step of decomposition are provided in Table 2.

From the obtained data, it was observed that up to 450 °C, the salts were completely decomposed on the surface of the materials, and this temperature was taken as a reference for thermal treatment. Thus, the calcination was carried out at this temperature under an air atmosphere flowing at 100 mL/min, in which the Co and Mo salts undergo complete decomposition, generating the CoMo/SBA-15 and CoMo/AlSBA-15 HDS catalysts.

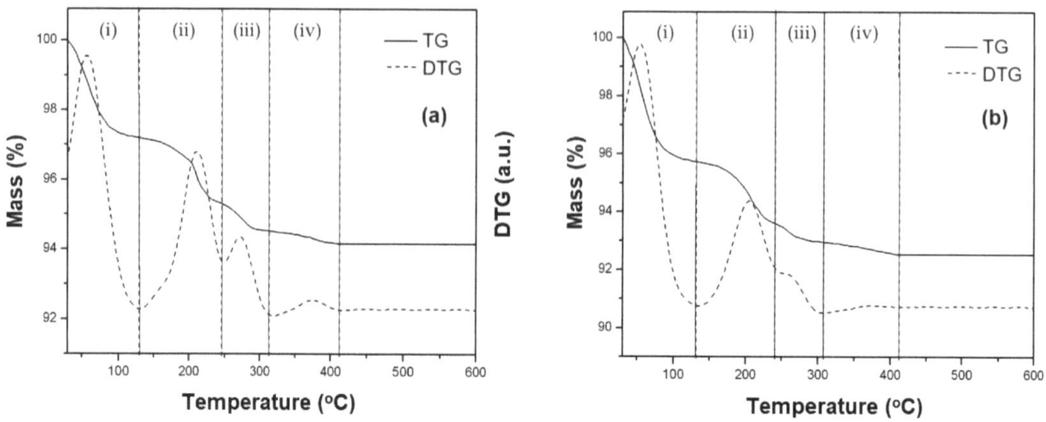

Figure 3. Thermogravimetry (TG) and derivative thermogravimetry (DTG) curves, obtained at a heating rate of 10 °C/min, showing the decomposition steps of Co and Mo salts for obtaining the supported mesoporous HDS catalysts: (**a**) CoMo/SBA-15 and (**b**) CoMo/AlSBA-15.

Table 2. Percentage of mass losses and respective temperature ranges for each step of decomposition of Co and Mo salts for obtaining CoMo/SBA-15 and CoMo/AlSBA-15.

Mesoporous Material	Temperature Range (°C) and Mass Loss (%)				
	(i) 30–130	(ii) 130–240	(iii) 240–310	(iv) 310–410	Total 30–410
CoMo/SBA-15	2.8	1.85	0.82	0.37	5.84
CoMo/AlSBA-15	3.14	2.76	0.95	0.51	7.36

2.2. X-ray Diffraction

The X-ray diffractograms (XRDs) of the materials obtained in calcined form were used to identify the hexagonal structure characteristic of mesoporous materials type SBA-15 [50,51]. Emphasis was placed on observing the obtaining of the three main diffraction peaks, referring to the crystalline planes, whose Miller indices are (100), (110), and (200). Two more peaks are observed, whose Miller indices are (210) and (300), indicating excellent textural uniformity of the material [52]. The first three peaks are characteristic of a two-dimensional p6 mm hexagonal symmetry, common to SBA-15-type materials [53,54]. Figure 4a,b show the X-ray diffractograms of mesoporous materials of SBA-15 and AlSBA-15 (Si/Al = 50), respectively.

The XRD analysis of CoMo-supported catalysts was carried out in two steps: low angle (0.5 to 5.0 degrees) and high angle (5.0 to 60.0 degrees), for observation of the ordered hexagonal phase and the presence of CoMo metal oxides, as shown in Figures 5 and 6, for CoMo/SBA-15 and CoMo/AlSBA-15, respectively.

From the diffractograms presented, the presence of the five main diffraction peaks was observed, whose Miller indices are (100), (110), (200), (210), and (300), indicating that high-quality mesoporous materials with defined structure were obtained, with well-ordered hexagonal arrangement [55]. Diffraction peaks for non-supported materials present a better definition in relation to CoMo-supported; this fact is due to the presence of heteroatom on the ordered structures. The mesoporous hexagonal arrangement parameter a_o (lattice parameter) of the SBA-15 structure is obtained from the reflection peak for the (100) plane, which is the most characteristic in the X-ray diffractogram, whose values are summarized in Table 3.

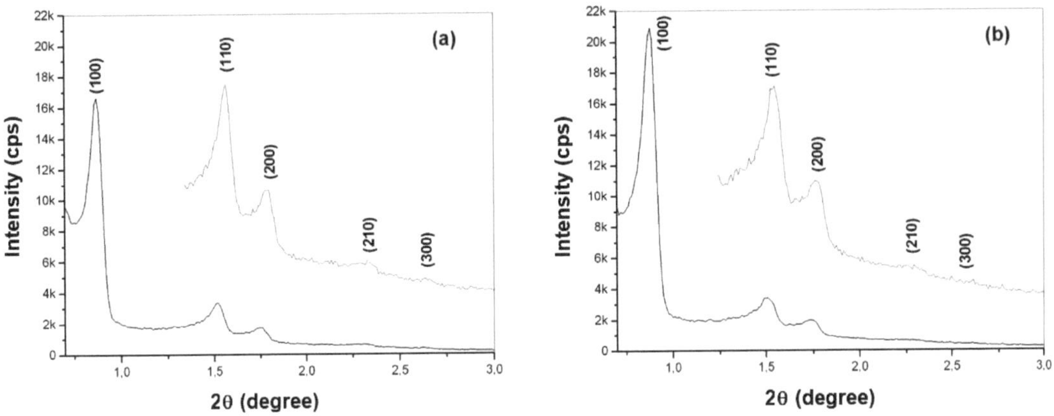

Figure 4. X-ray diffractograms of the calcined mesoporous materials: (**a**) SBA-15 and (**b**) AlSBA-15, showing the Miller indices.

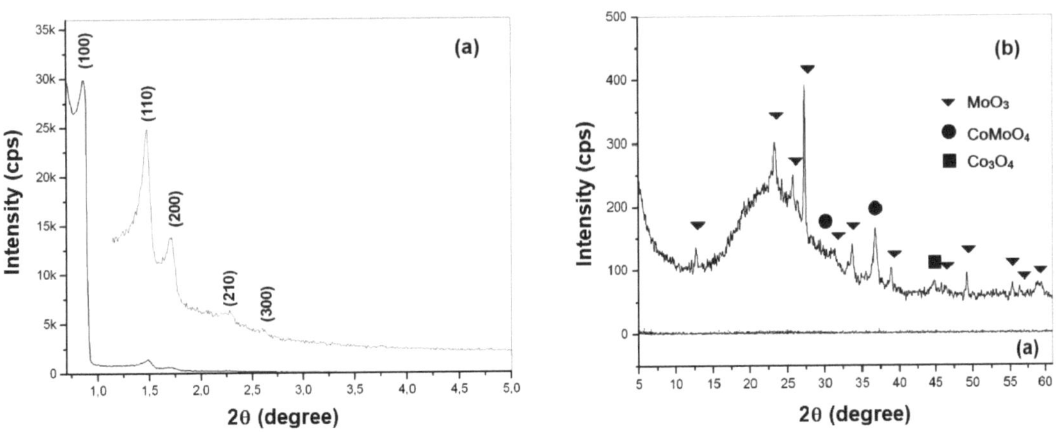

Figure 5. X-ray diffractograms of the calcined CoMo/SBA-15 obtained at low (**a**) and high (**b**) diffraction angles.

Analyzing the data on the mesoporous parameter (a_o) of the SBA-15 and AlSBA-15 supports from Table 3, it can be noted that in all cases, there was a decrease in the value of this parameter. That value (a_o) represents the sum of the pore diameter (dp) and the silica wall thickness (wt). This decrease may have occurred due to the deposition of nanoparticles of cobalt and molybdenum oxides inside the mesopores of the materials.

The crystalline phases of cobalt and molybdenum oxides were identified through the crystallographic charts of these oxides found in the JCPDS (Joint Committee on Powder Diffraction Standards)/ICDD (International Center for Diffraction Data) library [56]. Research on crystallographic chartsverified the presence of MoO_3 (JCPDS Chart: 05-0508) with an orthorhombic structure, Co_3O_4 (JCPDS Chart: 43-1003) with a cubic structure, and mixed oxides of cobalt and molybdenum in the form of $CoMoO_4$ (JCPDS Registration: 21-0868) with a monoclinic structure [56]. The main peaks identified based on JCPDS were as follows: MoO_3 (2⊖ = 12.79, 23.32, 25.88, 27.32, 33.12, 33.72, 35.46, 38.96, 39.66, 38.96, 39.66, 45.76, 46.3, 49.26, 52.22, 54.13, 55.12, 56.36, 57.59, and 58.75), $CoMoO_4$ (2⊖ = 26.40, 28.34, 31.98, and 36.63) and Co_3O_4 (2⊖ = 18.93, 31.38, 36.92, 38.52, 44.97, 55.57, and 59.49). In all samples studied, the predominance of the crystalline phases MoO_3 and $CoMoO_4$ was observed. The presence of a diffraction peak at 44.97 degrees, corresponding to Co_3O_4, was identified in all samples. Other cobalt and molybdenum oxides may also have occurred, but in very small quantities not identified from XRD due to interference with background radiation or because they are present in amorphous form.

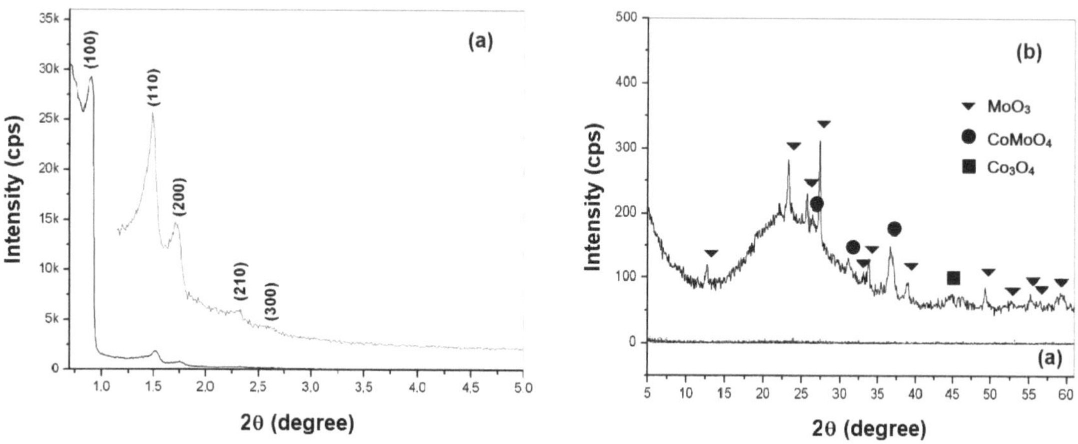

Figure 6. X-ray diffractograms of the calcined CoMo/AlSBA-15 obtained at low (**a**) and high (**b**) diffraction angles.

Table 3. Crystallographic properties of calcined SBA-15, AlSBA-15, CoMo/SBA-15, and CoMo/AlSBA-15 materials.

Materials	(hkl)	2q (°)	$d_{(hkl)}$ (nm)	a_o (nm)
SBA-15	(100)	0.868	10.18	11.76
	(110)	1.516	5.83	
	(200)	1.748	5.05	
	(210)	2.313	3.82	
	(300)	2.630	3.36	
AlSBA-15(Si/Al = 50)	(100)	0.892	9.91	11.44
	(110)	1.503	5.88	
	(200)	1.738	5.08	
	(210)	2.282	3.87	
	(300)	2.610	3.38	
CoMo/SBA-15	(100)	0.873	10.12	11.69
	(110)	1.489	5.94	
	(200)	1.714	5.16	
	(210)	2.275	3.88	
	(300)	2.560	3.45	
CoMo/AlSBA-15(Si/Al = 50)	(100)	0.907	9.74	11.25
	(110)	1.528	5.78	
	(200)	1.733	5.10	
	(210)	2.284	3.87	
	(300)	2.623	3.37	

2.3. Nitrogen Adsorption

The adsorption and desorption isotherms, as well as the distribution of pore diameters obtained for samples SBA-15 and AlSBA-15, are presented in Figures 7 and 8, respectively. It can be observed that type IV isotherms were obtained in the samples, according to the IUPAC classification, which is characteristic of mesoporous materials. Accordingly, the hysteresis found was type I, which is characteristic of materials with a cylindrical pore system or made from aggregates or clusters of spheroidal particles with pores of uniform size [57].

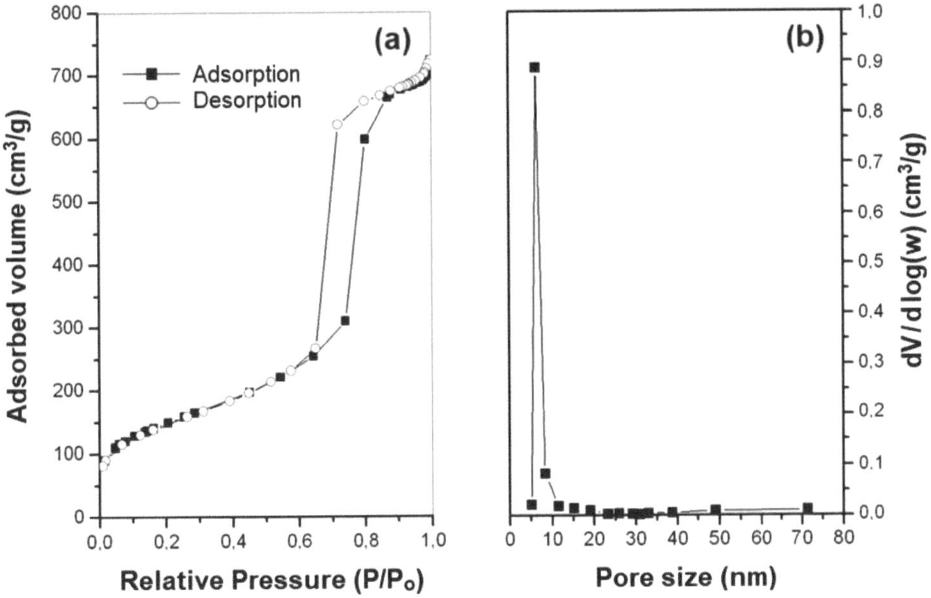

Figure 7. (**a**) N_2 adsorption isotherms and (**b**) pore size distribution of the SBA-15 mesoporous support.

SBA-15 and AlSBA-15 are types of mesoporous materials known for their well-defined pore structures. These materials contain both micropores and mesopores, which contribute to their unique properties and applications. The micropores are primarily formed within the walls of the silica framework, arising from the arrangement of silica units during the synthesis process. They provide additional surface area and can influence the adsorption and diffusion of small molecules and ions. The presence of micropores enhances the overall surface area and adsorption capacity of the material. However, mesopores are the dominant type of pore and are responsible for the ordered and uniform pore structures characteristic of these materials. Mesopores are formed between the silica walls, creating channels or void spaces that allow for the efficient diffusion of molecules within the material. These larger pores enable the transport of larger molecules and facilitate mass transfer processes. The presence of both micropores and mesopores in SBA-15 and AlSBA-15 contributes to their high surface area, large pore volume, and uniform pore size distribution. This combination of pore characteristics makes them attractive materials for HDS applications, where precise control over pore structure and surface properties is essential.

The surface areas of the mesoporous materials obtained were determined from data on nitrogen adsorption isotherms at 77 K using the BET model [58] in the P/Po range of 0.05–0.20. The pore diameter distributions of mesoporous materials were obtained by the BJH method [59] correlating the desorbed volume values as a function of relative pressure (P/Po) in the algorithms in a pore range of 1–80 nm. The average pore diameters were estimated through the pore distribution curves obtained by the BJH method and revealed values of 6.84 and 6.91 nm for SBA-15 and AlSBA-15 (see Table 4), with low variation. After the impregnation of Co and Mo metals, these values decreased to 6.11 and 6.82, respectively. As summarized in Table 4, the obtained materials presented pore volumes in the range of 0.84 to 1,12 cm^3/g. Using the BET method, it was observed that the samples had surface areas in the range of 402 to 602 m^2/g. These values are compatible with those found in the literature for SBA-15 containing aluminum [60,61].

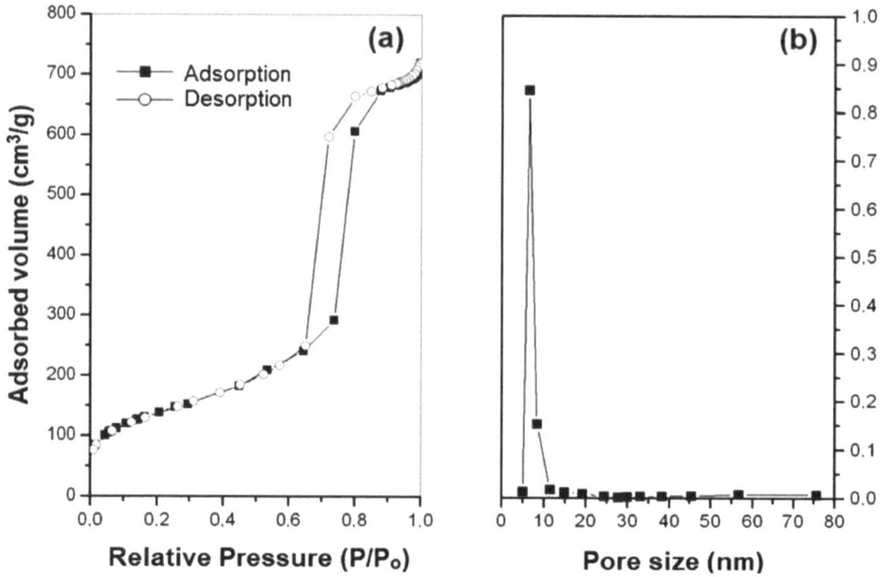

Figure 8. (**a**) N_2 adsorption isotherms and (**b**) pore size distribution of the AlSBA-15 mesoporous support.

Table 4. Surface properties of the mesoporous materials, determined from XRD and nitrogen adsorption.

Sample	a_o (nm)	Dp (nm)	Wt (nm) *	Vp ($cm^3\ g^{-1}$)	S_{BET} ($m^2\ g^{-1}$)
SBA-15	11.76	6.84	4.92	1.12	602
AlSBA-15	11.44	6.91	4.53	1.07	495
CoMo/SBA-15	11.69	6.11	5.58	0.95	406
CoMo/AlSBA-15	11.25	6.82	4.43	0.84	402

a_o: mesoporous parameter; Dp: pore diameter; Wt: wall thickness * (Wt = a_o − Dp); Vp: pore volume, obtained from the BJH method.

It was observed that there was a decrease in the mesoporous parameter, the sum of the average pore diameter (Dp), and the silica wall thickness (Wt) with the introduction of Al on the support. The average pore diameter did not vary significantly with the introduction of Al. The average values of the silica wall thickness (Wt) can give these materials high mechanical resistance and the possibility for application as catalytic supports in processes of oil refining, where catalysts are often subjected to operating conditions with high temperatures and pressures [62,63]. The structure and pore system of the materials, showing hexagonal ordering with micropores and mesopores, is shown in Figure 9.

The determination of the wall thickness (Wt) of SBA-15 and AlSBA-15 mesoporous materials can be achieved through various experimental techniques and characterization methods. Small-angle X-ray diffraction provided structural insights by analyzing the intense peak at an interplanar distance of $d_{(100)}$, while nitrogen adsorption–desorption isotherms indirectly inferred the wall thickness (W_t) through BET analysis. The nitrogen physisorption offered additional means to determine pore size distribution and local atomic connectivity, contributing to wall thickness estimation. The combination of these methods is often utilized to comprehensively characterize the structure and properties of SBA-15 materials, considering factors such as sample preparation requirements and thermal treatments.

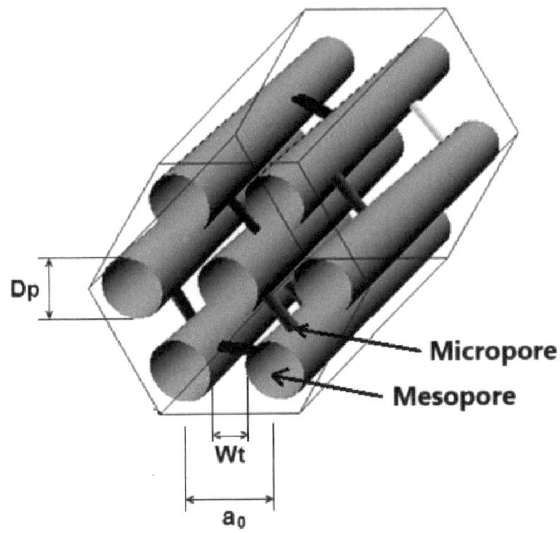

Figure 9. Hexagonal arrangement of the mesoporous structure of the SBA-15, showing the mesoporous parameter (a_o), wall thickness (Wt), and pore size diameter (Dp).

After the impregnation of cobalt and molybdenum oxides on the supports, there were no changes in the shapes of the adsorption and desorption isotherms, continuing to be type IV, thus maintaining the mesoporous structure. The average pore diameter decreased with the introduction of Al into SBA-15. With the introduction of Co and Mo metals, the pore volumes were 0.95 and 0.84 $cm^3\ g^{-1}$, and the surface areas were 406 and 402 m^2g^{-1} for CoMo/SBA-15 and CoMo/AlSBA-15, respectively, and a decrease in the total surface area in relation to the mesopore supports was observed.

2.4. Scanning Electron Microscopy

Scanning Electron Microscopy (SEM) provided surface topography images of the materials, and they were obtained at magnifications of 15,000× of the SBA-15 and AlSBA-15 materials, as shown in Figure 10. The micrographs of CoMo/SBA-15 and CoMo/AlSBA-15 with details of the pore systems are shown in Figures 10 and 11, respectively. The SEM analyses were carried out with the aim of observing the morphology of the synthesized nanostructured materials. It can be seen in the figures that silica fibers with micrometric dimensions are formed from the linear adhesion of nodules of submicrometric particles. The morphology of the AlSBA-15 (samples with Si/Al = 50) was similar to the SBA-15 sample, even after the impregnation of Co and Mo. In all cases, non-uniform fibers were observed, giving the appearance of "intertwined bead necklaces" [64–66], indicating that this is probably the phase corresponding to SBA-15 since XRD and nitrogen adsorption analyses showed that these samples are pure and have a high degree of ordering and suitable porosity, as shown in Figures 11 and 12.

In hydrodesulfurization reactions, MoO_3 and $CoMoO_4$ species can, during the HDS and sulfidation steps, transform into the MoS_2 and "CoMoS" phases, which are active and stable for the reaction. The presence of Co_3O_4 can give rise to Co_9S_8, a phase that is very inactive for HDS catalysts but can also be reduced to metallic cobalt, which, properly accommodated at the ends of MoS_2 crystals, gives rise to active phases, known as "CoMoS" [67–69].

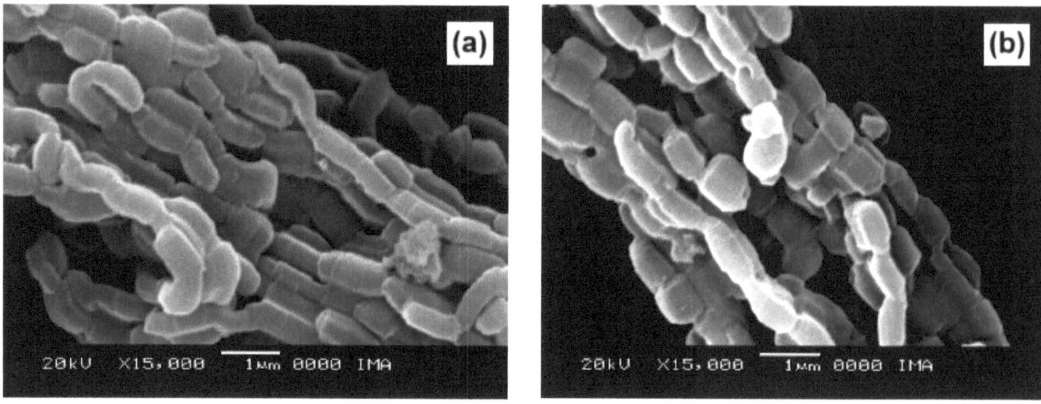

Figure 10. Scanning electron micrographs of the mesoporous supports: (**a**) SBA-15 and (**b**) AlSBA-15 (Si/Al = 50).

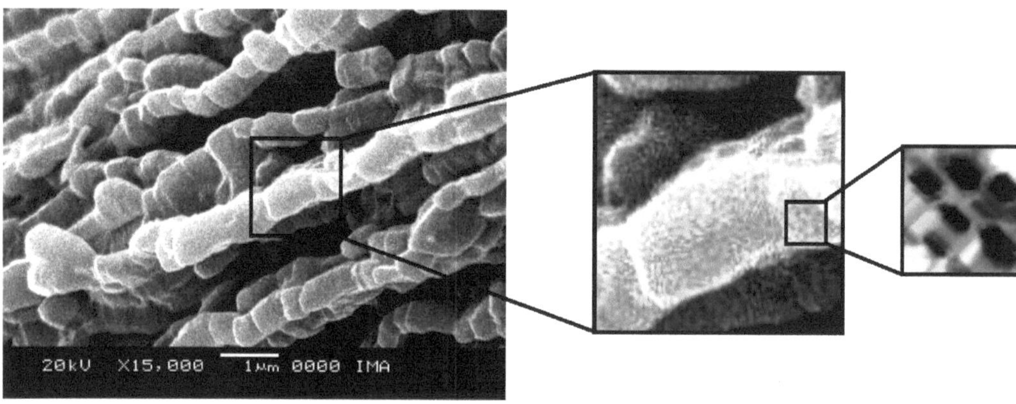

Figure 11. Scanning electron micrograph of the mesoporous CoMo/SBA-15 catalyst showing details of porosity.

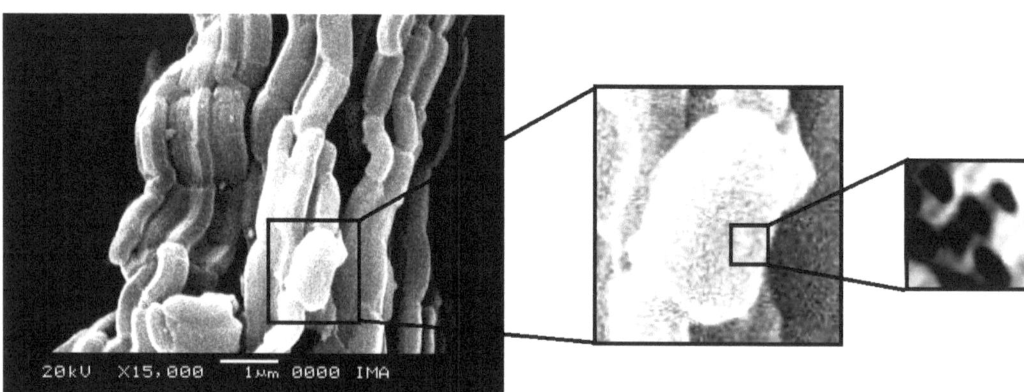

Figure 12. Scanning electron micrograph of the mesoporous CoMo/AlSBA-15 catalyst showing details of porosity.

2.5. Catalytic Activities of CoMo/SBA-15 and CoMo/AlSBA-15

Before starting the catalytic tests for hydrodesulfurization of thiophene, some preliminary tests were carried out with the aim of verifying the occurrence of thermal cracking reactions under the operating conditions of the tests. The first test consisted of a mixture of 13,500 ppm of thiophene (ca. 5100 ppm of sulfur) in n-heptane steam through the reactor at 350 °C without a catalyst, with the aim of observing the occurrence of thermal degradation of the mixture at this temperature, which was not observed, where the chromatogram showed only the peaks relating to thiophene and n-heptane. The second test was conducted to verify the occurrence of catalytic cracking of pure n-heptane on the catalytic bed at 350 °C containing the obtained catalysts. Once again, no peaks other than n-heptane were detected, proving that there was no catalytic cracking under these conditions. After preliminary tests, catalytic hydrodesulfurization tests (HDS) of the mixture of 13,500 ppm of thiophene in n-heptane (ca. 5100 ppm of sulfur) were carried out with the objective of evaluating the conversion and selectivity of 15% CoMo/SBA-15 and 15% CoMo/AlSBA-15 catalysts (Si/Al = 50). According to chromatograms, H_2S and C_4-hydrocarbon compounds were typically obtained in the following elution order: isobutane, 1-butene, n-butane, trans-2-butene, and cis-2-butene. The presence of butadiene or isobutene was not observed. In thiophene HDS reactions, butadiene can occur as a primary reaction product or act as an intermediate to obtain butenes through a hydrogenation reaction, thus having a very short lifetime in the catalytic cycle and not appearing in appreciable quantities in product distribution [70]. In the case of isobutene, this product is thermodynamically unfavorable, with the conversion of linear butenes being preferred.

Figure 13a,b show the conversion and paraffin/olefin ratio, respectively, for HDS reactions on catalysts with 15% cobalt and molybdenum metal phase supported on SBA-15, AlSBA-15 (Si/Al = 50). Figure 14a,b show the selectivity for the products. The conversion and selectivity were defined according to Equations (1) and (2), respectively.

$$\text{Conversion}(\%) = \frac{\text{mol(i)Thiophene} - \text{mol(f)Thiophene}}{\text{mol(i)Thiophene}} \quad (1)$$

$$\text{Seletivity to P}(\%) = \frac{\text{mol of Thiophene converted to P}}{\text{mol of Thiophene converted}} \quad (2)$$

where "mol(i)" and "mol(f)" represent the initial and final molar quantities obtained by analyzing the chromatograms at the beginning and end of the reaction, respectively. The letter "P" represents the reaction product measured at the reactor outlet, which may be butane, isobutane, 1-butene, cis-2-butene, or trans-2-butene. The paraffin/olefin ratio was determined considering the selectivity of (n-butane + isobutane) divided by the selectivity of (1-butene + cis-2-butene + trans-2-butene).

It was observed that in the first 15 min of reaction, the highest conversion values were obtained for all catalysts studied. These conversions progressively decreased until reaching stability normally after 60 min of reaction. Considering the conversion values obtained in 120 min of reaction, the CoMo/SBA-15 was more active than the CoMo/AlSBA-15. Also, it was observed that after 60 min of reaction, the paraffin/olefin ratio increased for CoMo/SBA-15, whereas with the CoMo/AlSBA-15, this ratio was almost constant, with values below 0.1. Through X-ray diffraction analyses, no diffractions of amorphous phases of MoO_3 or other diffractions related to Co_3O_4 were observed other than that found for all samples. According to a previous report [71], cobalt appears as a promoter for the hydrogenolysis reactions of the C-S bonds of thiophene, but it can also act as a promoter for other reactions, such as isomerization and hydrogenation of butadiene after the HDS catalytic cycle. This promoting effect of cobalt can be attributed to the transfer of electrons to molybdenum oxide, reducing its oxidation state from Mo^{6+} to Mo^{4+}.

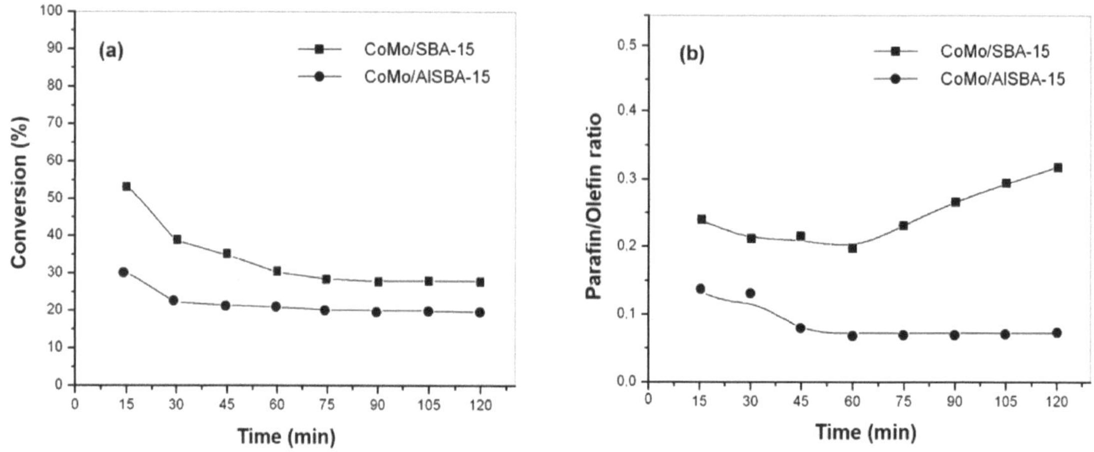

Figure 13. Catalytic activity for the CoMo/AlSBA-15 and CoMo/AlSBA-15 catalysts: (**a**) conversion as a function of reaction time and (**b**) paraffin/olefin ratio as a function of reaction time.

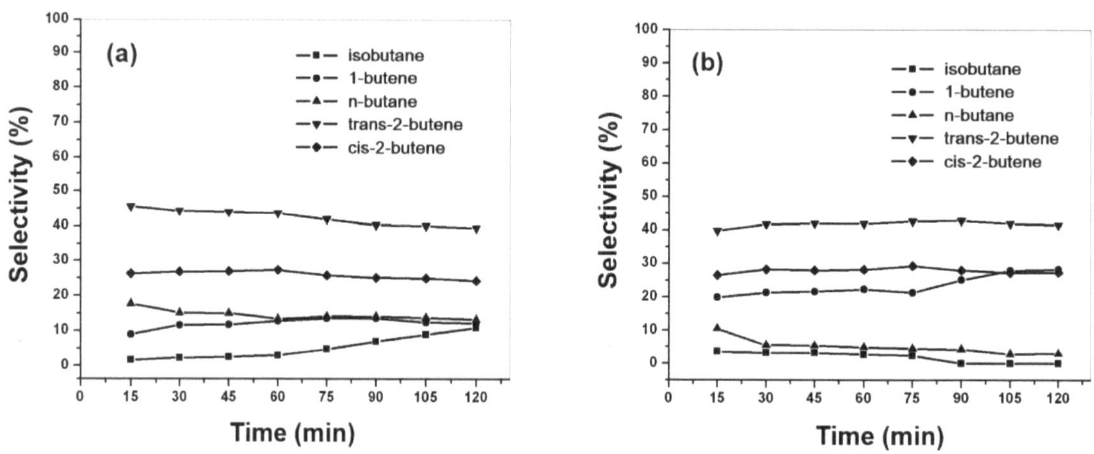

Figure 14. Product selectivity to paraffins and olefins as a function of reaction time for the mesoporous catalysts: (**a**) CoMo/SBA-15 and (**b**) CoMo/AlSBA-15.

A proposed reaction mechanism of thiophene on MoO_3 considers that the formation of butenes and n-butane can occur directly through one or two butadiene hydrogenation steps, forming 1-butene or n-butane, respectively [3]. Through the isomerization of 1-butene, it is possible to obtain cis-2-butene and trans-2-butene by displacing the double bond. Isobutane can be obtained through chain isomerization of n-butane. The presence of 1,3-butadiene and tetrahydrothiophene (THT) molecules was not detected. Thus, it is supposed that these compounds suffer interconversion reactions inside the mesopores of the CoMo/SBA-15 and CoMo/AlSBA-15 catalysts, as schematized in Figure 15.

It is known that some transition metals are effective catalysts for hydrodesulfurization (HDS) reactions, and molybdenum (Mo) and cobalt (Co) are most commonly used. The Co and Mo metals play different roles in the HDS process, and sometimes, the catalysts are composed of combinations of these metals for enhanced performance. In general, Mo facilitates the breaking of carbonsulfur bonds in sulfur-containing compounds, promoting desulfurization. It has a high affinity for sulfur, enabling it to form stable sulfide compounds on the catalyst surface. Mo also helps in hydrogenation reactions, which are often coupled with desulfurization in HDS processes. Co-based catalysts are less common than Mo-based catalysts but are still used in certain HDS applications. Co facilitates desulfurization reactions by breaking carbonsulfur bonds. Co-based catalysts also exhibit suitable selectivity toward thiophene conversion. Catalysts used in industrial HDS processes contain combinations of these metal species, such as Co-Mo, to leverage their synergistic effects. Thus, Mo facilitates the initial breaking of carbonsulfur bonds, while Co promotes further desulfurization and hydrogenation reactions. Additionally, the presence of mesoporous silica support (SBA-15 or AlSBA-15) can enhance the catalyst ability, improving the sulfur removal efficiency.

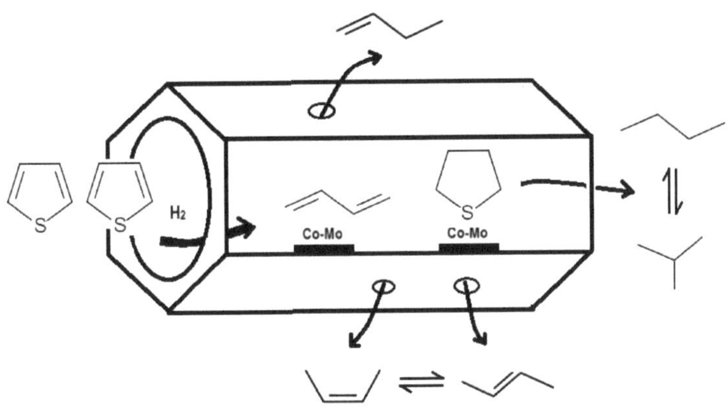

Figure 15. Possible HDS thiophene reaction inside the mesopore system of the CoMo/SBA-15 and CoMo/AlSBA-15 catalysts and desorption of the products.

According to the obtained results, it was proposed that thiophene molecules suffer desulfurization in the presence of hydrogen to form 1,3-butadiene and hydrogenation to form tetrahydrothiophene (THT). In the presence of Co and Mo metals, 1,3-butadiene reacts with hydrogen to obtain 1-butene and 2-butene, and this suffers isomerization to cis- and trans-2-butene. The THT species are adsorbed on Co-Mo metals, and a new step of desulfurization is suggested for the formation of paraffins, n-butane, and a subsequent isomerization to isobutane. The selectivity for n-butane suggests that it forms via secondary reactions of primary products. Therefore, after one hour of reaction, using the CoMo/SBA-15 catalyst, the isobutane selectivity increased with n-butane decreasing with reaction time, suggesting a step of isomerization.

For the CoMo/AlSBA-15 catalyst, this behavior was not observed, indicating that the presence of aluminum, generating Bronsted acid sites, stabilizes the structure and inhibits the paraffin isomerization. From the conversion, selectivity, and paraffin/olefin ratio, a mechanistic scheme for thiophene HDS is proposed in Figure 16, showing a sequence of primary and secondary reactions.

One probable explanation for the presence of acid sites on the CoMo/SBA-15 catalyst can be attributed to silanol groups (≡Si-O-H) inside the micropores (Si-OH bulk), with acid character due to the formation of hydrogen bonds, in addition to moderate acidity on the surface of the mesopores (associated Si-OH), forming hydrogen bonds on the external surface of silicon oxide [72]. Another possibility that must be considered is the reactions of H2S generated in the HDS reaction, with oxygen from the silicate groups forming sulfated silica with surface proton sites [73]. In this case, we would have the formation of Bronsted super-acidic sites on the internal surface of the mesopores, as proposed in Figure 17, with the sulfated CoMo/SBA-15 catalyst presenting silanol groups capable of donating or accepting hydrogen bonds with different strengths, consequently promoting an increase in the concentration of isobutane from the isomerization process of n-butane.

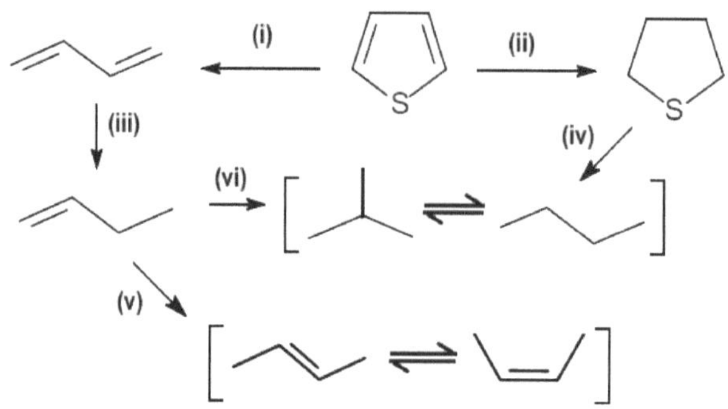

Figure 16. Proposed mechanism reaction for hydrodesulfurization of thiophene using the mesoporous CoMo/SBA-15 and CoMo/AlSBA-15 catalysts, showing steps: (i) desulfurization; (ii) hydrogenation; (iii) skeletal hydrogenation; (iv) desulfurization and decyclization; (v) cis- and trans-isomerization; (vi) hydrogenation and C_4-paraffins isomerization.

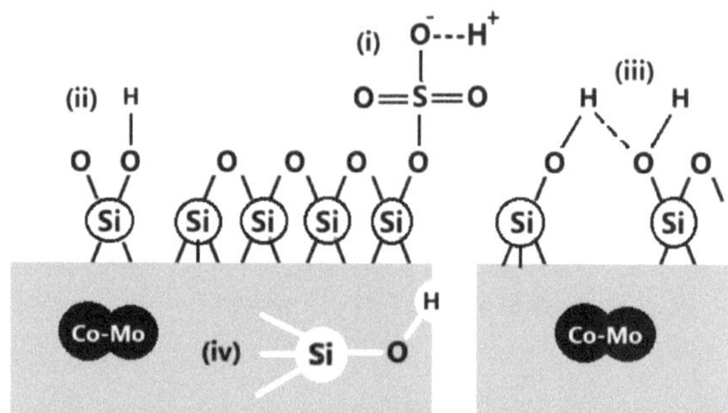

Figure 17. Proposed scheme for the silanol groups and acidity on the CoMo/SBA-15 catalyst, evidencing the formation of different hydrogen bonds and acid sites: (i) Bronsted superacid site due to presence of sulfate on the SBA-15 surface; (ii) isolated silanol; (iii) associated silanol; (iv) bulk silanol, showing an acid site on the microporous surface.

3. Materials and Methods

3.1. Synthesis of the SBA-15 and AlSBA-15 Supports

The CoMo/SBA-15 and CoMo/AlSBA-15 catalysts were synthesized in two steps: (i) hydrothermal synthesis and calcination of the supports and (ii) impregnation of the Co and Mo on the obtained supports.

The hydrothermal synthesis of mesoporous supports of types SBA-15 and AlSBA-15 with molar ratio Si/Al = 50 were synthesized using the following reagents: tetraethylorthosilicate—TEOS, Sigma-Aldrich, 98%—$Si(OC_2H_5)_4$, (St. Louis, MO USA), Pseudobohemite—AlOOH, Vista Chemical Company, 70% Al_2O_3 and 30% water (Westlake, LA, USA), Pluronic P123 (Triblock Copolymer, BASF Co., average PM = 5750 g/mol, (Houston, TX, USA), hydrochloric acid (Merck, HCl, 37% vol. (Sao Paulo, SP Brazil), and distilled water. The hydrothermal syntheses were carried out using 250 mL Teflon autoclaves wrapped in a stainless steel protection manufactured by Parr Instruments.

The reagents were mixed to obtain a reactive hydrogel with a molar composition of 0.017P123:1.0TEOS:5.7HCl:193H_2O [74]. First, the P123 template was dissolved in distilled water and HCl, with stirring and heating to 35 °C. Once the temperature was reached, the silica source, tetraethylorthosilicate (TEOS), was added. The mixture was kept under stirring and heated at 35 °C for 24 h (pH = 0–1) to obtain a homogeneous gel. Then, it was transferred to the autoclave and stored in an oven for 48 h, previously heated to 100 °C.

For the AlSBA-15 support, the reagents were mixed to obtain a reactive hydrogel with the following molar composition: 0.017P123:1.0 TEOS:xAl_2O_3:5.7HCl:193H_2O. The value of "x" was used to maintain the molar ratio Si/Al = 50.

Once the hydrothermal syntheses were completed, the materials obtained were vacuum filtered and washed with 50 mL of a 2% solution by volume of hydrochloric acid in ethanol. This procedure facilitates the removal of the organic director from the pores of the material, reducing calcination time. After this procedure, each material was placed to dry at room temperature for 24 h. To completely remove P123 from the pores of the mesoporous molecular sieves, the calcination technique was used. In this procedure, each sample was subjected to heating from room temperature to 500 °C under a dynamic nitrogen atmosphere with a flow of 100 mL min^{-1} and a heating rate of 10 °C min^{-1}. Upon reaching 500 °C, each material remained for one hour under nitrogen in the same flow. After this time, the gas was changed to synthetic air (reactive gas) and heated at the sample temperature for another hour with a flow of 100 mL min^{-1}. The supports were called SBA-15 and AlSBA-15.

3.2. Preparation of the CoMo/SBA-15 and CoMo/AlSBA-15

Co and Mo metals were deposited on mesoporous supports using the impregnation technique with excess solvent using absolute ethanol: C_2H_5OH (99.5%, Merck, Sao Paulo, SP, Brazil) as solvent, cobalt nitrate hexahydrate: $Co(NO_3)_2.6H_2O$ (99%, Merck, Sao Paulo, SP, Brazil) as a source of cobalt, and ammonium heptamolybdate tetrahydrate: $(NH_4)_6Mo_7O_{24}.4H_2O$ (Ecibra, 82.5% in MoO_3, Santo Amaro, SP, Brazil) as a source of molybdenium. Before impregnation, all mesoporous supports were subjected to a TG run in a nitrogen atmosphere at a heating rate of 20 °C min^{-1} from 30 to 900 °C, with the aim of determining the relative humidity levels of each support, starting from mass losses in the range of 30 to 130 °C, and using these data to correct the dry mass of the supports in order to minimize weighing errors during the deposition stage of the cobalt and molybdenum precursor salts.

The metal impregnation procedure consisted of weighing ca. 0.5 g of the support (SBA-15 or AlSBA-15), considering the relative humidity. The amounts of cobalt nitrate, $Co(NO_3)_2.6H_2O$, and ammonium hepatmolybidate, $(NH_4)_6Mo_7O_{24}.4H_2O$, were weighed in a porcelain crucible and solubilized in 20 mL of absolute ethanol using a glass rod. After the solubilization of the salts, the support was slowly added, stirring with the glass rod. The crucible with the suspension was transferred to the heating mantle at 70 °C, homogenizing periodically to evaporate the excess solvent. After evaporation of the excess ethanol, the crucible was transferred to the oven and dried at 100 °C for 6 h. The depositions of the metallic phases were carried out to obtain a loading of 15% by weight of the active phase, with a [Co/(Co + Mo)] atomic ratio of 0.45.

The synthesis process allows the cobalt and molybdenum ions from the solution to be absorbed into the pores of the support material. After impregnation, the solvent was evaporated, leaving behind the metals distributed on the surface and within the pores of the support materials. The impregnated materials were dried to remove any remaining moisture. The dried catalysts were subjected to calcination at 500 °C under an atmosphere of air flowing at 100 mL/min. This step serves to decompose the cobalt and molybdenum salts and convert them into cobalt and molybdenum oxides on the support material. Finally, before the catalytic evaluation, the calcined materials were subjected to a reduction step in a hydrogen atmosphere. This reduces the cobalt and molybdenum species to their metallic form, the active metal phase required for catalysis.

3.3. Physicochemical Characterization of the Obtained Materials

3.3.1. Thermogravimetry (TG)

Thermal analysis using TG was used to carry out studies to determine the best calcination conditions for eliminating P123 from the pores of mesoporous materials SBA-15 and AlSBA-15, as well as verifying the best calcination conditions for the decomposition of the precursor salts of the metallic phases of cobalt and molybdenum. Thermogravimetric analyses of the as-synthesized mesoporous materials (SBA-15 and AlSBA-15) were obtained in a thermobalance with a horizontal furnace model TA/SDTA 951 from Mettler (Switzerland, GmbH). The thermogravimetric curves of the non-calcined samples were obtained by heating the sample from room temperature to 900 °C in a dynamic nitrogen atmosphere at three heating rates of 5, 10, and 20 °C min^{-1}, with the aim of carrying out a series of kinetic studies regarding the best conditions for removing the P123 template from the pores of the materials and thus establishing the best calcination conditions. For each test, alumina crucibles with a mass of around 10 mg were used.

By using TG, it was possible to study the best conditions for calcining mesoporous materials impregnated with cobalt and molybdenum salts. In all cases, approximately 10 mg of each non-calcined sample was heated from room temperature to 900 °C at a heating rate of 10 °C min^{-1}. The curves were obtained in a dynamic synthetic air atmosphere of 25 mL min^{-1}.

3.3.2. X-ray Diffraction (XRD)

XRD analyses using the powder method were carried out on materials obtained in calcined form, with the aim of verifying whether the mesoporous hexagonal structure had formed. In the samples impregnated after the calcination process, new XRD analyses were carried out to verify variations in the hexagonal mesoporous structure and to identify the crystalline phases of the cobalt and molybdenum oxides formed.

The X-ray diffractograms of the SBA-15 and AlSBA-15 samples were obtained in an angular scan of 0.5 to 5.0 degrees on Shimadzu model XRD 6000 equipment (Nakagyo-ku, Kyoto, Japan). The tests were conducted using CuKα radiation and a nickel filter with a tube voltage of 30 kV and a current of 30 mA. The slit had an opening of 0.15 degrees, and the X-ray beam was phased in relation to the sample with a speed of 0.5 degrees/min and a step of 0.01 degrees. For samples containing deposited cobalt and molybdenum oxides, XRD was carried out in an angular range of 5 to 60 degrees.

3.3.3. Nitrogen Adsorption

The specific surface area, determined by the BET method, total pore volume, distribution, and average pore size diameter, was determined through N_2 adsorption at the temperature of liquid N_2 (77 K). The experiments of the adsorption isotherms of the calcined samples were carried out on Micromeritics ASAP2010 equipment (Norcross, GA, USA). To this end, approximately 100 mg of each sample was previously treated at 170 °C for 12 h under vacuum and then subjected to nitrogen adsorption at 77 K. The adsorption and desorption isotherms were obtained in a relative pressure (P/P_o) range of 0.01 to 0.95. The data relating to the volume of adsorbed gas as a function of partial pressure were correlated using mathematical models to determine the BET surface area [58] and BJH to volume and distribution of pores [59].

3.3.4. Scanning Electron Microscopy (SEM)

Scanning electron micrographs of the mesoporous supports SBA-15 and AlSBA-15 with Si/Al = 50, as well as the supported cobalt and molybdenum catalysts, were carried out with the aim of observing the morphology of the synthesized mesoporous materials and some change in the morphology after impregnation of the CoMo metals. The analysis was obtained using a Jeol equipment model JSM-5610 LV (Miami, MA, USA). Before analysis, the samples were adhered to the sample holder using a thin carbon tape and subjected to a pretreatment that consisted of the deposition of a thin nanolayer of gold, with the aim of making the sample a suitable electron conductor and thus be able to provide suitable quality and resolution of the images. The analyses were carried out with magnifications ranging from 100 to 25,000 times.

3.4. Thiophene Hydrodesulfurization (HDS)

The thiophene HDS catalytic tests were carried out in a fixed-bed continuous-flow reactor under atmospheric pressure, according to the scheme shown in Figure 18. Thiophene was chosen as a probe molecule, which is characterized as the most common sulfur contaminant present in middle petroleum distillates. To carry out the tests, approximately 50 mg of sample was introduced into the Pyrex glass "U" reactor heated from room temperature to 450 °C at a heating rate of 5 °C min^{-1} in a dynamic atmosphere of N_2 containing 10% of H_2 with a total flow of 30 mL min^{-1}. After reaching 450 °C, the sample remained under these conditions for another 1 h and was then cooled to the reaction temperature of 350 °C, still maintaining the reducing atmosphere. Then, a mixture of n-heptane containing 12,070 ppm of thiophene (ca. 5100 ppm of sulfur) was drawn from a saturator maintained at room temperature through a line heated at 120 °C to the catalytic bed with a flow of 30 mL min^{-1}, maintaining the molar ratio H_2/(thiophene and n-heptane) of 8.2. The composition of the standard mixture of thiophene in n-heptane was confirmed through chemical analysis using EDX-700 equipment Shimadzu (Nakagyo-ku, Kyoto, Japan).

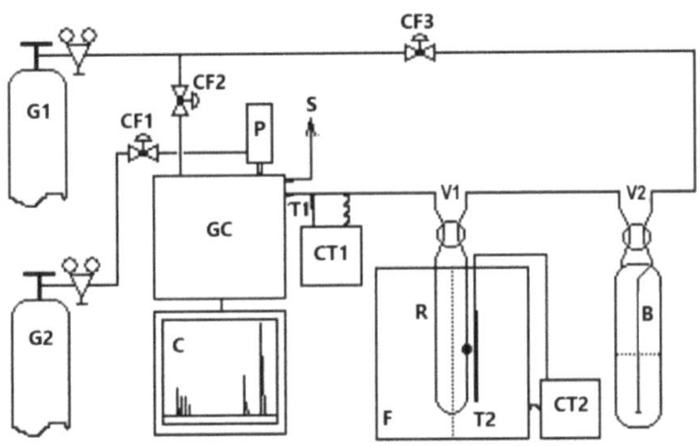

Figure 18. Catalytic evaluation unit. Where G1 = hydrogen; G2 = nitrogen; V1, V2, = 4-way valves; CF1, CF2, CF3 = flow control valves; B = saturator; P = 10-way pneumatic valve; CT1, CT2 = temperature controllers; T1, T2 = thermocouples; S = gas output; C = chromatogram; GC = gas chromatograph; F= furnace; and R = reactor with catalyst.

During the reaction, the catalytic bed was maintained at a constant temperature of 350 °C using a COEL HW1500 temperature controller (Sao Paulo, SP, Brazil). The reactor effluent products were successively injected "on-line" through a ten-way valve into a Varian CP3800 gas chromatograph (Palo Alto, CA, USA) with a thermal conductivity detector at 15 min intervals until reaching the pseudo-stationary state. The products were separated and analyzed in a 60 m fused silica column. The identification of products was carried out by comparing the retention times of the analytes of each chromatogram with the retention times of thiophene, n-heptane, and natural gas standards, considering the elution orders of the substances through the stationary phase used in the column (separation based on boiling points) as proposed by the manufacturer. Quantification of chromatogram peaks was performed using the method of external standards analyzed in the linearity range of the detector as recommended for thermal conductivity detectors. The tests were conducted with all catalysts in powder form to minimize the effects arising from internal mass transport. The following aspects were also taken into consideration: isothermal reaction in a fixed bed, vapor phase in an ideal gas state, uniform porosity, and negligible pressure drop in the bed without the presence of axial dispersion effects.

4. Conclusions

The effect of mesoporous supports such as silica (SBA-15) and aluminosilicate (AlSBA-15) on the catalytic activities of cobalt and molybdenum (CoMo) catalysts was demonstrated for thiophene hydrodesulfurization in a n-heptane stream as a model reaction. The characterization of the mesoporous supports by nitrogen adsorption–desorption analysis showed that SBA-15 and AlSBA-15 (Si/Al = 50) materials possessed surface area, pore diameter, and pore volume appropriate for the impregnation and dispersion of the Co and Mo metals on the surface. The crystallization properties by X-ray diffraction analysis showed that the Co and Mo metals were well dispersed in the catalytic supports. The catalytic activities indicated that the ordering and open pore channel of the CoMo/SBA-15 and CoMo/AlSBA-15 mesoporous catalysts are appropriate for thiophene conversion and selectivity to paraffin butane and olefins 1-butene, cis-, and trans-2-butene. The 1,3-butene and tetrahydrothiophene (THT) molecules were not de-detected in the products, evidencing that these compounds are strongly adsorbed on the Co-Mo active sites and undergo hydrogenation and desulfurization, respectively, with subsequent formation of C_4-paraffins, such as n-butane and isobutane.

The catalytic activity of CoMo/SBA-15 for thiophene HDS reaction was higher than that of CoMo/AlSBA-15, reaching ca. 20 and 30% conversion, respectively, after 1 h of reaction. The same trend was observed for the paraffin/olefin ratio. Therefore, using the CoMo/SBA-15 catalyst, after 1 h of reaction, the paraffin/olefin ratio increased from 0.2 to 0.3 up to 2 h of reaction. With the CoMo/AlSBA-15 catalyst, it was found that this ratio remained constant between 1 and 2 h of reaction, with a value lower than 0.1, showing that the AlSBA-15 support stabilized the structure. Regarding product selectivity, in general, the two catalysts were selective for the olefins 1-butene and trans- and cis-2-butenes. In relation to paraffins, for the CoMo/SBA-15 catalyst, ca. 10% selectivity was obtained for isobutane and n-butane. On the other hand, using the CoMo/AlSBA-15 catalyst, low concentrations of paraffins were observed, with a subsequent increase in the concentration of 1-butene, showing that the n-butane dehydrogenation reaction caused by metallic sites probably occurred, as well as isomerization to iso-butane, due to the presence of Bronsted acid sites generated by Al in the CoMo/AlSBA-15 catalyst.

Author Contributions: Conceptualization, A.C.S.L.S.C. and J.M.F.B.; methodology, A.C.S.L.S.C. and M.J.B.S.; formal analysis, M.D.S.A. and J.B.S.; investigation and data curation, A.C.S.L.S.C.; data curation, writing—original draft preparation, A.S.A. and R.C.O.B.D.; writing—review and editing, M.D.S.A., A.S.A. and J.B.S.; project administration, A.S.A. and V.J.F.J.; funding acquisition, A.S.A. All authors have read and agreed to the published version of the manuscript.

Funding: This research was funded by CNPq, grant number 312461/2022-4.

Data Availability Statement: Data are contained within the article.

Acknowledgments: The authors thank the Brazilian Agency of Petroleum, Natural Gas and Biofuel (ANP) and the National Council for Scientific and Technological Development (CNPq Brazil) for supporting this research.

Conflicts of Interest: The authors declare no conflicts of interest.

References

1. Ohtsuka, T. Catalyst for hydrodesulfurization of petroleum residua. *Catal. Rev. Sci. Eng.* **1997**, *16*, 291–325. [CrossRef]
2. Delmon, B.; Li, Y.W. Modeling of hydrotreating catalysts based on the remote control: HYD e HDS. *J. Mol. Catal. A Chem.* **1997**, *127*, 163–190.
3. Topsoe, H.; Clausen, B.S. Importance of Co-Mo type structures in hydrodesulfurization. *Catal. Rev. Sci. Eng.* **1984**, *26*, 395–420. [CrossRef]
4. Topsoe, H. Characterization of the structures and active sites in sulfided Co-Mo/Al_2O_3 and Ni-Mo/Al_2O_3 catalysts by NO chemiosorption. *J. Catal.* **1983**, *84*, 386–401. [CrossRef]
5. Grande, P.; Vanhaeren, X. Hydrotreating catalysts, an old story with new challenges. *Catal. Today* **1997**, *36*, 375–391. [CrossRef]
6. Morais, C.G.D.P.; Silva, J.B.; Almeida, J.S.; Oliveira, R.R.; Araujo, M.D.S.; Fernandes, G.J.T.; Delgado, R.C.O.B.; Coriolano, A.C.F.; Fernandes, V.J., Jr.; Araujo, A.S. Catalytic Distillation of Atmospheric Residue of Petroleum over HY-MCM-41 Micro-Mesoporous Materials. *Catalysts* **2023**, *13*, 296. [CrossRef]
7. Yue, L.; Li, G.; Zhang, F.; Chen, L.; Li, X.; Huang, X. Size-dependent activity of unsupported Co–Mo sulfide catalysts for the hydrodesulfurization of dibenzothiophene. *Appl. Catal. A Gen.* **2016**, *512*, 85–92. [CrossRef]

8. Zepeda, T.A.; de Leon, J.N.D.; Alonso, G.; Infantes-Molina, A.; Galindo-Ortega, Y.I.; Huirache-Acuna, R.; Fuentes, S. Hydrodesulfurization activity of Ni-containing unsupported Ga(x)WS_2 catalysts. *Catal. Commun.* **2019**, *130*, 105760. [CrossRef]
9. Morales-Ortuno, J.C.; Ortega-Domínguez, R.A.; Hernandez-Hipolito, P.; Bokhimi, X.; Klimova, T.E. HDS performance of NiMo catalysts supported on nanostructured materials containing titania. *Catal. Today* **2016**, *271*, 127–139. [CrossRef]
10. Shang, H.; Liu, C.; Xu, Y.; Qiu, J.; Wei, F. States of carbon nanotube supported Mo based HDS catalysts. *Fuel Process. Technol.* **2007**, *88*, 117–123. [CrossRef]
11. Liu, Z.; Zhang, L.; Jiang, J.; Bian, C.; Zhang, Z.; Gao, Z. Advancement of hydrodesulfurization catalyst and discussion of its application in coal tar. *Adv. Chem. Eng.* **2013**, *3*, 36–46. [CrossRef]
12. Umar, M.; Abdulazeez, I.; Tanimu, A.; Ganiyu, S.A.; Alhooshani, K. Modification of ZSM-5 mesoporosity and application a catalyst support in hydrodesulfurization of dibenzothiophene: Experimental and DFT studies. *J. Environ. Chem. Eng.* **2021**, *9*, 106738. [CrossRef]
13. Yu, Q.; Zhang, L.; Guo, R.; Sun, J.; Fu, W.; Tang, T.; Tang, T. Catalytic performance of CoMo catalysts supported on meso rous ZSM-5 zeolite-alumina composites in the hydrodesulfurization of 4,6-dimethyldibenzothiophene. *Fuel Process. Technol.* **2017**, *159*, 76–87. [CrossRef]
14. Qi, L.; Zheng, P.; Zhao, Z.; Duan, A.; Xu, C.; Wang, X. Insights into the intrinsic kinetics for efficient hydrodesulfurization of 4,6-dimethyldibenzothiophene over mesoporous $CoMoS_2$/ZSM-5. *J. Catal.* **2022**, *408*, 279–293. [CrossRef]
15. Zhou, W.; Zhou, A.; Zhang, Y.; Zhang, C.; Chen, Z.; Liu, L.; Zhou, Y.; Wei, Q.; Tao, X. Hydrodesulfurization of 4,6-dimethyldibenzothiophene over NiMo supported on Ga-modified Y zeolites catalysts. *J. Catal.* **2019**, *374*, 345–359. [CrossRef]
16. Song, H.; Wang, J.; Wang, Z.; Song, H.; Li, F.; Jin, Z. Effect of titanium content on dibenzothiophene HDS performance over Ni_2P/Ti-MCM-41 catalyst. *J. Catal.* **2014**, *311*, 257–265. [CrossRef]
17. Souza, M.J.B.; Marinkovic, B.A.; Jardim, P.M.; Araujo, A.S.; Pedrosa, A.M.G.; Souza, R.R. HDS of thiophene over CoMo/AlMCM-41 with different Si/Al ratios. *Appl. Catal. A Gen.* **2007**, *316*, 212–218. [CrossRef]
18. Calderon-Magdaleno, M.A.; Mendoza-Nieto, J.A.; Klimova, T.E. Effect of the amount of citric acid used in the preparation of NiMo/SBA-15 catalysts on their performance in HDS of dibenzothiophene-type compounds. *Catal. Today* **2014**, *220*, 78–88. [CrossRef]
19. Soni, K.K.; Chandra Mouli, K.; Dalai, A.K.; Adjaye, J. Effect of Ti loading on the HDS and HDN activity of KLGO on NiMo/TiSBA-15 catalysts. *Microp. Mesop. Mater.* **2012**, *152*, 224–234. [CrossRef]
20. Alonso-Perez, M.O.; Pawelec, B.; Zepeda, T.A.; Alonso-Núnez, G.; Nava, R.; Navarro, R.M.; Huirache-Acuna, R. Effect of the titanium incorporation method on the morphology and HDS activity of supported ternary Ni–Mo–W/SBA-16 catalysts. *Microp. Mesop. Mater.* **2021**, *312*, 110779. [CrossRef]
21. Dominguez Garcia, E.; Chen, J.; Oliviero, E.; Oliviero, L.; Mauge, F. New insight into the support effect on HDS catalysts: Evidence for the role of Mo-support interaction on the MoS_2 slab morphology. *Appl. Catal. B Environ.* **2020**, *260*, 117975. [CrossRef]
22. Roy, T.; Rousseau, J.; Daudin, A.; Pirngruber, G.; Lebeau, B.; Blin, J.-L.; Brunet, S. Deep hydrodesulfurization of 4,6-dimethydibenzothiophene over CoMoS/TiO_2 catalysts: Impact of the TiO_2 treatment. *Catal. Today* **2021**, *377*, 17–25. [CrossRef]
23. Mazurelle, J.; Lamonier, C.; Lancelot, C.; Payen, E.; Pichon, C.; Guillaume, D. Use of the cobalt salt of the heteropolyanion $[Co_2Mo_{10}O_{38}H_4]_6$ for the preparation of CoMo HDS catalysts supported on Al_2O_3, TiO_2 and ZrO_2. *Catal. Today* **2008**, *130*, 41–49. [CrossRef]
24. Zhang, L.; Chen, Z.; Zheng, S.; Cai, G.; Fu, W.; Tang, T.; He, M. Effect of the Co/Mo Ratio on the Morphology and Activity of the CoMo Catalyst Supported on MgO Nanosheets in Dibenzothiophene Hydrodesulfurization. *Ind. Eng. Chem. Res.* **2020**, *59*, 12338–12351. [CrossRef]
25. Bing, L.C.; Tian, A.X.; Li, J.J.; Yi, K.F.; Wang, F.; Wu, C.Z.; Wang, G.J. The effects of chelating agents on CoMo/TiO_2-Al_2O_3 hydrodesulfurization catalysts. *Catal. Lett.* **2018**, *148*, 1309–1314. [CrossRef]
26. Li, G.; Li, W.; Zhang, M.; Tao, K. Characterization and catalytic application of homogeneous nano-composite oxides ZrO_2–Al_2O_3. *Catal. Today* **2004**, *93*, 595–601. [CrossRef]
27. Tavizon Pozos, J.A.; Esquivel, G.C.; Cervantes Arista, I.; de los Reyes Heredia, J.A.; Suarez Toriello, V.A. Co-processing of hydrodeoxygenation and hydrodesulfurization of phenol and dibenzothiophene with NiMo/Al_2O_3–ZrO_2 and NiMo/TiO_2–ZrO_2 catalysts. *Int. J. Chem. React. Eng.* **2022**, *20*, 47–60. [CrossRef]
28. Kazakov, M.O.; Kazakova, M.A.; Vatutina, Y.V.; Larina, T.V.; Chesalov, Y.A.; Gerasimov, E.Y.; Prosvirin, I.P.; Klimov, O.V.; Noskov, A.S. Comparative study of MWCNT and alumina supported CoMo hydrotreating catalysts prepared with citric acid as chelating agent. *Catal. Today* **2020**, *357*, 221–230. [CrossRef]
29. Saleh, T.A. Carbon nanotube-incorporated alumina as a support for MoNi catalysts for the efficient hydrodesulfurization of thiophenes. *Chem. Eng. J.* **2021**, *404*, 126987. [CrossRef]
30. Saleh, T.A.; Al-Hammadi, S.A.; Abdullahi, I.M.; Mustageem, M. Synthesis of molybdenum cobalt nanocatalysts supported on carbon for hydrodesulfurization of liquid fuels. *J. Mol. Liq.* **2018**, *272*, 715–721. [CrossRef]
31. Prajapati, Y.N.; Verma, N. Hydrodesulfurization of thiophene on activated carbon fiber supported NiMo catalysts. *Energy Fuel* **2018**, *32*, 2183–2196. [CrossRef]
32. Abubakar, U.C.; Alhooshani, K.R.; Adamu, S.; Al Thagfi, J.; Saleh, T.A. The effect of calcination temperature on the activity of hydrodesulfurization catalysts supported on mesoporous activated carbon. *J. Clean. Prod.* **2019**, *211*, 1567–1575. [CrossRef]

33. Ali, I.; Al-Arfaj, A.A.; Saleh, T.A. Carbon nanofiber-doped zeolite as support formolybdenum based catalysts for enhanced hydrodesulfurization of dibenzothiophene. *J. Mol. Liq.* **2020**, *304*, 112376. [CrossRef]
34. Araujo, A.S.; Quintella, S.A.; Coutinho, A.C.S.L.S. Synthesis monitoring of SBA-15 nanostructured materials. *Adsorption* **2009**, *15*, 306–311. [CrossRef]
35. Antochshuk, V.; Araujo, A.S.; Jaroniec, M. Functionalized MCM-41 and CeMCM-41 materials synthesized via interfacial reactions *J. Phys. Chem. B* **2000**, *104*, 9713–9719. [CrossRef]
36. Araujo, A.M.M.; Queiroz, G.S.; Maia, D.O.; Gondim, A.D.; Souza, L.D.; Fernandes, V.J., Jr.; Araujo, A.S. Fast Pyrolysis of Sunflower Oil in the Presence of Microporous and Mesoporous Materials for Production of Bio-Oil. *Catalysts* **2018**, *8*, 261. [CrossRef]
37. Araujo, A.S.; Jaroniec, M. Determination of the surface area and mesopore volume for lanthanide-incorporated MCM-41 materials by using high resolution thermogravimetry. *Thermochim. Acta* **2000**, *345*, 173–177. [CrossRef]
38. Barros, J.M.F.; Fernandes, G.J.T.; Araujo, M.D.S.; Melo, D.M.A.; Gondim, A.D.; Fernandes, V.J., Jr.; Araujo, A.S. Hydrothermal Synthesis and Properties of Nanostructured Silica Containing Lanthanide Type Ln–SiO_2 (Ln = La, Ce, Pr, Nd, Eu, Gd, Dy, Yb, Lu). *Nanomaterials* **2023**, *13*, 382. [CrossRef] [PubMed]
39. Araujo, A.S.; Jaroniec, M. Synthesis and properties of lanthanide incorporated mesoporous molecular sieves. *J. Colloid. Interf. Sci.* **1999**, *218*, 462–467. [CrossRef] [PubMed]
40. Coutinho, A.C.S.L.S.; Quintella, S.A.; Araujo, A.S.; Barros, J.M.F.; Pedrosa, A.M.G.; Fernandes, V.J., Jr.; Souza, M.J.B. Thermogravimetry applied to characterization of SBA-15 nanostructured material. *J. Therm. Anal. Calorim.* **2007**, *87*, 457–461. [CrossRef]
41. Souza, M.J.B.; Silva, A.O.S.; Aquino, J.M.F.B.; Fernandes, V.J., Jr.; Araujo, A.S. Kinetic study of template removal of MCM-41 nanostructured material. *J. Therm. Anal. Calorim.* **2004**, *75*, 693–698. [CrossRef]
42. Araujo, A.S.; Fernandes, V.J., Jr.; Souza, M.J.B.; Silva, A.O.S.; Aquino, J.M.F.B. Model free-kinetics applied to CTMA+ removal of AlMCM-41 molecular sieves. *Thermochim. Acta* **2004**, *413*, 235–2408. [CrossRef]
43. Araujo, S.A.; Ionashiro, M.; Fernandes, V.J., Jr.; Araujo, A.S. Thermogravimetric investigations during the synthesis of silica-based MCM-41. *J. Therm. Anal. Calorim.* **2001**, *64*, 801–805. [CrossRef]
44. Souza, M.J.B.; Lima, S.H.; Araujo, A.S.; Pedrosa, A.M.G.; Coutinho, A.C.S.L.S. Determination of the acidity of MCM-41 with different Si/A1 ratios by the temperature programmed desorption of pyridine. *Ads. Sci. Techn.* **2007**, *25*, 751–756. [CrossRef]
45. Araujo, A.S.; Souza, C.D.R.; Souza, M.J.B.; Fernandes, V.J., Jr.; Pontes, L.A.M. Acid properties of ammonium exchanged A1MCM-41 with different Si/Al ratio. *Stud. Surf. Sci. Catal.* **2002**, *141*, 467–472.
46. Farias, M.F.; Domingos, Y.S.; Fernandes, G.J.T.; Castro, F.L.; Fernandes, V.J., Jr.; Costa, M.J.F.; Araujo, A.S. Effect of acidity in the removal-degradation of benzene in water catalyzed by Co-MCM-41 in medium containing hydrogen peroxide. *Microp. Mesop. Mater.* **2018**, *258*, 33–40. [CrossRef]
47. Coriolano, A.C.F.; Barbosa, G.F.S.; Alberto, C.K.D.; Delgado, R.C.O.B.; Castro, K.K.V.; Araujo, A.S. Catalytic processing of atmospheric residue of petroleum over AlSBA-15 nanomaterials with different acidity. *Petrol. Sci. Techn.* **2016**, *34*, 627–632. [CrossRef]
48. Gajardo, J.; Colmenares-Zerpa, J.; Peixoto, A.F.; Silva, D.S.A.; Silva, J.A.; Gispert-Guirado, F.; Llorca, J.; Chimentão, R.J. Revealing the effects of high Al loading incorporation in the SBA-15 silica mesoporous material. *J. Porous Mat.* **2023**, *5*, 1687–1707. [CrossRef]
49. Fernandes, V.J., Jr.; Araujo, A.S.; Fernandes, G.J.T. Thermal analysis applied to solid catalysts. Acidity, activity and regeneration. *J. Therm. Anal. Calorim.* **1999**, *56*, 275–285. [CrossRef]
50. Zhao, D.; Hou, Q.; Feng, J.; Chmelka, B.F.; Stucky, G.D. Nonionic triblock and star diblock copolymer and oligomeric surfactant syntheses of highly ordered, hydrothermally stable, mesoporous silica structures. *J. Am. Chem. Soc.* **1998**, *120*, 6024–6036. [CrossRef]
51. Zhao, D.; Feng, J.; Hou, Q.; Melosh, N.; Fredrickson, G.H.; Chmelka, B.F.; Stucky, G.D. Triblock copolymer syntheses of mesoporous silica with periodic 50 to 300 angstrom pores. *Science* **1998**, *279*, 548–552. [CrossRef]
52. Han, Y.J.; Kim, J.M.; Stucky, G.D. Preparation of nobel metal nanowires using hexagonal mesoporous silica SBA-15. *Chem. Mater.* **2000**, *12*, 2068–2069. [CrossRef]
53. Dhar, G.M.; Kumaran, G.M.; Kumar, M.; Rawat, K.S.; Sharma, L.D.; Raju, B.D.; Rao, K.S.R. Physico-chemical characterization and catalysis on SBA-15 supported molybdenum hydrotreating catalysts. *Catal. Today* **2005**, *99*, 309–314. [CrossRef]
54. Kumaran, G.M.; Garg, S.; Soni, K.; Kumar, M.; Sharma, L.D.; Dhar, G.M.; Rao, K.S.R. Effect of Al-SBA-15 support on catalytic functionalities of hydrotreating catalysts I. Effect of variation of Si/Al ratio on catalytic functionalities. *Appl. Catal. A Gen.* **2006**, *305*, 123–129.
55. Tuel, A.; Hubert-Pfalzgraf, L.G. Nanometric monodispersed titanium oxide particles on mesoporous silica: Synthesis, characterization and catalytic activity in oxidation reactions in the liquid phase. *J. Catal.* **2003**, *217*, 343–353. [CrossRef]
56. ICDD (International Center for Diffraction Data) and JCPDS (Joint Committee on Powder Diffraction Standards). Available online: http://www.icdd.com/ (accessed on 20 February 2024).
57. Thommes, M.; Kaneko, K.; Neimark, A.V.; Olivier, J.P.; Rodriguez-Reinoso, F.; Rouquerol, J.; Sing, K.S.W. Physisorption of Gases, with Special Reference to the Evaluation of Surface Area and Pore Size Distribution (IUPAC Technical Report). *Pure Appl. Chem.* **2015**, *87*, 1051–1069. [CrossRef]
58. Brunauer, S.; Emmett, P.H.; Teller, E. Adsorption of gases in multimolecular layers. *J. Am. Chem. Soc.* **1938**, *60*, 309–319. [CrossRef]

59. Barrett, E.P.; Joyner, L.G.; Halenda, P.P. The determination of pore volume and area distributions in porous substances. I. Computations from nitrogen isotherms. *J. Am. Chem. Soc.* **1951**, *73*, 373–380. [CrossRef]
60. Eswaramoorthi, I.; Dalai, A.K. Synthesis, characterization and catalytic performance of boron substituted SBA-15 molecular sieves. *Microp. Mesop. Mat.* **2006**, *93*, 1–11. [CrossRef]
61. Vinu, A.; Kumar, G.S.; Ariga, K.; Murugesan, V. Preparation of highly ordered mesoporous AlSBA-15 and its application to isopropylation of m-cresol. *J. Mol. Catal. A* **2005**, *235*, 57–66. [CrossRef]
62. Gédéon, A.; Lassoued, A.; Bonardet, J.L.; Fraissard, J. Surface acidity diagnosis and catalytic of AlSBA materials obtained by direct synthesis. *Microp. Mesop. Mat.* **2001**, *801*, 44–45. [CrossRef]
63. Ooi, Y.S.; Zakaria, R.; Mohamed, A.R.; Bhatia, S. Hydrothermal stability and catalytic activity of mesoporous aluminum-containing SBA-15. *Catalysis Comm.* **2004**, *5*, 441–445. [CrossRef]
64. Chao, M.C.; Lin, H.P.; Sheu, H.S.; Mou, C.Y. A study of morphology of mesoporous silica SBA-15. *Stud. Surf. Sci. Catal.* **2002**, *141*, 387–394.
65. Choi, D.G.; Yang, S.M. Effect of two-step sol-gel reaction on the mesoporous silica structure. *J. Colloid Interf. Sci.* **2003**, *261*, 127–132. [CrossRef] [PubMed]
66. Katiyar, A.; Yadav, S.; Smirniotis, P.; Pinto, N.G. Synthesis of ordered large pore SBA-15 spherical particles for adsorption of biomolecules. *J. Chromatogr. A* **2006**, *1*, 13–20. [CrossRef]
67. Wang, E.; Yang, F.; Song, M.; Chen, G.; Zhang, Q.; Wang, F.; Bing, L.; Wang, G.; Han, D. Recent advances in the unsupported catalysts for the hydrodesulfurization of fuel. *Fuel Proc. Techn.* **2022**, *235*, 107386. [CrossRef]
68. Lauritsen, J.V.; Besenbacher, F. Atom-resolved scanning tunneling microscopy investigations of molecular adsorption on MoS2 and CoMoS hydrodesulfurization catalysts. *J. Catal.* **2015**, *328*, 49–58. [CrossRef]
69. Tétényi, P.; Szarvas, T.; Ollár, T. Experimental proof of thiophene hydrodesulfurization reaction steps by isotope (14C) labeled thiophene. *React. Kinet. Mechan. Catal.* **2021**, *134*, 697–710. [CrossRef]
70. Desikan, P.; Amberg, C.H. Catalytic hydrodesulphurization of thiophene. Selective poisoning plus acidity of catalyst surface. *Can. J. Chem.* **1964**, *42*, 843–850.
71. Iwamoto, R.; Inamura, K.; Nozaki, T.; Lino, A. Effect of cobalt on the sulfiding temperature of CoO-MoO_3/Al_2O_3 studied by temperature programmed sulfiding. *Appl. Catal. A Gen.* **1997**, *163*, 217–225. [CrossRef]
72. Tsyganenko, A.A.; Storozheva, E.N.; Manoilova, O.V.; Lesage, T.; Daturi, M.; Lavalley, J.-C. Brønsted acidity of silica silanol groups induced by adsorption of acids. *Catal. Lett.* **2000**, *70*, 159–163. [CrossRef]
73. Wijaya, K.; Lammaduma Malau, M.L.; Utami, M.; Mulijani, S.; Patah, A.; Wibowo, A.C.; Chandrasekaran, M.; Rajabathar, J.R.; Al-Lohedan, H.A. Synthesis, Characterizations and Catalysis of Sulfated Silica and Nickel Modified Silica Catalysts for Diethyl Ether (DEE) Production from Ethanol towards Renewable Energy Applications. *Catalysts* **2021**, *11*, 1511. [CrossRef]
74. Yamada, T.; Zhou, H.; Asai, K.; Honma, I. Pore size controlled mesoporous silicate powder prepared by triblock copolymer templates. *Mater. Lett.* **2002**, *56*, 93–96. [CrossRef]

catalysts

MDPI

Review

Porous Aerogel Structures as Promising Materials for Photocatalysis, Thermal Insulation Textiles, and Technical Applications: A Review

Kang Hoon Lee [1], Zafar Arshad [2,*], Alla Dahshan [3], Mubark Alshareef [4], Qana A. Alsulami [5,*], Ayesha Bibi [6], Eui-Jong Lee [7], Muddasir Nawaz [8], Usman Zubair [2] and Amjed Javid [2]

1 Department of Energy and Environmental Engineering, The Catholic University of Korea, 43 Jibong-ro, Bucheon-si 14662, Republic of Korea; diasyong@catholic.ac.kr
2 School of Engineering and Technology, National Textile University Faisalabad, Faisalabad 37610, Pakistan; usmank980@gmail.com (U.Z.); amjedljno907@gmail.com (A.J.)
3 Department of Physics, Faculty of Science, King Khalid University, P.O. Box 9004, Abha 62529, Saudi Arabia; adahshan73@gmail.com
4 Department of Chemistry, Faculty of Applied Science, Umm Al Qura University, Makkah 24230, Saudi Arabia; mmshreef@uqu.edu.sa
5 Chemistry Department, Faculty of Science, King Abdulaziz University, Jeddah 21589, Saudi Arabia
6 Department of Human Nutrition and Dietetics, Women University Mardan, Mardan 23200, Pakistan; ayeshabb2009@yahoo.com
7 Department of Environmental Engineering, Daegu University, 201 Daegudae-ro, Jillyang, Gyeongsan-si 38453, Republic of Korea; lujong@daegu.ac.kr
8 Centre for Advance Materials, Qatar University, Doha P.O. Box 2713, Qatar; m.nawaz@qu.edu.qa
* Correspondence: zafarnubii@gmail.com (Z.A.); qalselami@kau.edu.sa (Q.A.A.)

Citation: Lee, K.H.; Arshad, Z.; Dahshan, A.; Alshareef, M.; Alsulami, Q.A.; Bibi, A.; Lee, E.-J.; Nawaz, M.; Zubair, U.; Javid, A. Porous Aerogel Structures as Promising Materials for Photocatalysis, Thermal Insulation Textiles, and Technical Applications: A Review. *Catalysts* **2023**, *13*, 1286. https://doi.org/10.3390/catal13091286

Academic Editor: Narendra Kumar

Received: 19 June 2023
Revised: 29 August 2023
Accepted: 3 September 2023
Published: 8 September 2023
Corrected: 7 December 2023

Abstract: Aerogels, due to their unique features like lightweight, ultra-low thermal conductivity, and design variations, have gotten a lot of interest in thermal insulation, photocatalysis, and protective areas. Besides their superior thermal properties, aerogel thermal insulation and photocatalyst materials also possess many inherent flaws, such as handling issues, high manufacturing costs, and low strength as well as toughness. The most persuasive and successful ways to improve photocatalytic and thermal insulating qualities while lowering costs are composition optimization and microstructure reconstruction. Their high surface area and porosity make them ideal for enhancing the efficiency and capacity of these devices. Research may lead to more efficient and longer-lasting energy storage solutions. This review describes the characteristics, microstructural reconstruction, design variation, and properties of all aerogel fabrication techniques and provides a comprehensive overview of scientific achievements linked to them. The effectiveness of raw material compositions, properties, and mechanical parameters are also discussed. The major goal of this review is to highlight the aerogel-based materials and design variations and to explore the most potential development trends for photocatalysis and thermal applications. The industrial as well as technical applications of silica aerogels are also highlighted. This review highlights futuristic applications of aerogel-based textile materials to alleviate the CO_2 burden on our atmosphere, either by providing next-level thermal insulation or by employing them in CO_2 mitigating technologies such as CO_2 capture.

Keywords: thermal insulation; photocatalysis; specific heat; aerogel porosity; mesoporous layer; thermal conductivity

1. General Aerogel Overview

Aerogels have been discovered as the most promising thermally insulated material for textile, industrial, and other protective applications. These materials, having the highest porosity, are also highly recommended for thermal insulation, purification, energy storage, and catalysis [1,2]. It emphasizes the recent contribution of aerogels with their surface modification, material variation, and fascinating optical characteristics [3,4]. The global energy crisis and environmental degradation are getting increasingly critical as national economies develop rapidly [5]. Energy demand has increased at a rate of 1.8% per year over the past 40 years, rising from 2790 Mtoe in 2010 to nearly 4400 Mtoe by 2050 [6]. Energy conservation utilizes all natural resources—gas, oil, coal, and nuclear energy—as a potential resource for energy conservation. During the era of 2016–2020, China was very much intended to meet its domestic economic lines by all possible means. China invested almost 1.2 trillion RMB in its energy conservation sector and environmental protection [7]. The European Union also anticipated that all energy-efficient buildings toward green conservation by the end of 2020 [8]. United Union planned to maintain all commercial buildings with energy-efficient infrastructure to meet the current energy demands of the sector. The introduction of new and innovative thermal insulation materials has been of greater interest to researchers involved in maintaining environmental structures and thermal equipment [9,10].

Conventional thermal insulation materials and fabrications demand high-cost raw materials, multiple fabrication layers, and high costs [11]. As aerogels are lightweight and highly thermal insulators, they have the capability to replace conventional ones with superior insulating characteristics [12]. Aerogels have shown tremendous potential for various applications due to their unique properties. These ultralight materials are known for their extremely low density, high surface area, and excellent thermal insulating properties, as shown in Figure 1 [13,14]. Aerogels' porosity varies in the nanometer-scale range, with pores of 1/3000th the diameter [3,12]. It has a series of networks with an air volume of around 80–99.8% and a surface area of 100–1600 m^2g^{-1}. Due to the low refractive index value (1.007–1.240), these materials are highly preferred for insulation purposes, laser experiments, nuclear particle detection, and ultrasonic gas sensors. It can behave like a superheat insulator in thermal applications [15]. Aerogels have a unique and intricate structure that contributes to their exceptional properties. They are often referred to as "frozen smoke" due to their translucent appearance and extremely low density. There are three basic parts: two solid parts and one porous part. The porous part is sandwiched, preserving the aerogel's structure [16,17]. In aerogels, the liquid is replaced by air with shrinkage of the solid network, achieving the highest porosity. When the liquid component of gel is replaced with air, higher porosity provides superior results. Its porosity lies at both meso and micro levels [18].

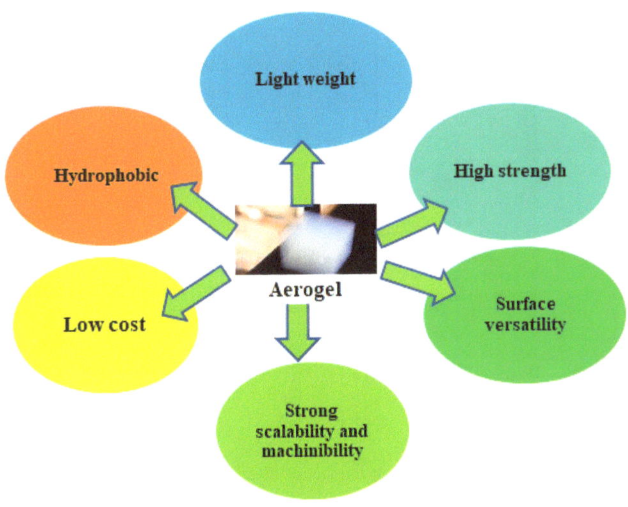

Figure 1. Aerogel characteristics.

2. History and Marketing

Aerogel was first discovered by Samuel Stephens Kistler in 1931. It was also named solid smoke, blue air, or frozen smoke. In the 1990s, it was a promising candidate for thermal insulation, flameproofing, and other lighter-weight applications. In the 21st century aerogel materials were a potential research area [19]. Around 2013–2020, aerogel marketing jumped up to US$ 1896.6 million from $221.8 million, and now it is expected to boom up to $1395 million by 2027. The growth of aerogel marketing is mainly due to the increased development of the build-tech and geo-tech sectors [20]. The top leading manufacturers of aerogel are Aerogel Technologies Leading Large Crystallographic Aerogels (LLCA), Armacell International Silica Aerogels (SA), Thermablok Aerogels Limited, JIOS Aerogel, Cabot Corporation, and Aspen Aerogel Corporation [21].

Factors Affecting the Aerogel Market

There are some limiting factors due to which the aerogel market is badly affected. Mainly, these factors are the expensive processing parameters of aerogel synthesis and higher production costs. Aerogel synthesis requires higher cost rates of raw materials to produce a market-competent product, thereby negatively impacting the market growth [22]. Scientists are highly focused on reducing production costs and making more customer-attractive products for aerogel marketing. Increasing product applications in multiple areas is an attractive way to improve marketing [23].

Public awareness about the advanced applications of aerogel-based materials in the field of insulation and other purposes can increase the market growth rate. The aerogel market is influenced by a combination of technological, economic, and societal factors. These factors can impact the demand, production, and adoption of aerogels in various industries [24,25]. The marketing and applications of aerogels are also categorized as exotic areas because of their peculiar characteristics and natural accessibility [26]. Silica is the most versatile aerogel material being used worldwide for thermal isolation, as shown in Figure 2.

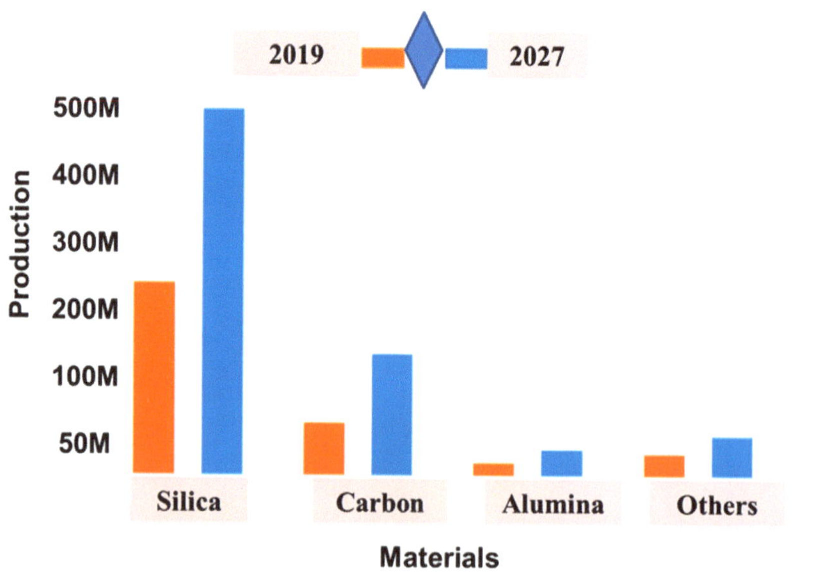

Figure 2. Growth rate market of different aerogel-based materials [14].

3. Aerogel Chemistry

Aerogel is a nanoporous polymeric material (Figure 3) with a backbone of linked nanoparticles (15 nm) that expand throughout its volume due to gas expansion, forming a 3D continuous network. Aerogel materials are 15 times heavier as compared to air, which has a very low dielectric constant [27]. Aerogel materials possess higher sound absorption but a low sound velocity [28] with damping > 50 dB and air velocity around ≈100 m/s. Aerogel possesses a very low thermal conductivity of 0.005–0.1 W/m·K. Aerogel's physical structure can be transparent or opaque, or it can be colored and have a variable refractive index (1001 to 2.1) [29]. Aerogel materials have internal surface areas ranging from 10–2000 m^2/g. Aerogel surfaces can be functionalized and made hydrophobic by polymeric residues with the help of chemical vapor deposition (CVD) of magnetic layers. Gel structure is basically functionalized by embedded particles (e.g., dyes, ferroelectrics) or with interpenetrating hybrid aerogel networks [21,30,31].

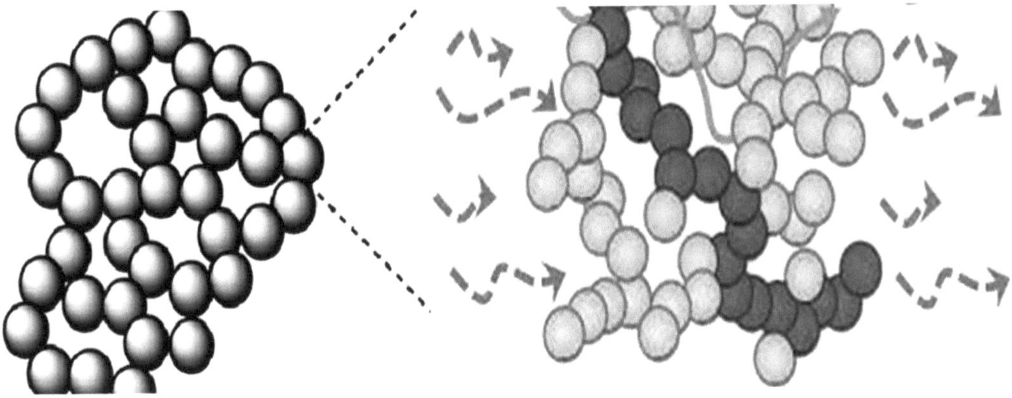

Figure 3. Aerogel structure. Reproduced from [32].

4. Aerogel Porosity and Calculations

Per International Union of Pure and Applied Chemistry (IUPAC) standards, pore size < 2 nm is termed as microporous, pore size 2–50 nm is termed as mesoporous, and pores > 50 nm are macroporous. Synthesized aerogels are varied from microporous to mesoporous structure; however, most of them are mesoporous, as shown in Figure 4. In aerogel pore networking, the pore size should be in that range where liquid can easily flow from pore to pore. The liquid should be able to flow through the whole structure without any restriction. Aerogels are unique materials with the highest porosity and smaller pore size diameters [18,33]. For the calculation of the pore size, the BET/nitrogen adsorption method is utilized. In this technique, adsorbed gas is measured to detect pore size. Ball milling was used to reduce the particle size [34]. Figure 4a presents the comparison of aerogels loaded with silica and titania through BET analysis. The presence of silica concentration affects the pore size of aerogels, thereby enhancing the mechanical properties of aerogels as $P/P_0 = 0.99$, where P is the pressure of the substrate and P_0 is the inherent pressure. From the observation, titania-based aerogel surface area is 529–587 m^2/g, whereas pore diameter ranges from 11.6 to 12.3 nm. For silica-based aerogels, the surface area equals 572.1 m^2/g, whereas the pore diameter is 12.1 nm. According to IUPAC standards and desorption hysteresis, both of these are mesoporous aerogels. However, there is a significant difference in both densities, as carbon-based materials exhibited more density due to their inherent characteristics [35,36].

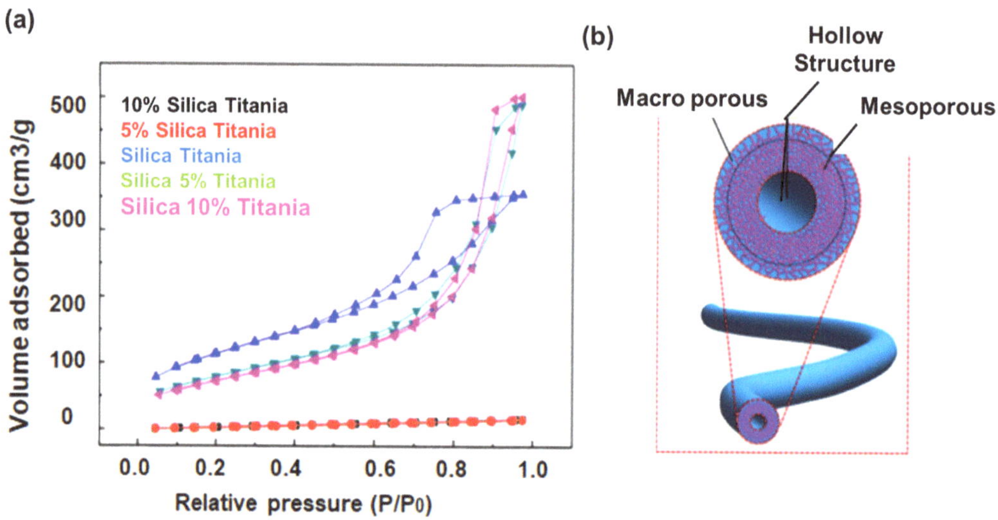

Figure 4. (**a**) BET Comparison between silica and titania-based aerogels; (**b**) Variations in aerogel as macrostructure and microstructures. Open access, reproduced from [37].

Volume shrinkage is calculated from the aerogel volume. There are two terms for understanding aerogel porosity: bulk density and skeletal density.

$$\text{Bulkdensity} = \frac{\text{Aerogelmass}}{\text{Aerogelvolume}} \tag{1}$$

$$\text{Skeletaldensity} = \frac{\text{Massofsolidmaterial}}{\text{Sumofthevolumeofsolidmaterial}} \tag{2}$$

Skeletal density value is almost closer to the bulk density. These two values are gained by the helium pycnometry [38].

Volume shrinkage percentage, porosity, and pore volume for aerogel are obtained by following formulae:

$$\text{Vs}(\%\text{age}) = \left(1 - \frac{\text{Va}}{\text{Vg}}\right) \times 100 \tag{3}$$

$$\text{Porosity} = \left(1 - \frac{\text{Pb}}{\text{Ps}}\right) \times 100 \tag{4}$$

$$\text{Porevolume} = \left(\frac{1}{\text{Pb}} - \frac{1}{\text{Ps}}\right) \tag{5}$$

In the above equations, Va is the volume of aerogels, Vg is the volume of air, P_b is the density of aerogels, whereas P_s is the density of base material, i.e., (cellulose).

5. Aerogels as Catalysis

Catalysts are essential components for the treatment of air and water pollutants on the way to a sustainable and clean environment. As active heterogeneous catalysts for a number of catalytic and photocatalytic environmental remediation processes, aerogels made from diverse molecular precursors are well known [39]. Because of this, aerogels have been viewed as a bridge bridging the nano and macroworlds, where the building blocks can both maintain their original features and develop new ones through their 3D interaction. Aerogels are particularly promising for photocatalytic applications due to their specific qualities [40]. Aerogels have an incredibly high surface area, which provides a large number of active sites for photocatalytic reactions to occur. This increased surface area allows for more interactions between the catalyst and the reactants, enhancing the efficiency of the reaction. In addition, the development of novel aerogel materials like graphene has encouraged the search for more mechanically stable and versatile aerogel photocatalysts [41]. The use has expanded to include solar energy conversion in addition to conventional environmental cleanup [42]. Aerogels have been explored and applied in various catalytic processes due to their unique properties and larger surface area [43]. One notable example of aerogels used in catalysis is their application as catalyst supports for heterogeneous catalysis. Heterogeneous catalysis involves the use of a solid catalyst to accelerate a chemical reaction between gaseous or liquid reactants. This type of aerogel-supported catalyst can be used not only in hydrogenation reactions but also in a wide range of other catalytic processes such as oxidation, hydrogenolysis, and more. The versatility of aerogels as catalyst supports enables their application in various industries, including fine chemicals, pharmaceuticals, petrochemicals, and environmental remediation [44].

It is important to note that the specific conditions, metal catalyst, and reactants used in such catalytic processes will vary depending on the target reaction and the desired outcome. As research in the field of aerogel-based catalysis continues, we can expect to see more tailored and efficient catalyst systems developed for a variety of chemical transformations as shown in Figure 5 [45].

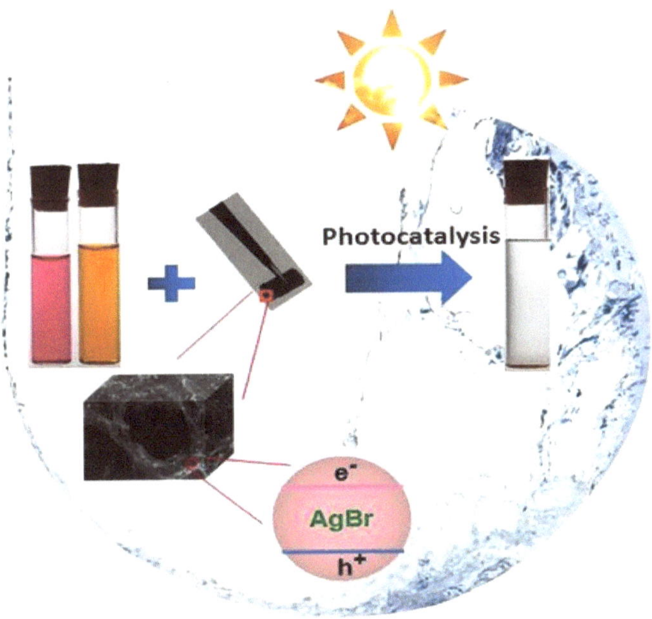

Figure 5. Aerogels as catalysis.

6. Aerogel Specific Heat

Specific heat is related to internal energy and is a very important thermodynamic property. The specific heat of aerogels is around 1900 J/g·K. Specific heat for volume and pressure is given in the following formulae:

$$Cv = \left(\frac{\partial u}{\partial T}\right)v \tag{6}$$

$$Cp = \left(\frac{\partial h}{\partial T}\right)p \tag{7}$$

In the above equations, u is internal energy, and h is enthalpy. T and P represent temperature and pressure [36]. Its SI units are J/mol·K or J/Kg·K [46]. In aerogel structures free molecules are allowed to move randomly in the structure. On heating, the movement of free molecules increases and the volume of gas increases, thereby increasing the acceleration of aerogel-free molecules, as shown in Figure 6.

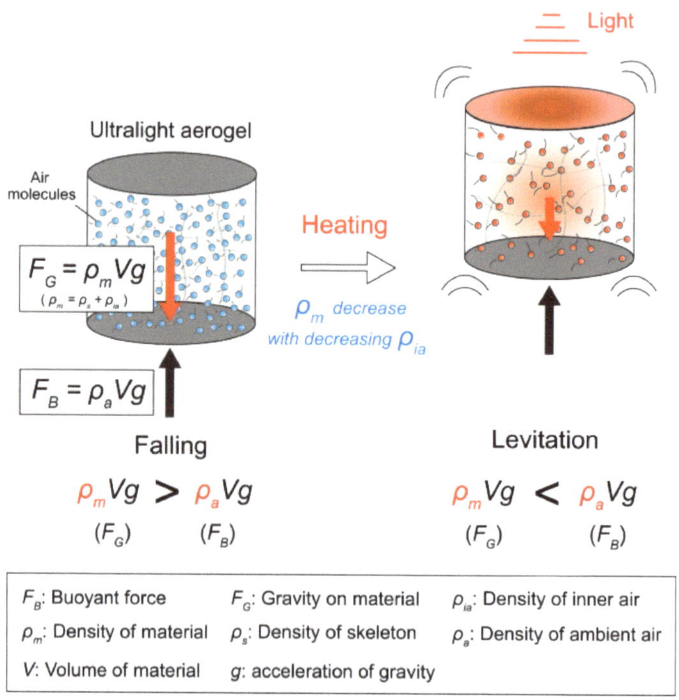

Figure 6. Aerogels specific heat measurements. Open access, Reproduced from [47].

7. Aerogel Heat Thermal Transfer Mechanism and Calculations

The heating transfer mechanism through hollow and aerogel fibers is termed "thermal conductivity". It is measured in W/m·K. It is measured with the basic principle of Fourier's law, which is applicable to all fluids, gases, and liquids. Thermal conductivity varies with temperature, and it is termed as follows [48]:

$$k = k(\bar{r}, T(r,t) = \frac{qx}{\frac{\partial T}{\partial x}} \tag{8}$$

As most of the materials are homogenous, k = k (T).

As shown in Figure 7a, thermal conductivity via hollow fibers is as follows:

$$\lambda_1 = \lambda_{conv1} + \lambda_{cond1} + \lambda_{rad1} \tag{9}$$

In the above equation, λ_{conv} is basically the heat transfer through convection; λ_{cond} is basically the heat transfer via conduction; whereas λ_{rad} is the heat transfer through radiation. Hollow fibers possess a porosity of around 85%, so heat transfer is quite easy and vast. Thermal convection is also simultaneously reduced this way. As the thermal conduction of air is much lower than that of solids, so thermal conduction of hollow fibers is reduced [48,49].

Aerogels possess thermal conductivity in different ways. Thermal conductivity via the aerogel core is expressed as follows:

$$\lambda_2 = \lambda_{conv1} + \lambda_{cond1} + \lambda_{rad1} \quad (10)$$

In aerogel, there is a cellular networking structure with pores diameter of tens of micrometers. These two factors suppress air circulation, thereby decreasing thermal convection. Air possesses the least value of thermal conduction as compared to solid structures. Aerogels also possess almost 99% porosity, thereby reducing thermal conduction. Aerogel porosity enhances its thermal properties and also prohibits all types of infrared radiation [50]. Hollow fibers can capture the aerogel precursor and also help to restrain air convection. Due to the above superior properties, aerogel behaves like a superior thermal insulating material with its microstructure and sheath. However, there are several factors that can modify its microstructure and sheath, like improvements in the design of the sheath and microstructure, enclosing air pockets in its structure, and further decreasing the pore size [51]. However, the volume ratio of the core microstructure and sheath should be uniform to tune all these characteristics. Aerogels are the best insulators because they do not favor all heat transfer modes. If the gas molecular free path diameter is more than its pore diameter (KN > 1), the gas molecule will not strike with other gas molecules; rather it will strike with the walls. Due to this, thermal conductivity decreases. This effect is known as Knudsen effect. Knudsen equation is as follows [52]:

$$\lambda g = \frac{\lambda g0}{1 + \beta . Kn} \quad (11)$$

Here, $\lambda g0$ is the thermal conductivity of the gas at standard conditions (lair 1/4 26 mW m·K1), β is a factor for the energy transfer between gas molecules and the surrounding cell walls (E2 for air) and Kn is the Knudsen number. The thermal conductivity of silica Aerogels granules and composites is evaluated by Lee's Disc method (ASTM C177-13). The apparatus consists of three plates made of copper (75 × 45 × 3 mm) [53]. The heater is also connected with 5 levels of input power 1.18, 1.92, 3.16, 4.75, and 6.84 W. Thermal probes are fixed at first, second, and third copper plates. Basically, the silica aerogel granules are fixed between these copper plates at ambient temperature and pressure. The required time is 1–2 h after that thermal probe temperature is at a steady state [54].

Thermal conductivity is basically calculated from the following equation:

$$Q = (\lambda Ay)\frac{(T2 - T3)}{t} \quad (12)$$

In this equation, "Q" represents the quantity of heat flowing in watts; "λ" is the thermal conductivity having units W/K-m. "A" is the area of the surface, and "t" is the required time. It was found that porosity and structural orientation highly affect thermal conductivity. It was also observed that thermal conductivity has a direct relationship with density. Heat transfer through conduction decreases in the same manner as density decreases, as shown in Figure 7b.

Thermal conductivity has an inverse relationship with porosity, as shown in Figure 7c. As porosity increases, it enhances the amorphous regions in the structure, which also hinders heat conduction. Heat conduction is very much decreased in amorphous regions because heat cannot find any trap states to flow through the material [55].

Thermal conductivity can be tailored using different precursors. TEOS is the most useful substrate; SS causes many impurities, whereas MTMS is also utilized for aging purposes. The effect of different substrates on density, processing temperature, porosity, and thermal conductivity is listed in Table 1.

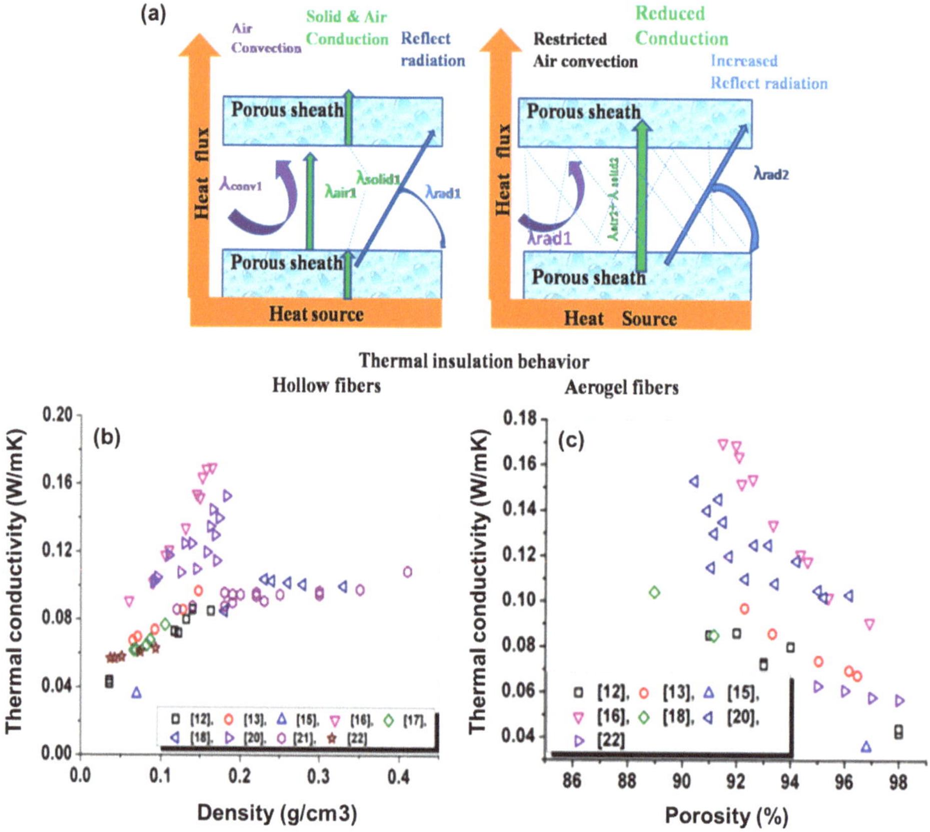

Figure 7. (**a**) Comparison between thermal insulation through hollow fibers and aerogel fibers; (**b**) Thermal conductivity with respect to density; (**c**) Thermal conductivity with respect to porosity. Reproduced from [56].

Table 1. Comparison of properties for different drying technologies [57].

Precursor	Drying Type	Thermal Stability °C	Thermal Conductivity	Porosity	Density	Ref
TEOS	APD	550	0.08–0.1	84.21–91.16	0.23–0.33	[58]
MTMS	SCD	480	0.09–0.098	----	0.1–0.35	[59]
MTMS	SCD	490	0.057–0.063	95–98	0.037–0.093	[59]
TEOS	SCD	300	0.068–0.099	92–96	0.06-0.15	[60]
TEOS	APD	200–520	0.042–0.086	91–98	0.036–0.163	[61]
SS	APD	320				[61]
SS	APD	325	0.091–0.170	92–96	0.152–0.06	[62]
TEOS	SCD	100–300	0.103–0.355	72–96.8	0.036–0.417	[62]

8. Heat Transfer Calculation

Many heat transmission techniques make use of composite systems, and some even combine conduction and convection. When working with these composite systems, it's generally easier to work with a U-factor or overall heat transfer coefficient. The U-factor is determined by a formula that is similar to Newton's cooling law [63]:

$$q = U\Delta T \tag{13}$$

q is the total heat flux density in units (Wm^{-2}). U is the overall heat transfer coefficient ($Wm^{-2}K$) whereas ΔT is the temperature difference. From Figure 8a, assuming one-dimensional heat transfer through the plane wall and disregarding radiation, the overall heat transfer coefficient can be calculated as follows:

$$U = \frac{1}{1/h1 + L1/k1 + 1/h2} \tag{14}$$

U is the overall heat transfer coefficient ($Wm^{-2}K$); K is the materials conductivity ($Wm^{-1}K^{-1}$); and h is the convection heat transfer coefficient ($Wm^{-2}K$) [64].

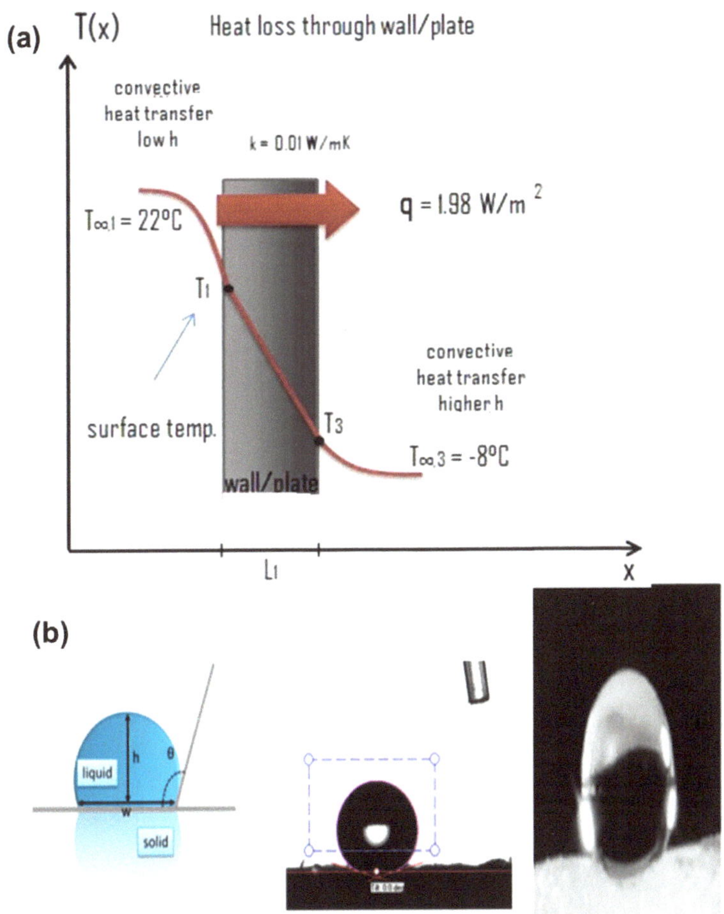

Figure 8. (**a**) Heat Transfer; (**b**) Hydrophobicity between different media.

9. Hydrophobicity

Aerogels can be hydrophilic or hydrophobic. Hydrophobicity and hydrophilicity depend on the substrate, precursor, silylating agent, and drying method. Hydrophilic aerogels did not get much emergence because when the water drop absorbs it, it scatters into the structure, colloids with the walls, and deforms the structure [65]. There was one main application of hydrophilic silica in drug delivery because of its hydrophobic character. Most of the applications fall under the hydrophobic nature of aerogels. Over 350 °C, the hydrophobicity of aerogel is converted into hydrophilicity due to oxidation. Hydrophobicity is basically the characteristics of the contact angle (θ) of a water droplet with an aerogel surface. It is the wetting of a solid by a liquid. When the liquid forms an angle at the three-phase boundary, the solid, liquid, and gas phases intersect, as shown in Figure 8b [66]. The contact angle is measured by the following equation:

$$\theta = 2\tan^{-1} 2h/w \tag{15}$$

In the above equation, w is the width and h is the height of the water droplet. If the angle is less than 90°, liquid will wet the surface. Complete wetting occurs at 0°. Materials having a contact angle greater than 90° are called hydrophobic. For superhydrophobic, the contact angle should be greater than 150°. Contact angles are subclassified as dynamic and static angles [67]. When a droplet is standing on the surface and the three-phase boundary is not moving, dynamic and static contact angles are measured (Scientific 2015). Dynamic contact angles are referred to as advancing and receding angles when the three-phase boundary is moving [68,69].

10. Classifications of Aerogel

Aerogels are classified according to chemical nature organic, inorganic, or composite, and further classified mentioned in Figure 9. Classifications of aerogels on the basis of appearance, preparation method, microstructure, and chemical structure are shown in Figure 10.

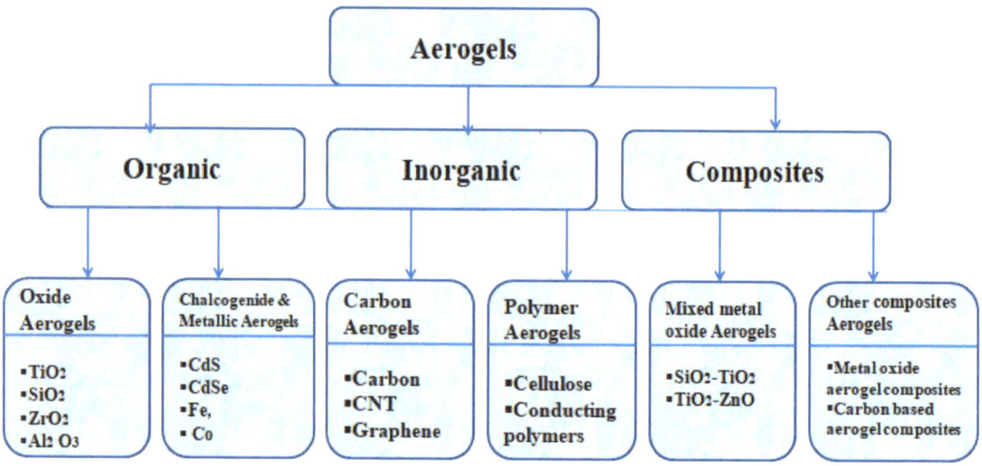

Figure 9. Aerogels classification. Open access, Reproduced from [70].

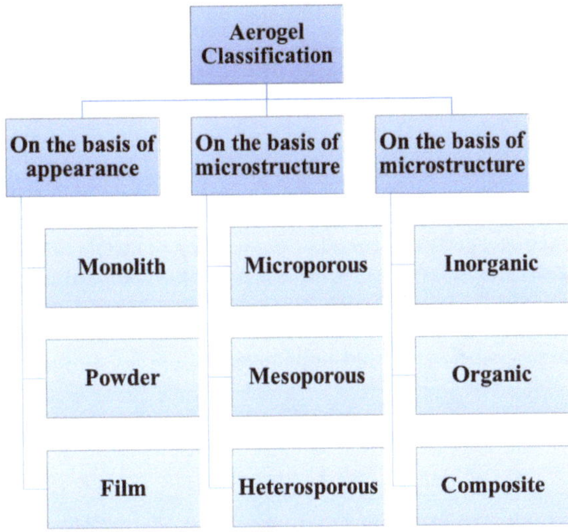

Figure 10. Aerogel classifications according to characteristics, Reproduced from [71].

10.1. Silica Aerogels

Silica aerogels are comprised of basic silica networking partial filler air packets, along with 3D networking structures. It has (-Si-) atoms at its backbone structure, forming siloxane bridges. Native silica aerogels cannot be handled alone and remain monolithic so there is a need to reinforce silica for enhanced mechanical properties (Figure 11). Silica aerogel properties are mainly affected due to their intrinsic fragility, thereby producing strains on the surface [72]. Scientists discovered many strategies, mainly to reduce its densities for high shock energy absorption. Low-density aerogels are highly preferred for bending and compression applications. For load-bearing applications, aging of wet gel is recommended, in which inorganic networks are formed, thereby increasing the strength of the final particles [73]. By this, the modulus of elasticity is also increased by a factor of two. Hybridization is also a good option for increasing mechanical properties by promoting the co-gelation of the silicon alkoxide with hybrid precursors such as poly (dimethylsiloxane) (PDMS) [74]. The gels formed by this are termed "ORMOSIL" (Organic modified silica) hybrids. Its flexibility is similar to rubber, but it can be elastically compressed up to 30% (by volume) with no damage [75,76].

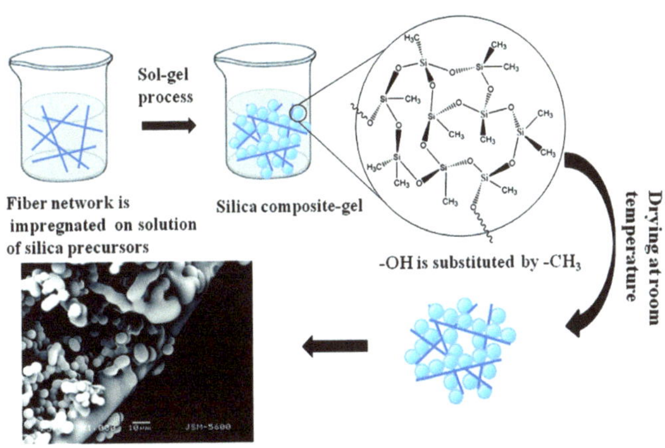

Figure 11. Silica aerogel synthesis and SEM morphology. Open access, Reproduced from [77].

10.2. Metal Oxide Aerogels

It is the most abundant class of aerogels structured with metal-oxygen bonds. These have unique characteristics; they behave as catalysts for chemical transformations and precursors for carbon nanotubes. Alumina oxide (Al_2O_3), zinc oxides (ZnO), titanium oxide (TiO_2), and iron oxide (Fe_2O_3) are major metal oxide materials used in this area. Their porosity ranges from 1 to 25 nm and possesses a low density of 0.06 g/cm^3; for Al_2O_3/SiO_2, the density is slightly higher, 0.54 $g{\cdot}cm^{-3}$, and porosity is varied between 77% and 96% [78,79]. These types of aerogels are prepared via the sol-gel route. In Figure 12, the hydroxyl group is formed by the reaction of metal aloxide with water. This step is known as hydrolysis. After this, aloxide is hydrolyzed partly, and then condensation occurs [80]. There may be a dehydration or dealcoholization reaction between aloxy and hydroxyl groups or with two separate hydroxyl groups. The nucleation rate is very low as compared to propagation to avoid agglomeration.

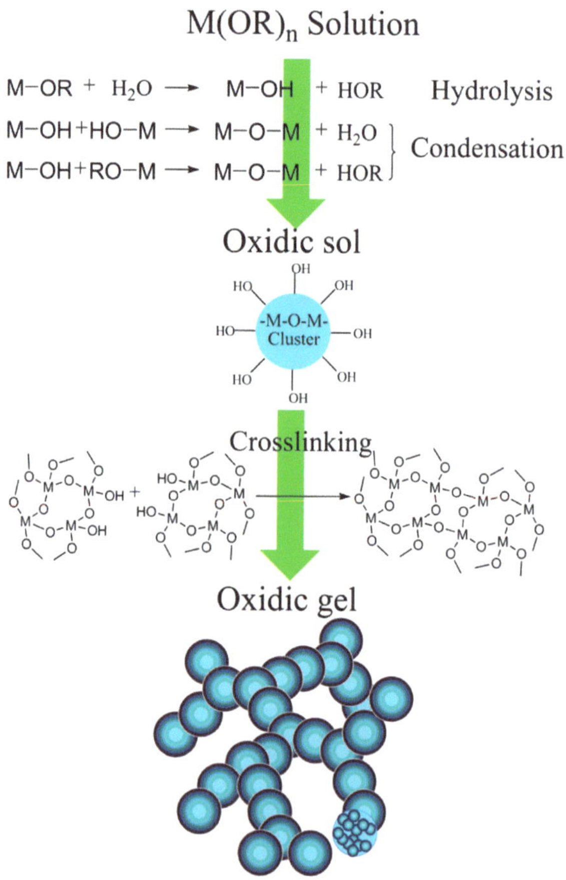

Figure 12. Metal oxide aerogels. Open access, Reproduced from [15].

This also affects the cluster size as the lowest nucleation rate also results in smaller particle sizes. Growth rate is more related to diffusion rate and degree of supersaturation. At the growth stage, the hydrolysis rate is important. A high hydrolysis rate leads to the formation of supersaturated solution so there is a need to decrease it. To decrease the hydrolysis rate, the reaction should be catalyzed under an acidic or basic environment to get a stable colloid [80].

10.3. Organic and Carbon Aerogels

Organic aerogels basically contain cellulose and resin in their backbone structure. Organic aerogels are synthesized with both monomeric and polymeric precursors such as polyvinyl chloride, phenol melamine, polymeric isocyanate, phenol-furfural cresol formaldehyde, and melamine–formaldehyde. Precursor is carefully chosen as end product applications are certainly dependent on it. These types of aerogels are highly recommended for membrane-based gas separation, acoustic insulators, dielectrics, and catalyst supports [81,82].

Its polymerization includes the following two steps:

1. Addition reaction between formaldehyde and resorcinol to form hydroxymethyl resorcinol monomers (Equation (16)):
2. -CH_2- or -CH_2OCH_2- bridging polymerization between monomers, producing formaldehyde or water (Equations (17) and (18)).

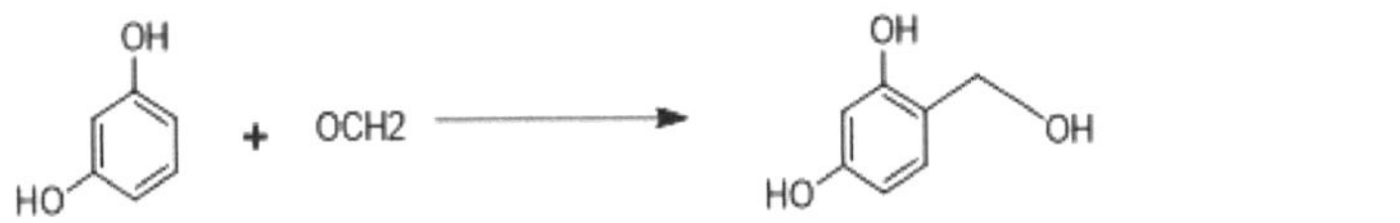

(16)

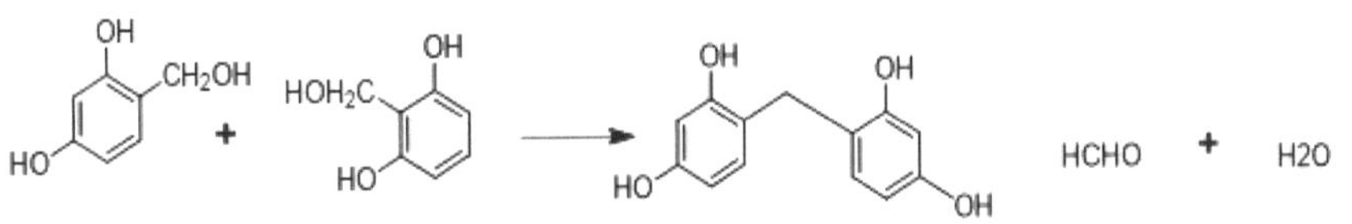

(17)

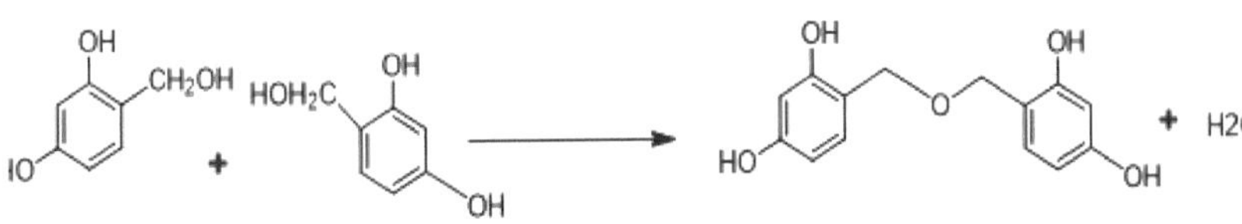

(18)

When clusters crosslink, gelation occurs. RF clusters are the result of continuous polymerization. These are used for high mechanical strength applications [51,83].

Carbon aerogel (Figure 13) is a branch of organic aerogel possessing a higher porosity rate and interconnected clusters of carbon nanoparticles having 3–20 nm diameter. To obtain carbon aerogels from organic aerogels, temperatures above 600 °C are required for carbonization or pyrolysis [84]. Additionally, there should be an inert gas atmosphere (nitrogen, helium, or argon). These are highly porous, graphite-based aerogels. Carbon aerogels are preferred for electric storage and hydrogen production devices because of their higher porosity and corrosion resistance. Carbon aerogels were further modified into diamond aerogels by laser-heated diamond anvil cells. These are superactive media for catalyst support and acoustic insulators [85–87].

Zhang et al. published a rather novel method for synthesizing graphene aerogel in 2011. L-ascorbic acid is used to self-crosslink graphene oxide, which is subsequently dried [19,88]. The benefit of this approach is that no extra pyrolysis treatment is required. Carbon nanotube (CNT) aerogel is another intriguing type of carbon aerogel. In 2007, it was originally produced by sonicating CNT into a surfactant solution, followed by gelation and drying. Polyvinyl alcohol could improve aerogel even further [89]. In 2009, Aliev et al. reported synthesizing CNT aerogel fibers by extracting them from multiwall nanotube forests' straight sidewalls [90]. Since its raw material (CNT forests) was created by catalytic chemical vapor deposition, it differs from practically all previous aerogels. Another "dry synthetic method" for creating carbon-based aerogel (aerographite) involves chemical vapor deposition of nanostructured graphite onto ZnO network templates, hydrogen atmosphere reduction of ZnO to metallic Zn, and high-temperature sublimation of Zn. The end product has an extremely low density [91].

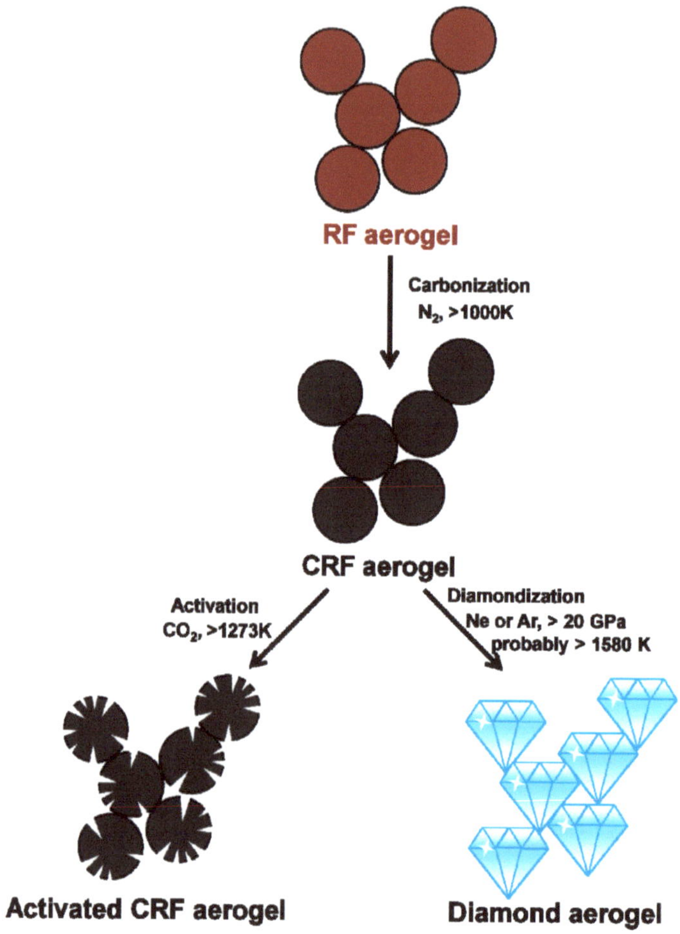

Figure 13. Carbon aerogels. Open access, Reproduced from [15].

There are some additional agents, like biopolymers, metal clusters, and some organic molecules, that can be added into SA matrix for enhanced mechanical properties. Powders, beads, rods, fibers, blankets, and boards are all examples of SA-TIMs. It has incredible potential to meet the current needs of fire prevention, thermal insulation, adsorption properties, medical, aerospace, and home building sectors [91], due to its unique combination of the above-mentioned properties. There are still numerous technical challenges, particularly in mass production and distribution.

10.4. Hybrid Aerogels and Composite Aerogels

The most researched hybrid aerogels incorporate both inorganic and organic phases, nanoparticles, and cross-linked polymers into the gel matrix. They are then reinforced with fibers and filler to create interpenetrating networks [92]. These aerogels were acquired in order to enhance their processing as well as their mechanical, physical, and other qualities. The co-pyrolysis of soft fibers such as polyacrylonitrile (PAN) with an RF aerogel matrix yields composite aerogels as carbon fiber-reinforced composites. The aerogels' toughness, strength, and thermal conductivity are all increased by this technique [93]. The aerogels' toughness, strength, and thermal conductivity are all increased by this technique. Examples of this kind of aerogels include cross-linked polymer aerogels, functional nanocomposite aerogels that contain metal or metal oxide nanoparticles in the aerogel matrix with special electrical, catalytic, optical, and magnetic properties, and nanocomposite carbon aerogels containing carbon nanotubes (CNT) [94,95]. Other examples include carbon aerogels containing nanoparticles of various metals, including Cr, Zr, Mo, Fe, W, Co, Ni, Pd, Pt, Cu, and Ag, as well as silica aerogels containing transition metal oxide nanoparticles, such as NiO [96], ZnO [97], TiO_2 [98], Al_2O_3 [99], CuO, and $CoFe_2O_4$ [100,101].

10.5. Carbon Nanotube Aerogels

Critical-point drying and freeze-drying were used to make carbon nanotube (CNT) aerogels from aqueous-gel precursors, according to M.B. Bryning et al. [102]. The CNT aerogels were robust and conducted electricity to enhanced electrical characteristics. Small amounts of polyvinyl alcohol can be used to strengthen the aerogels. This type of aerogel can withstand 8000 times its weight. Because of their high strength-to-weight and surface-area-to-volume ratios, these materials frequently offer unique features. These aerogels were created by constructing a free-standing, three-dimensional network of carbon nanotubes in the air directly from CNT networks in suspension. Because CNT aerogels can be back-filled with a polymer or other material, they can be used as the foundation for composite structures made of a wide range of materials [103].

10.6. Polyurethane Aerogels

Polyurethane aerogels are a type of aerogel material that is derived from polyurethane polymers. They belong to the broader category of polymer-based aerogels and offer unique properties and applications due to their composition and structure. Polyurethane aerogels are typically formed through a sol-gel process similar to other types of aerogels [104]. The process involves creating a solution (sol) of polyurethane precursors, often consisting of diisocyanates and polyols. These precursors undergo a polymerization reaction to form a three-dimensional network structure within a liquid phase. A solvent exchange and drying process, often involving supercritical drying, removes the solvent while preserving the porous structure, forming polyurethane aerogels [105].

11. Aerogels Photocatalysts Synthesis

Aerogel photocatalysts can currently be made using two different methods: direct synthesis using a molecular-based sol-gel strategy, assembling using nanoscale units as templates, and immobilizing photocatalysts in aerogel frameworks. The former is influenced by traditional sol-gel chemistry, which uses molecular precursors, condensation, and hydrolysis to create SiO_2 aerogels. Only a few different types of metal oxides, including SiO_2, TiO_2, Al_2O_3, and ZrO_2, as well as chalcogenides (ZnS, WSx, and GeS2), are present in the molecular precursor [106]. Additionally, the prepared aerogel has generally low crystallinity. It is preferable to use the assembly method to investigate high-crystallinity aerogels. The range of aerogel photocatalysts has been greatly expanded by nanocrystalline colloids such as chalcogenides (CdS, CdSe, PbS, ZnS), metal oxides (MnO_2, Fe_3O_4, SnO_2), metal-free compounds, and their composites. Recent advancements in this subject have been made possible by the construction of 1D and 2D crystalline building components like nanowires and nanosheets. More potential photocatalytic applications are provided by the availability of these premade building blocks to attain the requisite characteristics prior to producing aerogels [44].

Immobilization of Aerogels as Photocatalysts

Although directly synthesized aerogel photocatalysts have been extensively studied, only a small number of photocatalysts can be converted into aerogel photocatalysts using the sol-gel or self-assembly processes, severely restricting their practical application [107].

Immobilizing photocatalysts on the already-prepared aerogel framework to create composite aerogel photocatalysts has thus attracted a lot of attention recently in an effort to streamline the fabrication process and increase the variety of aerogel photocatalysts. In order to create composite aerogel photocatalysts, it should be remembered that practically all photocatalysts can be added to the aerogel framework as shown in Figure 14 [108]. Additionally, the composite aerogel photocatalyst has the benefits of the aerogel framework as well as the ability to operate as a photocatalyst. When a photocatalyst is exposed to light, photons are absorbed, and excitons are generated. Higher excitons can enhance photocatalytic activity by increasing the probability of electron-hole separation and, thus, the production of ROS. These excitons can then be separated into their constituent electrons and holes, which can participate in redox reactions with adsorbed molecules on the photocatalyst surface. This photocatalysis in the mobile phase facilitates the generation of ROS and increases the efficiency of the process [109]. Compared to the mobile phase, the photocatalysis is slower in the passivated state, i.e., coated on fabric. However, the easier recovery of active materials and careful design, as well as optimization of the photocatalyst's properties with minimal secondary pollution, make the proposed strategy closer to practical applications [110,111].

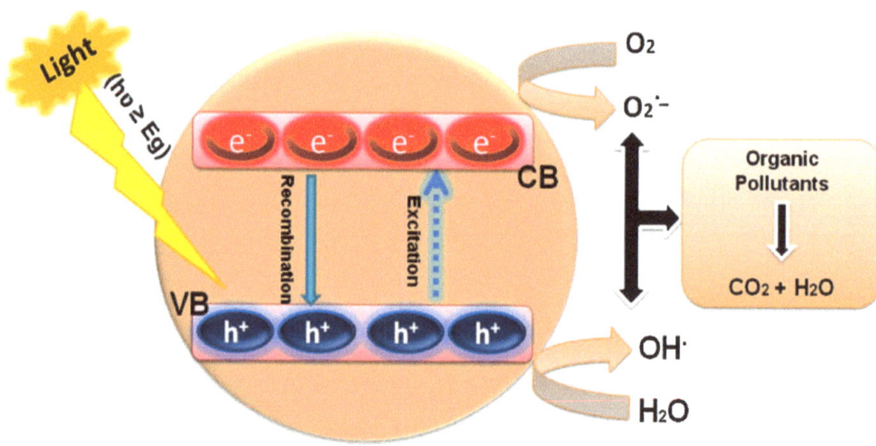

Figure 14. Aerogels photocatalysis mechanism.

12. Preparation Process

To synthesize the aerogel structures, it is quite essential to control the microstructure for preparation of silica aerogel thermal insulation microstructures SA-TIMs because their end application is strictly based on microstructure and processing parameters. Generally, SA-TIMs synthesis includes the following three strategies, as also shown in Figure 15.

12.1. Solution-Sol Transition

Nanoparticles of the sol formation occurred; precursor solution is prepared at any solvent and may be catalyzed by hydrolysis or through other assisted reactions, and finally, aloxysilanes are added [32]. The required solvent should be properly miscible and able to prepare the homogenous mixture. Preferred solvents are ethanol, alcohol, acetone, tetrahydroflourane, etc. The phase diagram of aloxylane-water-solvent is shown in Figure 16a. If the solvent is alcohol, it can decrease the hydrolysis rate by esterification reaction. The choice of solvent is necessary to synthesize the desired product. If the solvent is not able to dissolve the required amount of aloxysilanes, it can cause agglomeration [112].

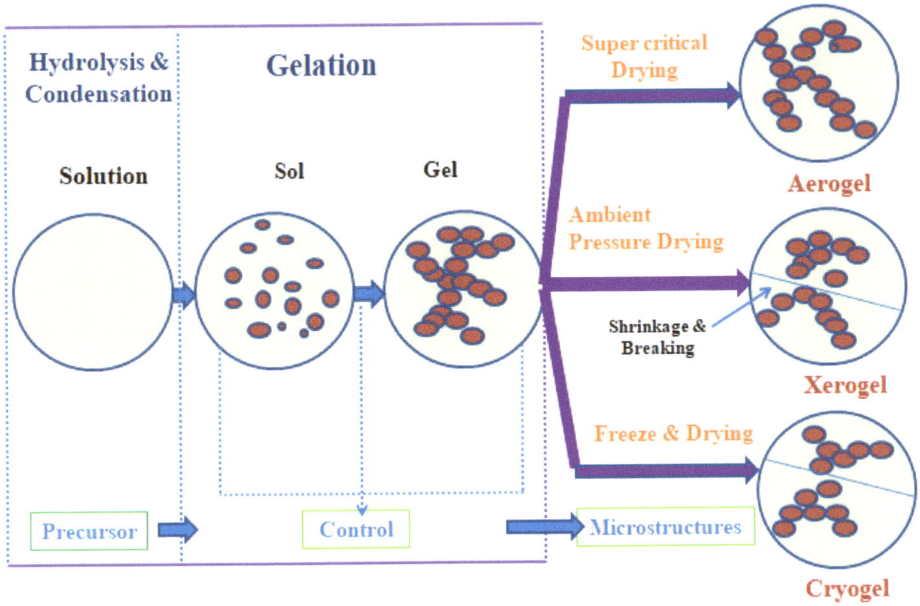

Figure 15. Synthesis strategies of silica aerogels.

12.2. Sol-Gel Transition (i.e., Gelation)

In this stage, the sol is converted into a gel by forming a 3D networking structure cross-linked with each other to form a metastable state. It can be chemically of any nature or physical (temperature, pressure, pH). This state is achieved by uniform stirring for 5 to 6 h with optimized parameters. Stirring rate and temperature are two confined parameters affecting its shape and structural parameters. Uniform stirring is necessary to avoid agglomeration [113].

12.3. Gel-Aerogel Transition (i.e., Drying)

Now sol-gel solution needs to be dried to replace the water and alcohol particles by air without cracking and subsequent drying of the dried aerogel. The existing nanoporous microstructure should not be collapsed; otherwise, it will deteriorate the final properties. Mainly, there are three techniques for drying the sol-gel [36,114].

i. Supercritical drying, in which the sol-gel is dried above the critical point of solvent without the capillary effect of surface tension between the vapor and liquid phases, leads to uniformity in the structure, higher porosity, and optimal textural characteristics [115]. Basically, this drying is generally used to convert the gels into Aerogels as shown in the Figure 16b. In the initial stage, wet gel is dried and expanded liquid is formed due to increased capillary movements of atoms [115]. Now its atoms are more interactive with each other. In the next step, a supercritical fluid mixture is formed which has the ability to effuse through a solid like a gas. At this stage, almost 50% of the solvent is removed. In the next stage, diffusion-controlled drying proceeds in which solvent is removed up to 98% without the action of capillary forces. Now the wet gel is completely converted into aerogel.
ii. Ambient pressure drying, in this drying, is achieved with silylation treatment thereby increasing the structural strength. Capillary forces are much reduced during solvent evaporation. It can reduce the production cost and safety risks [116].
iii. Freeze-drying, in which the solvent is evaporated by decreasing the wet gel temperature below the crystalline temperature of the solvent, like sublimation. Most

xerogels and cryogels are produced by this drying [117]. It allows complete removal of solvents while achieving 95% porosity without any shrinkage. This process also has some disadvantages, like health hazards and irreversible shrinkages. The network of a silica cryogel takes a considerable aging period to solidify, and the network is easily damaged by the crystallization of the solvent in the pores. As a result, the majority of silica cryogel products are powders, and producing monolithic silica cryogels is extremely challenging [118].

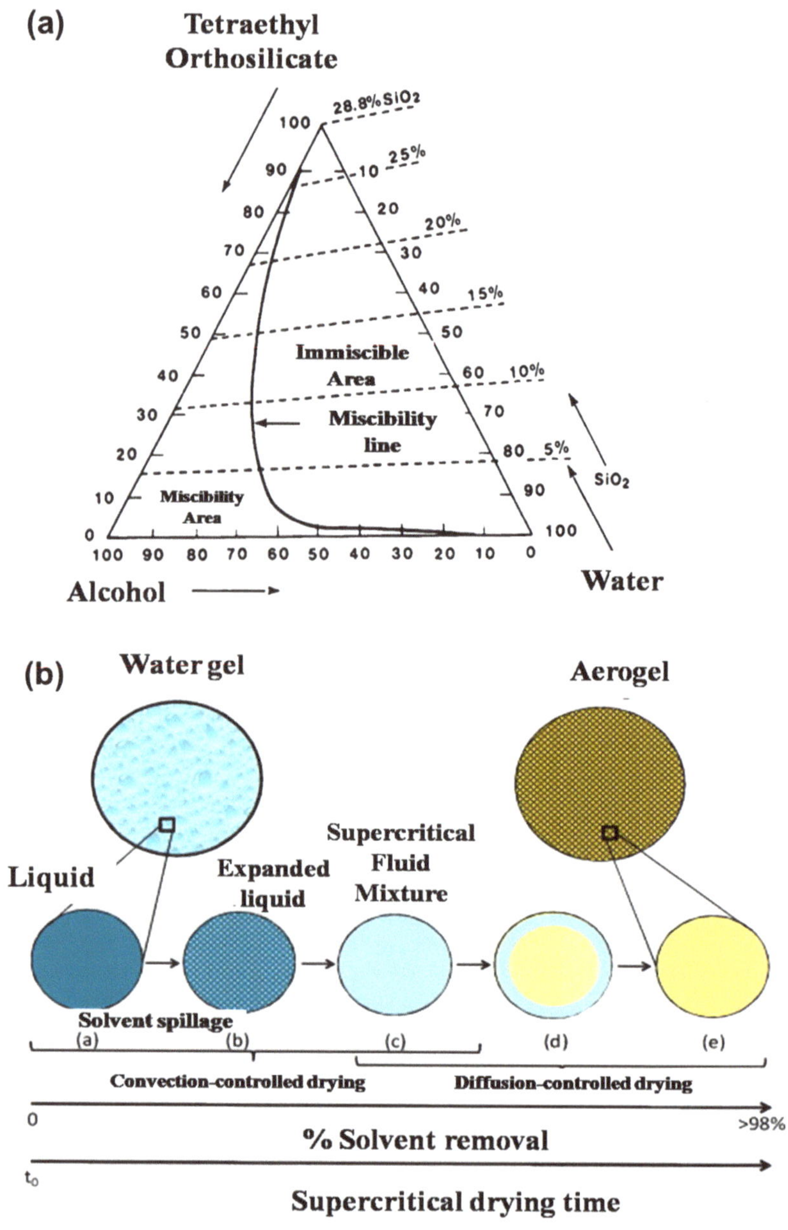

Figure 16. (a) Phase Diagram of aloxylane-water-solvent, (b) Supercritical drying.

13. Design Principles

There are some challenges of SA-TIMs, like enhancing hydrophobic functionalities, strong scalability, high-temperature performance, and mechanical strengths for desired aspects. There are many factors like material and process design, synthesis route, composition, and environmental factors that ultimately affect its performance, as shown in Figure 17. If we have a simulative control of the microstructure and morphology of nanoparticles, it will enhance thermal conductivity and mechanical strength. Some other properties like density, processability, and temperature resistance can also be tuned by optimizing synthesis parameters [119,120].

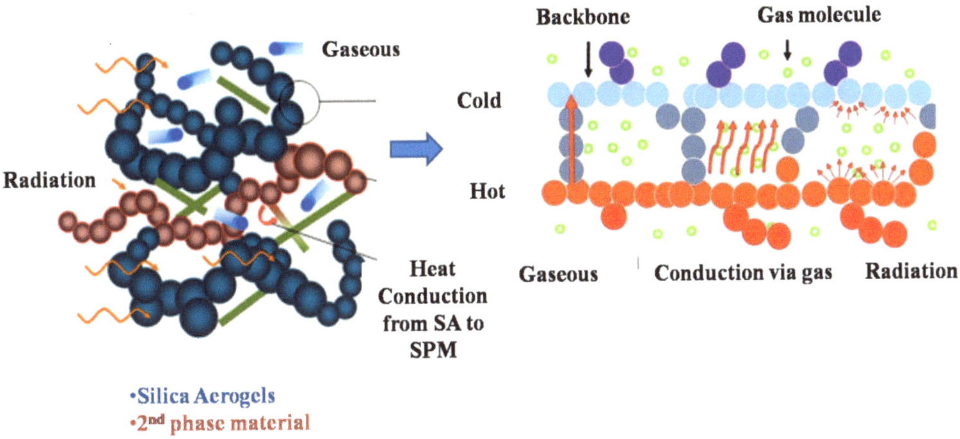

Figure 17. Design principles.

Thermal transportation of aerogel includes heat conduction through a solid backbone, heat conduction with gas molecules, and then there occurred thermal radiation through the backbone structure. Thermal conductivity basically depends on the type, composition, shape, and hybrid structure of SA Skelton and nanoparticles. Morphology, packing arrangement structure, and structure ordering also have as strong impact on thermal conductivity [121,122]. If there is an addition of second phase material in the structure, it exhibits the radiation collapse at the structure. If the surrounding air produces a microclimate with an even temperature difference, a net amount of heat flows because of the air. The thermal conductivity of SA-TIMs is basically the sum result of solid, gas, and radiation conductivity as follows:

$$(\lambda_{total}) \text{ of SA-TIMs} = (\lambda_{solid}) + (\lambda_{gas}) + (\lambda_{rad}) \quad (19)$$

Thermal conductivity of solid is limited due to porous and nanoporous structure so it does not affect highly. Thermal conductivity of radiations is controlled by porosity, pores sizes, and refracting and reflecting surfaces [123]. The remaining space is obviously filled by gas and the thermal conductivity of gas is as:

$$\lambda gas = \lambda g0/(1 + \alpha T/\delta P) \quad (20)$$

In the above equation T, P, and δ represent the temperature, pressure, and pore size of gas, respectively, whereas α is a specific constant.

SA-TIMs λ_{tot} depends on all these external and internal parameters: morphology, presence f side chain, functional atoms, weaker or stronger bonds, pendant group, as well as segments [122,123]. Due to the difference in water vapor pressure at the surface and in the surroundings, moisture is exchanged between the two surfaces. The net amount of moisture exchange between these two surfaces affects vapor pressure as well as thermal conductivity [124,125]. Furthermore, during the service life of SA-TIMs, vapor phase transport (convection and diffusion), liquid phase transport (wicking), heat transfer (convection, conduction, and radiation), liquid evaporation and condensation, and sorption/diffusion of water vapor and liquid through the solid phase occur. Because water vapor (0.025 W/(m·K)) has a lower heat capacity than water (0.620 W/(m·K)) and ice (2.220 W/(m·K)), reducing the number of adsorption sites (Si-OH) and siloxane (Si-O-Si) on the surface is critical [126,127].

14. Drying Technologies for Aerogels

Different drying technologies of aerogels i.e., ambient drying, freeze drying, supercritical drying parameters, conditions, limitations, costs, and preparation steps are mentioned in Table 2.

Table 2. Drying Technologies comparison.

Drying	Conditions	Preparation Steps Prior to Drying	Limitations for the Gel Matrix	Main Energy Costs/Risks
Ambient drying	• Room T • Ambient P	Hydrophobization of the matrix is essential	Not preferable for hydrophilic and fragile matrices. Density > 0.1 g/cm^3 achieved	Low-energy cost, safe and less hazardous.
Freeze drying	• P < 100 mBar • −70 °C < T < −20 °C	Structure compaction, use modifiers.	Pores are somehow destroyed. Density below 0.03 g/cm^3	High-energy cost, because of low-temperature batch process
Direct supercritical drying	• T > 100 °C • P > 30 bar	Direct conversion of solvent to critical parameters. No solvent conversion.	Side reactions occur Temperature > 100 °C is not suitable with organic gels.	Moderate-energy cost, toxic
Super critical drying by CO_2 extraction	• P > 74 bar • T > 31 °C	Hydrogel solvent exchange occurs, so solvent should be compatible with CO_2.	CO_2/solvent transportation affects the properties	Energy cost is high because of compressed CO_2, lesser explosion risks

15. Aerogel Fibers

Producing aerogel-based fibers with unique properties like high porosity, lighter weight, and optimum density has been of scientific interest for a century. These fibers possess promising applications in hygiene products, drug deliveries, and filter materials [128]. Cellulose-based aerogels are synthesized with the dissolution of cellulose into suitable media like urea solution, sodium hydroxide, or any other ionic liquids. After this washing is performed before coagulation, and then drying is performed with any of the special drying methodologies, i.e., supercritical drying, freeze drying, ambient drying, etc. [90,129].

The first cellulose-based aerogel was reported by Tan et al., in which cellulosic aerogel was synthesized by salt hydrate as a unique dissolving agent. Ester with toluene-2,4-diisocyanate was a crosslinking agent. In this unique method, dissolution was carried out with $Ca(SCN)_2$ because of a similar coordination number. Toxic cyanates are also avoided in this technique. By dissolving $Ca(SCN)_2$ salt hydrate, porous aerogels were synthesized with gelation, coagulation, and supercritical drying [130]. Different ratios of cellulosic content up to 6% were added, and their properties were compared. In this way, 3D structures of cellulosic fibers at the nanoscale are achieved with envelop density from 0.5 to 6 wt% having cellulosic content between 0.009 and 0.137 g/cm^3. The surface area ranged from 120 to 230 m^2/g calculated by BET. Synthesized aerogels showed excellent thermal conductivity as well as mechanical strength at room temperature [131]. Synthesized aerogels obtained by coagulation, washing, and extrusion are also shown in the following Figure 18. It was found that the optimum extrusion range is 95–110 °C. SEM images of monolith fibers are also shown to have a pore size of 10–100 nm and a diameter range of 10–25 nm. Both monolith and micro-extruded fibers have the same morphology except core structure difference. This difference is due to evaporation in the coagulation bath whereas the melt solution submerged into the regeneration bath. The fibers' unique architecture, which includes a meso and macroporous structure constructed at the top of a nanoporous network, opens up new application possibilities [132].

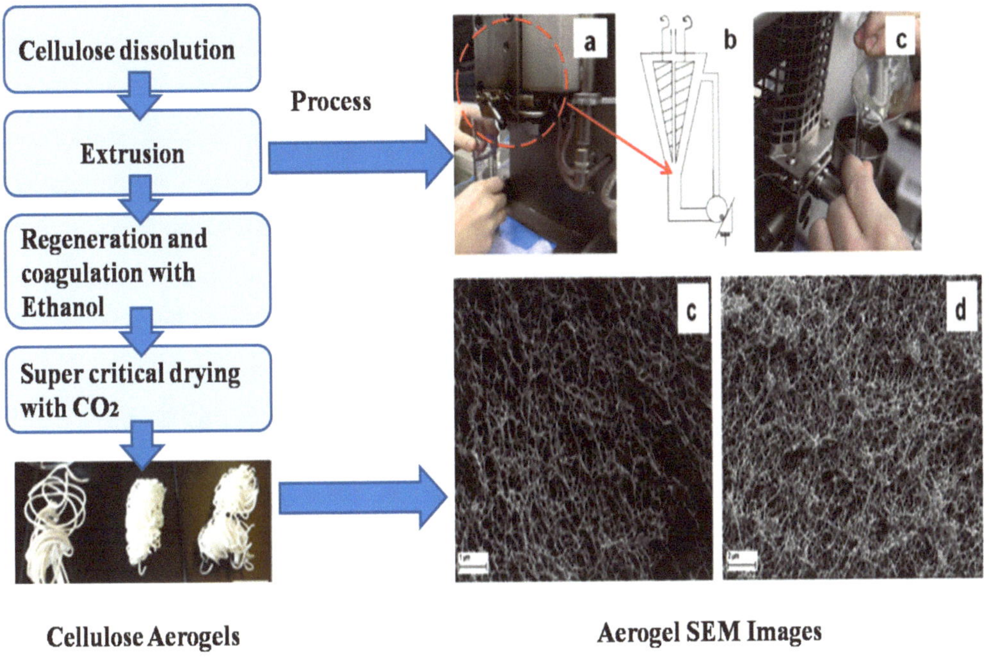

Figure 18. Aerogel fibers (**a–c**) Synthesis routes and design principles, (**c**,**d**) SEM morphology. Reproduced from [129].

Porous CA/PAA-wrapped SF aerogel fibers were synthesized and reported for high thermal insulation and high mechanical properties in applications [133]. Silk fibers were degummed, and dissolution gel was prepared. Then, the silk fibers were added into dialysis bag for preparation of silk fiber aqueous solution. At the final stage, freeze-drying process was performed. The CA/PAA hollow fibers having porous sheath not only possess high mechanical properties but also facilitate the formation of SF aerogel core. These also possess high porosity (86%), low density (0.21 g/cm^3), and high tensile strength (2.6 ± 0.4 MPa). Additionally, it is suitable for thermal insulation in both hot (100 °C) and cold (−20 °C) conditions. Thermal insulation properties can be tuned by core and sheath proportion and by numerically tuning other parameters. As a result of the delicate core–shell structure of aerogel fiber, it offers a new way to manufacture high-performance wearable thermal insulation materials, as shown in Figure 19.

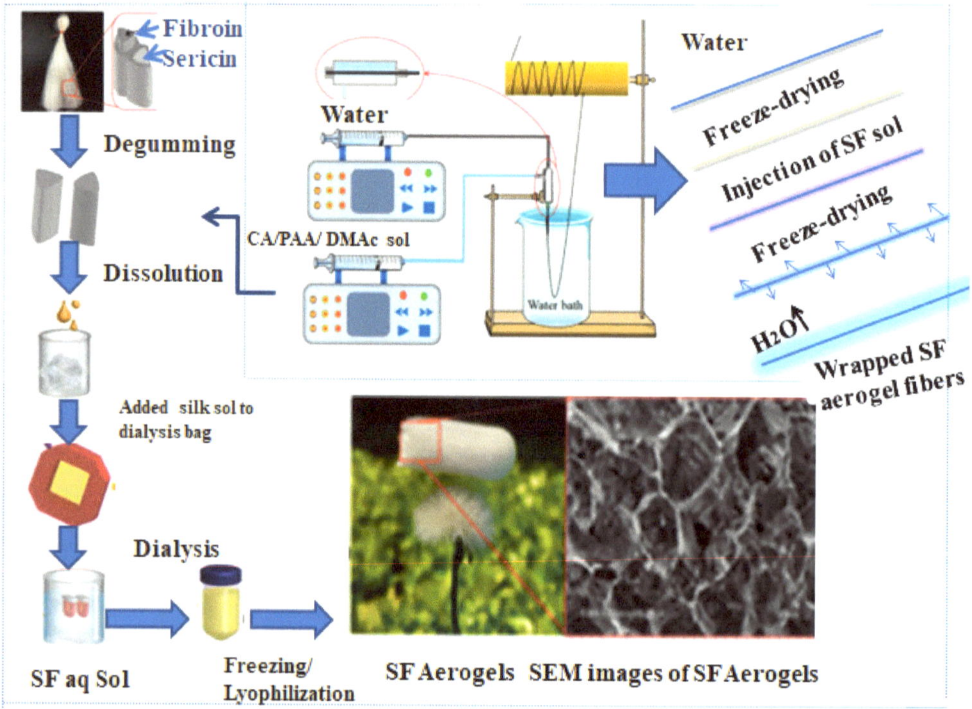

Figure 19. Aerogel fibers synthesis, Open access, Reproduced from [130].

16. Aerogel Fabrics

Aerogel fabrics (woven or non-woven) are manufactured to have high insulation and comfortable properties. Cellulosic or synthetic fibers are woven or non-woven together to assemble aerogel fabrics. Different synthetic or natural fibers were spun on an industrial scale by a compact melt spinning machine and subsequently processed into nonwoven fabrics on a laboratory-scale needling line, and sound absorption properties of blankets made of silica aerogel/polyester (PET) were examined [134]. The silica aerogel blankets were made by synthesizing silica aerogel on nonwoven fabrics in situ using a two-step tetraethoxysilane sol-gel technique and drying at ambient pressure. Various synthesis parameters were employed to generate aerogel particles with varying pore structures and characteristics [135,136]. As coarser fibers are used, there are fewer fibers per unit area and wider pores [137]. Figure 20 shows SEM micrographs of the nonwoven textile structures and aerogel blanket samples synthesized by nanofibrication. The SEM images clearly revealed the effectiveness of nanoaerogels on fibers. These coated fibers show enhanced thermal and heat regulation properties by providing sufficient moisture management capability to the wearer. As previously mentioned, the μCT device used for this study has a resolution of 1–3 μm. Therefore, this technique cannot be used for the characterization of aerogels in the internal structure of blankets [137,138].

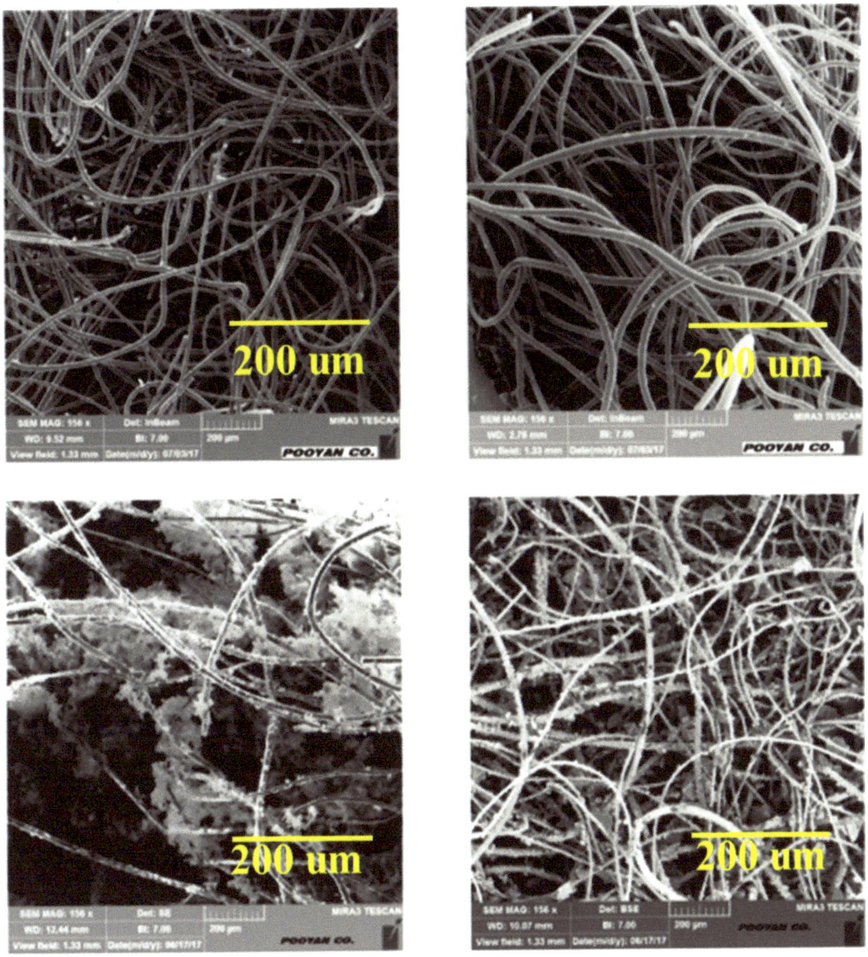

Figure 20. SEM images of neat nonwoven fabrics and aerogel blankets. Reproduced from [139].

The thermos-physiological comfort of a newly created fabric, which could be used for firefighter protective equipment, was discussed [140]. Super hydrophobic silica aerogel nanoparticles were incorporated into a 65/35 wool-aramid blended fabric to create thermally comfortable garments [141]. The performance of air, moisture, and heat transport was then used to assess thermo-physiological comfort. It has been discovered that a 2% aerogel nanoparticle coating can boost thermal resistance by up to 68.64% and reduce air permeability by up to 45.46%, while a 4% aerogel coating can reduce air permeability by up to 61.76%. The aerogel-coated fabric's moisture management properties have also been examined and described in depth [142]. The coating thickness has a positive effect on moisture transportation and overall moisture management, in conclusion. Again, the next-to-skin layer coated with aerogel and the outside layer coated with aerogel had different effects on the identical fabric. When the coating is applied to a surface that is near the skin, it has been found that the coated fabric behaves more like a moisture management fabric [143].

Using an acrylic glue, varying amounts of "NANOGEL" (superhydrophobic aerogel nanoparticle from Cabot) were applied to a 65/35 wool/Aramid blend fabric that weighs around 230 g/m^2. 2%, 4%, and 8% aerogel-coated fabrics were created with coating pastes that ranged in viscosity from 20,000 to 30,000 centipoise as shown in Figure 21. After coated fabrics were cured for 10 min at 105 °C, their thermophysiological characteristics were examined [144].

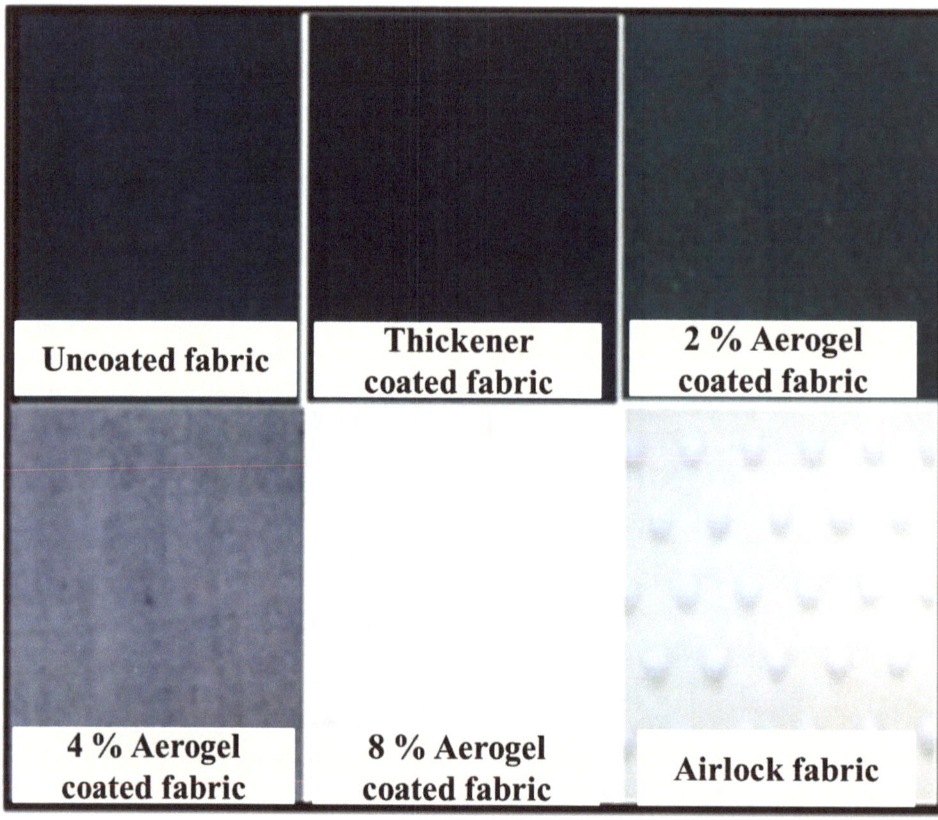

Figure 21. Aerogel coated fabrics microscopic images. Open access, Reproduced from [144].

In order to increase the wearer's thermophysiological comfort, the research looked into the potential use of aerogel in firefighter protective gear. The ability of aerogel-coated cloth to transmit heat, moisture, and air was examined in terms of thermophysiology [145]. These three qualities are thoroughly discussed [144]. The test results can be summarized as follows: the aerogel-coated fabric has good air resistance, the right coating thickener can reduce the hydrophobicity of the aerogel and the base fabric, improving the ability to manage moisture, and the aerogel-coated fabric provides better thermal insulation. In addition to being thermally insulative, aerogel can also significantly resist airflow, according to recent studies [146].

These two taken together show a technique for making textiles for clothing that offer thermally better insulation for a variety of uses in extremely hot and cold weather. Aside from the encouraging results, the research was hampered by a few inevitable limitations brought on by the short time restriction, such as the fact that not all varieties of made-up cloth samples had their heat resistance evaluated [147]. Again, the moisture management tests conducted call for a further in-depth investigation of several coating types with different Aerogel and thickener compositions on different fabric structures in the border range. However, the report offers general recommendations for further study on the use of Aerogel in protective apparel, based on successful research outcomes [148,149].

Figure 22 also shows that when a coating of 2% aerogel paste was applied, the aerogel particles did not completely cover the surface, whereas a coating of 4% showed partial coverage. The aerogel powder thoroughly coated the fabric surface for an 8% coating. The fabric's thickness and mass per unit area were examined to check for physical changes. Aerogel coating resulted in a 4–6% increase in fabric thickness and a 6–9% increase in fabric weight. The barrier of heat and moisture transport rises with thickness. When two fabrics are of the same thickness, previous research has shown that the lower-density cloth provides better thermal insulation [150]. Coating resulted in a progressive increase in insulation in our situation, as indicated in Table 3.

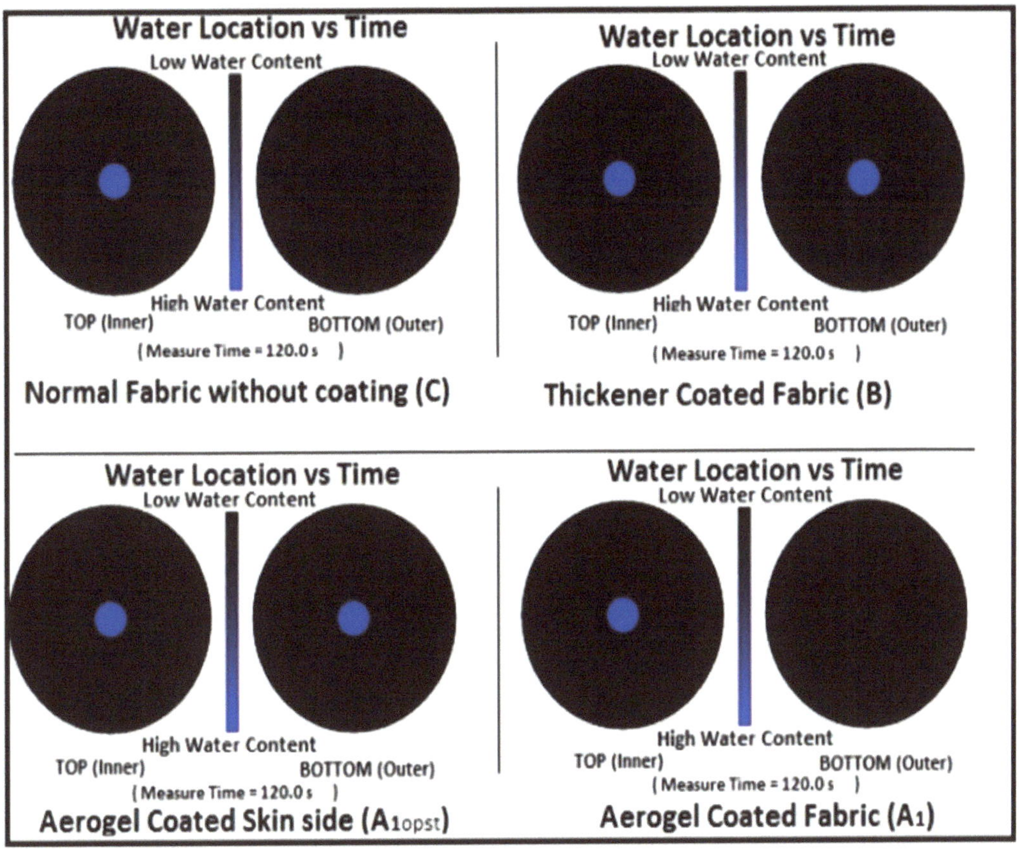

Figure 22. Maximum wetted radius lie at both top, and bottom surface. Open access, Reproduced from [144].

Table 3. Aerogel fabrics parameters. Open access, [144].

Fabric Construction	Surfaces Wetting Time (s)		Absorption (%/s)		Fabris OMMC
	Top, WT_1	Bottom, WT_b	Top, MAR_1	Bottom, MAR_b	
Aerogel Coated (Al)	5.88	119.95	58.66	0	−478.4
Thickener coated (B)	7.08	57.64	350.66	29.12	−118.5
Plain fabric (C)	6.28	119.95	332.73	0	−391.1
Aerogel skin	7.16	84.44	380.75	4.15	−393.4
Airlock (D)	7.19	119.95	363.39	0	−427.7

17. Aerogel Finishes and Coatings

To prevent air pollution from ultrafine particles, hydrophobic silica aerogel rather than its powder form is employed for aerogel superhydrophobic coatings. It is used to modify materials' surfaces so they are extremely hydrophobic while still being able to endure physical abrasion, something a regular aerogel cannot. The hydrophobic gel was combined with DOW CORNING® 2405 resin as a binder and different amounts of DOWSILTM Z-6137 silane and tetraethyl orthosilicate (TEOS) to create the superhydrophobic silica aerogel coating. Static contact angles (CA) and abrasion testing were used to describe the coating. Investigated were scanning electron micrographs of various coating formulations. According to the results, the hydrophobic gels combined with resin and Z-6137 silane have a contact angle of >179 degrees. Glass, fiber, polymers, and other materials can be coated with superhydrophobic silica aerogel [151,152].

Aerogel composites fabricated with epoxy with carbon fiber as a substrate and dip coating were utilized for the one-hour and one-and-a-half-hour marks of the epoxy cure. It was then compared with an undoped epoxy coating; there was a remarkable difference in heat conductivity values of 39% and 47%, respectively. Further proof that the aerogel particles contained nanopores came from the reflectance spectra of the coatings. Finite element techniques evaluated the aerogel coating using material parameters. The performance of the coating under cyclic thermal stresses was then predicted using the model, which was first validated using experimental data. Additionally, top and bottom coatings on a single surface were modeled and compared to the double coating system. It was discovered that the double coating system had the lowest rate of temperature change and fluctuations at steady-state, in contrast to the bottom coating, which displayed the fastest temperature drop and the highest fluctuations at steady-state conditions. The top coating's performance was average [153].

18. Aerogels Photocatalysis Applications

Photocatalysis typically involves a photocatalyst material that absorbs light and generates electron-hole pairs, which can initiate various chemical reactions. Aerogels can serve as an excellent support material for photocatalysts due to their porous nature and large surface area, allowing for high photocatalyst loading and efficient utilization of light. Aerogels can be used as a three-dimensional support matrix for anchoring photocatalytic nanoparticles (Figure 23). The high porosity of aerogels enables the dispersion of nanoparticles, providing a large active surface area for photocatalytic reactions. The highly porous structure of aerogels can trap and scatter light, increasing light-matter interactions within the photocatalyst (Figure 23). This enhances the absorption of light and promotes efficient utilization of the energy for photocatalytic reactions. Aerogels can act as protective coatings for photocatalysts, preventing their degradation and enhancing their stability. The porous structure of aerogels can also prevent the agglomeration of photocatalytic nanoparticles, maintaining their high surface area and activity. Combining aerogels with photocatalytic materials can lead to synergistic effects. For example, incorporating metal nanoparticles into aerogel structures can enhance photocatalytic activity by promoting charge separation and facilitating electron transfer processes. The choice of aerogel material depends on factors such as the desired photocatalyst, target reactions, and environmental conditions. Overall, aerogels offer promising opportunities for improving the efficiency and performance of photocatalytic systems. However, it's worth noting that the field of aerogels in photocatalysis is still evolving, and further research is needed to optimize the design and synthesis of aerogels for specific photocatalytic applications [107].

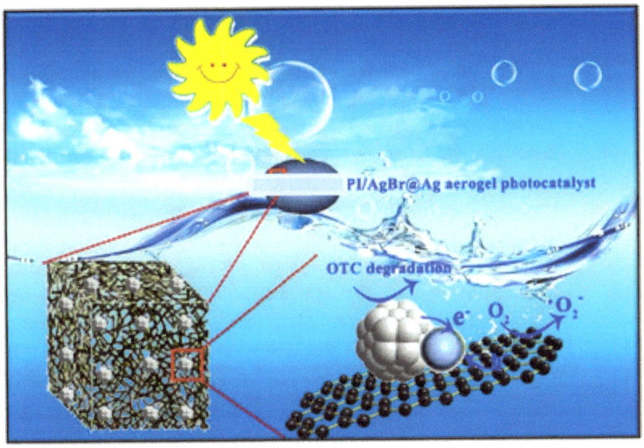

Figure 23. Aerogels applications in photocatalysis.

19. Aerogels Technical Applications

Aerogel has the most diverse applications in all technical and functional applications because of its unique physical and chemical properties (Figure 24). In the beginning, aerogel was discovered as spacesuit fabrication material for NASA projects. It can bear extreme temperatures and weather parameters and show a wispy appearance [153,154].

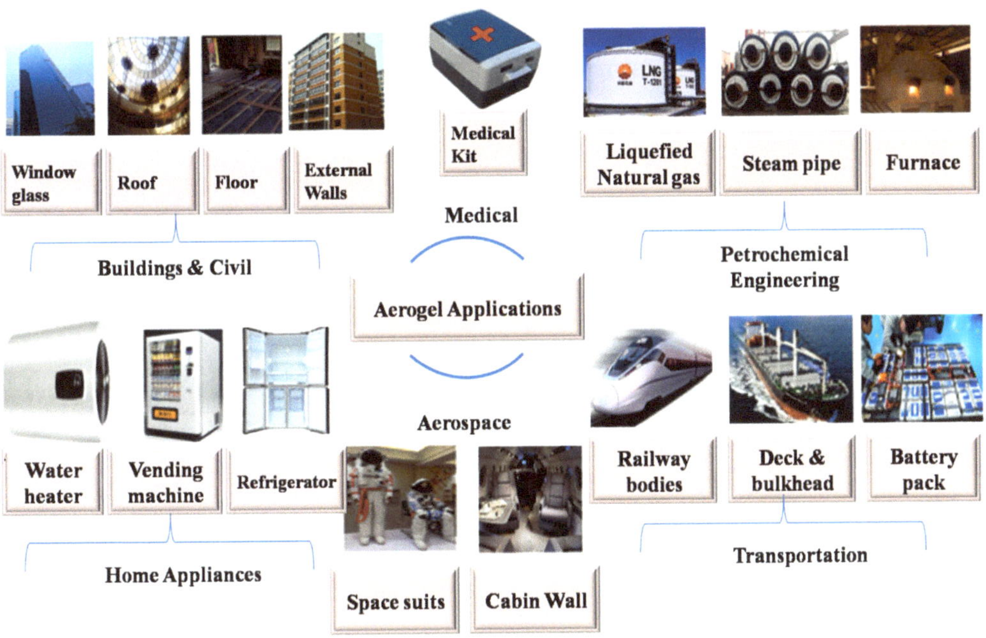

Figure 24. Aerogels applications.

In the early 21st century, aerogel started being used in astronaut applications by the 'Stardust' mission, with the motivation of extracting the back particles from beyond the moon. Basically, before this, dust was collected from the comet 'Wild 2'. Aerogel was the first material to capture physical particles of dust without physically reacting with them. Basically, when the dust particle hits aerogel, its velocity speeds up to 6 times as compared to a rifle bullet and removes it from the surface without reacting [3,155,156]. The dust particle loses its momentum when interacts with the aerogel due to the formation of more porous regions. So, it also plays a vital role in the cleaning of astronaut's parts and enhances its life. As aerogel provides insulation in advanced engineering fabrications it is utilized for space suits where insulation and flexibility are demanded [157]. Aerogel can be molded into different forms and shapes so it can be utilized as shape memory material. Carbon Aerogel materials are preferred for energy storage devices due to the networking structure of carbon atoms in them. In supercapacitors, electrode lining is made by aerogel because of its textural and unique chemical properties. Electrode lining should be porous for better transportation so aerogel porous structure is very much feasible for this [158,159]. Due to its porosity, it is also applicable in rechargeable batteries and catalyst support systems. As the graphene aerogel structure provides the highest porosity and charge mobility it was very attractive to the batteries market. In 2009, graphene aerogel anode materials were first used in battery support systems. It was further decided that graphene-based nano-porous aerogel was the best material for low-cost and high-performance energy applications. For electric vehicle electrode structures, graphene aerogel with doped cations was the best-optimized material to reduce the dependency on conventional fuel usage and to enhance aerogel market growth in the future [160].

There is an innovative and emerging application of aerogel used through the 1980s in laser fusion experiments. For the laser fusion experiment, the material should be homogenous, low dense, and pure and present in the near-spherical shells. For this, the average pore size should be smaller than 1 mm so that it must keep the liquid fuel and ensure its homogeneity [161]. Other material characteristics for this are hydrophilic in liquid deuterium–tritium, highly stable, and should be able to bear mechanical stresses and coating processes. Silica Aerogels are used for laser target materials and machined from monolithic pieces. Materials used for this are boron, silica, carbon, and organic based. Scientists are highly focused on the development of microsphere surfaces for aerogel fiber extrusions. Microporous polymeric thin doped nanomaterials are also used for this purpose [162,163]. Aerogels are widely used in hydrogen and energy applications because of their high surface area [164]. There is a very exciting application in energy loss using radiography application using Cherenkov radiation. In this setup, a monoenergetic charge particle was the source material to illuminate the imaging object. When the particle interacts with the object, it reduces its velocity and causes energy losses. The emitted photons are quite linear and it will measure the uniform velocity and density of the object. Image is the formed on the screen. It is very high speed and based on X-ray technique. Both electrons and proton beams can be utilized to form the image and it is quite economical application [165].

In aerocapacitor applications, aerogels are widely used due to their higher conductivity rate and large surface area. It provides a double layer for charge separation and a porous sheet for charge mobility. It also possesses high energy density and high power conversion as compared to traditional aerocapacitors. Electrode material provides continual charge production and high surface per unit volume due to that stored energy released with the highest power densities, e.g., 7.5 kW. Carbide-derived carbon aerogels and nitrogen-doped carbon aerogels are the best materials for this application [166]. Aerogel materials are very attractive for energetic applications. A tremendous amount of energy can be stored on its surface because of the highest porosity and low refractive index. There is much variety of its products i.e., aerogel hybrid composites, direct conversion of aerogels from energetic molecules and compound aerogels containing both reducer and oxidizer together. Lithium-ion-doped silica nanocomposite aerogel and single-metal organic framework aerogels are its most applicable products [167]. Aerogels have been used for optical applications since the 1900s because of high-temperature processes. Lanthanide-doped aerogels were considered the best material for lasing applications. Silica aerogel was not favorable for optical transmission applications. Doping of silica with radioactive phosphor and tritium converts it into the more efficient radioluminescent light source. Translucent Aerogels are very active photocatalytic materials for solar cell applications. Solar cell applications require less dense material and with higher conductivity rate and photon absorption ability. Ni/C Aerogels have the highest conductivity among all other members so highly selective for this purpose [120]. All the aerogel varieties are recognized best insulating material for industrial and home-tech applications. It possesses high damping rates, low thermal conductivity, and low density suitable for acoustic applications. Acoustic impedance and high absorption are also characteristics. Aerogels have also been projected as a shock-absorbing material in industrial applications [168].

19.1. Thermal Insulator Applications

Energy conversion is decreased by thermal insulation. For regulation of indoor climate and to maintain proper thermal heat management, aerogel products are highly preferred for build-tech applications. Very lightweight materials (1-person structural panel) having higher fire protection class are used for rooftop materials construction. There is no colonization of fungi at the top surface of aerogel thereby enhancing its life [169]. These are also very low-cost materials. Mostly silica aerogel is utilized for domestic and industrial insulation due to its easy process but iron oxide, carbon, and organic polymer aerogel can also be used for these applications. It is also named 'frozen smoke' because its structure utilizes 99.8% of the amorphous region nothing but air [170]. Aerogel materials also lead to many applications like rooftop insulation, windows insulation, cold storage applications, and piping insulation. Silica-based aerogels, polyaniline/pectin aerogels, and Hydroxyethyl aerogels are among the best support materials for aerogel applications in the field of insulation. Carbon-silica composite Aerogels provide the best insulation with long-term durability, surface modification, and functional properties [171].

Aerogel materials with improved insulating properties have been utilized for fire retardant applications, flexible blankets, and high-performance fabrics. Aerogel-treated non-woven flexible fabrics show better thermal performance for eco-friendly garments. The "pyrogel" insulation was developed by Aspen Aerogel Inc. to be used as an effective thermal barrier in extreme weather for hikers and climbers on high peaks, where their light weight and flexibility are extremely useful [24]. Flame retardant fabrics hinder/stop the growth of flame onto the fabrics. To achieve this, aerogel coating is applied on the surface of fabric for protective clothing. Silica aerogel reduces the thermal conductivity by evacuation. As silica-biopolymers are highly hydrophobic in nature, these materials can be used to build the anti-wetting, flame-retardant surface on the textile. Silica structure can be converted into hydrophobic form by silane agents [172–174].

Green buildings, households, industrial applications, and aerogel fibers/composite structures have been explored to overcome energy consumption. To prevent heat losses and to harvest more and more solar energy, a layer of aerogel composites is embedded into two layers of PTFE that can boost insulation value (R) to 12. Aerogel blankets reduce the energy consumption rate from cooling and heating the fabric by 30–70%. It has large-scale applications in water parks, sports stadiums, and green shopping centers on the industrial scale. These also have been utilized for thermal awnings and window replacements with new energy codes. Researchers were focused on developing the insulative blanket [31]. It is produced by inserting aerogels interstitially within a fiber matrix. Matrix played an important role in enhancing the thermal properties of the final product. Aerogel composites reinforced with double-layer low-density fibers are very much suitable for thermal resistance [175]. It was comprised of two phases. Matrix was composed of low-density aerogel whereas reinforcing phase is sandwiched in between. The reinforcing is basically composed of low-density fibers, low thickness and a lesser aspect ratio. A preferred combination of composed fibrous system is an aerogel matrix bounded by short, high-aspect microfibers in which one fibrous material is continuous batting and the other one is dispersed throughout an Aerogel matrix [176].

19.2. Innovative and Medical Textile Applications

Aerogel materials also have the potential to be used for active smart textile materials and performance clothing. Piezoelectric materials based on alumina-silica are widely used for energy harvesting carpets and convert the stresses into suitable energy forms, stresses mainly come from the people walking onto it [102]. It is also used for military helmets for thermal insulation and to utilize wind and thermal energy in suitable form. Wall coverings and curtains are also mainly comprised of these aerogel structures for indoor applications. Aerogel-based smart garments are also composed of silica to map the body areas body such as the spine, kidneys, and torso. If we lead aerogel structure towards brittleness and fragility, its applications will be opened up from clothing to heatproofing and smart textiles [177]. To enhance antibacterial activity in the textile, aerogel materials are very attractive to be utilized. As the aerogel structure is porous, it hinders the growth of bacteria hence utilized for antibacterial properties. TiO_2, ZnO aerogels solution gives higher antibacterial activity and remains active many times also showing durability. Composite aerogels having titania and zinc oxide combination show enhanced antibacterial characteristics due to incubation of bacterial zone. Aerogel solution processed with Nylon and polypropylene shows enhanced antibacterial activity. The main antibacterial property is basically due to the inherent raw material used [149,178]. For medical textiles, silica hybrid biopolymers are used for membranes and coatings. Biocompatible materials, bone substituent, and cardio-joints are fabricated by hybrid materials. Collagen, chitosan, and gelatin proteins are obtained by silica sol mixture with proteins, these sols are further processed to make biocomposite layers and hence biomaterials are fabricated. These biomaterials are highly dense, biodegradable, and have multifunctional properties. Joints are mainly fabricated by these strategies having enhanced medical applications where high strength and flexibility are the key requirements [179,180].

19.3. Aerogels in Environmental Applications

Aerogels are primarily composed of a porous network of solid materials, typically derived from silica, metal oxides, polymers, or carbon-based materials. The excellent properties make aerogels suitable for a wide range of environmental applications, thanks to their ability to address various challenges related to energy efficiency, pollution control, and sustainability [181]. Aerogels have shown promise in addressing oil spills. Their high surface area and porous structure enable them to absorb and capture oil, making them useful for cleaning up oil spills in water bodies. The absorbed oil can potentially be recovered and reused. Aerogels can be tailored to have specific pore sizes and surface chemistries, allowing them to selectively adsorb and remove contaminants from water, such as heavy metals, organic pollutants, and even microplastics. This makes them valuable for water purification and treatment [182]. Aerogels can be functionalized with chemical groups that have an affinity for capturing carbon dioxide (CO_2) from the atmosphere. These modified aerogels can potentially play a role in carbon capture and sequestration technologies to mitigate climate change. The large surface area and porous nature of aerogels also make them excellent catalyst supports. They can enhance catalytic reactions by providing a high surface area for active sites, leading to improved efficiency in various industrial processes, such as chemical production and emissions control [183]. Aerogels have been explored as materials for energy storage applications, including supercapacitors and batteries. Aerogels can be incorporated into solar panels and solar absorbers to improve their energy capture and conversion efficiency. They can also be used to create transparent insulating materials for windows that allow light in while minimizing heat transfer.

20. Future Works and Conclusions

Research is likely to focus on refining the manufacturing processes of aerogels to make them more cost-effective, scalable, and environmentally friendly. This could involve innovations in supercritical drying methods, precursor materials, and gelation techniques. One of the challenges with aerogels is their brittleness and low mechanical strength. Future research may focus on enhancing their mechanical properties by incorporating reinforcing fibers and nanoparticles or by structurally modifying the aerogel's network. Currently, most aerogels are made from silica-based precursors. In the future, researchers might develop new types of aerogels using different precursor materials, such as polymers, metal oxides, and organic compounds. This could lead to aerogels with tailored properties for specific applications. After decades of development, a wide range of aerogel materials have been in the interest of scientists to achieve superior mechanical and thermal performance. Up to now, aerogels are still a research dimension for scientists to meet the desired thermal insulation, photocatalytic, and performance needs. Composition optimization of used raw materials, and designing novel assembly structures and hybrid structures will give a new dimension to aerogels having superior characteristics [27]. Tailoring the microstructure and composition of aerogels, and fabrication simulation of aerogels will meet the target application requirements. These scientific inquiries prompt us to investigate the underlying processes of each necessary mechanical and physicochemical property and consider the techniques of implementation in order to comprehend how to improve and combine the necessary features in a single material. Furthermore, due to preparation technological limitations, the development of aerogel products with a substantial and full structure may take years or even decades. Generally speaking, the goal of future research is to create Aerogels that are more cost-effective, perform better, and have more features.

The most persuasive and successful tactics for improving the photocatalytic and thermal insulating capabilities of aerogels, as well as promoting functionalization and industrialization, are composition optimization and microstructure reconstruction. In recent years, a wide range of aerogels with different compositions and multi-component microstructures have been developed for designers and practical life applications. The aerogel field will continue to grow at a rapid rate in the coming years, with effective raw material composition and tuned hybrid structures being key factors. With significant collaboration between academic and industrial partners, aerogels with enhanced efficiencies and tuned properties can be developed more quickly.

Author Contributions: Conceptualization, Z.A., M.A. and A.J.; methodology, Z.A., K.H.L. and A.D.; software, A.D. and M.A.; validation, M.A., Q.A.A. and M.N.; formal analysis, Z.A., M.A. and Q.A.A.; investigation, A.B., E.-J.L. and M.N. resources, K.H.L., A.J. and M.A.; writing—original Z.A. draft preparation, Z.A. and A.B.; writing—review and editing, M.A. and U.Z.; supervision, M.A.; project administration, K.H.L.; funding acquisition, K.H.L. and M.A. All authors have read and agreed to the published version of the manuscript.

Funding: This research received no external funding.

Acknowledgments: The Authors extend their appreciation to the Deanship of Scientific Research at King Khalid University for funding this research through a large group Research project under grant number RGP.2/158/44. This work was supported by the Korea Environment Industry and Technology Institute (KEITI) through the Environmental R&D Project on the Disaster Prevention of Environmental Facilities Project, funded by the Korea Ministry of Environment (MOE) (2020002870004).

Conflicts of Interest: The authors declare no conflict of interest. We declare that we have no commercial or associative interest in the work submitted that would create a conflict of interest.

References

1. Guo, J.; Fu, S.; Deng, Y.; Xu, X.; Laima, S.; Liu, D.; Zhang, P.; Zhou, J.; Zhao, H.; Yu, H. Hypocrystalline ceramic aerogels for thermal insulation at extreme conditions. *Nature* **2022**, *606*, 909–916. [CrossRef]
2. Ganesamoorthy, R.; Vadivel, V.K.; Kumar, R.; Kushwaha, O.S.; Mamane, H. Aerogels for water treatment: A review. *J. Clean. Prod.* **2021**, *329*, 129713. [CrossRef]
3. Jones, S.M. Aerogel: Space exploration applications. *J. Sol-Gel Sci. Technol.* **2006**, *40*, 351–357. [CrossRef]
4. Chen, Y.; Zhang, L.; Yang, Y.; Pang, B.; Xu, W.; Duan, G.; Jiang, S.; Zhang, K. Recent progress on nanocellulose aerogels: Preparation, modification, composite fabrication, applications. *Adv. Mater.* **2021**, *33*, 2005569. [CrossRef]
5. Ghalehkhondabi, I.; Ardjmand, E.; Weckman, G.R.; Young, W.A. An overview of energy demand forecasting methods published in 2005–2015. *Energy Syst.* **2017**, *8*, 411–447. [CrossRef]

6. Sorrell, S. Reducing energy demand: A review of issues, challenges and approaches. *Renew. Sustain. Energy Rev.* **2015**, *47*, 74–82. [CrossRef]
7. Van Ruijven, B.J.; De Cian, E.; Sue Wing, I. Amplification of future energy demand growth due to climate change. *Nat. Commun.* **2019**, *10*, 2762. [CrossRef]
8. Bertoldi, P. Overview of the European Union policies to promote more sustainable behaviours in energy end-users. In *Energy and Behaviour*; Elsevier: Amsterdam, The Netherlands, 2020; pp. 451–477.
9. Economidou, M.; Todeschi, V.; Bertoldi, P.; D'Agostino, D.; Zangheri, P.; Castellazzi, L. Review of 50 years of EU energy efficiency policies for buildings. *Energy Build.* **2020**, *225*, 110322. [CrossRef]
10. Lucarelli, C.; Mazzoli, C.; Rancan, M.; Severini, S. Classification of sustainable activities: EU taxonomy and scientific literature. *Sustainability* **2020**, *12*, 6460. [CrossRef]
11. Jelle, B.P. Traditional, state-of-the-art and future thermal building insulation materials and solutions—Properties, requirements and possibilities. *Energy Build.* **2011**, *43*, 2549–2563. [CrossRef]
12. Feng, J.; Su, B.-L.; Xia, H.; Zhao, S.; Gao, C.; Wang, L.; Ogbeide, O.; Feng, J.; Hasan, T. Printed aerogels: Chemistry, processing, and applications. *Chem. Soc. Rev.* **2021**, *50*, 3842–3888. [CrossRef] [PubMed]
13. Sen, S.; Singh, A.; Bera, C.; Roy, S.; Kailasam, K. Recent developments in biomass derived cellulose aerogel materials for thermal insulation application: A review. *Cellulose* **2022**, *29*, 4805–4833. [CrossRef]
14. Li, C.; Chen, Z.; Dong, W.; Lin, L.; Zhu, X.; Liu, Q.; Zhang, Y.; Zhai, N.; Zhou, Z.; Wang, Y. A review of silicon-based aerogel thermal insulation materials: Performance optimization through composition and microstructure. *J. Non-Cryst. Solids* **2021**, *553*, 120517. [CrossRef]
15. Du, A.; Zhou, B.; Zhang, Z.; Shen, J. A special material or a new state of matter: A review and reconsideration of the aerogel. *Materials* **2013**, *6*, 941–968. [CrossRef] [PubMed]
16. Liu, Q.; Yan, K.; Chen, J.; Xia, M.; Li, M.; Liu, K.; Wang, D.; Wu, C.; Xie, Y. Recent advances in novel aerogels through the hybrid aggregation of inorganic nanomaterials and polymeric fibers for thermal insulation. *Aggregate* **2021**, *2*, e30. [CrossRef]
17. Wang, L.; Xu, H.; Gao, J.; Yao, J.; Zhang, Q. Recent progress in metal-organic frameworks-based hydrogels and aerogels and their applications. *Coord. Chem. Rev.* **2019**, *398*, 213016. [CrossRef]
18. Schaefer, D.W.; Keefer, K.D. Structure of random porous materials: Silica aerogel. *Phys. Rev. Lett.* **1986**, *56*, 2199. [CrossRef]
19. Worsley, M.A.; Pauzauskie, P.J.; Olson, T.Y.; Biener, J.; Satcher, J.H., Jr.; Baumann, T.F. Synthesis of graphene aerogel with high electrical conductivity. *J. Am. Chem. Soc.* **2010**, *132*, 14067–14069. [CrossRef]
20. An, L.; Wang, J.; Petit, D.; Armstrong, J.N.; Hanson, K.; Hamilton, J.; Souza, M.; Zhao, D.; Li, C.; Liu, Y. An all-ceramic, anisotropic, and flexible aerogel insulation material. *Nano Lett.* **2020**, *20*, 3828–3835. [CrossRef]
21. Wittwer, V. Development of aerogel windows. *J. Non-Cryst. Solids* **1992**, *145*, 233–236. [CrossRef]
22. Baktash, A.; Amiri, O.; Sasani, A. Improve efficiency of perovskite solar cells by using magnesium doped ZnO and TiO_2 compact layers. *Superlattices Microstruct.* **2016**, *93*, 128–137. [CrossRef]
23. Joly, M.; Bourdoukan, P.; Ibrahim, M.; Stipetic, M.; Dantz, S.; Nocentini, K.; Aulagnier, M.; Caiazzo, F.G.; Fiorentino, B. Competitive high performance Aerogel-Based Composite material for the European insulation market. *Energy Procedia* **2017**, *122*, 859–864. [CrossRef]
24. Berardi, U. Aerogel-enhanced insulation for building applications. In *Nanotechnology in Eco-Efficient Construction*; Elsevier: Amsterdam, The Netherlands, 2019; pp. 395–416.
25. Herrmann, G.; Iden, R.; Mielke, M.; Teich, F.; Ziegler, B. On the way to commercial production of silica aerogel. *J. Non-Cryst. Solids* **1995**, *186*, 380–387. [CrossRef]
26. Rahmanian, V.; Pirzada, T.; Wang, S.; Khan, S.A. Cellulose-Based Hybrid Aerogels: Strategies toward Design and Functionality. *Adv. Mater.* **2021**, *33*, 2102892. [CrossRef] [PubMed]
27. An, L.; Wang, J.; Petit, D.; Armstrong, J.N.; Li, C.; Hu, Y.; Huang, Y.; Shao, Z.; Ren, S. A scalable crosslinked fiberglass-aerogel thermal insulation composite. *Appl. Mater. Today* **2020**, *21*, 100843. [CrossRef]
28. Arshad, Z.; Khoja, A.H.; Shakir, S.; Afzal, A.; Mujtaba, M.A.; Soudagar, M.E.M.; Fayaz, H.; Saleel C, A.; Farukh, S.; Saeed, M. Magnesium doped TiO2 as an efficient electron transport layer in perovskite solar cells. *Case Stud. Therm. Eng.* **2021**, *26*, 101101. [CrossRef]
29. Zhao, S.; Siqueira, G.; Drdova, S.; Norris, D.; Ubert, C.; Bonnin, A.; Galmarini, S.; Ganobjak, M.; Pan, Z.; Brunner, S. Additive manufacturing of silica aerogels. *Nature* **2020**, *584*, 387–392. [CrossRef]
30. Koebel, M.; Rigacci, A.; Achard, P. Aerogel-based thermal superinsulation: An overview. *J. Sol-Gel Sci. Technol.* **2012**, *63*, 315–339. [CrossRef]
31. Mazrouei-Sebdani, Z.; Begum, H.; Schoenwald, S.; Horoshenkov, K.V.; Malfait, W.J. A review on silica aerogel-based materials for acoustic applications. *J. Non-Cryst. Solids* **2021**, *562*, 120770. [CrossRef]
32. Maleki, H.; Durães, L.; Portugal, A. An overview on silica aerogels synthesis and different mechanical reinforcing strategies. *J. Non-Cryst. Solids* **2014**, *385*, 55–74. [CrossRef]
33. Sehaqui, H.; Zhou, Q.; Berglund, L.A. High-porosity aerogels of high specific surface area prepared from nanofibrillated cellulose (NFC). *Compos. Sci. Technol.* **2011**, *71*, 1593–1599. [CrossRef]
34. Gesser, H.; Goswami, P. Aerogels and related porous materials. *Chem. Rev.* **1989**, *89*, 765–788. [CrossRef]

35. Zheng, Q.; Fang, L.; Guo, H.; Yang, K.; Cai, Z.; Meador, M.A.B.; Gong, S. Highly porous polymer aerogel film-based triboelectric nanogenerators. *Adv. Funct. Mater.* **2018**, *28*, 1706365. [CrossRef]
36. Dorcheh, A.S.; Abbasi, M. Silica aerogel; synthesis, properties and characterization. *J. Mater. Process. Technol.* **2008**, *199*, 10–26. [CrossRef]
37. Dai, H.; Jun, Z.; Yin, Y.; Shao, G.; Yu, C. Synthesis of Ag doped SiO_2-TiO_2 aerogels with nano-sized microcrystalline anatase structure through IL control. *IOP Conf. Ser Shangai, China.: Mater. Sci. Eng.* **2019**, *587*, 012016. [CrossRef]
38. Koyuncu, D.D.E.; Okur, M. Investigation of dye removal ability and reusability of green and sustainable silica and carbon-silica hybrid aerogels prepared from paddy waste ash. *Colloids Surf. A Physicochem. Eng. Asp.* **2021**, *628*, 127370. [CrossRef]
39. Wan, W.; Zhang, R.; Ma, M.; Zhou, Y. Monolithic aerogel photocatalysts: A review. *J. Mater. Chem. A* **2018**, *6*, 754–775. [CrossRef]
40. Wong, K.J.; Foo, J.J.; Siang, T.J.; Ong, W.J. Shining Light on Carbon Aerogel Photocatalysts: Unlocking the Potentials in the Quest for Revolutionizing Solar-to-Chemical Conversion and Environmental Remediation. *Adv. Funct. Mater.* **2023**, 2306014. [CrossRef]
41. Schreck, M.; Kleger, N.; Matter, F.; Kwon, J.; Tervoort, E.; Masania, K.; Studart, A.R.; Niederberger, M. 3D Printed Scaffolds for Monolithic Aerogel Photocatalysts with Complex Geometries. *Small* **2021**, *17*, 2104089. [CrossRef]
42. Ge, B.; Ren, G.; Yang, H.; Yang, J.; Pu, X.; Li, W.; Jin, C.; Zhang, Z. Fabrication of BiOBr-silicone aerogel photocatalyst in an aqueous system with degradation performance by sol-gel method. *Sci. China Technol. Sci.* **2020**, *63*, 859–865. [CrossRef]
43. Ferreira-Neto, E.P.; Worsley, M.A.; Rodrigues-Filho, U.P. Towards thermally stable aerogel photocatalysts: $TiCl_4$-based sol-gel routes for the design of nanostructured silica-titania aerogel with high photocatalytic activity and outstanding thermal stability. *J. Environ. Chem. Eng.* **2019**, *7*, 103425. [CrossRef]
44. Hasanpour, M.; Hatami, M. Photocatalytic performance of aerogels for organic dyes removal from wastewaters: Review study. *J. Mol. Liquids* **2020**, *309*, 113094. [CrossRef]
45. Zhao, X.; Yi, X.; Wang, X.; Chu, W.; Guo, S.; Zhang, J.; Liu, B.; Liu, X. Constructing efficient polyimide (PI)/Ag aerogel photocatalyst by ethanol supercritical drying technique for hydrogen evolution. *Appl. Surf. Sci.* **2020**, *502*, 144187. [CrossRef]
46. Sleator, T.; Bernasconi, A.; Posselt, D.; Kjems, J.; Ott, H. Low-temperature specific heat and thermal conductivity of silica aerogels. *Phys. Rev. Lett.* **1991**, *66*, 1070. [CrossRef] [PubMed]
47. Yanagi, R.; Takemoto, R.; Ono, K.; Ueno, T. Light-induced levitation of ultralight carbon aerogels via temperature control. *Sci. Rep.* **2021**, *11*, 12413. [CrossRef]
48. Scheuerpflug, P.; Hauck, M.; Fricke, J. Thermal properties of silica aerogels between 1.4 and 330 K. *J. Non-Cryst. Solids* **1992**, *145*, 196–201. [CrossRef]
49. Ebert, H.-P. Thermal properties of aerogels. In *Aerogels Handbook*; Springer: Berlin/Heidelberg, Germany, 2011; pp. 537–564.
50. Bernasconi, A.; Sleator, T.; Posselt, D.; Ott, H. Dynamic technique for measurement of the thermal conductivity and the specific heat: Application to silica aerogels. *Rev. Sci. Instrum.* **1990**, *61*, 2420–2426. [CrossRef]
51. Wiener, M.; Reichenauer, G.; Braxmeier, S.; Hemberger, F.; Ebert, H.-P. Carbon aerogel-based high-temperature thermal insulation. *Int. J. Thermophys.* **2009**, *30*, 1372–1385. [CrossRef]
52. Strzałkowski, J.; Garbalińska, H. Thermal and strength properties of lightweight concretes with the addition of aerogel particles. *Adv. Cem. Res.* **2016**, *28*, 567–575. [CrossRef]
53. Li, D.; Zhang, C.; Li, Q.; Liu, C.; Arıcı, M.; Wu, Y. Thermal performance evaluation of glass window combining silica aerogels and phase change materials for cold climate of China. *Appl. Therm. Eng.* **2020**, *165*, 114547. [CrossRef]
54. Bellini, T.; Clark, N.A.; Muzny, C.D.; Wu, L.; Garland, C.W.; Schaefer, D.W.; Oliver, B.J. Phase behavior of the liquid crystal 8CB in a silica aerogel. *Phys. Rev. Lett.* **1992**, *69*, 788. [CrossRef] [PubMed]
55. Zeng, S.; Hunt, A.; Greif, R. Theoretical modeling of carbon content to minimize heat transfer in silica aerogel. *J. Non-Cryst. Solids* **1995**, *186*, 271–277. [CrossRef]
56. Hasan, M.A.; Sangashetty, R.; Esther, A.C.M.; Patil, S.B.; Sherikar, B.N.; Dey, A. Prospect of thermal insulation by silica aerogel: A brief review. *J. Inst. Eng. (India) Ser. D* **2017**, *98*, 297–304. [CrossRef]
57. Du, A.; Wang, H.; Zhou, B.; Zhang, C.; Wu, X.; Ge, Y.; Niu, T.; Ji, X.; Zhang, T.; Zhang, Z. Multifunctional silica nanotube aerogels inspired by polar bear hair for light management and thermal insulation. *Chem. Mater.* **2018**, *30*, 6849–6857. [CrossRef]
58. Wang, F.; Dou, L.; Dai, J.; Li, Y.; Huang, L.; Si, Y.; Yu, J.; Ding, B. In situ synthesis of biomimetic silica nanofibrous aerogels with temperature-invariant superelasticity over one million compressions. *Angew. Chem.* **2020**, *132*, 8362–8369. [CrossRef]
59. Wilson, S.M.; Gabriel, V.A.; Tezel, F.H. Adsorption of components from air on silica aerogels. *Microporous Mesoporous Mater.* **2020**, *305*, 110297. [CrossRef]
60. Tian, X.; Liu, J.; Wang, Y.; Shi, F.; Shan, Z.; Zhou, J.; Liu, J. Adsorption of antibiotics from aqueous solution by different aerogels. *J. Non-Cryst. Solids* **2019**, *505*, 72–78. [CrossRef]
61. Cheng, H.; Fan, Z.; Hong, C.; Zhang, X. Lightweight multiscale hybrid carbon-quartz fiber fabric reinforced phenolic-silica aerogel nanocomposite for high temperature thermal protection. *Compos. Part A Appl. Sci. Manuf.* **2021**, *143*, 106313. [CrossRef]
62. Xie, T.; He, Y.-L. Heat transfer characteristics of silica aerogel composite materials: Structure reconstruction and numerical modeling. *Int. J. Heat Mass Transf.* **2016**, *95*, 621–635. [CrossRef]
63. Wei, G.; Liu, Y.; Zhang, X.; Du, X. Radiative heat transfer study on silica aerogel and its composite insulation materials. *J. Non-Cryst. Solids* **2013**, *362*, 231–236. [CrossRef]
64. Yokogawa, H.; Yokoyama, M. Hydrophobic silica aerogels. *J. Non-Cryst. Solids* **1995**, *186*, 23–29. [CrossRef]

65. Korhonen, J.T.; Kettunen, M.; Ras, R.H.; Ikkala, O. Hydrophobic nanocellulose aerogels as floating, sustainable, reusable, and recyclable oil absorbents. *ACS Appl. Mater. Interfaces* **2011**, *3*, 1813–1816. [CrossRef] [PubMed]
66. He, S.; Huang, Y.; Chen, G.; Feng, M.; Dai, H.; Yuan, B.; Chen, X. Effect of heat treatment on hydrophobic silica aerogel. *J. Hazard. Mater.* **2019**, *362*, 294–302. [CrossRef] [PubMed]
67. Schwertfeger, F.; Frank, D.; Schmidt, M. Hydrophobic waterglass based aerogels without solvent exchange or supercritical drying. *J. Non-Cryst. Solids* **1998**, *225*, 24–29. [CrossRef]
68. Ge, D.; Yang, L.; Li, Y.; Zhao, J. Hydrophobic and thermal insulation properties of silica aerogel/epoxy composite. *J. Non-Cryst. Solids* **2009**, *355*, 2610–2615. [CrossRef]
69. Alwin, S.; Sahaya Shajan, X. Aerogels: Promising nanostructured materials for energy conversion and storage applications. *Mater. Renew. Sustain. Energy* **2020**, *9*, 7. [CrossRef]
70. Akhter, F.; Soomro, S.A.; Inglezakis, V.J. Silica aerogels; a review of synthesis, applications and fabrication of hybrid composites. *J. Porous Mater.* **2021**, *28*, 1387–1400. [CrossRef]
71. Tabata, M.; Adachi, I.; Kawai, H.; Sumiyoshi, T.; Yokogawa, H. Hydrophobic silica aerogel production at KEK. *Nuclear Instrum. Methods Phys. Res. Sect. A Accel. Spectrom. Detect. Assoc. Equip.* **2012**, *668*, 64–70. [CrossRef]
72. Smirnova, I.; Suttiruengwong, S.; Arlt, W. Feasibility study of hydrophilic and hydrophobic silica aerogels as drug delivery systems. *J. Non-Cryst. Solids* **2004**, *350*, 54–60. [CrossRef]
73. Dai, S.; Ju, Y.; Gao, H.; Lin, J.; Pennycook, S.; Barnes, C. Preparation of silica aerogel using ionic liquids as solvents. *Chem. Commun.* **2000**, *3*, 243–244. [CrossRef]
74. Lee, K.-H.; Kim, S.-Y.; Yoo, K.-P. Low-density, hydrophobic aerogels. *J. Non-Cryst. Solids* **1995**, *186*, 18–22. [CrossRef]
75. Zeng, S.; Hunt, A.; Greif, R. Transport properties of gas in silica aerogel. *J. Non-Cryst. Solids* **1995**, *186*, 264–270. [CrossRef]
76. Shi, M.; Tang, C.; Yang, X.; Zhou, J.; Jia, F.; Han, Y.; Li, Z. Superhydrophobic silica aerogels reinforced with polyacrylonitrile fibers for adsorbing oil from water and oil mixtures. *RSC Adv.* **2017**, *7*, 4039–4045. [CrossRef]
77. Ayen, R.; Iacobucci, P. Metal oxide aerogel preparation by supercritical extraction. *Rev. Chem. Eng.* **1988**, *5*, 157–198. [CrossRef]
78. Gash, A.E.; Tillotson, T.M.; Satcher Jr, J.H.; Hrubesh, L.W.; Simpson, R.L. New sol–gel synthetic route to transition and main-group metal oxide aerogels using inorganic salt precursors. *J. Non-Cryst. Solids* **2001**, *285*, 22–28. [CrossRef]
79. Benad, A.; Jürries, F.; Vetter, B.; Klemmed, B.; Hübner, R.; Leyens, C.; Eychmüller, A. Mechanical properties of metal oxide aerogels. *Chem. Mater.* **2018**, *30*, 145–152. [CrossRef]
80. Li, J.; Wang, X.; Huang, Q.; Gamboa, S.; Sebastian, P. Studies on preparation and performances of carbon aerogel electrodes for the application of supercapacitor. *J. Power Sources* **2006**, *158*, 784–788. [CrossRef]
81. Meena, A.K.; Mishra, G.; Rai, P.; Rajagopal, C.; Nagar, P. Removal of heavy metal ions from aqueous solutions using carbon aerogel as an adsorbent. *J. Hazard. Mater.* **2005**, *122*, 161–170. [CrossRef]
82. Ying, T.-Y.; Yang, K.-L.; Yiacoumi, S.; Tsouris, C. Electrosorption of ions from aqueous solutions by nanostructured carbon aerogel. *J. Colloid Interface Sci.* **2002**, *250*, 18–27. [CrossRef]
83. Hwang, S.-W.; Hyun, S.-H. Capacitance control of carbon aerogel electrodes. *J. Non-Cryst. Solids* **2004**, *347*, 238–245. [CrossRef]
84. Hao, P.; Zhao, Z.; Tian, J.; Li, H.; Sang, Y.; Yu, G.; Cai, H.; Liu, H.; Wong, C.; Umar, A. Hierarchical porous carbon aerogel derived from bagasse for high performance supercapacitor electrode. *Nanoscale* **2014**, *6*, 12120–12129. [CrossRef]
85. Xu, P.; Drewes, J.E.; Heil, D.; Wang, G. Treatment of brackish produced water using carbon aerogel-based capacitive deionization technology. *Water Res.* **2008**, *42*, 2605–2617. [CrossRef]
86. Yang, K.-L.; Ying, T.-Y.; Yiacoumi, S.; Tsouris, C.; Vittoratos, E.S. Electrosorption of ions from aqueous solutions by carbon aerogel: An electrical double-layer model. *Langmuir* **2001**, *17*, 1961–1969. [CrossRef]
87. Zhu, C.; Han, T.Y.-J.; Duoss, E.B.; Golobic, A.M.; Kuntz, J.D.; Spadaccini, C.M.; Worsley, M.A. Highly compressible 3D periodic graphene aerogel microlattices. *Nat. Commun.* **2015**, *6*, 6962. [CrossRef]
88. Zhang, X.; Sui, Z.; Xu, B.; Yue, S.; Luo, Y.; Zhan, W.; Liu, B. Mechanically strong and highly conductive graphene aerogel and its use as electrodes for electrochemical power sources. *J. Mater. Chem.* **2011**, *21*, 6494–6497. [CrossRef]
89. Xu, Z.; Zhang, Y.; Li, P.; Gao, C. Strong, conductive, lightweight, neat graphene aerogel fibers with aligned pores. *ACS Nano* **2012**, *6*, 7103–7113. [CrossRef] [PubMed]
90. Yang, M.; Zhao, N.; Cui, Y.; Gao, W.; Zhao, Q.; Gao, C.; Bai, H.; Xie, T. Biomimetic architectured graphene aerogel with exceptional strength and resilience. *ACS Nano* **2017**, *11*, 6817–6824. [CrossRef] [PubMed]
91. El Kadib, A.; Bousmina, M. Chitosan bio-based organic–inorganic hybrid aerogel microspheres. *Chem.–Eur. J.* **2012**, *18*, 8264–8277. [CrossRef] [PubMed]
92. Hu, H.; Zhao, Z.; Wan, W.; Gogotsi, Y.; Qiu, J. Polymer/graphene hybrid aerogel with high compressibility, conductivity, and "sticky" superhydrophobicity. *ACS Appl. Mater. Interfaces* **2014**, *6*, 3242–3249. [CrossRef] [PubMed]
93. Shah, N.; Lin, D. Composite Aerogels for Biomedical and Environmental Applications. *Curr. Pharm. Des.* **2020**, *26*, 5807–5818. [CrossRef]
94. Zhao, X.; Yang, F.; Wang, Z.; Ma, P.; Dong, W.; Hou, H.; Fan, W.; Liu, T. Mechanically strong and thermally insulating polyimide aerogels by homogeneity reinforcement of electrospun nanofibers. *Compos. Part B Eng.* **2020**, *182*, 107624. [CrossRef]

95. Yue, Y.; Liu, N.; Ma, Y.; Wang, S.; Liu, W.; Luo, C.; Zhang, H.; Cheng, F.; Rao, J.; Hu, X. Highly self-healable 3D microsupercapacitor with MXene–graphene composite aerogel. *ACS Nano* **2018**, *12*, 4224–4232. [CrossRef] [PubMed]
96. Anderson, M.L.; Stroud, R.M.; Morris, C.A.; Merzbacher, C.I.; Rolison, D.R. Tailoring advanced nanoscale materials through synthesis of composite aerogel architectures. *Adv. Eng. Mater.* **2000**, *2*, 481–488. [CrossRef]
97. Nawaz, M.; Miran, W.; Jang, J.; Lee, D.S. One-step hydrothermal synthesis of porous 3D reduced graphene oxide/TiO_2 aerogel for carbamazepine photodegradation in aqueous solution. *App. Catal. B Environ.* **2017**, *203*, 85–95. [CrossRef]
98. Zu, G.; Shen, J.; Wei, X.; Ni, X.; Zhang, Z.; Wang, J.; Liu, G. Preparation and characterization of monolithic alumina aerogels. *J. Non-Cryst. Solids* **2011**, *357*, 2903–2906. [CrossRef]
99. Fan, W.; Zuo, L.; Zhang, Y.; Chen, Y.; Liu, T. Mechanically strong polyimide/carbon nanotube composite aerogels with controllable porous structure. *Compos. Sci. Technol.* **2018**, *156*, 186–191. [CrossRef]
100. Guo, W.; Liu, J.; Zhang, P.; Song, L.; Wang, X.; Hu, Y. Multi-functional hydroxyapatite/polyvinyl alcohol composite aerogels with self-cleaning, superior fire resistance and low thermal conductivity. *Compos. Sci. Technol.* **2018**, *158*, 128–136. [CrossRef]
101. Bryning, M.B.; Milkie, D.E.; Islam, M.F.; Hough, L.A.; Kikkawa, J.M.; Yodh, A.G. Carbon nanotube aerogels. *Adv. Mater.* **2007**, *19*, 661–664. [CrossRef]
102. Kim, K.H.; Oh, Y.; Islam, M. Graphene coating makes carbon nanotube aerogels superelastic and resistant to fatigue. *Nat. Nanotechnol.* **2012**, *7*, 562–566. [CrossRef]
103. Merillas, B.; Villafañe, F.; Rodríguez-Pérez, M.Á. Super-insulating transparent polyisocyanurate-polyurethane aerogels: Analysis of thermal conductivity and mechanical properties. *Nanomaterials* **2022**, *12*, 2409. [CrossRef]
104. Maiuolo, L.; Olivito, F.; Algieri, V.; Costanzo, P.; Jiritano, A.; Tallarida, M.A.; Tursi, A.; Sposato, C.; Feo, A.; De Nino, A. Synthesis, characterization and mechanical properties of novel bio-based polyurethane foams using cellulose-derived polyol for chain extension and cellulose citrate as a thickener additive. *Polymers* **2021**, *13*, 2802. [CrossRef] [PubMed]
105. Maleki, H.; Hüsing, N. Current status, opportunities and challenges in catalytic and photocatalytic applications of aerogels: Environmental protection aspects. *Appl. Catal. B Environ.* **2018**, *221*, 530–555. [CrossRef]
106. Zhao, X.; Zhang, J.; Wang, X.; Zhang, J.; Liu, B.; Yi, X. Polyimide aerogels crosslinked with MWCNT for enhanced visible-light photocatalytic activity. *Appl. Surf. Sci.* **2019**, *478*, 266–274. [CrossRef]
107. Yan, S.; Song, H.; Li, Y.; Yang, J.; Jia, X.; Wang, S.; Yang, X. Integrated reduced graphene oxide/polypyrrole hybrid aerogels for simultaneous photocatalytic decontamination and water evaporation. *Appl. Catal. B Environ.* **2022**, *301*, 120820. [CrossRef]
108. Bai, Y.; Yi, X.; Li, B.; Chen, S.; Fan, Z. Constructing porous polyimide/carbon quantum dots aerogel with efficient photocatalytic property under visible light. *Appl. Surf. Sci.* **2022**, *578*, 151993. [CrossRef]
109. Zhi, M.; Tang, H.; Wu, M.; Ouyang, C.; Hong, Z.; Wu, N. Synthesis and Photocatalysis of Metal Oxide Aerogels: A Review. *Energy Fuels* **2022**, *36*, 11359–11379. [CrossRef]
110. Jiang, G.; Wang, J.; Li, N.; Hübner, R.; Georgi, M.; Cai, B.; Li, Z.; Lesnyak, V.; Gaponik, N.; Eychmüller, A. Self-supported three-dimensional quantum dot aerogels as a promising photocatalyst for CO_2 reduction. *Chem. Mater.* **2022**, *34*, 2687–2695. [CrossRef]
111. Korkmaz, S.; Kariper, İ.A. Graphene and graphene oxide based aerogels: Synthesis, characteristics and supercapacitor applications. *J. Energy Storage* **2020**, *27*, 101038. [CrossRef]
112. Liu, W.; Herrmann, A.-K.; Bigall, N.C.; Rodriguez, P.; Wen, D.; Oezaslan, M.; Schmidt, T.J.; Gaponik, N.; Eychmüller, A. Noble Metal Aerogels Synthesis, Characterization, and Application as Electrocatalysts. *Accounts Chem. Res.* **2015**, *48*, 154–162. [CrossRef]
113. Kharissova, O.V.; Ibarra Torres, C.E.; González, L.T.; Kharisov, B.I. All-carbon hybrid aerogels: Synthesis, properties, and applications. *Ind. Eng. Chem. Res.* **2019**, *58*, 16258–16286. [CrossRef]
114. Zhi, D.; Li, T.; Li, J.; Ren, H.; Meng, F. A review of three-dimensional graphene-based aerogels: Synthesis, structure and application for microwave absorption. *Compos. Part B Eng.* **2021**, *211*, 108642. [CrossRef]
115. Campbell, L.; Na, B.; Ko, E. Synthesis and characterization of titania aerogels. *Chem. Mater.* **1992**, *4*, 1329–1333. [CrossRef]
116. Rao, A.V.; Bhagat, S.D.; Hirashima, H.; Pajonk, G. Synthesis of flexible silica aerogels using methyltrimethoxysilane (MTMS) precursor. *J. Colloid Interface Sci.* **2006**, *300*, 279–285.
117. Minisy, I.M.; Acharya, U.; Veigel, S.; Morávková, Z.; Taboubi, O.; Hodan, J.; Breitenbach, S.; Unterweger, C.; Gindl-Altmutter, W.; Bober, P. Sponge-like polypyrrole–nanofibrillated cellulose aerogels: Synthesis and application. *J. Mater. Chem. C* **2021**, *9*, 12615–12623. [CrossRef]
118. Aegerter, M.A.; Leventis, N.; Koebel, M.M. *Aerogels Handbook*; Springer Science & Business Media: Berlin, Germany, 2011.
119. Maleki, H.; Durães, L.; García-González, C.A.; Del Gaudio, P.; Portugal, A.; Mahmoudi, M. Synthesis and biomedical applications of aerogels: Possibilities and challenges. *Adv. Colloid Interface Sci.* **2016**, *236*, 1–27. [CrossRef] [PubMed]
120. Ratke, L.; Gurikov, P. *The Chemistry and Physics of Aerogels: Synthesis, Processing, and Properties*; Cambridge University Press: Cambridge, UK, 2021.
121. Stolarski, M.; Walendziewski, J.; Steininger, M.; Pniak, B. Synthesis and characteristic of silica aerogels. *Appl. Catal. A Gen.* **1999**, *177*, 139–148. [CrossRef]
122. Morris, C.A.; Anderson, M.L.; Stroud, R.M.; Merzbacher, C.I.; Rolison, D.R. Silica sol as a nanoglue: Flexible synthesis of composite aerogels. *Science* **1999**, *284*, 622–624. [CrossRef]

123. Esquivias, L.; Pinero, M.; Morales-Flórez, V.; de la Rosa-Fox, N. Aerogels synthesis by sonocatalysis: Sonogels. In *Aerogels Handbook*; Springer: Berlin/Heidelberg, Germany, 2011; pp. 419–445.
124. Hoepfner, S.; Ratke, L.; Milow, B. Synthesis and characterisation of nanofibrillar cellulose aerogels. *Cellulose* **2008**, *15*, 121–129. [CrossRef]
125. Pinelli, F.; Nespoli, T.; Rossi, F. Graphene Oxide-Chitosan Aerogels: Synthesis, Characterization, and Use as Adsorbent Material for Water Contaminants. *Gels* **2021**, *7*, 149. [CrossRef]
126. Jafari, S.; Dehghani, M.; Nasirizadeh, N.; Baghersad, M.H.; Azimzadeh, M. Label-free electrochemical detection of Cloxacillin antibiotic in milk samples based on molecularly imprinted polymer and graphene oxide-gold nanocomposite. *Measurement* **2019**, *145*, 22–29. [CrossRef]
127. Du, Y.; Zhang, X.; Wang, J.; Liu, Z.; Zhang, K.; Ji, X.; You, Y.; Zhang, X. Reaction-spun transparent silica aerogel fibers. *ACS Nano* **2020**, *14*, 11919–11928. [CrossRef] [PubMed]
128. Karadagli, I.; Schulz, B.; Schestakow, M.; Milow, B.; Gries, T.; Ratke, L. Production of porous cellulose aerogel fibers by an extrusion process. *J. Supercrit. Fluids* **2015**, *106*, 105–114. [CrossRef]
129. Yang, H.; Wang, Z.; Liu, Z.; Cheng, H.; Li, C. Continuous, strong, porous silk firoin-based aerogel fibers toward textile thermal insulation. *Polymers* **2019**, *11*, 1899. [CrossRef]
130. Li, X.; Dong, G.; Liu, Z.; Zhang, X. Polyimide Aerogel Fibers with Superior Flame Resistance, Strength, Hydrophobicity, and Flexibility Made via a Universal Sol–Gel Confined Transition Strategy. *ACS Nano* **2021**, *15*, 4759–4768. [CrossRef]
131. Meng, S.; Zhang, J.; Chen, W.; Wang, X.; Zhu, M. Construction of continuous hollow silica aerogel fibers with hierarchical pores and excellent adsorption performance. *Microporous Mesoporous Mater.* **2019**, *273*, 294–296. [CrossRef]
132. Mitropoulos, A.N.; Burpo, F.J.; Nguyen, C.K.; Nagelli, E.A.; Ryu, M.Y.; Wang, J.; Sims, R.K.; Woronowicz, K.; Wickiser, J.K. Noble metal composite porous silk fibroin aerogel fibers. *Materials* **2019**, *12*, 894. [CrossRef]
133. Venkataraman, M.; Mishra, R.; Jasikova, D.; Kotresh, T.; Militky, J. Thermodynamics of aerogel-treated nonwoven fabrics at subzero temperatures. *J. Ind. Text.* **2015**, *45*, 387–404. [CrossRef]
134. Arshad, Z.; Alharthi, S.S. Enhancing the Thermal Comfort of Woven Fabrics and Mechanical Properties of Fiber-Reinforced Composites Using Multiple Weave Structures. *Fibers* **2023**, *11*, 73. [CrossRef]
135. Qian, H.; Kucernak, A.R.; Greenhalgh, E.S.; Bismarck, A.; Shaffer, M.S. Multifunctional structural supercapacitor composites based on carbon aerogel modified high performance carbon fiber fabric. *ACS Appl. Mater. Interfaces* **2013**, *5*, 6113–6122. [CrossRef]
136. Bhuiyan, M.R.; Wang, L.; Shaid, A.; Shanks, R.A.; Ding, J. Polyurethane-aerogel incorporated coating on cotton fabric for chemical protection. *Progr. Org. Coat.* **2019**, *131*, 100–110. [CrossRef]
137. Xiong, X.; Yang, T.; Mishra, R.; Militky, J. Transport properties of aerogel-based nanofibrous nonwoven fabrics. *Fibers Polym.* **2016**, *17*, 1709–1714. [CrossRef]
138. Talebi, Z.; Soltani, P.; Habibi, N.; Latifi, F. Silica aerogel/polyester blankets for efficient sound absorption in buildings. *Constr. Build. Mater.* **2019**, *220*, 76–89. [CrossRef]
139. Venkataraman, M.; Mishra, R.; Kotresh, T.; Sakoi, T.; Militky, J. Effect of compressibility on heat transport phenomena in aerogel-treated nonwoven fabrics. *J. Text. Inst.* **2016**, *107*, 1150–1158. [CrossRef]
140. Altay, P.; Atakan, R.; Özcan, G. Silica aerogel application to polyester fabric for outdoor clothing. *Fibers Polym.* **2021**, *22*, 1025–1032. [CrossRef]
141. Jabbari, M.; Åkesson, D.; Skrifvars, M.; Taherzadeh, M.J. Novel lightweight and highly thermally insulative silica aerogel-doped poly (vinyl chloride)-coated fabric composite. *J. Reinf. Plast. Compos.* **2015**, *34*, 1581–1592. [CrossRef]
142. Shaid, A.; Fergusson, M.; Wang, L. Thermophysiological comfort analysis of aerogel nanoparticle incorporated fabric for fire fighter's protective clothing. *Chem. Mater. Eng.* **2014**, *2*, 37–43. [CrossRef]
143. Venkataraman, M.; Mishra, R.; Militky, J.; Hes, L. Aerogel based nanoporous fibrous materials for thermal insulation. *Fibers Polym.* **2014**, *15*, 1444–1449. [CrossRef]
144. Venkataraman, M.; Mishra, R.; Wiener, J.; Militky, J.; Kotresh, T.; Vaclavik, M. Novel techniques to analyse thermal performance of aerogel-treated blankets under extreme temperatures. *J. Text. Inst.* **2015**, *106*, 736–747. [CrossRef]
145. Han, Y.; Zhang, X.; Wu, X.; Lu, C. Flame retardant, heat insulating cellulose aerogels from waste cotton fabrics by in situ formation of magnesium hydroxide nanoparticles in cellulose gel nanostructures. *ACS Sustain. Chem. Eng.* **2015**, *3*, 1853–1859. [CrossRef]
146. Ahmad, F.; Ulker, Z.; Erkey, C. A novel composite of alginate aerogel with PET nonwoven with enhanced thermal resistance. *J. Non-Cryst. Solids* **2018**, *491*, 7–13. [CrossRef]
147. Lang, X.H.; Zhu, T.Y.; Zou, L.; Prakashan, K.; Zhang, Z.X. Fabrication and characterization of polypropylene aerogel material and aerogel coated hybrid materials for oil-water separation applications. *Prog. Org. Coat.* **2019**, *137*, 105370. [CrossRef]
148. Chakraborty, S.; Pisal, A.; Kothari, V.; Venkateswara Rao, A. Synthesis and characterization of fibre reinforced silica aerogel blankets for thermal protection. *Adv. Mater. Sci. Eng.* **2016**, *2016*, 2495623. [CrossRef]
149. Ibrahim, M.; Bianco, L.; Ibrahim, O.; Wurtz, E. Low-emissivity coating coupled with aerogel-based plaster for walls' internal surface application in buildings: Energy saving potential based on thermal comfort assessment. *J. Build. Eng.* **2018**, *18*, 454–466. [CrossRef]

150. Masera, G.; Wakili, K.G.; Stahl, T.; Brunner, S.; Galliano, R.; Monticelli, C.; Aliprandi, S.; Zanelli, A.; Elesawy, A. Development of a super-insulating, aerogel-based textile wallpaper for the indoor energy retrofit of existing residential buildings. *Procedia Eng.* **2017**, *180*, 1139–1149. [CrossRef]
151. Schuss, M.; Pont, U.; Mahdavi, A. Long-term experimental performance evaluation of aerogel insulation plaster. *Energy Procedia* **2017**, *132*, 508–513. [CrossRef]
152. Wakili, K.G.; Stahl, T.; Heiduk, E.; Schuss, M.; Vonbank, R.; Pont, U.; Sustr, C.; Wolosiuk, D.; Mahdavi, A. High performance aerogel containing plaster for historic buildings with structured façades. *Energy Procedia* **2015**, *78*, 949–954. [CrossRef]
153. Schmidt, M.; Schwertfeger, F. Applications for silica aerogel products. *J. Non-Cryst. Solids* **1998**, *225*, 364–368. [CrossRef]
154. Fesmire, J.E. Aerogel insulation systems for space launch applications. *Cryogenics* **2006**, *46*, 111–117. [CrossRef]
155. Smirnova, I.; Gurikov, P. Aerogel production: Current status, research directions, and future opportunities. *J. Supercrit. Fluids* **2018**, *134*, 228–233. [CrossRef]
156. Pekala, R.; Farmer, J.; Alviso, C.; Tran, T.; Mayer, S.; Miller, J.; Dunn, B. Carbon aerogels for electrochemical applications. *J. Non-Cryst. Solids* **1998**, *225*, 74–80. [CrossRef]
157. Jelle, B.P.; Baetens, R.; Gustavsen, A. Aerogel insulation for building applications. In *The Sol-Gel Handbook*; Levy, D., Zayat, M., Eds.; Wiley-VCH: Weinheim, Germany, 2015; pp. 1385–1412.
158. Pierre, A.C.; Pajonk, G.M. Chemistry of aerogels and their applications. *Chem. Rev.* **2002**, *102*, 4243–4266. [CrossRef] [PubMed]
159. Zheng, L.; Zhang, S.; Ying, Z.; Liu, J.; Zhou, Y.; Chen, F. Engineering of aerogel-based biomaterials for biomedical applications. *Int. J. Nanomed.* **2020**, *15*, 2363. [CrossRef] [PubMed]
160. Long, L.-Y.; Weng, Y.-X.; Wang, Y.-Z. Cellulose aerogels: Synthesis, applications, and prospects. *Polymers* **2018**, *10*, 623. [CrossRef]
161. Fricke, J. Aerogels and their applications. *J. Non-Cryst. Solids* **1992**, *147*, 356–362. [CrossRef]
162. Linhares, T.; de Amorim, M.T.P.; Durães, L. Silica aerogel composites with embedded fibres: A review on their preparation, properties and applications. *J. Mater. Chem. A* **2019**, *7*, 22768–22802. [CrossRef]
163. Maleki, H. Recent advances in aerogels for environmental remediation applications: A review. *Chem. Eng. J.* **2016**, *300*, 98–118. [CrossRef]
164. Buratti, C.; Merli, F.; Moretti, E. Aerogel-based materials for building applications: Influence of granule size on thermal and acoustic performance. *Energy Build.* **2017**, *152*, 472–482. [CrossRef]
165. Moreno-Castilla, C.; Maldonado-Hódar, F. Carbon aerogels for catalysis applications: An overview. *Carbon* **2005**, *43*, 455–465. [CrossRef]
166. Fricke, J.; Emmerling, A. Aerogels—Recent progress in production techniques and novel applications. *J. Sol-Gel Sci. Technol.* **1998**, *13*, 299–303. [CrossRef]
167. Cuce, E.; Cuce, P.M.; Wood, C.J.; Riffat, S.B. Toward aerogel based thermal superinsulation in buildings: A comprehensive review. *Renew. Sustain. Energy Rev.* **2014**, *34*, 273–299. [CrossRef]
168. Hrubesh, L.W.; Poco, J.F. Thin aerogel films for optical, thermal, acoustic and electronic applications. *J. Non-Cryst. Solids* **1995**, *188*, 46–53. [CrossRef]
169. Zhao, S.; Malfait, W.J.; Guerrero-Alburquerque, N.; Koebel, M.M.; Nyström, G. Biopolymer aerogels and foams: Chemistry, properties, and applications. *Angew. Chem. Int. Ed.* **2018**, *57*, 7580–7608. [CrossRef]
170. Biener, J.; Stadermann, M.; Suss, M.; Worsley, M.A.; Biener, M.M.; Rose, K.A.; Baumann, T.F. Advanced carbon aerogels for energy applications. *Energy Environ. Sci.* **2011**, *4*, 656–667. [CrossRef]
171. Santos-Rosales, V.; Alvarez-Rivera, G.; Hillgärtner, M.; Cifuentes, A.; Itskov, M.; García-González, C.A.; Rege, A. Stability studies of starch aerogel formulations for biomedical applications. *Biomacromolecules* **2020**, *21*, 5336–5344. [CrossRef] [PubMed]
172. Li, F.; Xie, L.; Sun, G.; Kong, Q.; Su, F.; Cao, Y.; Wei, J.; Ahmad, A.; Guo, X.; Chen, C.-M. Resorcinol-formaldehyde based carbon aerogel: Preparation, structure and applications in energy storage devices. *Microporous Mesoporous Mater.* **2019**, *279*, 293–315. [CrossRef]
173. Hu, L.; He, R.; Lei, H.; Fang, D. Carbon aerogel for insulation applications: A review. *Int. J. Thermophys.* **2019**, *40*, 39. [CrossRef]
174. Stahl, T.; Brunner, S.; Zimmermann, M.; Wakili, K.G. Thermo-hygric properties of a newly developed aerogel based insulation rendering for both exterior and interior applications. *Energy Build.* **2012**, *44*, 114–117. [CrossRef]
175. Randall, J.P.; Meador, M.A.B.; Jana, S.C. Tailoring mechanical properties of aerogels for aerospace applications. *ACS Appl. Mater. Interfaces* **2011**, *3*, 613–626. [CrossRef]
176. López-Iglesias, C.; Barros, J.; Ardao, I.; Monteiro, F.J.; Alvarez-Lorenzo, C.; Gómez-Amoza, J.L.; García-González, C.A. Vancomycin-loaded chitosan aerogel particles for chronic wound applications. *Carbohydr. Polym.* **2019**, *204*, 223–231. [CrossRef]
177. Fricke, J.; Tillotson, T. Aerogels: Production, characterization, and applications. *Thin Solid Films* **1997**, *297*, 212–223. [CrossRef]
178. Cheng, Y.; Zhao, H.; Lv, H.; Shi, T.; Ji, G.; Hou, Y. Lightweight and flexible cotton aerogel composites for electromagnetic absorption and shielding applications. *Adv. Electron. Mater.* **2020**, *6*, 1900796. [CrossRef]
179. Olivito, F.; Algieri, V.; Jiritano, A.; Tallarida, M.A.; Costanzo, P.; Maiuolo, L.; De Nino, A. Bio-Based Polyurethane Foams for the Removal of Petroleum-Derived Pollutants: Sorption in Batch and in Continuous-Flow. *Polymers* **2023**, *15*, 1785. [CrossRef] [PubMed]
180. García-González, C.A.; Budtova, T.; Durães, L.; Erkey, C.; Del Gaudio, P.; Gurikov, P.; Koebel, M.; Liebner, F.; Neagu, M.; Smirnova, I. An opinion paper on aerogels for biomedical and environmental applications. *Molecules* **2019**, *24*, 1815. [CrossRef]

181. Ferreira-Neto, E.P.; Ullah, S.; Da Silva, T.C.; Domeneguetti, R.R.; Perissinotto, A.P.; De Vicente, F.S.; Rodrigues-Filho, U.P.; Ribeiro, S.J. Bacterial nanocellulose/MoS_2 hybrid aerogels as bifunctional adsorbent/photocatalyst membranes for in-flow water decontamination. *ACS Appl. Mater. Interfaces* **2020**, *12*, 41627–41643. [CrossRef] [PubMed]
182. Novak, Z.; Horvat, G. Book of Abstracts. In Proceedings of the 3rd International Conference on Aerogels for Biomedical and Environmental Applications, Maribor, Slovenia, 5–7 July 2023.
183. Lakatos, Á.; Trník, A. Thermal diffusion in fibrous aerogel blankets. *Energies* **2020**, *13*, 823. [CrossRef]

catalysts

MDPI

Article

Dimethyl Ether to Olefins on Hybrid Intergrowth Structure Zeolites

Maria V. Magomedova [1,2,*], Anastasiya V. Starozhitskaya [1], Ilya A. Davidov [1], Dmitry E. Tsaplin [1,3] and Anton L. Maximov [1]

1 A.V. Topchiev Institute of Petrochemical Synthesis, RAS (TIPS RAS), 29 Leninsky Prospekt, 119991 Moscow, Russia
2 Lomonosov Institute of Fine Chemical Technologies, Russian Technological University (MIREA), 86 Vernadsky Prospekt, 119454 Moscow, Russia
3 Faculty of Chemistry, Lomonosov Moscow State University, 1 Leninskie Gory Str., 119234 Moscow, Russia
* Correspondence: podlesnaya@ips.ac.ru

Abstract: A series of catalysts based on hybrid intergrowth structure zeolites MFI-MEL, MFI-MTW, and MFI-MCM-41 are studied in the reaction of olefins synthesis from dimethyl ether at atmospheric pressure and a temperature of 340 °C. The total acidity of hybrid zeolite-based catalysts is shown to correlate with their activity. However, the use of zeolite with the structure MFI-MCM-41, which is characterized by a high content of medium acid sites, additionally catalyzes the methanol dehydration reaction, resulting in a decrease in the observed DME conversion. The obtained product distributions are brought into correlation with the texture of catalysts. It is shown that the use of hybrid zeolites does not change the mechanism of reaction, but the structural features of zeolites influence the priority of the competing MTO reactions: high ethylene yield is observed for catalysts with high micropore volume. The topology of the hybrid zeolite has been shown to influence the hydrogen transfer reaction rate, but not to change the isomerizing activity of the catalyst.

Keywords: hybrid zeolites; intergrowth structure; MTO reaction; dimethyl ether; light olefins; MEL; MTW; MCM-41; MFI

Citation: Magomedova, M.V.; Starozhitskaya, A.V.; Davidov, I.A.; Tsaplin, D.E.; Maximov, A.L. Dimethyl Ether to Olefins on Hybrid Intergrowth Structure Zeolites. *Catalysts* **2023**, *13*, 570. https://doi.org/10.3390/catal13030570

Academic Editor: Narendra Kumar

Received: 7 February 2023
Revised: 9 March 2023
Accepted: 10 March 2023
Published: 11 March 2023

1. Introduction

The reaction of oxygenates (methanol and dimethyl ether) to light olefins occurs with the involvement of microporous catalysts based on MFI zeolite or CHA silicoaluminophosphate [1–4]. A large number of studies using solid-state NMR spectroscopy, a pulse reactor, and the reaction with labeled ^{13}C atoms have been conducted to determine the MTO (methanol-to-olefins) reaction mechanism [5–11]. Despite the different structural characteristics of MFI and CHA microporous materials, the reaction is considered to follow a dual-cycle mechanism, which was proposed by a group of Norwegian researchers in 2007 [12–14]. This mechanism was subsequently confirmed by both theoretical studies and a number of experimental observations [15–24]. According to this mechanism, in the first step, hydrocarbon pool species (aromatic cations, alkyl cyclopentyl cations) are formed in the micropores of the catalyst. Then, hydrocarbon pool species contact with reagents and semi-products in a series of parallel methylation and dealkylation reactions to form light olefins. The resulting products participate in oligomerization, isomerization, and H-transfer reactions to form higher olefins, aromatic compounds, and alkanes [15,23,25,26]. Hydrocarbon pool species act as an organic catalyst for the conversion of oxygenates into hydrocarbons, while Lewis and Brønsted acid sites of zeolite or silicoalmoaphosphate stabilize hydrocarbon pool species in micropores. Moreover, due to the molecular-sieve properties, the structure of zeolite or silicoalamophosphate determines product selectivity.

However, the catalysts used in the MTO reaction are characterized by a low diffusion coefficient of the molecules inside the zeolite or silicoalamophosphate micropores [27,28].

The low diffusion rate of the reagents to the active sites leads to a decrease in the activity of the catalyst, while the low diffusion rate of the reaction products from the micropores leads to active secondary reactions, a decrease in the selectivity of target products, and rapid catalyst deactivation [29–34].

To improve mass transfer, micro-mesoporous catalysts are used. In the case of MFI-type zeolite, they can be obtained by mixing it with mesoporous materials (Al_2O_3, clay, galloisite) or by posttreatment with alkali or acid, removing silicon or aluminum atoms from the framework of the crystal lattice and forming the hierarchical structure [35–39]. The use of such catalysts reduces steric difficulties and increases the diffusion rate of reagents and products, which, in turn, leads to increased activity.

An alternative method of forming MFI-type zeolites with micro- and mesopores is the synthesis of hybrid zeolites—co-crystallites, which, in addition to the MFI structure, contain the structure of the other microporous material. Mesopores are formed directly during synthesis by the formation of additional cavities at the intersections of the channels and/or interparticle voids of zeolites [40–43]. Such hybrid zeolites in the reaction of hydrocarbon synthesis from oxygenates have not been sufficiently investigated [42–44].

In this paper, we conducted a study of the influence of textural and acid properties of micro- and mesoporous structure catalysts based on hybrid zeolites MFI-MEL/Al_2O_3, MFI-MTW/Al_2O_3, and MFI-MCM-41/Al_2O_3 on the distribution of reaction products in the synthesis of light olefins from DME. The results were compared with the standard catalyst MFI/Al_2O_3, which was studied in [45].

2. Results and Discussion

2.1. Characterization of the Catalysts

The characterization of hybrid zeolites is presented in Supplementary Materials (Table S1. Phase ratio, Si/Al ratio; Figure S1. ^{27}Al MAS NMR spectra; Figure S2. IR spectra of pyridine adsorbed on hybrid zeolites; Table S2. The distribution of Lewis and Brønsted acid sites; Figure S3. SEM and TEM micrographs).

According to XRD patterns of the catalysts, all samples contain the MFI topology structure, as evidenced by the presence of specific-to-MFI reflections at $2\theta = 7.9$, 8, 23.2, 23.9, and 24.4° [46] (Figure 1A).

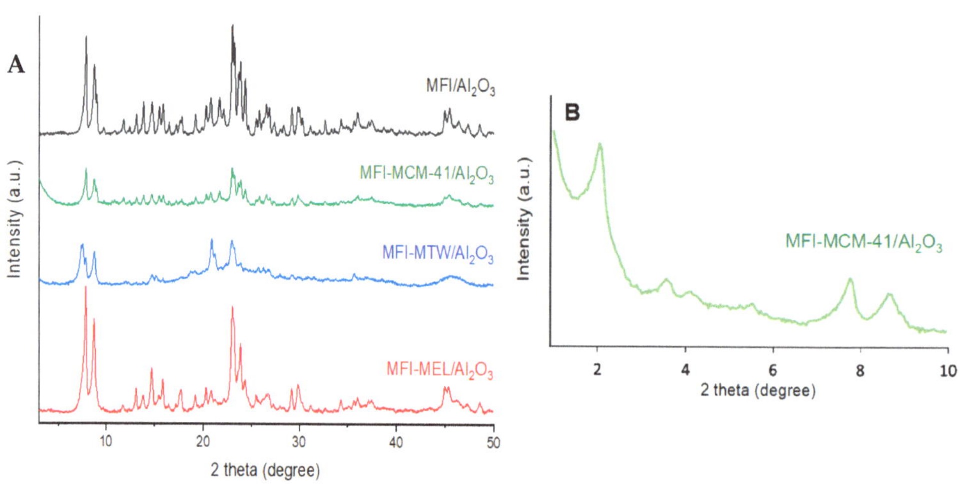

Figure 1. XRD patterns of the catalysts. (**A**) WAXRD-patterns; (**B**) SAXRD-pattern of MFI-MCM-41/Al_2O_3.

The presence of the MEL structure in the MFI-MEL/Al_2O_3 sample is confirmed by the presence of reflections at 2θ = 7.92, 8.78, 23.14, 23.98, and 45.2° [47]. The MFI-MTW/Al_2O_3 catalyst XRD pattern contains reflections that are characteristic of the MTW type structure at 2θ = 7.2, 8.8, 20.7, and 23.1° [48]. The presence of MCM-41 in the MFI-MCM-41/Al_2O_3 catalyst is confirmed by the characteristic reflection of the amorphous SiO_2 of MCM-41 at 2θ = 22.8° and the reflection at 2θ = 2.2° on the small-angle X-ray diffraction pattern (Figure 1B) [40].

Quantification of the hybrid zeolite phase ratio can be performed through an assessment of the contribution of different structures to the final XRD pattern [49]. On the catalysts MFI-MEL/Al_2O_3 and MFI-MTW/Al_2O_3, the zeolite phase ratio is 55/45 and 60/40, respectively [50]. The MFI/MCM-41 ratio is 80/20. All hybrid zeolites have an intergrowth structure and do not contain crystallites of individual phases, which is confirmed by SEM and TEM micrographs (Supplementary Materials Figure S3).

Table 1 shows the texture properties of the samples. External surface areas, micro- and mesopore surface areas, and the average diameters of micro- and mesopores for all catalysts are presented in Supplementary Materials (Table S3).

Table 1. Textural properties of the MFI-MEL/Al_2O_3, MFI-MTW/Al_2O_3, MFI-MCM-41/Al_2O_3 samples.

No	Catalyst	S_{BET}, $m^2/g(Cat)$	S_{micro}, $m^2/g(Cat)$	V(pore), $cm^3/g(Cat)$		
				Total	Micro	Meso
1	MFI-MEL/Al_2O_3	361	250	0.286	0.088 (30.8%)	0.198 (69.2%)
2	MFI-MTW/Al_2O_3	179	11	0.443	0.012 (2.7%)	0.431 (97.3%)
3	MFI-MCM-41/Al_2O_3	250	42	0.340	0.040 (11.8%)	0.300 (88.2%)
4	MFI/Al_2O_3	293	181	0.198	0.057 (28.7%)	0.142 (71.3%)

Catalysts based on hybrid zeolites MFI-MEL, MFI-MTW, and MFI-MCM-41 have a significantly larger mesopore volume (0.198–0.431 cm^3/g) than the standard catalyst based on MFI zeolite (0.142 cm^3/g).

The hybrid zeolite-based catalyst MFI-MEL/Al_2O_3 is characterized by the largest specific surface area among the studied samples—361 m^2/g, the average total pore volume—0.286 cm^3/g, and, at the same time, the largest micropore volume—0.88 cm^3/g (30.8%).

The hybrid zeolite-based catalyst MFI-MTW/Al_2O_3 is characterized by the smallest specific surface area—179 m^2/g, and the largest total pore volume—0.443 cm^3/g, while it almost does not have micropores—their volume is 0.012 cm^3/g (2.7%). The large total pore volume of the MFI-MTW/Al_2O_3 catalyst is created by mesopores, which have a volume of 0.431 cm^3/g.

The MFI-MCM-41/Al_2O_3 catalyst is characterized by an average specific surface area of 250 m^2/g, which is slightly lower than for the standard MFI/Al_2O_3 sample, but higher than for the catalyst based on the MFI-MTW/Al_2O_3 hybrid zeolite. This sample is in the middle in relation to the total pore volume—0.340 cm^3/g, and the micropore volume—0.040 cm^3/g (11.8%).

In terms of the reaction mechanism, MFI-MEL/Al_2O_3 should be considered the most promising sample, which has the largest volume of micropores necessary for stabilizing hydrocarbon pool species.

The results of the analysis of the ammonia TPD spectra for the samples are shown in Figure 2. It should be noted that the acidity determined by TPD NH_3 can be overestimated; however, these values can be compared with each other with a high degree of confidence. The TPD spectra are resolved to Gaussian peaks using the Peak Deconvolution tool of the Origin 2018 software package [51,52].

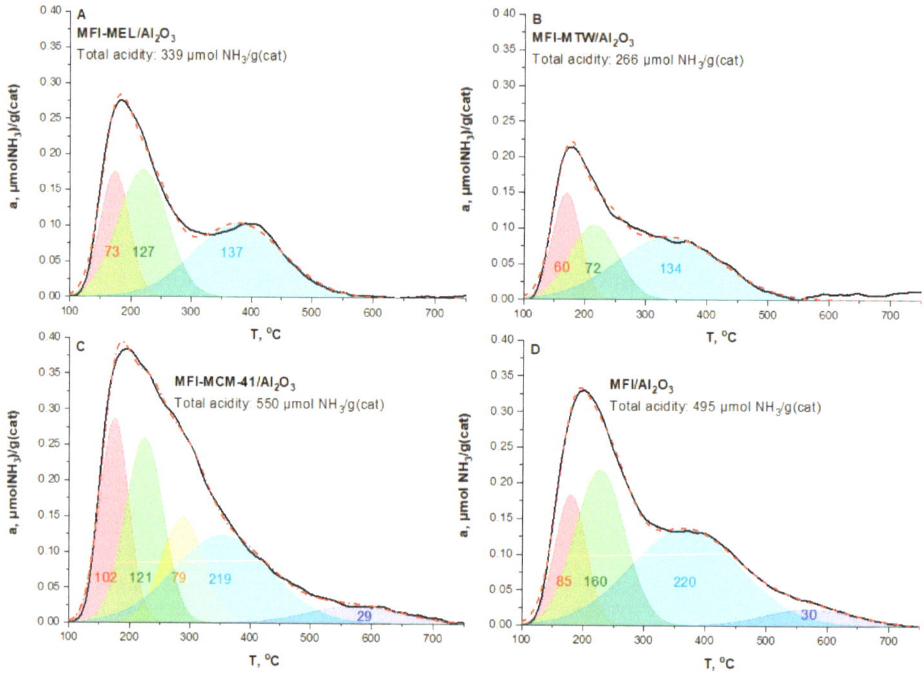

Figure 2. NH_3-TPD profiles of ammonia: (**A**) MFI-MEL/Al_2O_3, (**B**) MFI-MTW/Al_2O_3, (**C**) MFI-MCM-41/Al_2O_3, (**D**) MFI/Al_2O_3 and their deconvolution into Gaussian peaks. The black solid line is the experimental curve. The red dotted line is the cumulative curve of Gaussian peaks. Numbers indicate the amount of acid sites corresponding to the peak, μmol NH_3/g(cat).

For all samples, there are two peaks of ammonia desorption at the temperatures of 170–180 °C and 215–230 °C that correspond to the aluminum oxide and zeolite weak Lewis acid sites such as AlO^+ or charged $Al_xO_y^{n+}$ clusters, respectively [53].

All samples are characterized by the desorption peak at the temperature of 350–380 °C. These peaks correspond to strong Brønsted acid sites represented by bridging OH^+ groups.

For the MFI-MCM-41/Al_2O_3 and the standard MFI/Al_2O_3, superacid sites are observed at a desorption temperature of 570 °C. These peaks correspond to strong Lewis acid sites—extraframework aluminum atoms [53–57].

For MFI-MCM-41/Al_2O_3, there is no pronounced local minimum between the peaks of weak 170–180 °C and strong 350–380 °C acid sites in the TPD spectra, and a peak at the desorption temperature of 290 °C can be observed at deconvolution. This peak corresponds to the Brønsted acid sites of medium strength—the OH-group located on tetrahedral embedded aluminum atoms in the structure MCM-41 [58–62]. As the unmodified MCM-41 has no aluminum in the lattice, the observed medium-strength Brønsted acid sites confirm the formation of a hybrid structure MFI-MCM-41/Al_2O_3 [63].

The centers of the peaks corresponding to the weak and strong acid sites on different samples are slightly shifted along the temperature axis (within 10–30 °C), which is most likely due to the different diffusion restrictions on different zeolite structures [53].

The total acidity of samples decreases in the series MFI-MCM-41/Al_2O_3 > MFI/Al_2O_3 > MFI-MEL/Al_2O_3 > MFI-MTW/Al_2O_3 and is 550, 495, 339, and 267 μmol NH_3/g(cat), respectively.

Numerical values of acidity for samples are presented in Table 2.

Table 2. Acidic properties of catalysts.

No	Sample	Acidity of Fresh Catalyst, μmol NH_3/g(cat)				
		Total	Weak Sites	Medium Sites	Strong Sites	
			T_1 = 170–180 °C T_2 = 215–230 °C	T_3 = 290 °C	T_4 = 350–380 °C	T_5 = 570 °C
1	MFI-MEL/Al_2O_3	339	200 (59.0%)	-	139 (41.0%)	-
2	MFI-MTW/Al_2O_3	266	132 (49.6%)	-	134 (50.4%)	-
3	MFI-MCM-41/Al_2O_3	550	223 (40.5%)	79 (14.4 %)	219 (39.8%)	29 (5.3%)
4	MFI/Al_2O_3	495	245 (49.5%)	-	220 (44.4%)	30 (6.0%)

2.2. DME Conversion to Olefin

The dependence of DME conversion on the specified contact time for the samples studied is shown in Figure 3A. For all catalysts, the DME conversion increases with an increase in specified contact time. The resulting pattern of dependences can be described by an S-shaped curve, which reflects the autocatalytic nature of the MTO reaction [25,64].

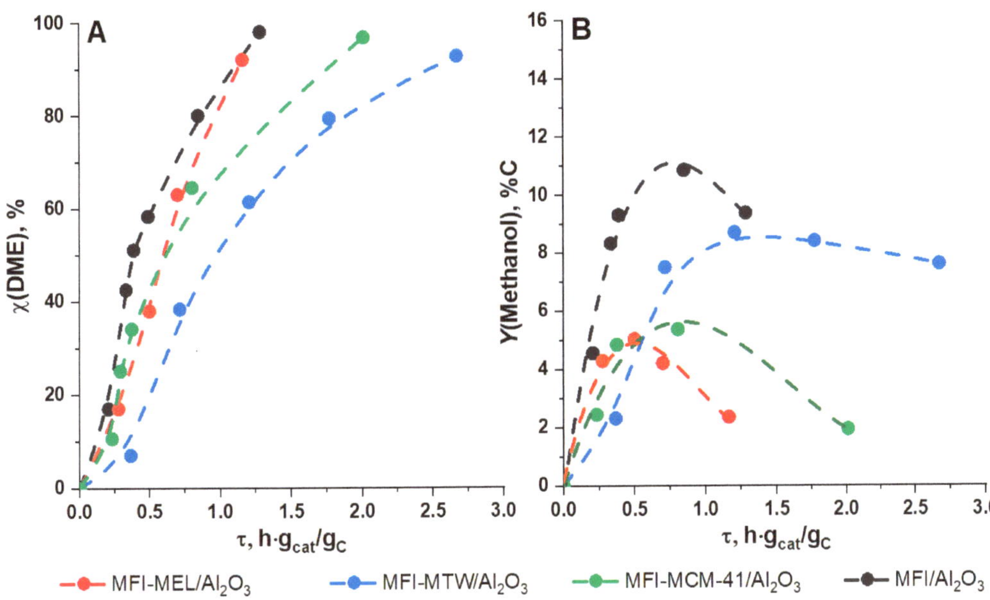

Figure 3. DME conversion (**A**) and methanol yield (**B**) as functions of specified contact time. p = 1 atm, T = 340 °C.

The activity of the studied catalysts decreases in the series MFI/Al_2O_3 > MFI-MEL/Al_2O_3 > MFI-MTW/Al_2O_3, which correlates with the total acidity of these samples: 495 > 339 > 266 μmol NH_3/g(cat).

At the same time, the sample based on the zeolite MFI-MCM-41/Al_2O_3, which is characterized by the highest total acidity of 550 μmol NH_3/g(cat), stands out from this sequence. The dependence of the DME conversion on the specified contact time for the MFI-MCM-41/Al_2O_3 catalyst lies below the analogous curve for the MFI/Al_2O_3 standard sample. Furthermore, on the MFI-MCM-41/Al_2O_3 sample, there is a significant decrease in the yield of methanol compared to the standard MFI/Al_2O_3 sample (Figure 3B). This is due to the fact that medium-strength acid sites (in the amount of 79 μmol NH_3/g(cat)) of MCM-41 do not participate in the reaction of hydrocarbon synthesis from oxygenates, but activate the dehydration of methanol with the formation of DME, as was shown in [65,66]. Therefore, for the sample based on the MFI-MCM-41 zeolite, not only are hydrocarbons formed due to the structure of the MFI, but the additional formation of the DME from methanol also occurs due to the presence of MCM-41.

On the MFI-MEL/Al_2O_3 catalyst, at low specified contact times (up to 0.25 h·g(cat)/g(C)), the DME conversion is lower than that of the standard MFI/Al_2O_3 sample. At specified contact times of more than 0.5 h·g(cat)/g(C), the DME conversion increases sharply and, at 1.2 h·g(cat)/g(C), it is close to DME conversion on MFI/Al_2O_3.

The MFI-MEL/Al_2O_3 sample is characterized by the largest micropore volume (0.088 cm^3/g). Micropores of zeolite are associated with the formation and stabilization of hydrocarbon pool species such as aromatic and polymethylcyclopentyl cations. The rate of this stage is low; therefore, an induction period is observed on the DME conversion curve at small specified contact times. However, when the hydrocarbon pool is already formed, it can be seen that DME conversion on MFI-MEL/Al_2O_3 increases sharply.

The interrelation between the micropores volume of the catalyst and its catalytic properties can be seen in the dependence of the methanol yield on the DME conversion. The maximum yield of methanol decreases in the series MFI-MTW/Al_2O_3 > MFI-MCM-41/Al_2O_3 > MFI-MEL/Al_2O_3 (Figure 3B). The minimum yield of methanol (5.0% C) is observed on MFI-MEL/Al_2O_3, while the maximum yield (8.5% C) is observed on MFI-MTW/Al_2O_3. In terms of catalyst micropore volume, the samples are in the inverse sequence MFI-MTW/Al_2O_3 < MFI-MCM-41/Al_2O_3 < MFI-MEL/Al_2O_3: 0.012 < 0.040 < 0.088 cm^3/g(cat).

The greater the number of micropores, the greater the number of diffusion restrictions. In the case of successive reactions, diffusion restrictions lead to a longer contact of intermediate products with the inner surface of the catalyst, which leads to more active secondary reactions. Methanol is an intermediate product. The priority pathway for its formation is the methylation of olefins and aromatics by DME [25,45]. Subsequently, methanol itself reacts as a methylating agent and is consumed. Therefore, the dependence of the methanol yield on the specified contact time passes through a maximum, then decreases. When methanol is formed in the system, it competes with DME in the diffusion rate into zeolite micropores and the participation in methylation reactions. The more micropores in the catalyst, the greater the methanol participation in methylation reactions due to smaller molecular size. That leads to the lower methanol yield observed in the gas phase.

The hypothesis about the determining role of micropores in the hydrocarbon pool formation and in the rate of products diffusion is supported by the graphical dependences of the products distribution (% mol) on DME conversion (Figure 4). In Figure 4, product selectivities are located one above the other along the *Y*-axis, and their sum is 100%. The product composition at a specified DME conversion is determined by the width of the bands corresponding to a specific component. For example, for a standard MFI/Al_2O_3 catalyst at a DME conversion of 43%, the selectivity for the formation of ethylene, propylene, butenes, and methanol is 9.3, 17.9, 7.7, and 48.5% mol, respectively.

The hybrid zeolite-based MFI-MEL/Al_2O_3 catalyst is characterized by the highest ethylene and propylene selectivity over the entire DME conversion range among the studied samples. The selectivity for ethylene and propylene is 19.4 and 24.2% mol at 50% DME conversion, and 26.8 and 22.2% mol at 95% DME conversion (Figure 4A), respectively. The ratio of ethylene/propylene ranges from 0.7 to 1.2 (Figure 5).

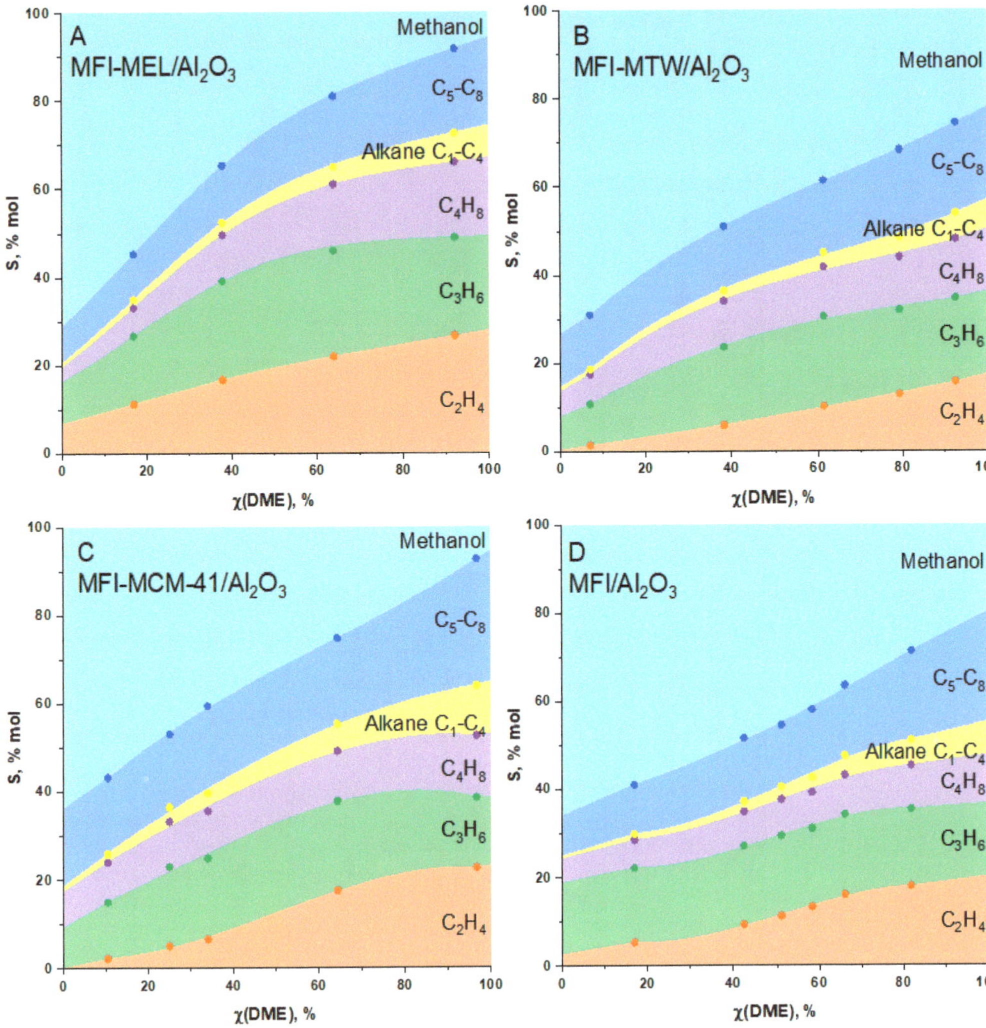

Figure 4. Dependence of the product selectivity (C_2-C_4 olefins, C_1-C_4 alkanes, C_5-C_8 hydrocarbons, and methanol) on DME conversion on catalysts (**A**) MFI-MEL/Al_2O_3, (**B**) MFI-MTW/Al_2O_3, (**C**) MFI-MCM-41/Al_2O_3, (**D**) MFI/Al_2O_3. $p = 1$ atm, T = 340 °C.

On the catalyst based on hybrid zeolite MFI-MTW/Al_2O_3, the selectivity for ethylene and propylene is 8.3 and 19.4% at 50% DME conversion and 16.4 and 18.8% at 95% DME conversion, respectively. In the range of low DME conversion values, the selectivity of ethylene is close to zero (Figure 4B). The ethylene/propylene ratio ranges from 0.15 to 0.75 (Figure 5).

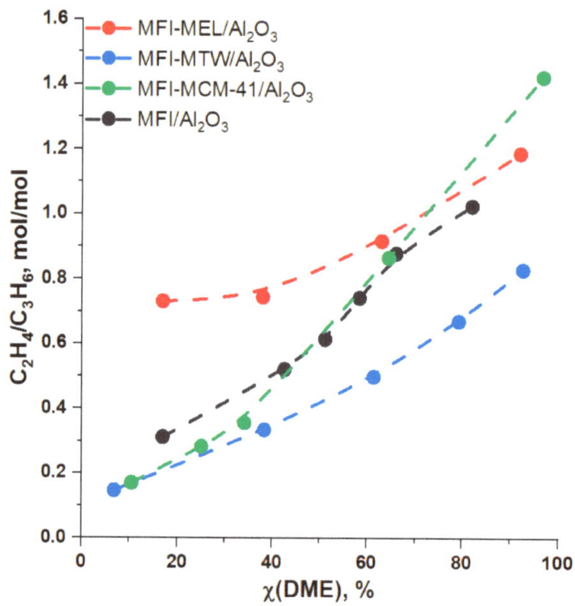

Figure 5. Ethylene/Propylene mole ratio as a function of DME conversion. p = 1 atm, T = 340 °C.

The MFI-MCM-41/Al_2O_3 catalyst is characterized by the product distribution, which can be defined as "average" between the two samples MFI-MEL/Al_2O_3 and MFI/Al_2O_3. Like MFI-MEL/Al_2O_3, it has a high ethylene and propylene selectivity—12.5 and 20.5% at 50% DME conversion, and 23.0 and 16.4% at 95% DME conversion (Figure 4C), respectively. At the same time, the selectivities of C_5–C_8 hydrocarbons and C_1–C_4 alkanes on the MFI-MCM-41/Al_2O_3 are the highest: up to 30% mol. and up to 12% mol, respectively. An additional feature of this catalyst is a sharp increase in the ethylene/propylene ratio from 0.18 to 1.4 with an increase in DME conversion. The ratio increases due to an increase in the ethylene selectivity from 1.7 to 12.3% mol, while the propylene selectivity remains approximately constant at 15–20% mol over the entire range of DME conversions (Figure 5).

As mentioned above, the MFI-MEL/Al_2O_3 catalyst has the highest micropore volume and the highest ethylene/propylene ratio. On the other hand, the MFI-MTW/Al_2O_3 catalyst has almost no micropores and has the lowest ethylene/propylene ratio.

The increase in the yield of ethylene on MFI-MEL/Al_2O_3 can be related to the low rate of diffusion of C_5–C_8 hydrocarbons from zeolite micropores. According to the dual-cycle mechanism, aromatic polymethyl-substituted intermediates are formed from C_5–C_8 hydrocarbons. Methylation and dealkylation of this type of hydrocarbon pool species lead to the formation of ethylene [12,23,67]. Thus, diffusion difficulties of C_5–C_8 hydrocarbons in zeolite micropores determine the yield of ethylene.

In addition to hydrocarbon selectivity, isomerizing activity and activity in hydrogen transfer reactions are important characteristics for zeolite catalysts.

The comparison of the hydrogen transfer index in C_2–C_5 hydrocarbons for the studied samples is shown in Figure 6. The hydrogen transfer index HTI is calculated according to the standard method as the ratio of the number of alkanes with n carbon atoms to the total number of alkanes and alkenes with n carbon atoms [68,69].

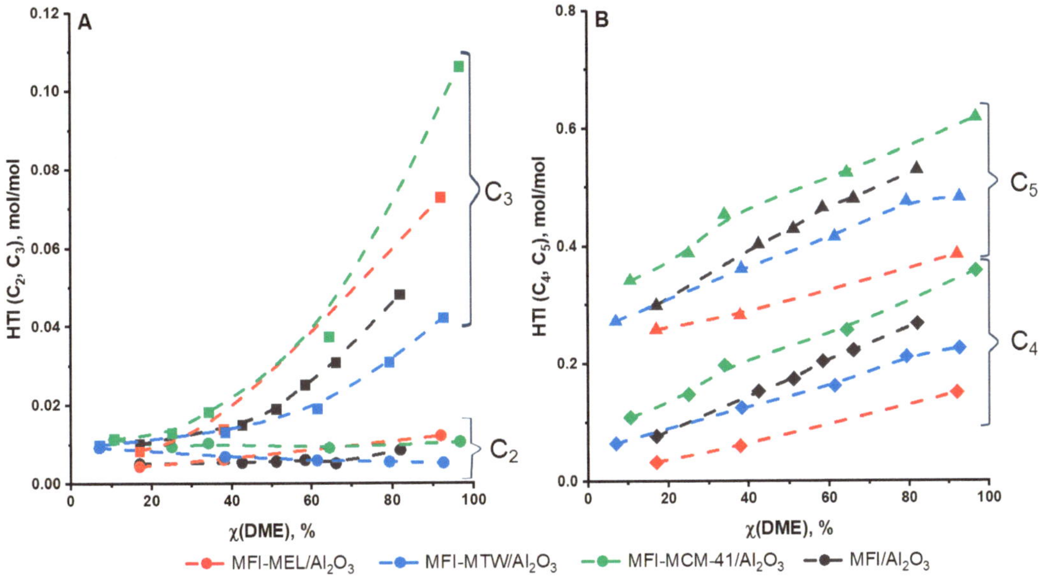

Figure 6. Hydrogen transfer index (HTI) for hydrocarbons with chain length (**A**) 2 (•), 3 (■), (**B**) 4 (◆), and 5 (▲) for MF-MEL/Al_2O_3, MFI-MTW/Al_2O_3, MFI-MCM-41/Al_2O_3, MFI/Al_2O_3 catalysts, respectively. p = 1 atm, T = 340 °C.

The hydrogen transfer index increases in the series $C_2 < C_3 < C_4 < C_5$ as shown in Figure 6. The HT-index for C_2 hydrocarbons is low (less than 0.012), and it almost does not depend on the DME conversion for all studied samples. The dependence of the HT-index for C_3 hydrocarbons on DME conversion can be approximated by an exponential function and for C_4–C_5 hydrocarbons by a linear one. This means that for C_3–C_5 hydrocarbons, the rate of the H-transfer reaction increases with an increase in DME conversion. Consequently, the contribution of hydrogen transfer reactions to the distribution of products also increases—the yield of alkanes rises.

In comparison to other samples, the MFI-MEL/Al_2O_3 is characterized by a high HT-index for C_3 hydrocarbons (0.075 at a DME conversion of 95%) and minimum HT-index for C_4 and C_5 hydrocarbons (0.15 and 0.38 at a DME conversion of 95%).

The maximum HT-index for C_3-C_5 hydrocarbons is observed for MFI-MCM-41/Al_2O_3. This catalyst has a large mesopore volume (0.300 cm^3/g) and it is not a catalyst with a low activity as MFI-MTW/Al_2O_3 is. As the hydrogen transfer reaction proceeds between the lower and higher olefins, the coordination of the reactants requires space, which is apparently provided by the mesopores of the hybrid structure.

To estimate the isomerizing activity of the catalysts, the dependences of the iso-/n-alkenes mole ratio on DME conversion were plotted (Figure 7).

It can be seen that the ratio of iso-/n-isomers does not depend on the DME conversion, and all the studied samples have the same isomerizing activity. Olefins with a longer carbon chain are more actively involved in isomerization: the iso-/n-ratio for C_4H_8 olefins is 0.08, and for C_5H_{10} olefins, it is 0.38.

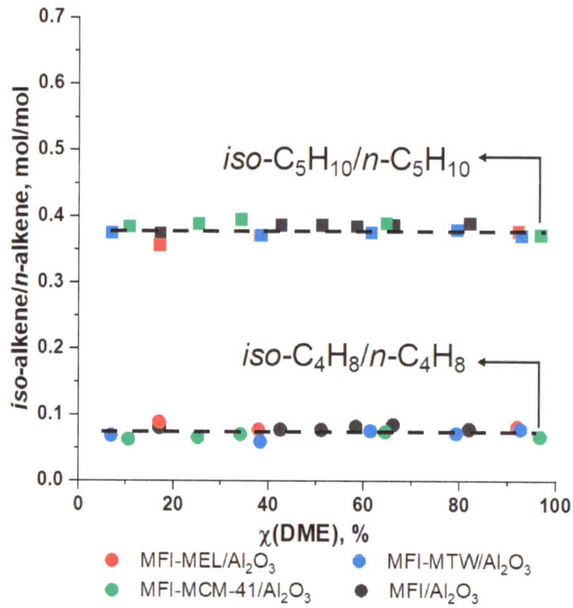

Figure 7. Iso-/n-olefins mole ratio as a function of DME conversion for MF-MEL/Al_2O_3, MFI-MTW/Al_2O_3, MFI-MCM-41/Al_2O_3, MFI/Al_2O_3 catalysts. p = 1 atm, T = 340 °C.

3. Materials and Methods

3.1. Catalyst Synthesis

Hybrid zeolites MFI-MEL, MFI-MTW, and MFI-MCM-41 were synthesized according to the method described in patents [50,70].

The industrial zeolite IK-17-1 (Novosibirsk Chemical Concentrates Plant (NCCP), Novosibirsk, Russia) was used as the MFI zeolite, which is the Russian analog of the ZSM-5 zeolite (Zeolyst, Conshohocken, PA, USA).

3.1.1. MFI-MEL Synthesis

Amounts of 3.73 g of aluminum decahydrate sulfate and 10.1 g of N^1,N^{10}-bis(2-hydroxyethyl)-N^1,N^1,N^{10},N^{10}-tetramethyldecane-1,10-diammonium bromide are dissolved in 12.6 g of distilled water (Solution 1).

Then, with stirring, 1.5 g of sodium hydroxide was added to Solution 1—the solution was stirred until the solid reagents were dissolved (Solution 2).

In a separate container, 25.2 g of a 40% (wt.) colloidal solution of silicon dioxide brand LUDOX HS-40 and 10.1 g of water were mixed until homogeneous (Solution 3).

Solution 2 was added dropwise to Solution 3 and mixed until a homogeneous gel-like mass was formed. For crystallization, the resulting gel was placed in a Teflon cup of the autoclave, sealed, and thermostated at 160 °C for 3 days. The resulting product was filtered off under reduced pressure (101.325 kPa) on a glass porous filter and washed with distilled water until the pH of the filtrate reached 9.0. The sample was transferred to a Petri dish and dried in an oven at 100 °C for 12 h; calcined in a muffle furnace, starting from room temperature, then with a heating step of 1 °C/min up to 520 °C, then at this temperature for 6 h. Ion exchange to obtain the NH_4 form of the zeolite was carried out with a 1.1 M aqueous solution of ammonium chloride for 15 h. The mixture was filtered under reduced pressure (98.5 kPa) on a porous glass filter, washed with distilled water until the pH of the filtrate reached 8.0, and dried in an oven at 100 °C for 12 h; calcined in a muffle furnace, starting from room temperature, then with a heating step of 1 °C/min up to 460 °C, then at this temperature for 4 h to obtain the H-form of the zeolite.

3.1.2. MFI-MTW Synthesis

Hybrid-structure zeolite MFI-MTW was prepared similarly to MFI-MEL with the difference that Solution 1 was prepared as follows: 3.73 g of aluminum decahydrate sulfate and 10.1 g of 8.95 g of N^1,N^6-bis(2) bromide-hydroxyethyl)-N^1,N^1,N^6,N^6-tetramethylhexane-1,6-diammonium were dissolved in 12.6 g of distilled water.

3.1.3. MFI-MCM-41 Synthesis

Amounts of 0.69 g of sodium hydroxide, 5.8 mL of a 1 M solution of tetrapropylammonium hydroxide, and 35.5 mL of water were stirred until all components were completely dissolved. An amount of 6.9 g of pyrogenic silicon dioxide was added to the resulting solution for an hour in portions with continued stirring, and the mixture was stirred at room temperature for another 1 h; then, the mixture was placed in a Teflon cup (Solution 1).

In a separate Teflon cup, 2.20 g of hexadecyltrimethylammonium bromide, 0.44 g of sodium hydroxide, and 43.3 mL of water were mixed until complete dissolution of the template. An amount of 0.51 g of sodium aluminate was added to the resulting solution and mixed. An amount of 4.92 g of silicon dioxide was added to the resulting solution in portions with stirring for one hour. The mixture was stirred for three hours at room temperature until complete homogenization (Solution 2).

Both Solution 1 and Solution 2 were placed in an autoclave and thermostated at 100 °C for 16 h. An amount of 6 g of Solution 1 was added to Solution 2 in a Teflon cup. The resulting mixture was stirred on a mechanical stirrer until homogeneous for one hour. For crystallization, the resulting mixture was placed in the autoclave, sealed, and thermostated at 180 °C for 3 days. The resulting product was filtered off under reduced pressure (101.3 kPa) on a glass porous filter and washed with distilled water until the pH of the filtrate reached 9.0. The sample was transferred to a Petri dish and dried in an oven at 90 °C for 6 h; calcined in a muffle furnace, starting from room temperature, then with a heating step of 1 °C/min up to 550 °C, then at this temperature for 6 h. Ion exchange to obtain the NH_4 form of the zeolite was carried out with a 1.1 M aqueous solution of ammonium chloride at 85 °C for 2 h. The mixture was filtered under reduced pressure (98.5 kPa) on a porous glass filter, washed with distilled water until the pH of the filtrate reached 8.0, and dried in an oven at 90 °C for 6 h; calcined in a muffle furnace, starting from room temperature, then with a heating step of 1 °C/min up to 550 °C, then at this temperature for 5 h to obtain the H-form of the zeolite.

3.1.4. Catalyst Preparation

An amount of 42.7 g of AlO(OH) was placed in a mixer and 20 mL of a peptizing solution (5 mL of 1.0 M aqueous HNO_3 + 15 mL of H_2O) was added. The mixture was stirred for 5 min. An amount of 76.5 g of dry zeolite was added. The mass was stirred for 20 min at 60 °C. After that, the catalytic mass was passed through an extruder with a die diameter of 2.5 mm. The extrudates were dried in an oven at 80 °C, 90 °C, 100 °C, 110 °C, and 120 °C for 3 h at each temperature, then calcined in a muffle furnace at 550 °C for 7 h. During calcination, AlO(OH) was converted to Al_2O_3. The final content of Al_2O_3 in the material obtained was 30% wt.

3.1.5. Hydrothermal Treatment

The MFI/Al_2O_3 standard sample was hydrothermally treated before catalytic experiments at 500 °C for 6 h at WHSV of steam 12 h^{-1}. The detailed description of the catalyst preparation method can be found in the patents [45,71,72].

3.2. Catalyst Characterization

X-ray powder diffraction (XRD) patterns of samples were recorded on a Rigaku Rotaflex RU-200 diffractometer with Cu Kα radiation (1.5418 Å, 50 kV, and 160 mA). The zeolites phase ratio was determined by the Reference Intensity Ratio (RIR) method using the MDI Jade 6 program [73].

The textural characteristics of the catalyst samples were studied by N_2 adsorption/desorption isotherms at −196 °C on a Belsorp mini X instrument (MICROTRAC MRB, Osaka, Japan). Prior to measurement, catalyst samples were degassed under a high vacuum at 350 °C and 1.36 10^{-6} atm for 3–6 h. The total specific surface area of the catalyst was determined by the BET method (Brunauer–Emmett–Teller). The total pore volume was determined according to the amount of adsorbed nitrogen at a relative pressure of p/p_0 = 0.95. The mesopore volume was determined as the difference between total pore volume and the micropore volume (t-plot). The micropore surface area was calculated as the difference of the total surface area (BET) and sum of the external surface area (t-plot) and the mesopore area (BJH method (Barrett–Joyner–Halenda), desorption curve). The average mesopore diameter was determined by the BJH method and the average micropore diameter was determined by the MP-plot.

The amount, strength, and distribution of acid sites were determined by NH_3-TPD on a chemical adsorption analyzer USGA (Moscow, Russia) with a thermal conductivity detector. All samples (0.15 g) before measurements were pretreated by heating in a helium flow at 500 °C (heating rate of 20 °C/min), calcination at 500 °C for 1 h, and cooling to 60 °C. Then, the samples were saturated with NH_3 for 15 min, flushed with helium at 100 °C for 1 h to remove physically adsorbed NH_3, and cooled to 60 °C. The final desorption of NH_3 was performed at 60–750 °C at a heating rate of 8 °C/min in a helium flow. The released ammonia was recorded using a thermal conductivity detector.

X-ray fluorescence elemental (XRF) analysis was conducted with the Thermo ARL Perform'x Sequential XFR instrument (Thermo Fisher Scientific, Waltham, MA, USA) using a 2500 W X-ray tube. Before the analysis, the samples weighing 200 mg were pressed into a tablet with boric acid.

Scanning electron microscopy (SEM). The micrographs of the samples were taken on a Hitachi (Chiyoda City, Tokyo, Japan) TM3030 desktop scanning electron microscope.

The structure and surface morphology of the synthesized samples were studied using transmission electron microscopy (TEM) on an LEO AB OMEGA instrument with the magnification from 80 to 500,000 and the image resolution of 0.2–0.34 nm.

Nuclear magnetic resonance spectra were recorded at a rotation magnetic angle on a Bruker AVANCE-II 400 WB spectrometer with a magnetic field of 9.4 T, which corresponds to operating frequencies of $\nu(^1H)$ = 400.13 MHz and $\nu(^{27}Al)$ = 104.2 MHz, using a MAS (magic-angle spinning), with the diameter of the MAS-rotor of 4 mm and the rotation frequency of 12,000 Hz. Before registration, air-dry samples were kept in a desiccator with a 25% aqueous ammonia solution for 24 h at room temperature. Spectra on ^{27}Al nuclei were recorded using a single-pulse technique (15° pulse) with the following parameters: the exciting 15° pulse width was 0.8 μs, the number of scans was 1024, and the interval between scans was 0.5 s. An 1 M $Al(NO_3)_3 \cdot H_2O$ aqueous solution was utilized as an external reference (0 ppm).

The concentration of acid sites in the samples was determined by the IR spectroscopy of adsorbed pyridine. IR spectra were recorded on a Nicolet Protégé 460 instrument with an optical resolution of 4 cm^{-1} and a range of 4000–400 cm^{-1}. Samples in the form of disks (diameter 1.6 cm, density ~10 mg cm^{-2}) were activated in an IR cell at 400 °C (heating rate of 7.5 deg min^{-1}) for 2 h at a pressure of 10^{-5} Torr. The adsorption of probe molecules was carried out at 150 °C and a pressure of 2 Torr of pyridine for 30 min. At the end of the adsorption cycle, pyridine was desorbed at 150 °C for 15 min. The concentrations of Bronsted acid sites (BAS) and Lewis acid sites (LAS) were determined from the intensity of the adsorbed pyridine bands (1545 and 1450 cm^{-1}, respectively); the molar extinction coefficients were ε(BAS) = 1.67 cm·μmol^{-1} and ε(LAS) = 2.22 cm·μmol^{-1}.

3.3. Catalytic Tests

Catalytic experiments were performed in a fixed-bed continuous flow quartz reactor with an inner diameter of 10 mm at 1 atm and 340 °C. Here, the catalyst, weighing 0.5 g (d_p = 0.4–0.63 mm), was mixed with quartz (d_p = 0.5–1.0 mm) with the volume ratio of 1:1. Prior to the experiment, the catalyst was preheated and purged with nitrogen flow at 400 °C for 1 h to remove physically adsorbed water. The dimethyl ether/nitrogen mixture with a concentration of DME of 10–13% vol was used as a feedstock. The GHSV was varied in the range of 2500–25,000 h^{-1}.

The specified contact time per carbon for DME (τ) varied from 0.2–3.0 g(cat)/g(C)·h, corresponding to DME conversion in the range of 5–95%. The results averaged over three parallel measurements with a relative error of 5–7% were used for the calculation of catalytic activity.

The reaction products were analyzed using an on-line gas chromatograph (Crystallux-4000 M) equipped with a flame ionization detector (FID) and a thermal conductivity detector (TCD). The capillary column with the CP-Poraplot Q phase (27.5 m × 0.32 mm × 10 μm) was used to determine the composition of hydrocarbons C_1-C_8, methanol, and DME. The packed column with a Porapak Q phase (3.0 m × 4 mm × 3 μm) was used to determine N_2. The analysis was performed at the programmed temperature increase from 90 to 250 °C with a heating rate of 30 °C/min, and the carrier gas was helium. Chromatograms were processed using the NetChromWin software (Version 2.0, JSC Scientific Production Company Meta-Chrome, Yoshkar-Ola, Russia, 2017) and hardware systems.

The reaction products observed were methanol, alkenes C_2–C_4, alkanes C_1–C_4, and hydrocarbons C_5–C_8 (alkanes, cyclo-, and aromatic compounds).

The DME conversion (1) and the selectivity of products (2) were used as the main indicators for characterizing the processes. The specified contact time was calculated by Equation (3)

$$X_{\mathrm{DME}} = \frac{n_{in}(\mathrm{DME}) - n_{out}(\mathrm{DME})}{n_{in}(\mathrm{DME})}, \ \% \tag{1}$$

$$S_i = \frac{n_{Ci}}{\sum_i n_{Ci}}, \ \% \text{ mol.} \tag{2}$$

$$\tau = \frac{m_{cat}}{m_{in,\,C}(\mathrm{DME})}, \ \frac{g_{cat} \times h}{g_C} \tag{3}$$

where $n_{in}(\mathrm{DME})$ and $n_{out}(\mathrm{DME})$ are moles of DME at the reactor inlet and outlet, respectively, n_{Ci} is mole of the *i*-th carbon-containing compound in products at the reactor outlet, mole; $\sum_i n_{Ci}$ is the sum of moles of all products at the reactor outlet, mole; m_{cat} is mass of the catalyst, g; $m_{in,C}(\mathrm{DME})$ is mass flow of carbon in the DME at the reactor inlet, g_C/h.

4. Conclusions

The catalysts based on hybrid zeolites MFI-MEL, MFI-MTW, and MFI-MCM-41 have been studied in the reaction of lower olefins synthesis from dimethyl ether. The texture-acid properties of the samples have been studied, and their influence on the activity and distribution of the reaction products has been established.

It has been shown that the activity of the zeolite catalyst is determined by the total acidity of weak (ammonia desorption temperature T = 170–230 °C) and strong (T = 350–380 °C) active sites, while the participation of medium-strength acid sites (T = 290 °C) in the reaction of hydrocarbon formation from oxygenates is unlikely.

It has been shown that the volume of micropores in the catalyst is responsible for the ethylene selectivity and the ethylene/propylene ratio in the reaction products. An increase in the volume of micropores contributes to an increase in the ethylene selectivity due to the stabilization of aromatic hydrocarbon pool species.

It has been shown that the topology of the hybrid zeolite affects the rate of hydrogen transfer reactions but does not affect the isomerizing activity of the catalyst.

Supplementary Materials: The following supporting information can be downloaded at: https://www.mdpi.com/article/10.3390/catal13030570/s1, Table S1: Composition of hybrid zeolites; Figure S1: ^{27}Al MAS NMR spectra of hybrid zeolites; Figure S2: IR spectra of pyridine adsorbed on hybrid zeolites; Table S2: Acid characteristics of hybrid zeolites; Figure S3: The micrographs of SEM- and TEM-hybrid zeolite; Table S3: Additional textural properties of the catalysts, Figure S4: BJH plots (desorption curve) for the catalysts, Figure S5: t-Plots for nitrogen adsorbed in the catalysts.

Author Contributions: Conceptualization, M.V.M. and A.V.S.; methodology, I.A.D. and D.E.T.; formal analysis, A.V.S.; resources, A.L.M.; writing—original draft preparation, A.V.S.; writing—review and editing, M.V.M.; visualization, A.V.S.; supervision, M.V.M.; project administration, A.L.M. All authors have read and agreed to the published version of the manuscript.

Funding: This work was financially supported by the Russian Science Foundation (grant No. 17-73-30046).

Data Availability Statement: The data presented in this study are available on request from the corresponding author.

Acknowledgments: The authors are grateful to Andrey Smirnov (M.V. Lomonosov Moscow State University), Vladislav Malyavin (A.V. Topchiev Institute of Petrochemical Synthesis, RAS), Ivan Levin (A.V. Topchiev Institute of Petrochemical Synthesis, RAS), and Sergey Sorokin (A.V. Topchiev Institute of Petrochemical Synthesis, RAS) for their help in physicochemical studies of catalysts and interpretation of the results. The authors are thankful to Svetlana Bychkova for her help in English editing.

Conflicts of Interest: The authors declare no conflict of interest.

References

1. Tian, P.; Wei, Y.; Ye, M.; Liu, Z. Methanol to olefins (MTO): From fundamentals to commercialization. *ACS Catal.* **2015**, *5*, 1922–1938. [CrossRef]
2. Yang, M.; Fan, D.; Wei, Y.; Tian, P.; Liu, Z. Recent Progress in Methanol-to-Olefins (MTO) Catalysts. *Adv. Mater.* **2019**, *31*, 1902181. [CrossRef]
3. Chang, C.D.; Silvestri, A.J.; Smith, R.L. (MTG) Production of Gasoline Hydrocarbons. U.S. Patent 3,928,483, 23 December 1975.
4. Chang, C.D.; Lang, W.H. Process for Manufacturing Olefins. U.S. Patent 566,166, 24 May 1977.
5. Li, J.; Wei, Z.; Chen, Y.; Jing, B.; He, Y.; Dong, M.; Jiao, H.; Li, X.; Qin, Z.; Wang, J.; et al. A route to form initial hydrocarbon pool species in methanol conversion to olefins over zeolites. *J. Catal.* **2014**, *317*, 277–283. [CrossRef]
6. Yarulina, I.; Chowdhury, A.D.; Meirer, F.; Weckhuysen, B.M.; Gascon, J. Recent trends and fundamental insights in the methanol-to-hydrocarbons process. *Nat. Catal.* **2018**, *1*, 398–411. [CrossRef]
7. Liu, Y.; Müller, S.; Berger, D.; Jelic, J.; Reuter, K.; Tonigold, M.; Sanchez-Sanchez, M.; Lercher, J.A. Formation mechanism of the first carbon–carbon bond and the first olefin in the methanol conversion into hydrocarbons. *Angew. Chem. Int. Ed.* **2016**, *55*, 5723–5726. [CrossRef]
8. Wu, X.; Xu, S.; Zhang, W.; Huang, J.; Li, J.; Yu, B.; Wei, Y.; Liu, Z. Direct mechanism of the first carbon–carbon bond formation in the methanol-to-hydrocarbons process. *Angew. Chem. Int. Ed.* **2017**, *56*, 9039–9043. [CrossRef]
9. Salehirad, F.; Anderson, M.W. Solid-state 13C MAS NMR study of methanol-to-hydrocarbon chemistry over H-SAPO-34. *J. Catal.* **1996**, *164*, 301–314. [CrossRef]
10. Ivanova, I.I.; Pomakhina, E.B.; Rebrov, A.I.; Hunger, M.; Kolyagin, Y.G.; Weitkamp, J. Surface species formed during aniline methylation on zeolite H-Y investigated by in situ MAS NMR spectroscopy. *J. Catal.* **2001**, *203*, 375–381. [CrossRef]
11. Van Der Mynsbrugge, J.; Moors, S.L.C.; De Wispelaere, K.; Van Speybroeck, V. Insight into the formation and reactivity of framework-bound methoxide species in h-zsm-5 from static and dynamic molecular simulations. *ChemCatChem* **2014**, *6*, 1906–1918. [CrossRef]
12. Svelle, S.; Joensen, F.; Nerlov, J.; Olsbye, U.; Lillerud, K.P.; Kolboe, S.; Bjørgen, M. Conversion of methanol into hydrocarbons over zeolite H-ZSM-5: Ethene formation is mechanistically separated from the formation of higher alkenes. *J. Am. Chem. Soc.* **2006**, *128*, 14770–14771. [CrossRef] [PubMed]
13. Bjørgen, M.; Svelle, S.; Joensen, F.; Nerlov, J.; Kolboe, S.; Bonino, F.; Palumbo, L.; Bordiga, S.; Olsbye, U. Conversion of methanol to hydrocarbons over zeolite H-ZSM-5: On the origin of the olefinic species. *J. Catal.* **2007**, *249*, 195–207. [CrossRef]
14. Bjørgen, M.; Lillerud, K.P.; Olsbye, U.; Svelle, S. Conversion of methanol to hydrocarbons: Hints to rational catalyst design from fundamental mechanistic studies on H-ZSM-5. *Stud. Surf. Sci. Catal.* **2007**, *167*, 463–468. [CrossRef]
15. Olsbye, U.; Svelle, S.; Bjrgen, M.; Beato, P.; Janssens, T.V.W.; Joensen, F.; Bordiga, S.; Lillerud, K.P. Conversion of methanol to hydrocarbons: How zeolite cavity and pore size controls product selectivity. *Angew. Chem. Int. Ed.* **2012**, *51*, 5810–5831. [CrossRef]
16. Lu, C.; Yang, H.; Wang, J.; Tan, Q.; Fu, L. Utilization of iron tailings to prepare high-surface area mesoporous silica materials. *Sci. Total Environ.* **2020**, *736*, 139483. [CrossRef]
17. Hunger, M.; Horvath, T. Adsorption of methanol on Bronsted acid sites in zeolite H-ZSM-5 investigated by multinuclear solid-state NMR spectroscopy. *J. Am. Chem. Soc.* **1996**, *118*, 12302–12308. [CrossRef]

18. Seiler, M.; Schenk, U.; Hunger, M. Conversion of methanol to hydrocarbons on zeolite HZSM-5 investigated by in situ MAS NMR spectroscopy under flow conditions and on-line gas chromatography. *Catal. Lett.* **1999**, *62*, 139–145. [CrossRef]
19. Seiler, M.; Wang, W.; Hunger, M. Local structure of framework aluminum in zeolite H-ZSM-5 during conversion of methanol investigated by in situ NMR spectroscopy. *J. Phys. Chem. B* **2001**, *105*, 8143–8148. [CrossRef]
20. Hunger, M.; Seiler, M.; Buchholz, A. In situ MAS NMR spectroscopic investigation of the conversion of methanol to olefins on silicoaluminophosphates SAPO-34 and SAPO-18 under continuous flow conditions. *Catal. Lett.* **2001**, *74*, 61–68. [CrossRef]
21. Wang, W.; Seiler, M.; Hunger, M. Role of surface methoxy species in the conversion of methanol to dimethyl ether on acidic zeolites investigated by in situ stopped-flow MAS NMR spectroscopy. *J. Phys. Chem. B* **2001**, *105*, 12553–12558. [CrossRef]
22. Wang, C.; Sun, X.; Xu, J.; Qi, G.; Wang, W.; Zhao, X.; Li, W.; Wang, Q.; Deng, F. Impact of temporal and spatial distribution of hydrocarbon pool on methanol conversion over H-ZSM-5. *J. Catal.* **2017**, *354*, 138–151. [CrossRef]
23. Yarulina, I.; De Wispelaere, K.; Bailleul, S.; Goetze, J.; Radersma, M.; Abou-Hamad, E.; Vollmer, I.; Goesten, M.; Mezari, B.; Hensen, E.J.M.; et al. Structure–performance descriptors and the role of Lewis acidity in the methanol-to-propylene process. *Nat. Chem.* **2018**, *10*, 804–812. [CrossRef]
24. Liu, C.; Uslamin, E.A.; Pidko, E.A.; Kapteijn, F. Direct discerning reaction pathways in methanol-to-hydrocarbons by transient operation–FASPA. *Chem. Eng. J.* **2023**, *453*, 139696. [CrossRef]
25. Magomedova, M.; Starozhitskaya, A.; Davidov, I.; Maximov, A.; Kravtsov, M. Dual-cycle mechanism based kinetic model for dme-to-olefin synthesis on hzsm-5-type catalysts. *Catalysts* **2021**, *11*, 1459. [CrossRef]
26. Dahl, I.M.; Kolboe, S. On the reaction mechanism for hydrocarbon formation from methanol over SAPO-34. I. Isotopic labeling studies of the co-reaction of ethene and methanol. *J. Catal.* **1994**, *149*, 458–464. [CrossRef]
27. Xiao, J.; Wei, J. Diffusion of Hydrocarbons Theory. *Chem. Eng. Sci.* **1992**, *47*, 1123–1141. [CrossRef]
28. Xiao, J.; Wei, J. Diffusion mechanism of hydrocarbons in zeolites-II. Analysis of experimental observations. *Chem. Eng. Sci.* **1992**, *47*, 1143–1159. [CrossRef]
29. Sharbini Kamaluddin, H.; Gong, X.; Ma, P.; Narasimharao, K.; Dutta Chowdhury, A.; Mokhtar, M. Influence of zeolite ZSM-5 synthesis protocols and physicochemical properties in the methanol-to-olefin process. *Mater. Today Chem.* **2022**, *26*, 101061. [CrossRef]
30. Weissenberger, T.; Machoke, A.G.F.; Bauer, J.; Dotzel, R.; Casci, J.L.; Hartmann, M.; Schwieger, W. Hierarchical ZSM-5 Catalysts: The Effect of Different Intracrystalline Pore Dimensions on Catalyst Deactivation Behaviour in the MTO Reaction. *ChemCatChem* **2020**, *12*, 2461–2468. [CrossRef]
31. Bai, R.; Song, Y.; Li, Y.; Yu, J. Creating Hierarchical Pores in Zeolite Catalysts. *Trends Chem.* **2019**, *1*, 601–611. [CrossRef]
32. Chen, L.H.; Sun, M.H.; Wang, Z.; Yang, W.; Xie, Z.; Su, B.L. Hierarchically structured zeolites: From design to application. *Chem. Rev.* **2020**, *120*, 11194–11294. [CrossRef] [PubMed]
33. Weckhuysen, B.M.; Fu, D.; Maris, J.J.E.; Stanciakova, K. Unravelling Channel Structure-Diffusivity Relationships in Zeolite ZSM-5 at the Single-Molecule Level. *Angew. Chem. Int. Ed.* **2022**, *61*, e202114388. [CrossRef]
34. Magomedova, M.V.; Afokin, M.I.; Starozhitskaya, A.V.; Galanova, E.G. Pilot Test of Olefin Synthesis from Dimethyl Ether in a Synthesis Gas Atmosphere. *Ind. Eng. Chem. Res.* **2021**, *60*, 4602–4609. [CrossRef]
35. Vu, X.H.; Armbruster, U.; Martin, A. Micro/mesoporous zeolitic composites: Recent developments in synthesis and catalytic applications. *Catalysts* **2016**, *6*, 183. [CrossRef]
36. Ivanova, I.I.; Knyazeva, E.E. Micro-mesoporous materials obtained by zeolite recrystallization: Synthesis, characterization and catalytic applications. *Chem. Soc. Rev.* **2013**, *42*, 3671–3688. [CrossRef] [PubMed]
37. Schwieger, W.; Machoke, A.G.; Weissenberger, T.; Inayat, A.; Selvam, T.; Klumpp, M.; Inayat, A. Hierarchy concepts: Classification and preparation strategies for zeolite containing materials with hierarchical porosity. *Chem. Soc. Rev.* **2016**, *45*, 3353–3376. [CrossRef] [PubMed]
38. Pérez-Ramírez, J.; Christensen, C.H.; Egeblad, K.; Christensen, C.H.; Groen, J.C. Hierarchical zeolites: Enhanced utilisation of microporous crystals in catalysis by advances in materials design. *Chem. Soc. Rev.* **2008**, *37*, 2530–2542. [CrossRef]
39. Shoinkhorova, T.; Dikhtiarenko, A.; Ramirez, A.; Dutta Chowdhury, A.; Caglayan, M.; Vittenet, J.; Bendjeriou-Sedjerari, A.; Ali, O.S.; Morales-Osorio, I.; Xu, W.; et al. Shaping of ZSM-5-Based Catalysts via Spray Drying: Effect on Methanol-to-Olefins Performance. *ACS Appl. Mater. Interfaces* **2019**, *11*, 44133–44143. [CrossRef] [PubMed]
40. Miao, C.; Shang, K.; Liang, L.; Chen, S.; Ouyang, J. Efficient and Stable Ni/ZSM-5@MCM-41 Catalyst for CO2 Methanation. *ACS Sustain. Chem. Eng.* **2022**, *10*, 12771–12782. [CrossRef]
41. Millward, G.R.; Ramdas, S.; Thomas, J.M.; Barlow, M.T. Evidence for semi-regularly ordered sequences of mirror and inversion symmetry planes in ZSM-5/ZSM-11 shape-selective zeolitic catalysts. *J. Chem. Soc. Faraday Trans. 2 Mol. Chem. Phys.* **1983**, *79*, 1075–1082. [CrossRef]
42. Li, P.; Zhang, W.; Han, X.; Bao, X. Conversion of methanol to hydrocarbons over phosphorus-modified ZSM-5/ZSM-11 intergrowth zeolites. *Catal. Lett.* **2010**, *134*, 124–130. [CrossRef]
43. Kim, J.J.; Jeong, D.J.; Jung, H.S.; Hur, Y.G.; Choung, J.W.; Baik, J.H.; Park, M.J.; Chung, C.H.; Bae, J.W. Dimethyl ether conversion to hydrocarbons on the closely interconnected FER@ZSM-5 nanostructures. *Microporous Mesoporous Mater.* **2022**, *340*, 112034. [CrossRef]

44. Liu, C.; Uslamin, E.A.; van Vreeswijk, S.H.; Yarulina, I.; Ganapathy, S.; Weckhuysen, B.M.; Kapteijn, F.; Pidko, E.A. An integrated approach to the key parameters in methanol-to-olefins reaction catalyzed by MFI/MEL zeolite materials. *Chin. J. Catal.* **2022**, *43*, 1879–1893. [CrossRef]
45. Magomedova, M.; Galanova, E.; Davidov, I.; Afokin, M.; Maximov, A. Dimethyl Ether to Olefins over Modified ZSM-5 Based Catalysts Stabilized by Hydrothermal Treatment. *Catalysts* **2019**, *9*, 485. [CrossRef]
46. Bin, F.; Song, C.; Lv, G.; Song, J.; Wu, S.; Li, X. Selective catalytic reduction of nitric oxide with ammonia over zirconium-doped copper/ZSM-5 catalysts. *Appl. Catal. B Environ.* **2014**, *150–151*, 532–543. [CrossRef]
47. Kokotailo, G.T.; Chu, P.; Lawton, S.L.; Meier, W.M. Synthesis and structure of synthetic zeolite ZSM-11. *Nature* **1978**, *275*, 119–120. [CrossRef]
48. Dugkhuntod, P.; Imyen, T.; Wannapakdee, W.; Yutthalekha, T.; Salakhum, S.; Wattanakit, C. Synthesis of hierarchical ZSM-12 nanolayers for levulinic acid esterification with ethanol to ethyl levulinate. *RSC Adv.* **2019**, *9*, 18087–18097. [CrossRef]
49. Conte, M.; Xu, B.; Davies, T.E.; Bartley, J.K.; Carley, A.F.; Taylor, S.H.; Khalid, K.; Hutchings, G.J. Enhanced selectivity to propene in the methanol to hydrocarbons reaction by use of ZSM-5/11 intergrowth zeolite. *Microporous Mesoporous Mater.* **2012**, *164*, 207–213. [CrossRef]
50. Maksimov, A.L.; Magomedova, M.V.; Afokin, M.I.; Tsaplin, D.E.; Kulikov, L.A.; Ionin, D.A. Method for Producing an HZSM-Type Zeolite (Variants) and Method for Producing Aromatic Hydrocarbons of the C6-C11 Fraction. RU Patent 2753263C1, 8 December 2021.
51. Lemus-Yegres, L.J.; Such-Basáñez, I.; Román-Martínez, M.C.; de Lecea, C.S.M. Catalytic properties of a Rh-diamine complex anchored on activated carbon: Effect of different surface oxygen groups. *Appl. Catal. A Gen.* **2007**, *331*, 26–33. [CrossRef]
52. Wang, Q.; Cui, Z.-M.; Cao, C.-Y.; Song, W.-G. 0.3 Å Makes the Difference: Dramatic Changes in Methanol-to-Olefin Activities between H-ZSM-12 and H-ZSM-22 Zeolites. *J. Phys. Chem. C* **2011**, *115*, 24987–24992. [CrossRef]
53. Karge, H.G. Comparative Measurements on Acidity of Zeolites. *Stud. Surf. Sci. Catal.* **1991**, *65*, 133–156. [CrossRef]
54. Lónyi, F.; Valyon, J. On the interpretation of the NH_3-TPD patterns of. *Microporous Mesoporous Mater.* **2001**, *47*, 293–301. [CrossRef]
55. Serrano, D.P.; García, R.A.; Vicente, G.; Linares, M.; Procházková, D.; Čejka, J. Acidic and catalytic properties of hierarchical zeolites and hybrid ordered mesoporous materials assembled from MFI protozeolitic units. *J. Catal.* **2011**, *279*, 366–380. [CrossRef]
56. Malinowski, P.J.; Derzsi, M.; Mazej, Z.; Jagličić, Z.; Gaweł, B.; ŁAsocha, W.; Grochala, W. AgIIso4: A genuine sulfate of divalent silver with anomalously strong one-dimensional antiferromagnetic Interactions. *Angew. Chem. Int. Ed.* **2010**, *49*, 1683–1686. [CrossRef] [PubMed]
57. Valyon, J.; Lo, F. A TPD and IR study of the surface species formed from ammonia on zeolite H-ZSM-5, H-mordenite, and H-beta. *Thermochim. Acta* **2001**, *373*, 53–57.
58. Corma, A.; Fornés, V.; Navarro, M.T.; Pérez-Pariente, J. Acidity and stability of MCM-41 crystalline aluminosilicates. *J. Catal.* **1994**, *148*, 569–574. [CrossRef]
59. Kosslick, H.; Lischke, G.; Parlitz, B.; Storek, W.; Fricke, R. Acidity and active sites of Al-MCM-41. *Appl. Catal. A Gen.* **1999**, *184*, 49–60. [CrossRef]
60. Jentys, A.; Pham, N.H.; Vinek, H. Nature of hydroxy groups in MCM-41. *J. Chem. Soc. Faraday Trans.* **1996**, *92*, 3287–3291. [CrossRef]
61. Hunger, M.; Schenk, U.; Breuninger, M.; Gläser, R.; Weitkamp, J. Characterization of the acid sites in MCM-41-type materials by spectroscopic and catalytic techniques. *Microporous Mesoporous Mater.* **1999**, *27*, 261–271. [CrossRef]
62. Chen, C.-Y.; Li, H.-X.; Davis, M.E. Studies on mesoporous materials. *Microporous Mater.* **1993**, *2*, 17–26. [CrossRef]
63. Silaghi, M.C.; Chizallet, C.; Raybaud, P. Challenges on molecular aspects of dealumination and desilication of zeolites. *Microporous Mesoporous Mater.* **2014**, *191*, 82–96. [CrossRef]
64. Pérez-Uriarte, P.; Ateka, A.; Aguayo, A.T.; Gayubo, A.G.; Bilbao, J. Kinetic model for the reaction of DME to olefins over a HZSM-5 zeolite catalyst. *Chem. Eng. J.* **2016**, *302*, 801–810. [CrossRef]
65. Naik, S.P.; Bui, V.; Ryu, T.; Miller, J.D.; Zmierczak, W. Al-MCM-41 as methanol dehydration catalyst. *Appl. Catal. A Gen.* **2010**, *381*, 183–190. [CrossRef]
66. Tang, Q.; Xu, H.; Zheng, Y.; Wang, J.; Li, H.; Zhang, J. Catalytic dehydration of methanol to dimethyl ether over micro-mesoporous ZSM-5/MCM-41 composite molecular sieves. *Appl. Catal. A Gen.* **2012**, *413–414*, 36–42. [CrossRef]
67. Sun, X.; Mueller, S.; Liu, Y.; Shi, H.; Haller, G.L.; Sanchez-Sanchez, M.; Van Veen, A.C.; Lercher, J.A. On reaction pathways in the conversion of methanol to hydrocarbons on HZSM-5. *J. Catal.* **2014**, *317*, 185–197. [CrossRef]
68. Zhang, S.; Gong, Y.; Zhang, L.; Liu, Y.; Dou, T.; Xu, J.; Deng, F. Hydrothermal treatment on ZSM-5 extrudates catalyst for methanol to propylene reaction: Finely tuning the acidic property. *Fuel Process. Technol.* **2015**, *129*, 130–138. [CrossRef]
69. Galanova, E.G.; Magomedova, M.V.; Afokin, M.I.; Starozhitskaya, A.V.; Maximov, A.L. Synthesis of olefins from dimethyl ether in a synthesis gas atmosphere. *Catal. Commun.* **2021**, *153*, 106297. [CrossRef]
70. Tsaplin, D.E.; Kulikov, L.A.; Maksimov, A.L.; Karakhanov, E.A. Method for Obtaining a Composite material with a Hierarchical Structure. RU Patent 2773945C1, 14 June 2022.
71. Khadzhiev, S.N.; Kolesnichenko, N.V.; Gorjainova, T.I.; Birjukova, E.N.; Kulumbegov, R.V. Catalyst and Method of Producing Olefins from Dimethyl Ether in Its Presence. RU Patent 2445158 C2, 20 May 2012.

72. Kolesnichenko, N.V.; Bukina, Z.M.; Jashina, O.V.; Zavalishin, I.N.; Markova, N.A.; Khadzhiev, S.N.; Lin, G.I.; Rozovskij, A.J.; Kitaev, L.E. Catalyst and Method to Manufacture Olefines from Dimethyl Ether in the Presence Thereof. RU Patent 2323777 C1, 10 May 2008.
73. Hubbard, C.R.; Snyder, R.L. RIR—Measurement and Use in Quantitative XRD. *Powder Diffr.* **1988**, *3*, 74–77. [CrossRef]

catalysts MDPI

Review

Hierarchical Zeolite Synthesis by Alkaline Treatment: Advantages and Applications

Daniele S. Oliveira [1], Rafael B. Lima [1], Sibele B. C. Pergher [1,*] and Vinícius P. S. Caldeira [2]

[1] LABPEMOL, Institute of Chemistry, Federal University of Rio Grande do Norte, Natal 59078-970, RN, Brazil
[2] LACAM, Chemistry Department, State University of Rio Grande do Norte, Mossoró 59610-210, RN, Brazil
* Correspondence: sibele.pergher@ufrn.br; Tel.: +55-(84)-991936083

Abstract: Zeolites are of great interest to the scientific and industrial communities due to their interesting catalytic properties, such as high specific area, shape selectivity, and thermal and hydrothermal stability. For this reason, zeolites have been intensively studied and applied in several reactions of great industrial interest. However, the size of zeolite micropores may hinder the diffusion of bulky molecules in the pore system, limiting the use of zeolites in some reactions/applications that use bulky molecules. One way to address this limitation is to generate secondary porosity (in the range of supermicropores, mesopores and/or macropores) in such a way that it connects with the existing micropores, creating a hierarchical pore system. There are different hierarchical approaches; however, most are not economically viable and are complicated/time-consuming. Alkaline treatment has been highlighted in recent years due to its excellent results, simplicity, speed and low cost. In this review, we highlight the importance of alkaline treatment in the generation of secondary porosity and the parameters that influence alkaline treatment in different zeolitic structures. The properties and catalytic performance of hierarchical zeolites prepared by alkaline treatment are extensively discussed. It is expected that this approach will be useful for understanding how alkaline treatment acts on different hierarchical structures and will thus open doors to achieve other hierarchical zeolites by this method.

Keywords: hierarchical zeolites; alkaline treatment; porous materials; synthesis strategies; post-synthesis methods

Citation: Oliveira, D.S.; Lima, R.B.; Pergher, S.B.C.; Caldeira, V.P.S. Hierarchical Zeolite Synthesis by Alkaline Treatment: Advantages and Applications. *Catalysts* **2023**, *13*, 316. https://doi.org/10.3390/catal13020316

Academic Editor: Narendra Kumar

Received: 15 December 2022
Revised: 23 January 2023
Accepted: 25 January 2023
Published: 1 February 2023

1. Hierarchical Zeolites

Conventional zeolites contain only micropores with a maximum size of 2 nm in their structure. The presence of only micropores provides greater selectivity in these materials; however, the uniquely microporous structure prevents molecules (especially bulky molecules) from accessing the interior of the pores and finding the active sites of the zeolite. This limited diffusion can minimize the performance of zeolites, for example, by enabling the formation of coke that, in turn, promotes the deactivation of the zeolite.

Due to diffusion limitations and possible rapid deactivation when large molecules enter microporous zeolites, several recent studies have focused on obtaining hierarchical zeolites to improve the accessibility and catalytic efficiency of zeolitic materials.

To better understand the advantages and use of hierarchical zeolites, it is necessary to understand their structural composition in relation to porosity. Hierarchical zeolites exhibit intrinsic microporosity and secondary porosity, which can be in the range of supermicropores (0.7–2 nm), mesopores (2–50 nm), and macropores (>50 nm). This secondary porosity can be classified into porosity with a narrow or wide range of pore sizes. Therefore, hierarchical zeolites exhibit selectivity arising from their intrinsic microporosity, together with improved mass transport arising from the introduction of supermicropores, mesopores, or macropores into the zeolitic structure [1].

The introduction of secondary porosity into zeolites depends on the desired application of the materials. Generally, secondary porosity is in the range of mesopores. However, the generated pore size range can be controlled and/or influenced by the chosen hierarchical synthesis method. This secondary porosity can be understood by two distinct mechanisms. Additional supermicropores, mesopores or macropores can be generated in the zeolitic crystals, resulting in intracrystalline porosity, or they can also be generated by the agglomeration of nanometric zeolite crystals, resulting in intercrystalline porosity. In both cases, the obtained hierarchical zeolites exhibit improved mass transfer due to the greater accessibility caused by the additional porosity. Due to the benefits of enhanced accessibility, great interest has been devoted to obtaining hierarchical zeolitic materials [2,3].

According to some studies [3–6], hierarchical zeolites can be classified into three categories:

- Hierarchical crystals: formed by the combination of the intrinsic microporosity of the predominant crystal in the zeolite with additional intracrystalline porosity that may be in the range of supermicropores, mesopores, or macropores. In addition, the formation of intercrystalline macropores is possible through spaces between the crystals. This secondary porosity (supermicropores, mesopores or macropores) can be introduced by either direct synthesis or post-synthesis.
- Nanosized crystals: formed by decreasing the size of the zeolite crystals, which generally reach dimensions lower than 100 nm. In addition to the well-defined microporous system characteristic of zeolites, these materials have a system of intercrystalline supermicropores, mesopores or macropores caused by the agglomeration/packing of the crystals. This secondary porosity is obtained by direct synthesis.
- Supported zeolite crystals: the zeolite crystals are supported and/or dispersed in another material/support. The material obtained is not a pure zeolite but a material consisting of the zeolite micropore system and a system of mesopores and/or intercrystalline macropores in the support (the size range of the additional pores is determined by the support material). This secondary porosity is obtained by post-synthesis or assisted direct synthesis.

Hierarchical zeolites obtained by different synthesis methods exhibit accessibility advantages for various catalytic reactions when compared to conventional zeolites. These advantages contribute to increasing mass transfer, minimizing catalytic deactivation and increasing activity with respect to bulky substrates in several chemical reactions.

Therefore, new, effective processes for the formation of hierarchical zeolites have been widely studied. Essentially, there are two types of approaches to obtaining hierarchical zeolites: top-down and bottom-up approaches [7–10]. The first strategy consists of a post-synthesis procedure, in which an already established zeolitic structure is subjected to processes to generate secondary porosity. The second approach is a direct synthesis procedure, in which the secondary porosity derives from the formation of the zeolitic structure.

The main synthesis methods, based on hierarchical approaches, used to obtain hierarchical zeolites are presented in Table 1.

Table 1. Main methods for the synthesis of hierarchical zeolites.

Approaches *Top-down*	Approaches *Bottom-up*
	Hard templating [18]
Desilication [11,12]	Soft templating [19]
Dealumination [13]	Template-free [20]
Irradiation [14]	Dual templating with surfactant [21]
Recrystallization from mixed methods [15–17]	Zeolitization of materials [22]
	Nanoparticle assembly [23]

2. Synthesis of Hierarchical Zeolites by Post-synthesis Procedures

The top-down methods of dealumination, desilication, irradiation, and recrystallization are the main methods of secondary porosity formation by post-synthesis. These methods have some advantages, such as applicability to different types of zeolites with different Si/Al ratios; low cost; high zeolitic crystallinity at the end of the process; and a high degree of secondary porosity. The main details of these post-synthesis methods are described below.

The dealumination method consists of the selective removal of aluminum atoms belonging to the zeolitic structure. This removal causes defects that are generated by the hydrolysis of Si-O-Al bonds [24–27]. Secondary porosity (usually mesopores) is introduced through the generation of vacancies caused by the removal of aluminum atoms [26–28]. However, this method exhibits some disadvantages, including the limitation of zeolites being rich in aluminum, difficulty in controlling the generated mesopores, poorly interconnected mesopores, and partial blockage of the pores (micropores and mesopores) caused by the deposition of amorphous material. In addition, the acidity of hierarchical zeolites can be significantly affected because the removal of aluminum atoms will reduce the number and strength of acid-active sites [9,25,29].

The irradiation method was developed by Valtchev [14]; in this method, macropores are introduced in a parallel orientation inside the zeolite crystals using uranium irradiation, and subsequently, etching with acid solution and washing with water are performed [1,30,31]. This process preserves the microporosity and crystallinity of the zeolite. Despite generating a uniform distribution of parallel macropores, this method has a disadvantage in the use of uranium to irradiate the zeolite crystals [1,30].

The desilication and recrystallization/restructuring methods are based on an alkaline treatment procedure. In desilication, silicon atoms are removed from the zeolitic structure, resulting in secondary porosity, usually in the range of mesopores [32–36]. Recrystallization/restructuring is characterized by the removal of atoms and subsequent reorganization of these atoms in the structure, which preserves the initial crystalline form and adds secondary porosity to make acid sites more accessible [37–40].

Desilication by alkaline treatment has become a topic of great relevance and interest in the preparation of hierarchical zeolites due to its success in obtaining zeolites with well-defined mesopores. This process introduces secondary porosity in a simple, fast, effective and low-cost way, which makes the method even more desirable. In addition, alkaline treatment can introduce secondary porosity while minimally affecting the acidic properties and microporous character of the zeolite, contributing to greater diversity in the application of these materials.

3. Alkaline Treatment

In recent years, alkaline treatment (a post-synthesis method usually involving NaOH solution) through desilication has become one of the most versatile procedures to generate mesoporosity in zeolites. Alkaline treatment can also lead to a reorganization in the zeolitic structure through the recrystallization/restructuring of the zeolite, generating secondary porosity and making acid sites more accessible.

The selective extraction of silicon from the structure by alkaline treatment, known as desilication or base leaching, is a top-down method widely used to prepare zeolites with hierarchical porosities. The controlled leaching of Si by OH^- ions forms intracrystalline mesopores, which facilitates molecular access and diffusion in the active sites of the zeolite [41]. These changes bring enormous benefits in catalysis associated with greater activity, selectivity and/or lifespan. Figure 1 illustrates the desilication process.

Alkaline treatment can be performed by not only conventional electric heating but also microwave heating. Microwave radiation induces rapid and uniform heating and has selective interactions with certain reagents or solvents [41,42]. Both methods are effective in the production of zeolites with secondary porosity. However, the microwave method can be considered more efficient because it leads to the formation of mesopores within a short treatment time [43,44].

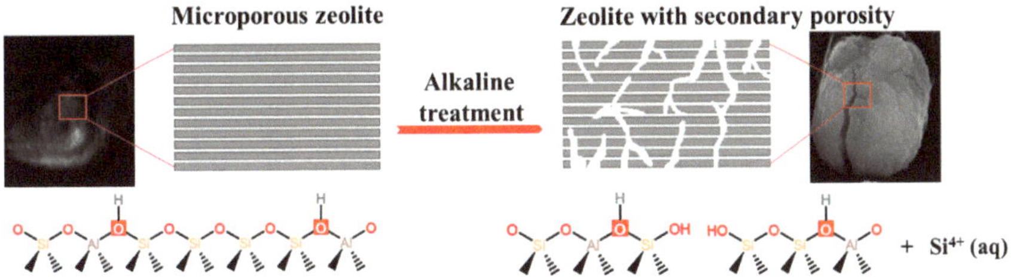

Figure 1. Method of desilication by alkaline treatment (SEM image of the ZSM-5 sample performed in TESCAN equipment, model MIRA3 FEG-SEM, magnification 2 μm).

The use of alkaline treatment (initially called caustic treatment) to improve the performance of zeolites in catalysis was patented by Dean Arthur Young and Yorba Linda in 1967 [45]. It was found that after treatment with solutions of alkali metal hydroxides, preferably NaOH, mordenite zeolite exhibited preserved crystallinity, a significantly increased benzene absorption capacity, and a threefold higher conversion to Pd/mordenite in oil and gas hydrocracking than standard mordenite zeolite (without alkaline treatment). Young and Linda speculated that the superior performance of the modified material could be due to better access to micropores.

In the patent "Caustic-treated zeolites" filed by Dean A. Young and Yorba Linda, 1968 [46], it was found that the properties of mordenite zeolite are improved by digestion with aqueous caustic solution (NaOH solution), in which a smaller proportion of structural silica is leached without significantly altering the crystal structure. It was found that caustic etching treatment increases the adsorption rate and effective capacity of zeolites, as well as their catalytic activity.

In 1977, Donald H. Rosback and Richard W. Neuzil [47] patented the use of an aqueous caustic solution (NaOH solution) in a precursor mass comprising zeolites X and Y in sodium form. They found that the treatment of this precursor mass under particular conditions produced an absorbent with increased capacity for olefins. The produced absorbent exhibits more efficient olefin separation due to its increased adsorption capacity and has a longer shelf life.

In 1979, Alan J. Rein, David D. Saperstein and Seemon H. Pines [48] patented a process for caustic washing of type 3A and/or 4A synthetic zeolites with an improved ability to eliminate acids. These materials can be used to prepare, for example, sodium 7-(2-thienylacetamido)-7-methoxy-3-carbamoyloxymethyl-3-cephem-4-carboxylate.

However, scientific articles on the modification of zeolite in alkaline media only began to be published approximately 25 years after the patent by Dean A. Young and Yorba Linda. In 1992, Dessau et al. [49] reported the dissolution of large ZSM-5 crystals in an attempt to identify Al gradients. They showed that treatment of ZSM-5 zeolite with an aqueous base (Na_2CO_3 solution) resulted in the partial dissolution of the sample with preferential removal of silicon. The treated zeolite had a lower silica/alumina ratio and exhibited a higher cation exchange capacity and higher catalytic activity.

In 1995, Le Van Mao et al. [50] analyzed the properties of zeolites Y, X, and ZSM-5 in alkaline media in more detail. It was concluded that treatment with aqueous sodium carbonate (Na_2CO_3) led to an increase in Al content and a higher ion exchange capacity without drastically altering the structure of the zeolites. This article reported the first nitrogen (N_2) adsorption/desorption isotherm of the mesoporous zeolite ZSM-5 obtained by alkaline treatment. Nevertheless, the main role of mesopores in increasing intracrystalline diffusion and/or access to micropore volume in reactions was not discussed.

In 1997, Čižmek et al. [51] focused on the dissolution mechanism of pentasyl zeolites with high Si contents (silicate-1 and ZSM-5 with different Al contents) in NaOH solution and confirmed the influence of aluminum on the dissolution kinetics.

However, it was only in 2000 that Ogura et al. [11] reported the importance of alkaline treatment and the remarkable porous changes it induced in ZSM-5 zeolite. They clearly demonstrated that the treatment of ZSM-5 in an alkaline solution (NaOH solution) drastically alters the morphology of ZSM-5, leading to the formation of mesopores with almost uniform size without destroying the microporous structure. They also observed by scanning electron microscopy (SEM) that after desilication, the morphology of ZSM-5 zeolite changed, with the appearance of cracks and holes, as shown in Figure 2.

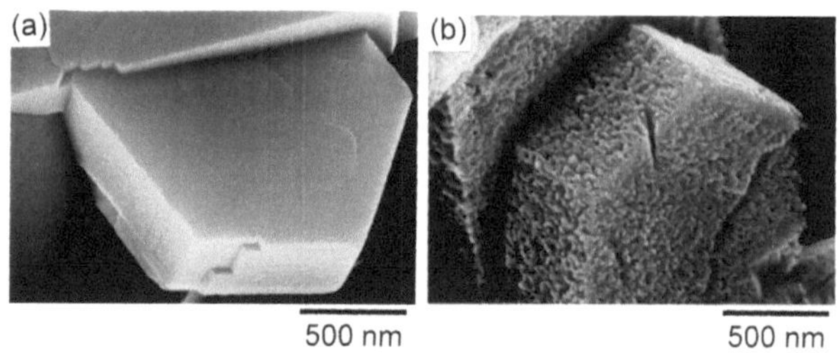

Figure 2. Desilication method: (**a**) SEM image of standard ZSM-5 and (**b**) SEM image of the ZSM-5 sample after alkaline treatment [11]. Copyright (2000), with permission from The Chemical Society of Japan.

Ogura et al. [11] performed N_2 adsorption/desorption at 77 K and observed a hysteresis loop in the desorption curve, a decrease in the microporous volume and an increase in the mesoporous volume, strongly suggesting the formation of mesopores in ZSM-5 zeolite. They concluded that the alkaline treatment of zeolite might be a promising procedure to create a uniform mesostructured material with zeolitic acidity.

Then, Groen, Pérez-Ramírez et al. [32,52–54] conducted several studies to investigate the potential of mesoporous zeolites (specifically ZSM-5) obtained by desilication. They established optimal conditions for the desilication method (0.2 mol L^{-1} NaOH for 30 min at 65 °C) and reported that the crystallinity and acidity of the zeolite remained practically unchanged during alkaline treatment; they further stated that the versatility of this alkaline treatment opens new ways to improve the diffusion characteristics in zeolite-catalyzed applications.

Groen, Pérez-Ramírez et al. also stated that alkaline treatment selectively extracts silicon atoms from the zeolitic lattice. According to these authors, the obtained porosity seems to result from the preferential extraction of silicon from the structure due to hydrolysis in the presence of OH^- ions. The aluminum in the structure controls the process of silicon extraction from the structure and makes the desilication process selective regarding the formation of intracrystalline mesopores. The presence of extra-lattice Al inhibits Si extraction and the formation of related mesopores; this is attributed to the re-alumination of extra-lattice Al species during alkaline treatment [32,52–56].

In recent years, the number of hierarchical zeolites prepared by alkaline treatment has increased significantly because this post-synthesis treatment method is effective, versatile and inexpensive and simply introduces secondary porosity into a wide variety of zeolitic structures. Table 2 shows some previously reported hierarchical zeolites prepared by alkaline treatment alone and by alkaline treatment combined with another approach.

It can be seen that different zeolites and treatments can be used to achieve hierarchical pores systems. Therefore, alkaline treatment can be used in different zeolites and can be easily extended to the industrial scale due to the simplicity and economy of the process as well as its compatibility with current technology. However, alkaline treatment can produce distinct effects on different zeolitic structures. This may occur due to differences in chemical and physical properties, such as composition, Si/Al molar ratio, symmetry, cell dimensions, structure, density, pore diameter and specific area. Therefore, they need to be better studied. In this article, we explore some of these properties in depth.

Table 2. Previously reported hierarchical zeolites prepared by alkaline treatment.

Material Name	Type of Structure [a]	Pore Size (nm)	Morphology [b]	Secondary Porosity (nm)	Reference
ZSM-5	MFI (3)	0.51 × 0.55; 0.53 × 0.56	Prismatic particles [c]	Intracrystalline, ~4	2000 [11]
ZSM-12	MTW (1)	0.56 × 0.60	Clusters of small crystalline particles [c]	Intracrystalline, 15–20	2006 [57]
Mordenite	MOR (2)	0.65 × 0.70; 0.26 × 0.57	Ellipsoidal particles (4 × 2 × 2)	Intracrystalline [d]	2007 [58]
Beta	*BEA (3)	0.66 × 0.67; 0.56 × 0.56	Truncated bi-pyramidal crystals [c]	Intracrystalline, ~3	2008 [59]
AlPO-16	AST (0)	Apertures formed by 6MR	Poorly faceted crystals (0.1–0.2)	Intercrystalline, ~5	2008 [5]
Ferrierite	FER (2)	0.42 × 0.54; 0.35 × 0.48	Plate-like crystals (0.3–0.8 × 0.05–0.15)	Inter- and intracrystalline [d]	2009 [60]
MCM-22	MWW (2)	0.40 × 0.55; 0.41 × 0.51	Pellet-like crystals (1–1.5)	Inter- and intracrystalline [d]	2009 [61]
ITQ-4	IFR (1)	0.62 × 0.72	Rod-shaped crystals (beamlike) (1.5 × 0.2 × 0.2)	Intracrystalline, 4–10	2010 [62]
SSZ-35	STF (1)	0.54 × 0.57	Agglomerates of small particles [c]	Not available	2010 [63]
SSZ-13	CHA (3)	0.38 × 0.38	Cubic crystals (10–15)	Intracrystalline, 2–10	2010 [64]
Zeolite Y	FAU (3)	0.74 × 0.74	Crystals [c]	Intracrystalline, ~2–10 and ~15–30	2010 [65]
Clinoptilolite	HEU (2)	0.31 × 0.55; 0.41 × 0.41; 0.28 × 0.34	Not available	Intracrystalline, ~11.7–17.8	2011 [66]
Silicalite-1	MFI (3)	0.51 × 0.55; 0.53 × 0.56	Prismatic crystals (2.5)	Intracrystalline, ~10	2011 [67]
ZSM-22	TON (1)	0.46 × 0.57	Nanorods (0.04 × 0.04 × 0.15)	Inter- and intracrystalline [d]	2011 [68]
Nu-10	TON (1)	0.46 × 0.57	Cylindrical particles (0.3–0.5 × ~0.05)	Intracrystalline, ~11.5 and ~20	2011 [69]
USY	FAU (3)	0.74 × 0.74	Octahedral crystals [c]	Intracrystalline, ~2.5–10 and ~2.5–20	2012 [70]
Zeolite X	FAU (3)	0.74 × 0.74	Octahedral crystals [c]	Intracrystalline, ~1000	2013 [71]
Zeolite L	LTL (3)	0.71 × 0.71	Clusters of crystalline particles [c]	Inter- and intracrystalline, ~3–50	2013 [72]
SSZ-74	-SVR (3)	0.55 × 0.57; 0.52 × 0.59; 0.52 × 0.56	Rectangular crystalline particles [c] (5 × 8 × 0.55)	Intracrystalline, 5–20	2014 [73]
ZSM-23	MTT (1)	0.45 × 0.52	Rod-like crystals (0.2–0.5)	Intracrystalline [d]	2014 [74]
SSZ-33	CON (3)	0.64 × 0.70; 0.45 × 0.51	Elliptical plates crystals (8 × 4 × 0.5)	Intracrystalline, 5–20	2015 [75]
EU-1	EUO (1)	0.41 × 0.54	Clusters of small particles [c]	Intracrystalline, 10–20 and >50	2015 [76]
ZSM-11	MEL (2)	0.53 × 0.54	Clusters of small crystalline particles (0.3–0.4)	Inter- and intracrystalline, 10–100	2015 [77]

Table 2. *Cont.*

Material Name	Type of Structure [a]	Pore Size (nm)	Morphology [b]	Secondary Porosity (nm)	Reference
IM-5	IMF (3)	0.55 × 0.56; 0.53 × 0.54; 0.53 × 0.59; 0.48 × 0.54; 0.51 × 0.53	Square aggregates of flakes (irregular cubes) [c]	Inter- and intracrystalline, ~4–10	2018 [78]

[a] Dimensionality in parenthesis. [b] Particle size in μm in parenthesis. [c] Particle size not available. [d] Range of secondary porosity not available.

3.1. Parameters That Influence Alkaline Treatment

Several parameters influence the generation of secondary porosity in hierarchical zeolites through alkaline treatment and need to be optimized, such as the Si/Al molar ratio, treatment time, temperature, alkaline agent concentration and zeolite structure. These parameters depend on the characteristics of the zeolitic structures that are used. In addition, the difficulty of controlling the size of the pores formed by this procedure represents an interesting field for investigation.

3.1.1. Si/Al Molar Ratio

Alkaline treatment can generate new pores or expand existing micropores through the partial dissolution or reorganization/restructuring of the structure, which often depends on the Si/Al molar ratio of the starting zeolite.

Dessau, Valyocsik and Goeke [49] reported the selective removal of silicon in ZSM-5 zeolites with different Si/Al molar ratios (13, 14, 16, 22, 24, 29, 35, 38, 70, 74, 76, 81, 94, 124, 150 and 223). It was observed that ZSM-5 samples with Si/Al molar ratios between 70 and 223 partially resisted prolonged alkaline treatment (16 to 20 h under reflux). In all cases, partial dissolution with preferential Si removal was observed. The recovered zeolites were significantly enriched in aluminum, of which very little entered the solution phase.

In 1995, Le Van Mao et al. [50] studied the selective removal of Si (desilication) in ZSM-5, Y and X zeolites with Si/Al molar ratios of 19.5, 2.5 and 1.2, respectively. They concluded that as a general rule for efficient desilication, the higher the Si/Al ratio is, the greater the Si removal and the less basic the solution required for the treatment.

In 1995 and 1997, Čižmek et al. [51,79] analyzed the dissolution of zeolites of the pentasyl family with high silica content (silicatelite-1 and ZSM-5 with different Si/Al ratios) in NaOH solution. The experimental results showed that the dissolution of zeolites with high silica content is controlled by two essential reactions: a direct reaction caused by the breaking of Si-O-Si and/or Si-O-Al bonds due to the action of OH^- ions in solution and a reverse process caused by the reaction between soluble species in the liquid phase or by reactions in/with the surface of the dissolved solid. The forward reaction rate decreased with increasing Al content in the ZSM-5 crystals, while the reverse reaction resulted in the formation of amorphous SiO_2 (ZSM-5) and/or different crystalline modifications of SiO_2 or even different hydrates of sodium silicate (silicatelite-1).

Groen, Pérez-Ramírez et al. [32,52,80–83] conducted extensive studies on the mesoporous ZSM-5 zeolite obtained by alkaline treatment. Commercial ZSM-5 zeolites with Si/Al ratios within the range of 15–1000 were used, and the usual desilication procedure was applied (0.2 mol L^{-1} NaOH for 30 min at 65 °C). These studies showed that mesoporosity clearly depends on the Si/Al molar ratio of the ZSM-5 zeolites. Furthermore, they emphasized that a Si/Al molar ratio between 25 and 50 is ideal for obtaining appropriate mesopores under fixed treatment conditions, while the lowest Si/Al ratio (<25, high Al content) results in limited mesoporosity, and a higher Si/Al ratio (>50, high Si content) leads to extra macropores due to uncontrolled Si extraction. Thus, ZSM-5 zeolites with high Al contents are relatively inert to silicon extraction. This is because most Si atoms are stabilized near $AlO_4{}^-$ tetrahedra. Consequently, these materials exhibit a relatively low degree of silicon dissolution and limited mesoporosity, as shown in Figure 3 for ZSM-5 zeolite.

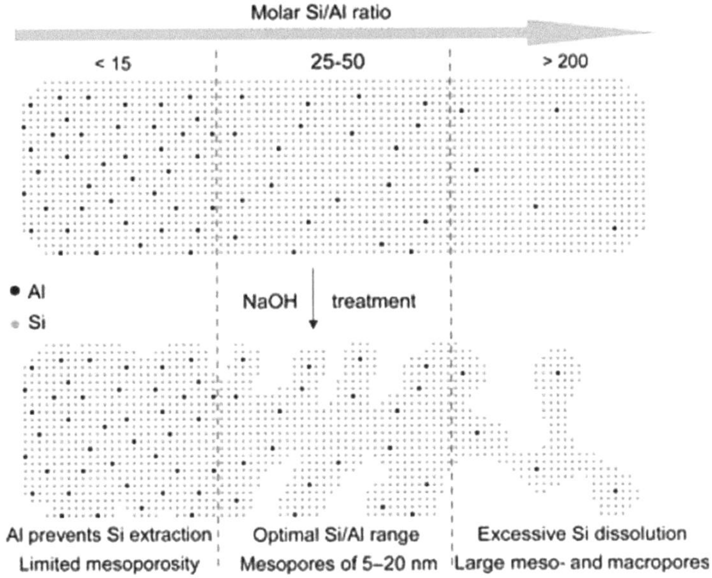

Figure 3. Influence of the Si/Al ratio on the desilication of ZSM-5 zeolite [52]. Copyright (2005), with permission from Wiley.

According to Groen et al. [52], the development of mesoporosity through Si extraction is fundamentally determined by the Si/Al ratio of the zeolite. Indeed, due to the negative charge of AlO_4 tetrahedra, the hydrolysis of the Si-O-Al bond in the presence of OH is impaired compared to the cleavage and relatively easy disruption of the Si-O-Si bond in the absence of neighboring aluminums. The number of aluminum atoms in the lattice is a determinant for the Si extraction process and, consequently, for mesopore formation [84].

According to Fernandez [85] and Verboekend and Pérez-Ramírez [86], virtually no aluminum is lost from the filtrate during alkaline treatment. From numerous analyses, the researchers found that in the process of creating mesopores, aluminum is removed from the structure and realuminated back on the outer surface of the zeolite. Therefore, the term "desilication" is not strictly accurate because although Si is leached in high amounts, both Si and Al are removed from the structure. This raises the question of whether realumination or the Al structure play the key role in the pore formation process. Other than aluminum being mainly a Lewis acid, very little is known about the nature and properties of aluminum deposited on the surface. Therefore, it is relevant to increase the understanding of these species, as they can significantly influence catalytic performance.

Verboekend and Pérez-Ramírez [67,86] recently extended the applicable Si/Al ratio range to 10–1000, which covers practically all compositions of ZSM-5 zeolite. For Al-rich zeolites (Si/Al = 10–20), alkaline treatment followed by subsequent acid washing removed amorphous Al-rich debris that blocked micro- and mesopores. For Si-rich zeolites (even silicic zeolites such as silicalite-1 and TS-1), deliberately added pore-directing agents (PDAs), such as $Al(OH)_4^-$ or tetrapropylammonium (TPA^+), are attracted to the surfaces of the zeolites, preventing excessive Si dissolution.

According to Verboekend [68] and Sadowska [87], intracrystalline mesoporosity is governed by the interaction of micro- and mesopores. However, it is also influenced by the Si/Al ratio of the structure. It was reported that alkaline leaching leads to the dissolution of Si and smaller amounts of Al species in the structure. However, most of these extracted Al species are able to realuminate at the surface of mesopores, resulting in a reduction in the Si/Al ratio of the hierarchical zeolite.

According to Verboekend, Vile and Pérez-Ramírez [88] and Tarach [89], the removal of silicon from the structure of some zeolites through alkaline treatment can cause the formation of large mesopores and a decrease in micropore volume and crystallinity. This shows that alkaline treatment, even under mild conditions, can affect the structural and acidic properties of some zeolites, such as beta zeolite (*BEA), and may be influenced by the content and position of Al atoms in the zeolite structure.

Zhang and Ostraat [2] performed pioneering work focused mainly on ZSM-5 zeolites and gradually applied post-synthesis alkaline treatment to other zeolite topologies, such as *BEA, FER, CHA, FAU, MTW, MWW, and MOR. The results revealed the challenges in obtaining the ideal mesoporosity with preserved microporosity/acidity for different zeolites through similar post-treatments with NaOH due to the differences in the stability of the Al structure and in the different crystallographies.

McGlone et al. [90] studied the alkaline treatment of ZSM-5 with SiO_2/Al_2O_3 ratios of 30 and 80. They concluded that desilication is more difficult at lower SiO_2/Al_2O_3 ratios due to repulsion by the negative charges on the ions imposed by the presence of Al in the structure and that it would be easier with SiO_2/Al_2O_3 ratios >50 due to uncontrolled desilication. In the case of SiO_2/Al_2O_3 ratio = 30, the rigid structure of the zeolite and the proximity of the Si and Al structures (Lowenstein's rule) prevent easy desilication to a certain extent.

These studies show that the secondary porosity generated by the alkaline treatment depends on the Si/Al molar ratio of the zeolite. A high Si/Al ratio can lead to the large formation of mesopores and extra macropores due to uncontrolled Si extraction. An average Si/Al ratio is considered ideal to obtain secondary porosity in the mesopore range. A low Si/Al ratio results in limited mesoporosity due to the greater amount of Al in the zeolite structure, which prevents greater dissolution. Al atoms have been shown to play a key role during alkaline treatment due to their influence on the stability of different crystallographic structures.

3.1.2. Alkaline Agents and Their Concentrations

The use of different alkaline agents and varying concentrations of solutions has been highlighted in recent years.

Suzuki and Okuhara [91] observed that alkaline treatment of ZSM-5 zeolite (Si/Al = 37) using lower-concentration NaOH solutions (0.05 mol L^{-1} dm^{-3}, 90 cm^3) for 0.5–30 h formed supermicropores (approximately 1.8 nm) instead of mesopores. Thus, the concentration of the alkaline solution is an important factor controlling the zeolitic properties and pore size distribution, which can be directed to improve the catalytic functions of zeolites.

According to Groen [80], the removal of Si from the zeolite structure in an alkaline medium, for example, with NaOH and Na_2CO_3, is the simplest and most economical way to introduce mesopores into different types of zeolites without altering the acidic properties of the zeolitic structures. The preservation of the acidic properties of zeolites after desilication is related to the realumination of the aluminum species extracted, which promotes accessibility to available acid centers.

Wei and Smirniotis [57] investigated the influence of NaOH solutions with different concentrations (0.05, 0.1, 0.2 and 0.4 mol L^{-1}) and various treatment times and temperatures on the desilication of ZSM-12 zeolite with different Si/Al ratios. The concentration of the NaOH solution was considered the most dominant factor affecting the desilication method.

Pioneering studies used the alkaline agents NaOH and Na_2CO_3 for desilication. However, studies have been conducted with other alkaline agents, such as the study performed by Groen, Moulijn and Pérez-Ramírez [83] using inorganic bases such as KOH and LiOH. These bases were shown to be less effective than NaOH in the development of mesopores.

According to Serrano, Escola and Pizarro [30], the most common and most studied desilication procedure involves the treatment of zeolite with 0.2 mol L^{-1} NaOH solution for 30 min at 65 °C using a zeolite-to-solution ratio of 33 g L^{-1}. Under these conditions, silica is preferentially removed from the crystal structure, which gives rise to mesoporosity, causing a decrease in the Si/Al atomic ratio of the desilicated zeolite.

Gackowski et al. [36] studied the mild alkaline treatment of ultrastable zeolite Y with a high silica content (Si/Al = 31) using 0.05 and 0.2 mol L^{-1} ammonia solutions and observed a major impact on the structure and properties of the desilicated samples, even in highly diluted ammonia solution (0.05 mol L^{-1}). The amount of silica extracted from the zeolite crystals under these conditions is quite low, leading, however, to significant structural changes in the solids. Thus, a high degree of amorphization was observed, as well as simultaneous changes in Al status and the creation of a high volume of mesopores. It was shown that the samples treated with dilute ammonia solutions exhibited short-range order associated with high Brønsted acidity. According to the authors, from an economic point of view, the treatment of zeolites with inexpensive ammonia solutions is more convenient than a hierarchical zeolite synthesis route based on expensive alkaline solutions, such as TPAOH.

According to Dai [92], in the alkaline treatment of the zeolite H-ZSM-5 using NaOH solution with different concentrations (0.1, 0.3, 0.5, 0.7 and 0.9 mol L^{-1}), when the concentration of the alkaline solution was low (<0.5 mol L^{-1}), the microporous area and microporous volume decreased slightly with the formation of mesopores. This indicates that moderate alkaline treatment increases the mesoporosity of the zeolite and preserves the micropore structure. However, the microporosity was severely destroyed after treatment with high concentrations (>0.5 mol L^{-1}). In addition, the mesoporous volume decreased after desilication with 0.9 mol L^{-1} alkaline solution. This may be a result of the collapse of the zeolite channels caused by the dissolution of zeolitic crystals.

According to Zhang [93], the hierarchical zeolite H-ZSM-5 with Si/Al ratio = 20 was synthesized by alkaline treatment using aqueous solutions of LiOH, NaOH, KOH and CsOH at the same concentration (0.2 mol L^{-1}). As expected, there was an increase in surface area and mesoporous volume after LiOH, NaOH and KOH treatment, which indicates that some of the micropores were destroyed and intracrystalline mesopores were formed. This indicates that OH^- ions can easily attack the internal siloxane groups within the zeolite channels. For CsOH-treated H-ZSM-5, the surface area did not increase dramatically due to more severe alkaline treatment (lower relative crystallinity), and its micropore surface area did not decrease, possibly due to crystal fragmentation (intracrystalline mesopores are covered by crystal fragments). This result indicates that treatment with CsOH is an effective and easy way to induce mesoporosity while maintaining microporosity. The hierarchical catalyst H-ZSM-5 treated with CsOH exhibited not only adequate acidity but also open interconnected mesopores and a smaller crystal size, resulting in greater catalytic activity and stability due to the presence of shorter diffusion paths (prolonging the usefulness of the catalyst). To better understand the main mechanism of enhanced diffusion, it is schematically demonstrated in Figure 4.

According to Tang [94], a series of hierarchical H-ZSM-5 zeolites were prepared by alkaline treatment using varying concentrations of NaOH solutions (0.1–0.5 mol L^{-1}) and the following solutions: 0.3 mol L^{-1} $NaAlO_2$, 0.3 mol L^{-1} Na_2CO_3 and 0.3 mol L^{-1} TPAOH. The acidic properties of H-ZSM-5 zeolite were less affected after treatment with NaAlO2, Na_2CO_3 and TPAOH than with NaOH. Fewer mesopores were introduced into the H-ZSM-5 treated with Na_2CO_3 than that treated with NaOH. Treatment with TPAOH did not have a significant effect on the introduction of mesoporosity into H-ZSM-5 because TPAOH acted as a template, helping to repair the crystal structure of the zeolite. Among the alkaline treatment conditions employed, 0.3 mol L^{-1} NaOH resulted in the best production of aromatic hydrocarbons.

Tanaka et al. [95] performed an alkaline treatment of zeolite H-ZSM-5 with Na_3PO_4 and NaOH, followed by acid treatment with H_3PO_4. H-ZSM-5 was kinetically more stable when treated with Na_3PO_4 than with NaOH at the same alkalinity. Thus, as in the treatment with NaOH, the yield and crystallinity decreased gradually with time under Na_3PO_4 treatment, and the volume of mesopores increased. In contrast to NaOH treatment, phosphorus species were introduced into the products by H_3PO_4 treatment. Acid treatment using H_3PO_4 was combined with NaOH treatment. In contrast to the alkaline treatments, the crystallinity, micropore volume and surface area increased slightly with $NaOH/H_3PO_4$ treatment time. While the yield of the solid product decreased, the Si/Al ratio increased, indicating the dealumination of the structure with $NaOH/H_3PO_4$. As in the treatment with H_3PO_4, phosphorus species were introduced into the products by treatment with $NaOH/H_3PO_4$.

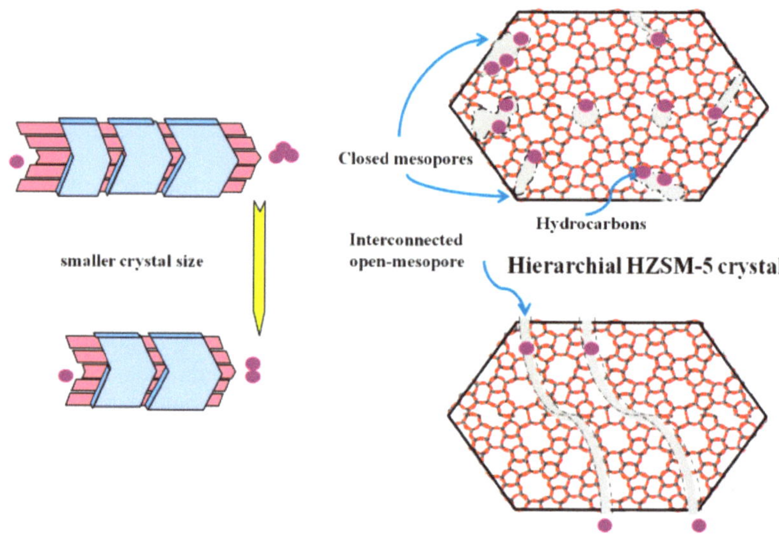

Figure 4. Schematic illustration of the main mechanism of enhanced diffusion [93]. Copyright (2018), with permission from MDPI.

The choice of the alkaline agent and the concentration of the alkaline solution are important factors controlling the zeolitic properties and pore size distribution. According to published studies, the more basic the alkaline agent is, the greater the volume and size of pores formed, and conversely, the lower the concentration is, the lower the volume and size of pores formed. By varying the concentration and type of the alkaline agent, the size and volume of the secondary porosity, the crystallinity and the acidity of the obtained zeolite can be controlled to some extent.

3.1.3. Templates as a Secondary Porosity Driver

The possibility of controlling the pore size in alkaline treatment with different pore-directing agents (PDAs) is currently being intensively researched.

Studies seeking to better control desilication introduced an additional compound into the solution to finalize the process, especially for less stable types of zeolites that are difficult to handle in NaOH solution. It was observed that the PDAs commonly used in the synthesis of zeolites could also be used as alternative compounds. Inspiration began with the use of aqueous solutions of tetraalkylammonium hydroxides (TAA, TPAOH, TBAOH, TMAOH) with the base medium [41].

Figure 5 compares alkaline treatment in conventional leaching with PDA treatment. The zeolite treated with PDAs exhibited the formation of more controlled mesopores.

Perez-Ramirez et al. [96] studied the desilication method involving NaOH treatment using ZSM-5 zeolite in the presence of quaternary ammonium cations. They found that these PDAs, such as TPA^+ and tetrabutylammonium hydroxide (TBA^+), act as a moderator of pore growth in zeolites by the extraction of silicon aided by OH^-, largely protecting the structure of zeolites during desilication and improving transport and catalytic performance.

This protective effect is not seen when cations capable of penetrating zeolite micropores, such as tetramethylammonium (TMA^+), are used.

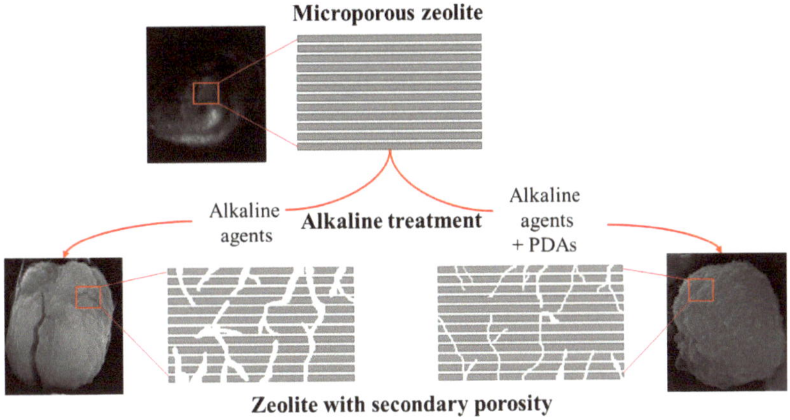

Figure 5. Illustration of alkaline treatment with and without PDAs.

Holm, Hansen and Christensen [97] investigated a "one-pot" desilication and ion exchange procedure, in which a subsequent ion exchange step that protonates the zeolite is avoided. This is possible because the TMA^+ charge compensation ions in the zeolite structure decompose during calcination to give protons. Additionally, according to the authors, the desilication of beta zeolite in a protonated form, i.e., without sodium in TMAOH, results in the limited formation of mesopores. This result indicates that Na^+ is required for efficient hydrolysis of the Si-O-Si bond, similar to the mineralization capacity during synthesis. Another potentially important consideration is the fact that in the absence of sodium during desilication, the structural charge will be compensated by only the TMA^+ ion, which can increase diffusion restrictions and thus prevent or simply delay the formation of pores.

Verboekend and Pérez-Ramirez [67] investigated the role of PDAs in the introduction of hierarchical porosity into silicalite-1 in an alkaline medium. They observed that the pore-directing role is not exerted directly by the trivalent cation metals of the structure but by species on the external surface of the zeolite. The inclusion of metal complexes ($Al(OH)_4{}^-$, $Ga(OH)_4{}^-$) and tetraalkylammonium cations (TMA^+, TPA^+) in the alkaline solution led to distinct mesopore surface areas and pore sizes centered in the range of 5 to 20 nm. All the aluminum partially integrated into the zeolite gave rise to Lewis and Brønsted acidity.

Verboekend and Pérez-Ramirez [67] also proposed a desilication model relating the affinity of the zeolite surface with the PDA and its mesopore formation efficiency (Figure 6). The illustration shows that the optimal formation of intracrystalline mesopores by controlled silicon leaching depends on a balance between the affinity of the PDA with the zeolite surface and the desilication of the zeolite crystal (the amount of PDA is exactly the same in the three scenarios). Evidently, when PDA does not show an affinity for (or is repelled by) the zeolite surface, there is no protection, which results in excessive dissolution (standard alkaline treatment). At the opposite extreme, when the affinity for the zeolite is very strong, as in the case of TMA^+, the surface is overprotected, and the dissolution process is completely inhibited, leading to a high yield of the solid and a lower formation of mesopores. In the medium, which is representative of $Al(OH)_4$ and TPA^+, the optimal balance between affinity for the zeolite and dissolution results in the formation of mesopores. In this case, TPA^+ proved to be the most effective PDA for the generation of mesopores in all silicic zeolites (causing partial protection of the zeolite surface).

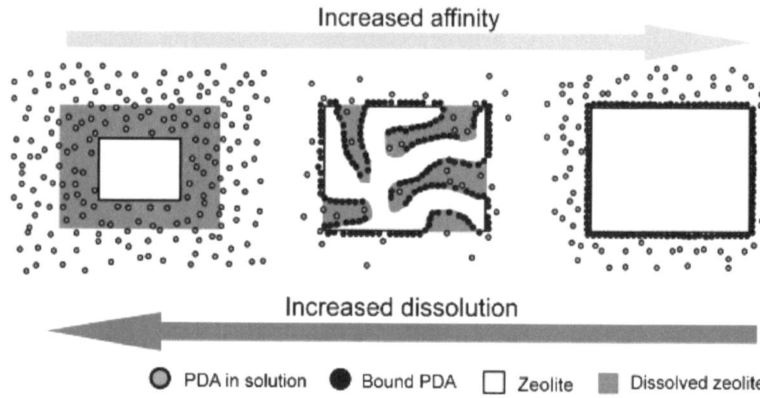

Figure 6. Illustration of how the affinity of PDA to the zeolite surface influences the rate of zeolite dissolution and the formation of mesopores [67]. Copyright (2011), with permission from Wiley.

Sadowska, Gora-Marek and Datka [98] investigated the acidity of ZSM-5 zeolite by desilication using NaOH and a mixture of NaOH/tetrabutylammonium hydroxide (TBAOH) with various concentrations, as well as different ratios of NaOH to TBAOH. It was demonstrated that the concentration of Brønsted acid centers increased as a result of the decrease in Si/Al due to desilication in dilute basic solutions (0.1 and 0.2 mol L^{-1}). Desilication in more concentrated bases (0.5 mol L^{-1}) resulted in the partial destruction of the zeolite, leading to the generation of weakly acidic protonic centers (groups other than SiOHAl). Studies have shown that desilication, which produces mesopores, improves the accessibility of acid centers. This effect was more apparent when desilication was performed with a NaOH/TBAOH mixture.

Sadowska et al. [87] found that NaOH/TBAOH mixtures produced mesopores with greater surface area and pore volume and smaller diameters in ZSM-5 zeolite than did NaOH. It is assumed that in the presence of TBAOH, a greater number of narrower pores are formed, or the narrow pores penetrate the zeolite crystal more deeply.

Tarach et al. [89] investigated the alkaline treatment of beta zeolite (Si/Al = 22) with NaOH and NaOH/TBAOH and found that desilication with NaOH/TBAOH resulted in more uniform intracrystalline mesoporosity with the formation of narrower mesopores while preserving the degree of crystallinity, resulting in catalysts with the most appropriate acidity and with better catalytic performance.

According to Raad [99], an alkaline treatment of H-ZSM-5 zeolite using NaOH with or without TBAOH leads to the formation of more or less structured intracrystalline mesopores. The presence of TBAOH allows the shaping of mesopores. Alkaline treatment preserves most of the Brønsted acid sites without any change in their concentration. A small number of new types of Lewis acid sites appear, particularly in the presence of TBAOH.

According to Verboekend, Vilé and Pérez-Ramírez [88], the desilication efficiency of a tetraalkylammonium cation strongly depends on its affinity with the zeolite surface. Nevertheless, only common quaternary ammonium cations used in the synthesis of zeolites have been explored for this purpose, for example, TPA^+ and TMA^+. Most likely, other molecules used in the preparation of zeolites exert the same or even a superior effect. For example, cetyltrimethylammonium (CTA^+), often used as a secondary template (mesopore template) during the synthesis of hierarchical zeolites, should exhibit a distinct effect on zeolites under alkaline conditions.

Additionally, according to Verboekend, Vilé and Pérez-Ramírez [88], the use of PDAs in NaOH leaching is a generic procedure to introduce intracrystalline mesoporosity into USY and beta zeolites while preserving the intrinsic properties of the zeolite, for example, microporosity, crystallinity and composition. It was found that a wide variety of quaternary ammonium cations and amines positively influence the desilication of these structures and that their impact depends on their charge and size. The type and concentration of PDAs affect the external (mesopore) surface and the size of mesopores, in addition to minimizing amorphization. They showed that TPA^+ produces mesoporous solids with greater zeolitic properties, while CTA^+ gives rise to the reassembly of dissolved species.

According to Zhang and Ostraat [2], a series of cationic, nonionic and anionic surfactants (PDAs) were combined with NaOH to optimize the formation of mesopores. An effective PDA for desilication requires a cationic charge and a mixture of alkyl compounds in the range of 10–20 carbon atoms, such as TPA^+ (tetrapropylammonium) and CTA^+ (cetyltrimethylammonium) cations. The use of TPA^+ results in the formation of highly mesoporous zeolites that retain intrinsic zeolitic properties; however, the use of CTA^+ facilitates the reassembly of dissolved species during alkaline treatment. Thus, both cations are ideal modifiers for desilication in terms of creating mesoporosity and preserving microporosity.

According to Ying and Garcia-Martínez [37] and Chal [38], alkaline treatment can also lead to the restructuring of the zeolite, generating secondary porosity and making acid sites more accessible.

Wang et al. [100] reported a new simple synthesis method to generate extra porosity by the recrystallization of MOR zeolite from NaOH solution in the presence of a mesoporous director. MOR/MCM-41 was successfully synthesized with more accessible acid sites.

Chal et al. [38] studied the recrystallization of zeolite Y using an organic base (TMAOH) in combination with a cationic surfactant (CTAB). The formation of mesoporosity was observed in the zeolitic structure, and the initial crystalline form was preserved.

Yoo et al. [101] presented a method for the preparation of mesoporous zeolite by a combination of desilication and reassembly methods, adopted from the concept of pseudomorphic transformations. Dissolved species containing silicates, aluminosilicates and fragments of ZSM-5 crystals can be deposited back into the zeolite structure by surfactant molecules (such as CTAB) through the formation of micelles under hydrothermal conditions (Figure 7). The final products prepared under specific alkaline conditions exhibited a bimodal mesopore size distribution (3 and 10–30 nm), increased external surface area and well-preserved crystallinity. The reassembly of the dissolved species by surfactant micellization produced small mesopores (3 nm), while desilication generated larger mesopores (10–30 nm).

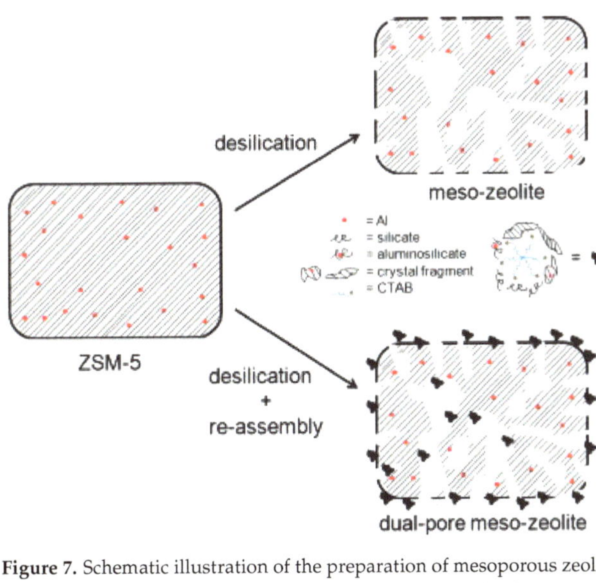

Figure 7. Schematic illustration of the preparation of mesoporous zeolite by means of desilication (upper right) and desilication/reassembly (lower right) [101]. Copyright (2012), with permission from Elsevier.

Garcia-Martínez et al. [15,37] described a new surfactant-based technique that allows precisely controlled mesoporosity to be introduced into a wide range of zeolite crystals, for example, FAU, MOR and MFI with various Si/Al ratios, while maintaining the chemical and physical properties of the zeolites. They suggested that the introduction of controlled mesoporosity in zeolites may occur through a surfactant-assisted crystal rearrangement mechanism. In summary, zeolite Y with well-controlled intracrystalline mesoporosity was prepared by means of surfactant-assisted synthesis (CTAB). According to the authors, this was the first time that well-controlled mesoporosity was introduced in Y zeolites with low Si/Al ratios, which are relevant for catalytic cracking.

Ivanova et al. [102] suggested a mechanism for the recrystallization of mordenite zeolite (Figure 8). According to the authors, in the first step of the process, Si-O-Si bonds are broken in an alkaline medium by desilication. The formation of negatively charged sites during desilication and ion exchange (sodium cations penetrate the interior of the crystals, and all zeolite protons and SiOH groups undergo rapid ion exchange with Na^+) stimulates the diffusion of CTA^+ cations attracted by negative charges in the intercrystalline and intracrystalline spaces of the zeolite. Hydrothermal treatment at 150 °C promotes the condensation of the species formed during desilication around the micelles, which results in the formation of fragments with uniform mesopores. The formation of such fragments occurs both inside the zeolite crystals and on the external surface of the crystallites. The resulting material is a zeolite with bimodal porosity (micro- and mesopores).

Silva, Ferracine and Cardoso [103] evaluated the effects of different concentrations of NaOH and the surfactant cetyltrimethylammonium bromide (CTA^+) on the textural, chemical and morphological characteristics of USY zeolite (Si/Al ratio = 15). The generation of mesoporosity in USY zeolite was enabled by the simultaneous presence of the surfactant and alkaline solution. Among the parameters studied, the concentration of the alkaline medium had the greatest influence on the textural properties of the zeolite. The presence of CTA^+ cations was fundamental in the process since the cations hindered the attack of hydroxyl groups (OH^-), preventing the dissolution of zeolite crystals during the treatment.

Gorzin et al. [104] evaluated the generation of mesoporosity in highly silicon-rich H-ZSM-5 zeolites (Si/Al ratio = 400) by a two-step approach: alkaline treatment using $NaAlO_2$ and $NaAlO_2$/TPAOH in different proportions, followed by washing treatment with acid. The results showed that the porosity of the desilicated samples was blocked mainly by deposits of sodium aluminate and silicon-containing debris. After a subsequent washing step with hydrochloric acid, the blocking species were removed, which resulted in improved mesoporosity. It was found that alkaline treatment in $NaAlO_2$/TPAOH solution followed by acid washing leads to the formation of narrow and uniform mesoporosity without severely destroying the crystal structure. The results showed that the desilication process leads to the considerable development of mesoporosity, while acid-washing treatment mainly influences the acidity of the catalyst. Therefore, the combination of alkali-acid treatment leads to the formation of a hierarchical H-ZSM-5 catalyst with a customized pore architecture and acidic surface properties.

Al-Ani et al. [105] concluded that surfactant species play a key role in the formation of regular intracrystalline mesopores and in the protection of the zeolite structure against desilication and excessive dealumination during the mesostructuring process; this process is in contrast to the treatment of zeolites by recrystallization, which can lead to the degradation and amorphization of the zeolite structure, generating a high degree of mesoporosity. According to the authors, zeolite treatment in the presence of an organic surfactant in a basic medium results in a much more controllable mesostructuring procedure than is possible with severe leaching using inorganic basic solutions (e.g., NaOH solution) without a surfactant, which generally leads to the significant degradation of the zeolite structure.

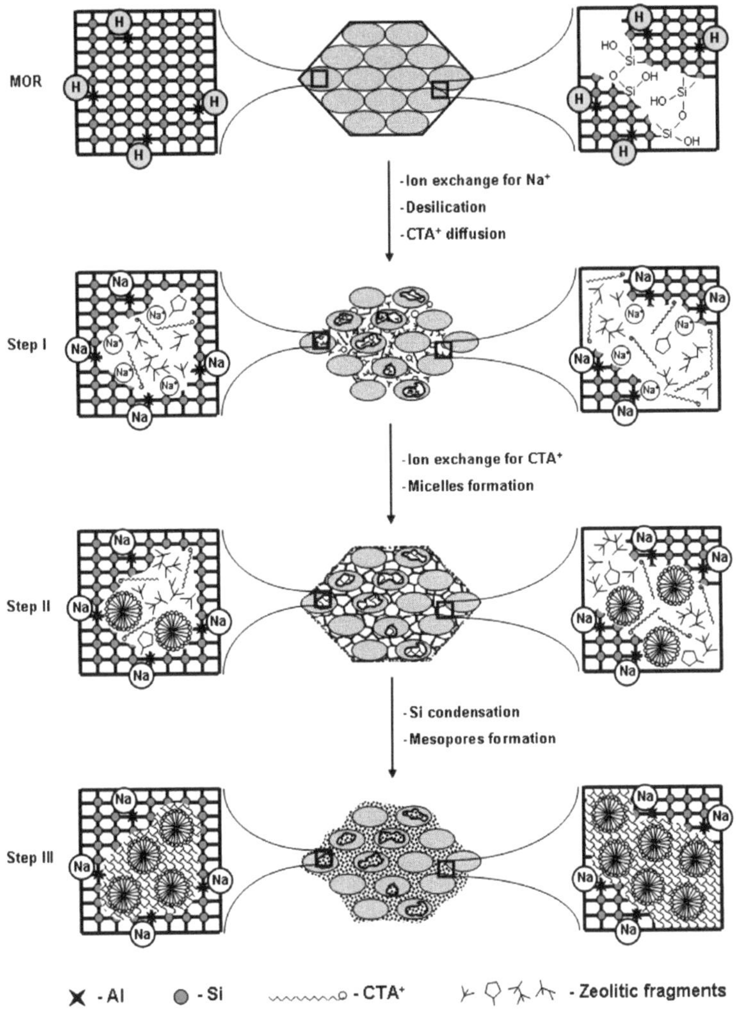

Figure 8. Proposed mechanism of the recrystallization of mordenite zeolite [102]. Copyright (2014), with permission from Elsevier.

3.1.4. Morphology and Crystal Size

According to Groen [52], information on how the porosity is distributed among the crystals and on the effect of crystal size is essential to rationally plan the potential applications of zeolites treated with an alkaline solution. It was shown that the controlled desilication of large and small crystals of ZSM-5 could modify and utilize the impact of Al zoning on the development of mesoporosity. The authors concluded that the Al gradient affects Si extraction throughout the zeolite particle and that the introduction of mesopores by desilication into large crystals may be inefficient due to Al zoning. Thus, for samples with a high degree of aluminum zoning, hollow crystals with an empty core and intact shell can be created.

Dessau, Valyocsik and Goeke [49] observed the nonuniform dissolution of large crystals of ZSM-5 (5–10 μm) after prolonged treatment with high concentrations of alkaline solution. This was speculatively attributed to the nonuniform concentration of Al in the crystals.

Verboekend et al. [68] suggested that the limited mesopore specific area and low desilication efficiency obtained in the alkaline treatment of ZSM-22 zeolite (one-dimensional) and ferrierite zeolite (two-dimensional) must be related to the morphology of the crystals. The introduction of mesoporosity into ZSM-22 crystals is not directly due to the unique characteristics of this zeolite: its nanorod (needle-like) morphology, unidimensional ellipsoidal microporous system and uneven distribution of Al. It was demonstrated that the desilication efficiency of ZSM-22 nanorods and ferrierite platelets is relatively low compared to that of ZSM-5, most likely due to the crystal morphology of the former two zeolites that facilitates the creation of intercrystalline mesoporosity.

Van Laak et al. [106] performed alkaline treatment on ZSM-5, ZSM-12 and beta zeolites. The ZSM-5 samples consisted of small crystallites between 20 and 100 nm in size that were agglomerated into larger particles of approximately 1 μm. SEM analysis (Figure 9A–C) indicated that the mesopores were intercrystalline and intracrystalline. Additional experiments on ZSM-12 yielded similar results and showed that the morphology of the standard zeolite (Figure 9D–F), i.e., crystallite/particle size, determines the amount of added mesoporosity, in which smaller crystallites give rise to larger pores. They also performed alkaline treatment on beta zeolite with 5 μm particles, where the deagglomeration of the particles was observed (Figure 9G–I). All of the tested zeolites (ZSM-5, ZSM-12 and beta) consisted of small crystallites (30–200 nm) that were agglomerated into larger particles between 1 and 5 μm. Intercrystalline mesopores were formed for all zeolites, but the treatment was more effective for zeolites with small crystallites.

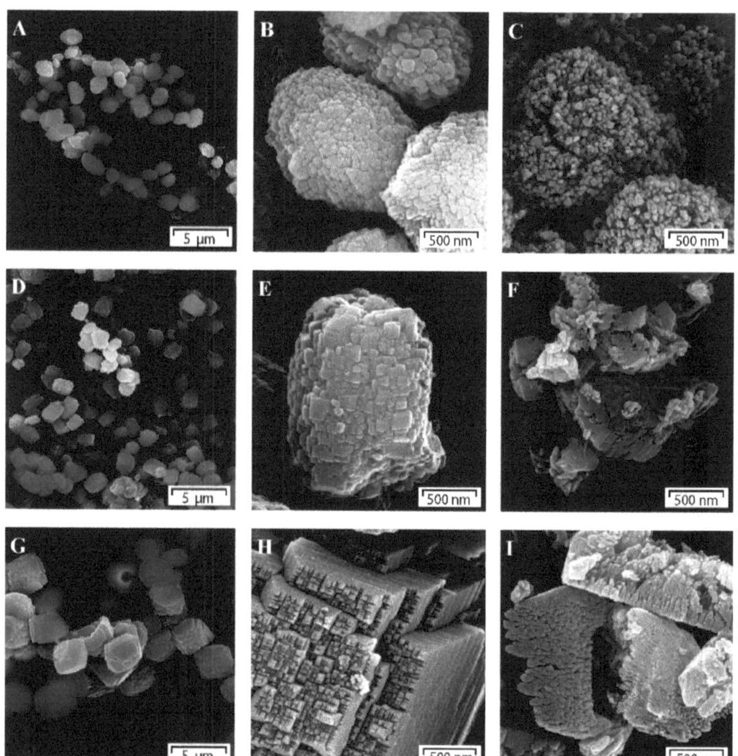

Figure 9. SEM images: (**A**) ZSM-5 standard; (**B**) ZSM-5 standard; (**C**) ZSM-5-at(60); (**D**) ZSM-12-(105) standard; (**E**) ZSM-12-(105) standard; (**F**) ZSM-12-(105)-at(60); (**G**) beta standard; (**H**) beta standard and (**I**) beta-at(120) [106]. Copyright (2011), with permission from Elsevier.

According to Svelle [107], the number of intergrowths in the ZSM-5 crystals determines the desilication characteristics. A large amount of intergrowth reduces the importance of the Si/Al ratio of the standard zeolite and leads to the formation of mesopores derived from the removal of defects. Figure 10 shows a schematic of the different mechanisms of mesopore formation, where small growths or defects require an ideal Si/Al ratio of 20–50 to introduce mesopores. Intermediate cases lead to a combination of the two mechanisms. In the case where many intergrowths are present, the Si/Al ratio is less important, and mesopores are formed mainly due to intergrowth/defect removal.

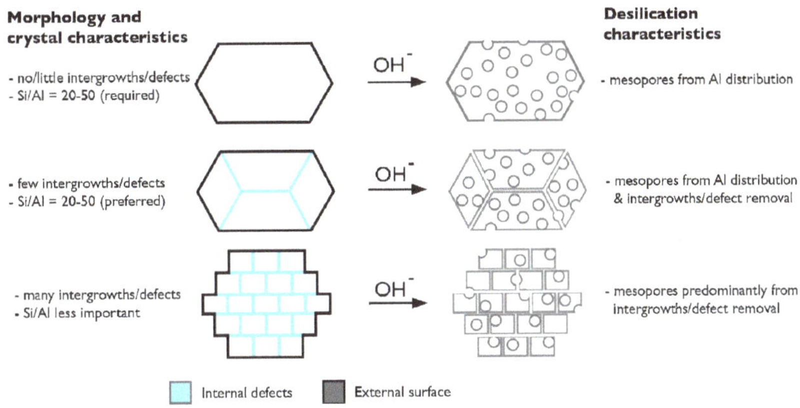

Figure 10. Schematic representation of different mechanisms of idealized mesopore formation [107]. Copyright (2011), with permission from Elsevier.

According to Verboekend [67], the optimal alkaline treatment strongly depends on the unique nature of each zeolite. For example, the properties that must be taken into account include the structure type, Si/Al ratio, Al distribution, crystal morphology and relative abundance of defects.

According to Tarach [89], compared to the previously reported desilication results for MFI, MOR and MTW zeolites, the relatively low stability of aluminum in the beta structure (caused by the presence of a high concentration of structural defects) negatively impacts the efficiency of the desilication process. High structural stability is a prerequisite for successful control of desilication, i.e., the coupling of mesoporosity development with the preservation of acidic properties.

Bi [108] successfully prepared the hierarchical zeolites H-ZSM-5 and Hβ (H-beta), which have interconnected micropores and mesopores, by alkaline treatment using different alkaline solutions (Na_2CO_3, NaOH, NaOH/TMAOH, NaOH/HCl). It was noted that the mesoporous surface areas of the hierarchical H-ZSM-5 zeolites were similar with different alkaline solutions. However, the surface area of the mesopores formed in the hierarchical Hβ zeolites was directly related to the basicity of the alkaline solution (more mesopores were produced in the NaOH solution). The main reason may be due to the different pore structures of the two zeolites. H-ZSM-5 contains 10-membered ring channels (0.51 nm × 0.55 nm and 0.53 nm × 0.56 nm), while Hβ contains 12-membered ring channels (0.66 nm × 0.67 nm and 0.56 nm × 0.56 nm). Hβ zeolite is more sensitive to alkaline post-treatment.

According to Al-Ani [105], a two-dimensional or three-dimensional pore system in some zeolites, for example, in medium-pore ZSM-5 (a zeolite with a three-dimensional pore system), may not be sufficient for the introduction of intracrystalline mesoporosity via the supramolecular modeling of ZSM-5 (alkaline treatment with NaOH/TPAOH + CTAB). These types of zeolites, with pores comprising 10-MR windows, are resistant to modification with surfactants; therefore, the potential routes for increasing their intracrystalline porosity are still under investigation. Additionally, according to the authors, the mechanism of the mesostructuring process is not yet completely understood. A detailed understanding of the mesostructuring processes would serve as a general guide to allow targeted post-synthesis modification and, thus, the production of improved materials with a direct impact on new catalytic applications, particularly those involving transformations of bulky molecules.

According to Li [109], the introduction of mesopores by desilication into large crystals may be inefficient due to Al zoning. The desilication process is affected by the local aluminum content. Basic leaching of zoned crystals leads to the dissolution of the aluminum-poor parts of the crystals without creating mesopores in the aluminum-rich parts. The results of base leaching indicate that aluminum tends to propagate outward along with the crystallization (crystal growth) of ZSM-5.

3.1.5. Temperature and Treatment Time

Groen et al. [81] investigated the alkaline treatment (0.2 mol L^{-1} NaOH) of ZSM-5 zeolite (Si/Al ratio = 37) with a focus on the evolution and optimization of the porous structure by varying the temperature (35–85 °C) and time (15–120 min) of treatment. They observed that the formation of mesoporosity during alkaline treatment is highly dependent on temperature and time. No substantial formation of mesopores was observed at temperatures up to 45 °C after 30 min of treatment. Only at 55 °C does some extra porosity begin to develop, and this porosity becomes substantial at 65 °C. Above 65 °C, the pore size distribution widens considerably, leading to a large volume of mesopores. A treatment temperature of 65 °C is considered optimal, combining extensive mesopore formation and almost completely preserved microporosity. Variations in treatment time in the range of 30–120 min at 65 °C lead to similar behavior but less spectacular changes. A longer treatment time yields a significantly wider pore size distribution, and a time of 30 min is the best to generate mesopores. The researchers concluded that the ideal temperature and time for the development of mesopores after the alkaline treatment of ZSM-5 are 65 °C and 30 min, respectively. Analysis of the influences of treatment temperature and time on the porous properties of the final material shows that the size and volume of mesopores can be adjusted to some extent by the optimization of these two variables.

Groen, Moulijn and Pérez-Ramirez [83] analyzed the effect of temperature and time in the alkaline treatment of ZSM-5 zeolite. The influence of temperature was studied in the temperature range 35–85 °C using 0.2 mol L^{-1} NaOH solution as the alkaline medium and a treatment time of 30 min. At lower temperatures in the range of 35–45 °C, limited mesoporosity developed. The increase in the surface area of mesopores becomes significant after treatment at 55 °C. Accordingly, a strong increase in silicon leaching was observed. A maximum mesopore surface area of 240 m^2 g^{-1} developed at a treatment temperature of 65 °C, which also led to a distinct decrease in microporosity. At higher temperatures, a decrease in the surface area of mesopores was observed. At these temperatures, excessive silicon extraction occurs, which leads to the formation of larger pores. Larger pores contribute strongly to the volume of mesopores but to a lesser extent to the surface area of mesopores. Consequently, a treatment temperature of 65 °C can be concluded to be optimal, considering the obtained surface area of mesopores in nanocrystalline zeolite crystals. The effect of time shows a similar trend. After exposure to NaOH at the ideal temperature of 65 °C for 15 min, the surface area of newly created mesopores already reached 180 m^2 g^{-1}, which was slightly lower than that after 30 min of treatment (240 m^2 g^{-1}). Similar to the effect of higher treatment temperatures, a longer treatment duration leads to a slight decrease in the surface area of mesopores, again due to excessive silicon dissolution and formation accompanied by larger pores.

Groen et al. [59] evaluated the effects of temperature (25–65 °C) and reaction time (10–60 min) in the alkaline treatment of beta zeolite with a Si/Al ratio = 35. They found that alkaline treatment at 25 °C hardly induced any new mesoporosity; the surface area of mesopores was only slightly larger than that of the standard sample; and treatment at 45 °C can be considered optimal in terms of combined micro- and mesoporous properties.

However, preferential silicon extraction can be substantially influenced by the treatment time, considering the large differences detected in the mesopore surface area of these samples and their progressively decreasing Si/Al ratios. At the ideal temperature (45 °C) for 10 min, mesopores are formed, and Brønsted acid sites remain. Extending the treatment time to 30 min causes the disappearance of /decrease in Brønsted acid sites and an increase in mesoporosity. Consequently, the marked influence of time on desilication was confirmed, not only in terms of porosity development but also in relation to the acidic properties associated with the presence of aluminum in the various materials. Shortening the treatment time may help retain the microporous and crystalline properties of purely microporous zeolite.

Santos et al. [110] studied the effect of time and temperature on the generation of secondary porosity in beta zeolite. The effect of time was evaluated using NaOH solution (0.2 mol L^{-1}) at 65 °C for a time period of 10–240 min. They observed that the formation of secondary porosity occurred under the selected experimental conditions, but the decrease in the microporous volume was drastic. The microporous volume decreased after a short treatment period of 20 min and then gradually increased. The mesoporous area initially increased with reaction time, reached a maximum value of 492 m^2 g^{-1} in 20 min and then exhibited a relatively small reduction. The mesoporous area plateaued at 413 m^2 g^{-1} after 240 min of treatment. The interesting results for beta zeolite as a function of treatment time led to the study of mesopore formation under a broader range of experimental conditions. Thus, the reaction temperature was varied from 30 to 100 °C, with the time set at 240 min. The mesopore area first increased with treatment temperature, reaching 499 m^2 g^{-1} at 50 °C and then decreasing to 267 m^2 g^{-1} at 100 °C. The microporous volume first decreased compared to the untreated sample and then increased. The trends observed with increasing treatment temperature were similar to those observed with respect to time.

Shah et al. [111] investigated the effects of alkaline treatment on the microporosity and acidity of the ZSM-5 zeolite structure by changing the temperature of the desilication method. The desilication reaction was performed at different temperatures (40, 60 and 80 °C). They observed that as the desilication temperature increased, the mesoporosity rate increased slightly after more Si was extracted. With increasing desilication temperature, the number of Lewis acid sites increased with the decreasing number of Brønsted acid sites. With increasing desilication temperature, a greater increase in pore size, acidity and mesoporosity was observed.

The temperature of the alkaline treatment is the key parameter in the generation of secondary porosity in zeolites with a well-defined structure. The pore size and microporosity distribution of zeolite can be controlled through temperature and time. According to the analyzed studies, it can be stated that temperature has a greater effect than time on the generation of secondary porosity.

4. Applications of Hierarchical Zeolites Prepared by Alkaline Treatment

The many efforts to obtain hierarchical zeolites with high thermal and hydrothermal stability and a high surface area without substantially losing their high micropore volume, acidity and crystallinity are mainly inspired by the industrial application of these materials in important reactions, for example, in catalysis, adsorption, optics, biomedicine and energy [108,109].

According to Hoff [34], the introduction of mesopores through alkaline treatment represents a possible strategy to improve intracrystalline diffusion and mass transport to promote the production of aromatics over unwanted coke formation.

According to Přech et al. [112], hierarchical zeolites exhibit improved the accessibility of active sites and faster mass transport and are generally more resistant to coke deactivation. Therefore, they often show higher catalytic activity than conventional zeolites, particularly in reactions that suffer from steric and/or diffusional limitations. In addition, the secondary porosity provides an ideal space for the incorporation and grafting of other components and phases, opening a wide diversity of routes for the preparation of multifunctional materials.

García-Martínez et al. [15] studied fluid catalytic cracking (FCC) catalysts made from mesoporous zeolite Y, which showed significantly improved selectivity in product yields (more transportation fuels, i.e., gasoline and LCO, and less coke, dry gases and uncracked bottoms). Mesoporous zeolites are not only superior FCC catalysts but also ideal materials for a wide variety of other applications where slow diffusion is a limiting factor in the reaction process, e.g., hydrocracking, catalytic pyrolysis of biomass, catalytic enhancement of bio-oil, transesterification of vegetable oil, conversion of methanol into hydrocarbons, water treatment and less energy-intensive adsorptive separations.

According to Vu [113], the introduction of mesoporosity improves the performance of zeolite ZSM-5 in triglyceride cracking under FCC conditions. Compared to commercial ZSM-5, which is known as the best zeolite-type catalyst for triglyceride cracking, the hierarchical ZSM-5 catalyst increases conversion, gasoline and light olefin yields by approximately 14%, 10% and 30%, respectively. The superior catalytic performance of hierarchical ZSM-5 catalysts stems from the improved accessibility and mass transfer provided by the created mesoporosity and simultaneous maintenance of the intrinsic catalytic properties of ZSM-5. This allows hierarchical ZSM-5 zeolites to effectively convert triglycerides, regardless of the degree of unsaturation, into gasoline hydrocarbons and light olefins.

Ding et al. [114] investigated the effects of hierarchical H-ZSM-5 zeolites on the improvement of aromatic hydrocarbon yields during the fast catalytic pyrolysis (FCP) of waste cardboard (WCB). The carbon yield of BTX using the hierarchical zeolite increased by up to 82% compared to that of the standard H-ZSM-5 zeolite, resulting in an increase of up to 44% in aromatic carbon yield. There was a decrease in the carbon yield of sugars, carbonyl compounds and other oxygenated compounds. This improvement in product yields was attributed to the generated mesoporosity, which shortened the diffusion path length of molecules, and the increase in weak acid sites contributed to the improved selectivity of hierarchical H-ZSM-5 zeolites for BTX. Thus, the coke yield decreased due to the increase in pore size and improved diffusion performance.

McGlone et al. [90] investigated the [4 + 2] Diels–Alder cycloaddition of 2,5-dimethylfuran with ethylene using a series of hierarchical H-ZSM-5 catalysts synthesized by alkaline treatment. An increase in conversion was observed for all hierarchical materials compared to untreated zeolite, and increases in temperature and ethylene pressure significantly improved both the conversion of dimethylfuran and the selectivity to p–xylene due to easier desorption from the zeolite surface and the increased cycloaddition rate, respectively. Additionally, according to the authors, a compromise between acidity and mesoporosity was considered the key to increasing the activity and maximizing the selectivity in the production of p-xylene from 2,5-dimethylfuran.

According to Chen, Xiong and Tao [115], mesoporous ZSM-5 prepared by alkaline treatment proved to be an efficient catalyst for the hydrolysis of cellulose in ionic liquid (IL), providing a high yield of reducing sugars. It was demonstrated that mesoporous ZSM-5 had 76.2% cellulose conversion and a 49.6% total reducing sugar (TRS) yield. In comparison, conventional ZSM-5 showed only 41.3% cellulose conversion with 33.2% TRS yield. The results indicated that the important role of mesopores in zeolites in increasing the TRS yield might be due to the diffusional alleviation of cellulose macromolecules. It was found that IL could enter the internal channel of mesoporous ZSM-5 to promote the generation of H^+ from Brønsted acid sites, which facilitated hydrolysis. In addition, mesoporous ZSM-5 showed excellent reuse for catalytic cycles, which is promising for practical applications in cellulose hydrolysis.

Zhang [93] studied butene oligomerization using a series of new types of hierarchical H-ZSM-5 zeolite catalysts. It was demonstrated that alkaline treatment could effectively modify the acidity properties and hierarchical structure of the H-ZSM-5 catalyst. The results showed that hierarchical catalysts with interconnected open mesopores, smaller crystal sizes and adequate acidity have a longer shelf life during butene oligomerization. A butene conversion rate of approximately 99% and C8+ selectivity of 85% were obtained at 12 h. Thus, an appropriate hierarchical catalyst can satisfy the requirements of the oligomerization process and has the potential to be used as a replacement for commercial ZSM-5 catalysts.

According to Tang [94], higher selectivity for aromatic hydrocarbons can be obtained in the FCP of lignin using the hierarchical zeolite catalyst H-ZSM-5. Alkaline treatment improved the catalytic performance of H-ZSM-5 zeolite for cracking bulky oxygenates released from lignin (such as guaiacol, syringol and their derivatives from lignin pyrolysis) to produce aromatic hydrocarbons.

According to Tanaka [95], the structure of hierarchical H-ZSM-5 zeolites modified with phosphorus by sequential alkaline/acid treatment reduced the residence time of light olefin products within the catalyst pores in methanol-to-olefin reactions, leading to a longer catalyst lifetime and higher methanol conversion and propylene selectivity.

Rac et al. [116] studied the possibility of improving the drug (atenolol, sodium diclofenac and salicylic acid) adsorption capacity of hierarchical ZSM-5 by using an alkaline treatment. Among the drugs tested, atenolol and diclofenac were most effectively adsorbed onto the hierarchical ZSM-5 zeolite. It was shown that, in the case of atenolol, the superior adsorption capacities of the hierarchical ZSM-5 may be even more pronounced at low initial concentrations. Therefore, the formation of mesopores can significantly improve the accessibility of active sites within the ZSM-5 structure.

Zhao [117] applied hierarchical H-MOR zeolite in the selective synthesis of ethylenediamine (EDA) via condensation amination of monoethanolamine (MEA) for the first time. It was observed that the diffusion conditions and the reactivities of the catalysts were improved by the generated secondary porosity and resulted in excellent catalytic performance under relatively mild reaction conditions. The MEA conversion was 52.8%, and the selectivity for EDA increased to 93.6% (according to the authors, this was the highest value ever reported). In addition, hierarchical H-MOR showed excellent catalytic stability in the selective synthesis of EDA. According to the authors, in the future, researchers can use hierarchical H-MOR as a support and still modify the structural and acidic properties of the catalyst, which will be one of the most promising methods to overcome the disadvantages of conventional methods of selective synthesis of EDA.

Our research group recently published an article on the application of hierarchical ZSM-5 zeolite in the catalytic cracking of polyethylene [118]. The hierarchical zeolite ZSM-5 was synthesized by alkaline treatment with NaOH in the presence and absence of CTAB using conventional processes of electric heating and microwave irradiation. It was observed that both forms of heating, regardless of the presence or absence of CTAB, efficiently formed secondary porosity in ZSM-5. The hierarchical samples exhibited efficient degradation of LDPE with a lower degradation temperature. This improvement in LDPE cracking is due to the introduction of secondary porosity, which consequently provides greater accessibility of LDPE molecules to the acid sites of hierarchical zeolites.

Smoliło-Utrata et al. [119] showed that modification of zeolite HY via alkaline treatment/desilication with vanadium impregnation could be an effective method to adjust the lattice oxygen basicity (a key parameter that plays a particularly important role in the process of oxidative dehydrogenation–ODH). The significantly increased mesopore surface ensures the binding of vanadium species to the silanol groups and the formation of the isolated $(SiO)_2(HO)V{=}O$ and $(SiO)_3\ V{=}O$ sites or highly dispersed polymeric forms located in the zeolite. The higher basicity of lattice oxygen in deSi V-HY compared to V-HY, resulting from the presence of the Al-rich shell, aids in the activation of the C−H bond and greater selectivity to propylene.

Table 3 shows the summary of applications of hierarchical zeolites pre-pared by alkaline treatment cited in these items.

Table 3. Summary of applications of hierarchical zeolites prepared by alkaline treatment cited in these items.

Hierarchical Zeolites	Applications	Reference
Y	Fluid catalytic cracking (FCC)	2012 [15]
ZSM-5	Triglyceride cracking	2014 [113]
H-ZSM-5	Fast catalytic pyrolysis (FCP) of waste cardboard	2017 [114]
H-ZSM-5	Diels–Alder cycloaddition of dimethylfuran and ethylene	2018 [90]
ZSM-5	Hydrolysis of cellulose in ionic liquid	2018 [115]
H-ZSM-5	Butene oligomerization	2018 [93]
H-ZSM-5	Catalytic pyrolysis of lignin	2019 [94]
H-ZSM-5	Methanol-to-olefin	2020 [95]
ZSM-5	Removal of pharmaceutically active substances	2020 [116]
H-MOR	Selective synthesis of ethylenediamine via condensation amination of monoethanolamine	2020 [117]
ZSM-5	Polyethylene catalytic cracking	2021 [118]
V-HY	Adjust the lattice oxygen basicity (oxidative dehydrogenation–ODH).	2022 [119]

5. Outlook and Final Considerations

Alkaline treatment has been proven to be an effective, inexpensive, versatile and simple post-synthesis procedure to generate secondary porosity in a wide variety of zeolite structures.

However, a disadvantage of alkaline treatment is that the use of a base alone can generate a wide range of mesopores and decrease crystallinity and acidity, which is undesirable.

The use of PDAs can help to overcome this limitation. The use of external PDAs allows the porosity, acidity and composition of hierarchical zeolites to be controlled. The presence of PDA on the outer surface induces partial protection that controls the dissolution process and leads to the formation of more uniform intracrystalline mesopores.

However, the use of PDAs during alkaline treatment can cause simultaneous desilication/reassembly. In this case, there is no loss of crystallinity, and a high surface area is obtained, making the application of this procedure even more desirable.

Thus, the careful choice of alkaline treatment conditions, such as the Si/Al molar ratio, alkaline agent, zeolite structure and treatment temperature and time, can lead to the realization of a hierarchical structure with preserved zeolite intrinsic properties.

Hierarchical zeolites synthesized by alkaline treatment have exceptional properties, such as improved accessibility of active sites, faster mass transport, a lower deactivation rate and better conversion capacity, thus making them promising for application in various industrial reactions.

However, to synthesize a specific hierarchical zeolite, it is necessary to carefully research the best alkaline treatment conditions to determine the ideal base to stabilize the zeolite structure during treatment. It is expected that this approach will be useful in achieving hierarchical forms of other zeolites by using an alkaline treatment.

Author Contributions: Conceptualization, S.B.C.P. and V.P.S.C.; methodology, D.S.O. and R.B.L.; formal analysis, V.P.S.C.; investigation, D.S.O. and R.B.L.; resources, S.B.C.P.; writing—original draft preparation, D.S.O. and R.B.L.; writing—review and editing, S.B.C.P. and V.P.S.C.; visualization, V.P.S.C.; supervision, S.B.C.P.; project administration, V.P.S.C.; funding acquisition, S.B.C.P. All authors have read and agreed to the published version of the manuscript.

Funding: This research was funded by CAPES, grant number 88887.370976/2019-00.

Data Availability Statement: Not applicable.

Conflicts of Interest: The authors declare no conflict of interest.

References

1. Wang, S.; Dou, T.; Li, Y.; Zhang, Y.; Li, X.; Yan, Z. A Novel Method for the Preparation of MOR/MCM-41 Composite Molecular Sieve. *Catal. Commun.* **2005**, *6*, 87–91. [CrossRef]
2. Wei, Y.; De Jongh, P.E.; Bonati, M.L.M.; Law, D.J.; Sunley, G.J.; De Jong, K.P. Enhanced Catalytic Performance of Zeolite ZSM-5 for Conversion of Methanol to Dimethyl Ether by Combining Alkaline Treatment and Partial Activation. *Appl. Catal. A Gen.* **2015**, *504*, 211–219. [CrossRef]
3. Zhang, K.; Ostraat, M.L. Innovations in Hierarchical Zeolite Synthesis. *Catal. Today* **2016**, *264*, 3–15. [CrossRef]
4. Yang, Y.; Xu, R.; Zheng, C.; Long, Y.; Tang, S.; Sun, Z.; Huang, B.; Chen, J.P. Hierarchical Hollow Zeolite Fiber in Catalytic Applications: A Critical Review. *Chemosphere* **2022**, *307*, 135899. [CrossRef]
5. Egeblad, K.; Christensen, C.H.; Kustova, M.; Christensen, C.H. Templating Mesoporous Zeolites. *Chem. Mater.* **2008**, *20*, 946–960. [CrossRef]
6. Pérez-Ramírez, J.; Abelló, S.; Villaescusa, L.A.; Bonilla, A. Toward Functional Clathrasils: Size- and Composition-Controlled Octadecasil Nanocrystals by Desilication. *Angew. Chem. Int. Ed.* **2008**, *47*, 7913–7917. [CrossRef]
7. De Tarso Figueiredo Grecco, S.; Do Carmo Rangel, M.; Urquieta-González, E.A. Zeólitas Hierarquicamente Estruturadas. *Quim. Nova* **2013**, *36*, 131–142. [CrossRef]
8. Hartmann, M.; Machoke, A.G.; Schwieger, W. Catalytic Test Reactions for the Evaluation of Hierarchical Zeolites. *Chem. Soc. Rev.* **2016**, *45*, 3313–3330. [CrossRef]
9. Koohsaryan, E.; Anbia, M. Nanosized and Hierarchical Zeolites: A Short Review. *Cuihua Xuebao/Chin. J. Catal.* **2016**, *37*, 447–467. [CrossRef]
10. Schwieger, W.; Machoke, A.G.; Weissenberger, T.; Inayat, A.; Selvam, T.; Klumpp, M.; Inayat, A. Hierarchy Concepts: Classification and Preparation Strategies for Zeolite Containing Materials with Hierarchical Porosity. *Chem. Soc. Rev.* **2016**, *45*, 3353–3376. [CrossRef]
11. Mardiana, S.; Azhari, N.J.; Ilmi, T.; Kadja, G.T.M. Hierarchical Zeolite for Biomass Conversion to Biofuel: A Review. *Fuel* **2022**, *309*, 122119. [CrossRef]
12. Ogura, M.; Shinomiya, S.Y.; Tateno, J.; Nara, Y.; Kikuchi, E.; Matsukata, M. Formation of Uniform Mesopores in ZSM-5 Zeolite through Treatment in Alkaline Solution. *Chem. Lett.* **2000**, *29*, 882–883. [CrossRef]
13. Van Aelst, J.; Haouas, M.; Gobechiya, E.; Houthoofd, K.; Philippaerts, A.; Sree, S.P.; Taulelle, F. Hierarchization of USY zeolite by NH4OH. A postsynthetic process investigated by NMR and XRD. *J. Phys. Chem. C* **2014**, *118*, 22573–22582. [CrossRef]
14. Feng, A.; Yu, Y.; Mi, L.; Cao, Y.; Yu, Y.; Song, L. Synthesis and Characterization of Hierarchical Y Zeolites Using NH4HF2 as Dealumination Agent. *Microporous Mesoporous Mater.* **2019**, *280*, 211–218. [CrossRef]
15. Valtchev, V.; Balanzat, E.; Mavrodinova, V.; Diaz, I.; El Fallah, J.; Goupil, J.M. High Energy Ion Irradiation-Induced Ordered Macropores in Zeolite Crystals. *J. Am. Chem. Soc.* **2011**, *133*, 18950–18956. [CrossRef]
16. García-Martínez, J.; Johnson, M.; Valla, J.; Li, K.; Ying, J.Y. Mesostructured Zeolite y—High Hydrothermal Stability and Superior FCC Catalytic Performance. *Catal. Sci. Technol.* **2012**, *2*, 987–994. [CrossRef]
17. Rac, V.; Rakić, V.; Stošić, D.; Otman, O.; Auroux, A. Hierarchical ZSM-5, Beta and USY zeolites: Acidity assessment by gas and aqueous phase calorimetry and catalytic activity in fructose dehydration reaction. *Microporous Mesoporous Mater.* **2014**, *194*, 126–134. [CrossRef]
18. Gackowski, M.; Kuterasiński, Ł.; Podobiński, J.; Korzeniowska, A.; Sulikowski, B.; Datka, J. Hierarchical zeolite mazzite: Physicochemical properties and α-pinene isomerization. *Appl. Catal. A-Gen.* **2019**, *578*, 53–62. [CrossRef]
19. Han, S.; Wang, Z.; Meng, L.; Jiang, N. Synthesis of Uniform Mesoporous ZSM-5 Using Hydrophilic Carbon as a Hard Template. *Mater. Chem. Phys.* **2016**, *177*, 112–117. [CrossRef]
20. Cho, K.; Cho, H.S.; De Ménorval, L.C.; Ryoo, R. Generation of Mesoporosity in LTA Zeolites by Organosilane Surfactant for Rapid Molecular Transport in Catalytic Application. *Chem. Mater.* **2009**, *21*, 5664–5673. [CrossRef]
21. Wang, Z.; Li, C.; Cho, H.J.; Kung, S.C.; Snyder, M.A.; Fan, W. Direct, Single-Step Synthesis of Hierarchical Zeolites without Secondary Templating. *J. Mater. Chem. A* **2015**, *3*, 1298–1305. [CrossRef]
22. Xiao, F.S.; Wang, L.; Yin, C.; Lin, K.; Di, Y.; Li, J.; Xu, R.; Su, D.S.; Schlögl, R.; Yokoi, T.; et al. Catalytic Properties of Hierarchical Mesoporous Zeolites Templated with a Mixture of Small Organic Ammonium Salts and Mesoscale Cationic Polymers. *Angew. Chem. Int. Ed.* **2006**, *45*, 3090–3093. [CrossRef]
23. Lysenko, N.D.; Il'in, V.G.; Yaremov, P.S. Structural and Sorption Characteristics of the Products from Zeolitization of Sba-15 in the Presence of Tetraalkylammonium Hydroxides. *Theor. Exp. Chem.* **2011**, *47*, 257–263. [CrossRef]
24. Guo, Y.P.; Wang, H.J.; Guo, Y.J.; Guo, L.H.; Chu, L.F.; Guo, C.X. Fabrication and Characterization of Hierarchical ZSM-5 Zeolites by Using Organosilanes as Additives. *Chem. Eng. J.* **2011**, *166*, 391–400. [CrossRef]
25. Möller, K.; Bein, T. Mesoporosity—A New Dimension for Zeolites. *Chem. Soc. Rev.* **2013**, *42*, 3689–3707. [CrossRef] [PubMed]
26. Jia, X.; Khan, W.; Wu, Z.; Choi, J.; Yip, A.C.K. Modern Synthesis Strategies for Hierarchical Zeolites: Bottom-up versus Top-down Strategies. *Adv. Powder Technol.* **2019**, *30*, 467–484. [CrossRef]
27. Chen, L.H.; Sun, M.H.; Wang, Z.; Yang, W.; Xie, Z.; Su, B.L. Hierarchically Structured Zeolites: From Design to Application. *Chem. Rev.* **2020**, *120*, 11194–11294. [CrossRef]
28. Kerstens, D.; Smeyers, B.; Van Waeyenberg, J.; Zhang, Q.; Yu, J.; Sels, B.F. State of the Art and Perspectives of Hierarchical Zeolites: Practical Overview of Synthesis Methods and Use in Catalysis. *Adv. Mater.* **2020**, *32*, 2004690. [CrossRef]

29. Mumtaz, F.; Irfan, M.F.; Usman, M.R. Synthesis Methods and Recent Advances in Hierarchical Zeolites: A Brief Review. *J. Iran. Chem. Soc.* **2021**, *18*, 2215–2229. [CrossRef]
30. Hartmann, M.; Thommes, M.; Schwieger, W. Hierarchically-Ordered Zeolites: A Critical Assessment. *Adv. Mater. Interfaces* **2021**, *8*, 2001841. [CrossRef]
31. Serrano, D.P.; Escola, J.M.; Pizarro, P. Synthesis Strategies in the Search for Hierarchical Zeolites. *Chem. Soc. Rev.* **2013**, *42*, 4004–4035. [CrossRef] [PubMed]
32. Feliczak-Guzik, A. Hierarchical Zeolites: Synthesis and Catalytic Properties. *Microporous Mesoporous Mater.* **2018**, *259*, 33–45. [CrossRef]
33. Groen, J.C.; Moulijn, J.A.; Pérez-Ramírez, J. Decoupling Mesoporosity Formation and Acidity Modification in ZSM-5 Zeolites by Sequential Desilication-Dealumination. *Microporous Mesoporous Mater.* **2005**, *87*, 153–161. [CrossRef]
34. Jung, J.; Jo, C.; Mota, F.M.; Cho, J.; Ryoo, R. Acid Catalytic Function of Mesopore Walls Generated by MFI Zeolite Desilication in Comparison with External Surfaces of MFI Zeolite Nanosheet. *Appl. Catal. A Gen.* **2015**, *492*, 68–75. [CrossRef]
35. Hoff, T.C.; Gardner, D.W.; Thilakaratne, R.; Proano-Aviles, J.; Brown, R.C.; Tessonnier, J.P. Elucidating the Effect of Desilication on Aluminum-Rich ZSM-5 Zeolite and Its Consequences on Biomass Catalytic Fast Pyrolysis. *Appl. Catal. A Gen.* **2017**, *529*, 68–78. [CrossRef]
36. Qi, R.; Fu, T.; Wan, W.; Li, Z. Pore Fabrication of Nano-ZSM-5 Zeolite by Internal Desilication and Its Influence on the Methanol to Hydrocarbon Reaction. *Fuel Process. Technol.* **2017**, *155*, 191–199. [CrossRef]
37. Gackowski, M.; Tarach, K.; Kuterasiński, Ł.; Podobiński, J.; Jarczewski, S.; Kuśtrowski, P.; Datka, J. Hierarchical Zeolites Y Obtained by Desilication: Porosity, Acidity and Catalytic Properties. *Microporous Mesoporous Mater.* **2018**, *263*, 282–288. [CrossRef]
38. Ying, J.Y.; García-Martínez, J. Mesostructured Zeolitic Materials, and Methods of Making and Using the Same. U.S. Patent 7,589,041 B2, 15 September 2009.
39. Chal, R.; Cacciaguerra, T.; Van Donk, S.; Gérardin, C. Pseudomorphic Synthesis of Mesoporous Zeolite y Crystals. *Chem. Commun.* **2010**, *46*, 7840–7842. [CrossRef]
40. Sachse, A.; García-Martínez, J. Surfactant-Templating of Zeolites: From Design to Application. *Chem. Mater.* **2017**, *29*, 3827–3853. [CrossRef]
41. Kazakov, M.O.; Nadeina, K.A.; Danilova, I.G.; Dik, P.P.; Klimov, O.V.; Pereyma, V.Y.; Paukshtis, E.A.; Golubev, I.S.; Prosvirin, I.P.; Gerasimov, E.Y.; et al. Influence of USY Zeolite Recrystallization on Physicochemical Properties and Catalytic Performance of NiMo/USY-Al_2O_3 Hydrocracking Catalysts. *Catal. Today* **2019**, *329*, 108–115. [CrossRef]
42. Abelló, S.; Bonilla, A.; Pérez-Ramírez, J. Mesoporous ZSM-5 Zeolite Catalysts Prepared by Desilication with Organic Hydroxides and Comparison with NaOH Leaching. *Appl. Catal. A Gen.* **2009**, *364*, 191–198. [CrossRef]
43. Paixão, V.; Monteiro, R.; Andrade, M.; Fernandes, A.; Rocha, J.; Carvalho, A.P.; Martins, A. Desilication of MOR Zeolite: Conventional versus Microwave Assisted Heating. *Appl. Catal. A Gen.* **2011**, *402*, 59–68. [CrossRef]
44. Hasan, Z.; Jun, J.W.; Kim, C.U.; Jeong, K.E.; Jeong, S.Y.; Jhung, S.H. Desilication of ZSM-5 Zeolites for Mesoporosity Development Using Microwave Irradiation. *Mater. Res. Bull.* **2015**, *61*, 469–474. [CrossRef]
45. Liu, Y.; Zheng, D.; Li, B.; Lyu, Y.; Wang, X.; Liu, X.; Li, L.; Yu, S.; Liu, X.; Yan, Z. Isomerization of α-Pinene with a Hierarchical Mordenite Molecular Sieve Prepared by the Microwave Assisted Alkaline Treatment. *Microporous Mesoporous Mater.* **2020**, *299*, 110117. [CrossRef]
46. Young, D.A.; Yorba, L. Hydrocarbon Conversion Process and Catalyst Comprising a Crystalline Alumino-Silicate Leached with Sodium Hydroxide. U.S. Patent 3,326,797, 20 June 1967.
47. Young, D.A.; Yorba, L. Caustic-Treated Zeolites. U.S. Patent 3,374,182, 19 March 1968.
48. Rosback, D.H.; Neuzil, R.W. Method for Manufacturing an Adsorbent Useful for Olefin Separation. U.S. Patent 4,048,111, 13 September 1977.
49. Rein, A.J.; Saperstein, D.D.; Pines, S.H. Caustic Washed Molecular Sieves. U.S. Patent 4,134,965, 16 January 1979.
50. Dessau, R.M.; Valyocsik, E.W.; Goeke, N.H. Aluminum Zoning in ZSM-5 as Revealed by Selective Silica Removal. *Zeolites* **1992**, *12*, 776–779. [CrossRef]
51. Le Van Mao, R.; Ramsaran, A.; Xiao, S.; Yao, J.; Semmer, V. PH of the Sodium Carbonate Solution Used for the Desilication of Zeolite Materials. *J. Mater. Chem.* **1995**, *5*, 533–535. [CrossRef]
52. Čižmek, A.; Subotić, B.; Šmit, I.; Tonejc, A.; Aiello, R.; Crea, F.; Nastro, A. Dissolution of High-Silica Zeolites in Alkaline Solutions II. Dissolution of "activated" Silicalite-1 and ZSM-5 with Different Aluminum Content. *Microporous Mater.* **1997**, *8*, 159–169. [CrossRef]
53. Groen, J.C.; Peffer, L.A.A.; Moulijn, J.A.; Pérez-Ramírez, J. Mechanism of Hierarchical Porosity Development in MFI Zeolites by Desilication: The Role of Aluminium as a Pore-Directing Agent. *Chem. A Eur. J.* **2005**, *11*, 4983–4994. [CrossRef]
54. Groen, J.C.; Bach, T.; Ziese, U.; Paulaime-Van Donk, A.M.; De Jong, K.P.; Moulijn, J.A.; Pérez-Ramírez, J. Creation of Hollow Zeolite Architectures by Controlled Desilication of A1-Zoned ZSM-5 Crystals. *J. Am. Chem. Soc.* **2005**, *127*, 10792–10793. [CrossRef]
55. Groen, J.C.; Moulijn, J.A.; Pérez-Ramírez, J. Desilication: On the Controlled Generation of Mesoporosity in MFI Zeolites. *J. Mater. Chem.* **2006**, *16*, 2121–2131. [CrossRef]
56. Gackowski, M.; Podobiński, J.; Broclawik, E.; Datka, J. IR and NMR Studies of the status of Al and acid sites in desilicated zeolite Y. *Molecules* **2019**, *25*, 31. [CrossRef] [PubMed]
57. Gackowski, M.; Datka, J. Acid properties of hierarchical zeolites Y. *Molecules* **2020**, *25*, 1044. [CrossRef] [PubMed]

58. Wei, X.; Smirniotis, P.G. Development and Characterization of Mesoporosity in ZSM-12 by Desilication. *Microporous Mesoporous Mater.* **2006**, *97*, 97–106. [CrossRef]
59. Groen, J.C.; Sano, T.; Moulijn, J.A.; Pérez-Ramírez, J. Alkaline-Mediated Mesoporous Mordenite Zeolites for Acid-Catalyzed Conversions. *J. Catal.* **2007**, *251*, 21–27. [CrossRef]
60. Groen, J.C.; Abelló, S.; Villaescusa, L.A.; Pérez-Ramírez, J. Mesoporous Beta Zeolite Obtained by Desilication. *Microporous Mesoporous Mater.* **2008**, *114*, 93–102. [CrossRef]
61. Bonilla, A.; Baudouin, D.; Pérez-Ramírez, J. Desilication of Ferrierite Zeolite for Porosity Generation and Improved Effectiveness in Polyethylene Pyrolysis. *J. Catal.* **2009**, *265*, 170–180. [CrossRef]
62. Mokrzycki, Ł.; Sulikowski, B.; Olejniczak, Z. Properties of Desilicated ZSM-5, ZSM-12, MCM-22 and ZSM-12/MCM-41 Derivatives in Isomerization of α-Pinene. *Catal. Lett.* **2009**, *127*, 296–303. [CrossRef]
63. Verboekend, D.; Groen, J.C.; Pérez-Ramírez, J. Interplay of Properties and Functions upon Introduction of Mesoporosity in ITQ-4 Zeolite. *Adv. Funct. Mater.* **2010**, *20*, 1441–1450. [CrossRef]
64. Musilová-Pavlačková, Z.; Zones, S.I.; Čejka, J. Post-Synthesis Modification of SSZ-35 Zeolite to Enhance the Selectivity in p-Xylene Alkylation with Isopropyl Alcohol. *Top. Catal.* **2010**, *53*, 273–282. [CrossRef]
65. Sommer, L.; Mores, D.; Svelle, S.; Stöcker, M.; Weckhuysen, B.M.; Olsbye, U. Mesopore Formation in Zeolite H-SSZ-13 by Desilication with NaOH. *Microporous Mesoporous Mater.* **2010**, *132*, 384–394. [CrossRef]
66. De Jong, K.P.; Zečević, J.; Friedrich, H.; De Jongh, P.E.; Bulut, M.; Van Donk, S.; Kenmogne, R.; Finiels, A.; Hulea, V.; Fajula, F. Zeolite y Crystals with Trimodal Porosity as Ideal Hydrocracking Catalysts. *Angew. Chem. Int. Ed.* **2010**, *49*, 10074–10078. [CrossRef]
67. Akgül, M.; Karabakan, A. Promoted Dye Adsorption Performance over Desilicated Natural Zeolite. *Microporous Mesoporous Mater.* **2011**, *145*, 157–164. [CrossRef]
68. Verboekend, D.; Pérez-Ramírez, J. Desilication Mechanism Revisited: Highly Mesoporous All-Silica Zeolites Enabled through Pore-Directing Agents. *Chem. A Eur. J.* **2011**, *17*, 1137–1147. [CrossRef] [PubMed]
69. Verboekend, D.; Chabaneix, A.M.; Thomas, K.; Gilson, J.P.; Pérez-Ramírez, J. Mesoporous ZSM-22 Zeolite Obtained by Desilication: Peculiarities Associated with Crystal Morphology and Aluminium Distribution. *CrystEngComm* **2011**, *13*, 3408–3416. [CrossRef]
70. Matias, P.; Sá Couto, C.; Graa, I.; Lopes, J.M.; Carvalho, A.P.; Ramôa Ribeiro, F.; Guisnet, M. Desilication of a TON Zeolite with NaOH: Influence on Porosity, Acidity and Catalytic Properties. *Appl. Catal. A Gen.* **2011**, *399*, 100–109. [CrossRef]
71. Verboekend, D.; Vilé, G.; Pérez-Ramírez, J. Hierarchical y and USY Zeolites Designed by Post-Synthetic Strategies. *Adv. Funct. Mater.* **2012**, *22*, 916–928. [CrossRef]
72. Verboekend, D.; Keller, T.C.; Mitchell, S.; Pérez-Ramírez, J. Hierarchical FAU- and LTA-Type Zeolites by Post-Synthetic Design: A New Generation of Highly Efficient Base Catalysts. *Adv. Funct. Mater.* **2013**, *23*, 1923–1934. [CrossRef]
73. Verboekend, D.; Keller, T.C.; Milina, M.; Hauert, R.; Pérez-Ramírez, J. Hierarchy Brings Function: Mesoporous Clinoptilolite and L Zeolite Catalysts Synthesized by Tandem Acid-Base Treatments. *Chem. Mater.* **2013**, *25*, 1947–1959. [CrossRef]
74. Kubů, M.; Opanasenko, M.; Shamzy, M. Modification of Textural and Acidic Properties of -SVR Zeolite by Desilication. *Catal. Today* **2014**, *227*, 26–32. [CrossRef]
75. Muraza, O.; Bakare, I.A.; Tago, T.; Konno, H.; Taniguchi, T.; Al-Amer, A.M.; Yamani, Z.H.; Nakasaka, Y.; Masuda, T. Selective Catalytic Cracking of N-Hexane to Propylene over Hierarchical MTT Zeolite. *Fuel* **2014**, *135*, 105–111. [CrossRef]
76. Kubů, M.; Opanasenko, M.; Vitvarová, D. Desilication of SSZ-33 Zeolite-Post-Synthesis Modification of Textural and Acidic Properties. *Catal. Today* **2015**, *243*, 46–52. [CrossRef]
77. Ahmed, M.H.M.; Muraza, O.; Al Amer, A.M.; Sugiura, Y.; Nishiyama, N. Development of Desilicated EU-1 Zeolite and Its Application in Conversion of Dimethyl Ether to Olefins. *Microporous Mesoporous Mater.* **2015**, *207*, 9–16. [CrossRef]
78. Liu, H.; Liu, S.; Xie, S.; Song, C.; Xin, W.; Xu, L. Effect of Desilication on the Performance of Hierarchical ZSM-11 Catalysts for Alkylation of Benzene with Dimethyl Ether. *Catal. Lett.* **2015**, *145*, 1972–1983. [CrossRef]
79. Yang, C.; Xie, S.; Liu, H.; Xin, W.; Feng, C.; Li, X.; Liu, S.; Xu, L.; Zeng, P. IM-5 Zeolite Treated with Mixed Solution of NaOH and TPABr: Characterization and Application for Alkylation of Benzene with Ethanol. *Catal. Lett.* **2018**, *148*, 2030–2041. [CrossRef]
80. Čižmek, A.; Subotić, B.; Aiello, R.; Crea, F.; Nastro, A.; Tuoto, C. Dissolution of High-Silica Zeolites in Alkaline Solutions I. Dissolution of Silicalite-1 and ZSM-5 with Different Aluminum Content. *Microporous Mater.* **1995**, *4*, 159–168. [CrossRef]
81. Groen, J.C.; Jansen, J.C.; Moulijn, J.A.; Pérez-Ramírez, J. Optimal Aluminum-Assisted Mesoporosity Development in MFI Zeolites by Desilication. *J. Phys. Chem. B* **2004**, *108*, 13062–13065. [CrossRef]
82. Groen, J.C.; Peffer, L.A.A.; Moulijn, J.A.; Pérez-Ramírez, J. Mesoporosity Development in ZSM-5 Zeolite upon Optimized Desilication Conditions in Alkaline Medium. *Colloids Surf. A Physicochem. Eng. Asp.* **2004**, *241*, 53–58. [CrossRef]
83. Groen, J.C.; Peffer, L.A.A.; Moulijn, J.A.; Pérez-Ramírez, J. On the Introduction of Intracrystalline Mesoporosity in Zeolites upon Desilication in Alkaline Medium. *Microporous Mesoporous Mater.* **2004**, *69*, 29–34. [CrossRef]
84. Groen, J.C.; Moulijn, J.A.; Pérez-Ramírez, J. Alkaline Posttreatment of MFI Zeolites. From Accelerated Screening to Scale-Up. *Ind. Eng. Chem. Res.* **2007**, *46*, 4193–4201. [CrossRef]
85. Fernandez, C.; Stan, I.; Gilson, J.P.; Thomas, K.; Vicente, A.; Bonilla, A.; Pérez-Ramírez, J. Hierarchical ZSM-5 Zeolites in Shape-Selective Xylene Isomerization: Role of Mesoporosity and Acid Site Speciation. *Chem. A Eur. J.* **2010**, *16*, 6224–6233. [CrossRef]

86. Verboekend, D.; Nuttens, N.; Locus, R.; Van Aelst, J.; Verolme, P.; Groen, J.C.; Sels, B.F. Synthesis, characterisation, and catalytic evaluation of hierarchical faujasite zeolites: Milestones, challenges, and future directions. *Chem. Soc. Rev.* **2016**, *45*, 3331–3352. [CrossRef] [PubMed]
87. Verboekend, D.; Pérez-Ramírez, J. Design of Hierarchical Zeolite Catalysts by Desilication. *Catal. Sci. Technol.* **2011**, *1*, 879–890. [CrossRef]
88. Sadowska, K.; Wach, A.; Olejniczak, Z.; Kuśtrowski, P.; Datka, J. Hierarchic Zeolites: Zeolite ZSM-5 Desilicated with NaOH and NaOH/Tetrabutylamine Hydroxide. *Microporous Mesoporous Mater.* **2013**, *167*, 82–88. [CrossRef]
89. Verboekend, D.; Vilé, G.; Pérez-Ramírez, J. Mesopore Formation in Usy and Beta Zeolites by Base Leaching: Selection Criteria and Optimization of Pore-Directing Agents. *Cryst. Growth Des.* **2012**, *12*, 3123–3132. [CrossRef]
90. Tarach, K.; Góra-Marek, K.; Tekla, J.; Brylewska, K.; Datka, J.; Mlekodaj, K.; Makowski, W.; Igualada López, M.C.; Martínez Triguero, J.; Rey, F. Catalytic Cracking Performance of Alkaline-Treated Zeolite Beta in the Terms of Acid Sites Properties and Their Accessibility. *J. Catal.* **2014**, *312*, 46–57. [CrossRef]
91. McGlone, J.; Priecel, P.; Da Vià, L.; Majdal, L.; Lopez-Sanchez, J.A. Desilicated ZSM-5 Zeolites for the Production of Renewable p-Xylene via Diels–Alder Cycloaddition of Dimethylfuran and Ethylene. *Catalysts* **2018**, *8*, 253. [CrossRef]
92. Suzuki, T.; Okuhara, T. Change in Pore Structure of MFI Zeolite by Treatment with NaOH Aqueous Solution. *Microporous Mesoporous Mater.* **2001**, *43*, 83–89. [CrossRef]
93. Dai, G.; Wang, S.; Zou, Q.; Huang, S. Improvement of Aromatics Production from Catalytic Pyrolysis of Cellulose over Metal-Modified Hierarchical HZSM-5. *Fuel Process. Technol.* **2018**, *179*, 319–323. [CrossRef]
94. Zhang, L.; Ke, M.; Song, Z.; Liu, Y.; Shan, W.; Wang, Q.; Xia, C.; Li, C.; He, C. Improvement of the Catalytic Efficiency of Butene Oligomerization Using Alkali Metal Hydroxide-Modified Hierarchical Zsm-5 Catalysts. *Catalysts* **2018**, *8*, 298. [CrossRef]
95. Tang, S.; Zhang, C.; Xue, X.; Pan, Z.; Wang, D.; Zhang, R. Catalytic Pyrolysis of Lignin over Hierarchical HZSM-5 Zeolites Prepared by Post-Treatment with Alkaline Solutions. *J. Anal. Appl. Pyrolysis* **2019**, *137*, 86–95. [CrossRef]
96. Tanaka, S.; Fukui, R.; Kosaka, A.; Nishiyama, N. Development of Hierarchical and Phosphorous-Modified HZSM-5 Zeolites by Sequential Alkaline/Acid Treatments and Their Catalytic Performances for Methanol-to-Olefins. *Mater. Res. Bull.* **2020**, *130*, 110958. [CrossRef]
97. Peréz-Ramírez, J.; Verboekend, D.; Bonilla, A.; Abelló, S. Zeolite Catalysts with Tunable Hierarchy Factor by Pore-Growth Moderators. *Adv. Funct. Mater.* **2009**, *19*, 3972–3979. [CrossRef]
98. Holm, M.S.; Hansen, M.K.; Christensen, C.H. "One-Pot" Ion-Exchange and Mesopore Formation during Desilication. *Eur. J. Inorg. Chem.* **2009**, 1194–1198. [CrossRef]
99. Sadowska, K.; Góra-Marek, K.; Datka, J. Hierarchic Zeolites Studied by IR Spectroscopy: Acid Properties of Zeolite ZSM-5 Desilicated with NaOH and NaOH/Tetrabutylamine Hydroxide. *Vib. Spectrosc.* **2012**, *63*, 418–425. [CrossRef]
100. Raad, M.; Hamieh, S.; Toufaily, J.; Hamieh, T.; Pinard, L. Propane Aromatization on Hierarchical Ga/HZSM-5 Catalysts. *J. Catal.* **2018**, *366*, 223–236. [CrossRef]
101. Yoo, W.C.; Zhang, X.; Tsapatsis, M.; Stein, A. Synthesis of Mesoporous ZSM-5 Zeolites through Desilication and Re-Assembly Processes. *Microporous Mesoporous Mater.* **2012**, *149*, 147–157. [CrossRef]
102. Ivanova, I.I.; Kasyanov, I.A.; Maerle, A.A.; Zaikovskii, V.I. Mechanistic Study of Zeolites Recrystallization into Micro-Mesoporous Materials. *Microporous Mesoporous Mater.* **2014**, *189*, 163–172. [CrossRef]
103. Silva, J.F.; Ferracine, E.D.; Cardoso, D. Effects of Different Variables on the Formation of Mesopores in Y Zeolite by the Action of CTA^+ Surfactant. *Appl. Sci.* **2018**, *8*, 1299. [CrossRef]
104. Gorzin, F.; Towfighi Darian, J.; Yaripour, F.; Mousavi, S.M. Preparation of Hierarchical HZSM-5 Zeolites with Combined Desilication with $NaAlO_2$/Tetrapropylammonium Hydroxide and Acid Modification for Converting Methanol to Propylene. *RSC Adv.* **2018**, *8*, 41131–41142. [CrossRef]
105. Al-Ani, A.; Haslam, J.J.C.; Mordvinova, N.E.; Lebedev, O.I.; Vicente, A.; Fernandez, C.; Zholobenko, V. Synthesis of Nanostructured Catalysts by Surfactantlating of Large-Pore Zeolites. *Nanoscale Adv.* **2019**, *1*, 2029–2039. [CrossRef]
106. Van Laak, A.N.C.; Zhang, L.; Parvulescu, A.N.; Bruijnincx, P.C.A.; Weckhuysen, B.M.; De Jong, K.P.; De Jongh, P.E. Alkaline Treatment of Template Containing Zeolites: Introducing Mesoporosity While Preserving Acidity. *Catal. Today* **2011**, *168*, 48–56. [CrossRef]
107. Svelle, S.; Sommer, L.; Barbera, K.; Vennestrøm, P.N.R.; Olsbye, U.; Lillerud, K.P.; Bordiga, S.; Pan, Y.H.; Beato, P. How Defects and Crystal Morphology Control the Effects of Desilication. *Catal. Today* **2011**, *168*, 38–47. [CrossRef]
108. Bi, Y.; Lei, X.; Xu, G.; Chen, H.; Hu, J. Catalytic Fast Pyrolysis of Kraft Lignin over Hierarchical HZSM-5 and Hβ Zeolites. *Catalysts* **2018**, *8*, 82. [CrossRef]
109. Li, T.; Krumeich, F.; Chen, M.; Ma, Z.; Van Bokhoven, J.A. Defining Aluminum-Zoning during Synthesis of ZSM-5 Zeolites. *Phys. Chem. Chem. Phys.* **2020**, *22*, 734–739. [CrossRef]
110. Dos Santos, L.R.M.; Da Silva, M.A.P.; De Menezes, S.C.; Chinelatto, L.S.; Lam, Y.L. Creation of Mesopores and Structural Re-Organization in Beta Zeolite during Alkaline Treatment. *Microporous Mesoporous Mater.* **2016**, *226*, 260–266. [CrossRef]
111. Shah, A.K.; Maitlo, G.; Korai, R.M.; Unar, I.N.; Shah, A.A.; Khan, H.A.; Shah, S.F.A.; Ismail, U.; Park, Y.H. Citronellal Cyclisation to Isopulegol over Micro-Mesoporous Zsm-5 Zeolite: Effects of Desilication Temperature on Textural and Catalytic Properties. *React. Kinet. Mech. Catal.* **2019**, *128*, 507–522. [CrossRef]

112. Přech, J.; Pizarro, P.; Serrano, D.P.; Áejka, J. From 3D to 2D Zeolite Catalytic Materials. *Chem. Soc. Rev.* **2018**, *47*, 8263–8306. [CrossRef] [PubMed]
113. Vu, X.H.; Bentrup, U.; Hunger, M.; Kraehnert, R.; Armbruster, U.; Martin, A. Direct Synthesis of Nanosized-ZSM-5/SBA-15 Analog Composites from Preformed ZSM-5 Precursors for Improved Catalytic Performance as Cracking Catalyst. *J. Mater. Sci.* **2014**, *49*, 5676–5689. [CrossRef]
114. Ding, K.; Zhong, Z.; Wang, J.; Zhang, B.; Addy, M.; Ruan, R. Effects of Alkali-Treated Hierarchical HZSM-5 Zeolites on the Production of Aromatic Hydrocarbons from Catalytic Fast Pyrolysis of Waste Cardboard. *J. Anal. Appl. Pyrolysis* **2017**, *125*, 153–161. [CrossRef]
115. Chen, T.; Xiong, C.; Tao, Y. Enhanced Hydrolysis of Cellulose in Ionic Liquid Using Mesoporous ZSM-5. *Molecules* **2018**, *23*, 529. [CrossRef]
116. Rac, V.; Rakić, V.; Stošić, D.; Pavlović, V.; Bosnar, S.; Auroux, A. Enhanced Accessibility of Active Sites in Hierarchical ZSM-5 Zeolite for Removal of Pharmaceutically Active Substances: Adsorption and Microcalorimetric Study. *Arab. J. Chem.* **2020**, *13*, 1945–1954. [CrossRef]
117. Zhao, F.W.; Zhang, Q.; Hui, F.; Yuan, J.; Mei, S.N.; Yu, Q.W.; Yang, J.M.; Mao, W.; Liu, Z.W.; Liu, Z.T.; et al. Catalytic Behavior of Alkali Treated H-MOR in Selective Synthesis of Ethylenediamine via Condensation Amination of Monoethanolamine. *Catalysts* **2020**, *10*, 386. [CrossRef]
118. Lima, R.B.; Neto, M.M.S.; Oliveira, D.S.; Santos, A.G.D.; Souza, L.D.; Caldeira, V.P.S. Obtainment of Hierarchical ZSM-5 Zeolites by Alkaline Treatment for the Polyethylene Catalytic Cracking. *Adv. Powder Technol.* **2021**, *32*, 515–523. [CrossRef]
119. Smoliło-Utrata, M.; Tarach, K.A.; Samson, K.; Gackowski, M.; Madej, E.; Korecki, J.; Mordarski, G.; Śliwa, M.; Jarczewski, S.; Podobiński, J.; et al. Modulation of ODH Propane Selectivity by Zeolite Support Desilication: Vanadium Species Anchored to Al-Rich Shell as Crucial Active Sites. *Int. J. Mol. Sci.* **2022**, *23*, 5584. [CrossRef] [PubMed]

catalysts

MDPI

Article

Catalytic Distillation of Atmospheric Residue of Petroleum over HY-MCM-41 Micro-Mesoporous Materials

Camila G. D. P. Morais [1], Jilliano B. Silva [2], Josue S. Almeida [3], Rafaela R. Oliveira [3], Marcio D. S. Araujo [3], Glauber J. T. Fernandes [3], Regina C. O. B. Delgado [4], Ana C. F. Coriolano [5], Valter J. Fernandes, Jr. [3,*] and Antonio S. Araujo [5,*]

1 Federal Institute of Education, Science and Technology of Pará, Campus Maraba, Maraba 68501-000, Brazil
2 Post-Graduate Program in Petroleum Science and Engineering, Federal University of Rio Grande do Norte, Natal 59078-970, Brazil
3 Laboratory of Fuels and Lubricants, Institute of Chemistry, Federal University of Rio Grande do Norte, Natal 59078-970, Brazil
4 Department of Engineering and Technology, Federal Rural University of Semi-Arid, Mossoro 59625-900, Brazil
5 Laboratory of Catalysis and Petrochemistry, Institute of Chemistry, Federal University of Rio Grande do Norte, Natal 59078-970, Brazil
* Correspondence: valter.fernandes@ufrn.br (V.J.F.J.); antonio.araujo@ufrn.br (A.S.A.)

Abstract: Catalytic distillation is a technology that combines a heterogeneous catalytic reaction and the separation of reactants and products via distillation in a single reactor/distillation system. This process combines catalysis, kinetics, and mass transfer to obtain more selective products. The heterogeneous catalyst provides the sites for catalytic reactions and the porous surface for liquid/vapor separation. The advantages of catalytic distillation are energy savings, low waste streams, catalyst longevity, higher conversion, and product selectivity; these properties are interesting for petrochemical and petroleum industries. For this study, 100 mL of atmospheric residue of petroleum (ATR) was distilled in the presence of 1.0 g of a micro/mesoporous catalyst composed of a HY-MCM-41, and the reactor used was an OptiDist automatic distillation device, operating according to ASTM D-86 methodology. The products were collected and analyzed by gas chromatography. The samples of ATR, HY/ATR, and HY-MCM-41/ATR were analyzed by thermogravimetry (TG) to determine the activation energies (Ea) relative to the thermal decomposition of the process, using the Ozawa–Flynn–Wall (OFW) kinetic model. The obtained results show a potential catalytic distillation system for use in the reaction of heavy petroleum fractions and product separation from the HY/MCM-41 micro/mesoporous catalyst. The TG data revealed two mass loss events for ATR in the ranges of 100–390 and 390–590 °C, corresponding to volatilization and thermal cracking, respectively. The Ea determined for the thermal degradation of the ATR without a catalyst was in the range of 83–194 kJ/mol, whereas in the presence of the HY-MCM-41 catalyst, it decreased to 61–105 kJ/mol, evidencing the catalytic effect of the micro-mesoporous material. The chromatography analysis allowed for the identification of gasoline and a major production of diesel and gasoil when the HY-MCM-41 mixture was used as the catalyst, evidencing the synergism of the combined effect of the acid sites, the crystalline phase, and the microporosity of the HY zeolite with the accessibility of the hexagonal mesoporous structure of the MCM-41 material.

Keywords: hybrid material; HY/MCM-41; micro-mesoporous material; petroleum residue; thermogravimetry; catalysis

Citation: Morais, C.G.D.P.; Silva, J.B.; Almeida, J.S.; Oliveira, R.R.; Araujo, M.D.S.; Fernandes, G.J.T.; Delgado, R.C.O.B.; Coriolano, A.C.F.; Fernandes, V.J., Jr.; Araujo, A.S. Catalytic Distillation of Atmospheric Residue of Petroleum over HY-MCM-41 Micro-Mesoporous Materials. *Catalysts* **2023**, *13*, 296. https://doi.org/10.3390/catal13020296

Academic Editor: Narendra Kumar

Received: 26 December 2022
Revised: 17 January 2023
Accepted: 21 January 2023
Published: 28 January 2023

1. Introduction

Currently, with the increasing demand for and depletion of crude oil reserves around the world, more research regarding the catalytic cracking of heavier hydrocarbons, such as the atmospheric residue of petroleum (ATR), vacuum gas oil (VGO), and oil slurry to produce valuable fuels is necessary [1–4]. The fluidized catalytic cracking (FCC) of vacuum gas oil (VGO) is considered a promising process for enhancing the gasoline yield to fulfill the global energy demand. Recent research on new active catalysts suggests the modification of FCC to increase its efficiency [2]. The most suitable option considered is the development of hierarchical catalyst systems, focusing the micro/mesoporous materials, mainly HY, ZSM-5, MCM-41, and Y/MCM-41, composed of a particle of Y zeolite and a thin layer of MCM-41 [5–12]. Reactive distillation of heavy and extra-heavy oil, ATR, and VGO is considered a promising process for enhancing the diesel and gasoline yield to fulfill the energy demand [13–16]. Catalytic distillation is a process taking advantage of the synergy created when combining the catalyzed reaction and separation into a single unit, allowing for the production and removal of products with improved conversion and selectivity, saving energy and leading to highly-efficient systems [17,18].

In catalysis, however, the presence of micropores may impose internal diffusion limitations, resulting in low catalyst effectiveness, pore clogging by large products or oligomers, and ultimately coke formation and deactivation, resulting in a decrease in productivity with time under continuous use. To increase the accessibility of the reactants to the active sites, as well as the desorption of the products, the creation of additional mesopores in zeolite crystals has been proposed [19–28].

The oil refining process starts in the distillation units, first at atmospheric pressure, and then at reduced pressure (vacuum). In these two stages, bottom products are formed, known as atmospheric residue (ATR) and vacuum residue oil (VGO), which consist of high molecular weight fractions, composed of saturates, aromatics, resins, and asphaltenes [19]. The atmospheric distillation consists of heating the oil and separating the hydrocarbon products in specific temperature ranges in an efficient way to avoid thermal cracking. From the distillation tower are obtained light gases, liquefied petroleum gas, and liquid fractions of naphtha, gasoline, kerosene, diesel, and gas oil. The atmospheric residue of petroleum (ATR) is a viscous and dark liquid obtained at the end of the distillation tower, as shown in Figure 1.

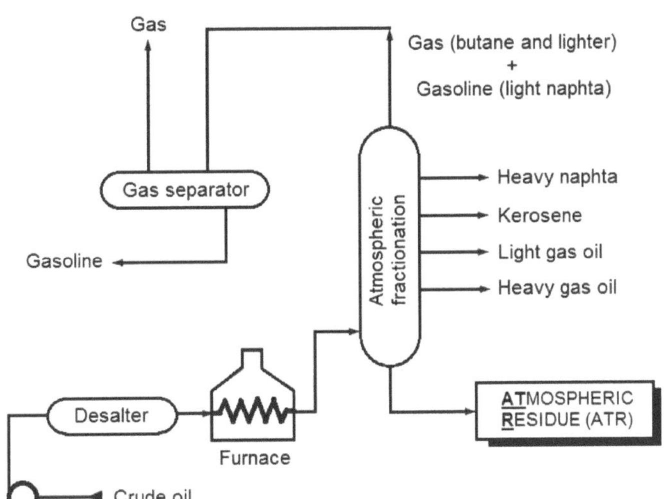

Figure 1. Scheme of the atmospheric distillation process in an oil refinery, showing the steps of the desalter, gas separator, and the raw hydrocarbon products, as well as the fraction of the atmospheric residue of petroleum at the bottom.

The ATR is a complex residue consisting of hydrocarbons, with carbon numbers predominantly greater than C_{11} and distillation above approximately 200 °C. In general, this residue contains ca. 5% or more by weight of four and six-membered polynuclear aromatic hydrocarbons. There is a growing need for the flexibility of conversion technologies, which leads to an emphasis on the processing of oil residues [20]. The processing consists of upgrading the residual fraction of heavy for high-value products, such as light distillates, fuels, and other raw materials for petrochemical industries [21]. Thus, distillation tests are useful tools to predict the yield of these fractions during refining. Catalytic or reactive distillation emerges as a promising technique, which can be used for processing even waste generated at refineries because it combines the reaction and separation of components in the same reactor [22], which is also beneficial for the environmental. The reactions include the cracking of heavy oil fractions, isomerization for increase in the octane of gasolines, and xylene isomerization for the synthesis of ethylbenzene [23]. The catalytic distillation offers some advantages, such as greater conversion, high selectivity, energy savings, and simple and easy equipment operation [24,25].

The use of heterogeneous catalysts in the degradation of petroleum residues has been a promising way to increase the yield and selectivity of products in the desired range of hydrocarbons. Among the oil improvement processes, catalytic distillation appears as an operation that can be used to refine the residues generated in refineries, since this process makes it possible to obtain lighter derivatives with greater added value for the consumer market. The catalyst most often used for petroleum refining is the microporous HY zeolite. The mesoporous materials are also of interest in catalysis, due to their beneficial properties such as uniform mesopores, a high surface area, and their high hydrocarbon sorption capacity [26–29].

The thermal degradation kinetics of ATR mixed with the AlSBA-15 catalyst have been evaluated using thermogravimetry [30,31]. Kinetic data were obtained by thermogravimetry and pyrolysis coupled with gas chromatography and mass spectrometry. Using AlSBA-15 as a catalyst, a better yield, with fractions in the range of gasoline and diesel, is obtained. This result demonstrated that the aluminum incorporated in the SBA-15 structure increased the acidity and consequently, allowed a better cracking activity of the ATR molecules. The use of the ZSM-5/MCM-41 hybrid catalyst for vacuum gas oil pyrolysis (VGO) was evaluated and compared to cracking using individual ZSM-5 and MCM-41 catalysts [32]. The use of this hybrid catalyst reduced the pyrolysis activation energy, and the products obtained in the process were hydrocarbons in the C_3–C_5 range (liquefied petroleum gas) and middle distillates, mainly C_6–C_{10} (gasoline) and C_{11}–C_{16} (diesel).

The aim of the current work is to evaluate the activity and selectivity of hybrid HY-MCM-41 micro-mesoporous material for the catalytic distillation of ATR in order to obtain high-value hydrocarbon products, such as gasoline and diesel. For this process, the large pores of hexagonal MCM-41 combined with the acidity of HY zeolite are useful for the catalytic cracking reactions of large molecules of hydrocarbons, as well as the subsequent separation of lower hydrocarbons, mainly in the range of natural gas, gasoline, and diesel.

2. Results and Discussion

2.1. Physicochemical Characterization of the Catalysts

Figure 2 presents the X-ray diffractograms of the HY and MCM-41 zeolite samples, along with the scanning electron micrograph for HY and the transmission electron micrograph for MCM-41, along with their pore systems.

For the HY zeolite diffractogram, the typical crystallinity of this material can be observed due to the presence of peaks at 2θ angles at 6.3, 10.1, 15.6, 18.2, 20.4, 23.7, 27.1, and 31.3 degrees. Similar data were found in the standard form of the JCPDS database 73-2310 [33]. The characteristic X-ray pattern of the mesoporous material of MCM-41 presents 3 to 5 peaks, referring to the reflection planes (100), (110), (200), (210), and (300), with the plane d(100) being the most intense, and the other planes of reflections being less intense. These reflections are characteristic of the hexagonal structure of the material. The absence of peaks at larger angles indicates that the material is not crystalline. However, it is known that there is an ordered network, which can also be observed by the high intensity of the peak corresponding to the reflection plane d(100) [34].

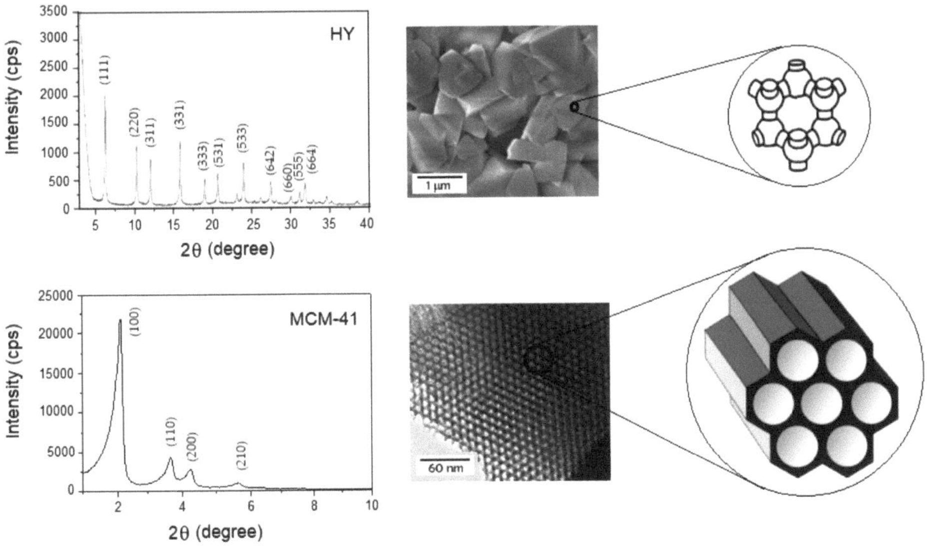

Figure 2. X-ray diffractogram and scanning electron micrograph of zeolite HY, and its transmission for MCM-41, with micro and mesopores systems, respectively.

The elemental analysis determined by X-ray fluorescence spectroscopy was carried out in order to determine the concentration of Si and Al present in the structure of zeolite HY and MCM-41, as well as to verify the presence of impurities. The sodium contents, in the form of Na_2O, determined for zeolite HY and MCM-41 were 0.93%wt. and 0.30%wt., respectively. Regarding the Si/Al ratio, the HY zeolite presented a value of 32, equivalent to that reported by the manufacturer. Regarding the pore diameter, it was observed that the HY zeolite presented a Dp of around 1.6 nm (<2 nm), while for MCM-41, the Dp was twice as high, around 3.4 nm (>2 nm), confirming the characteristics of the micros and mesopores of the materials. The specific area and pore volume values determined for the materials were considered satisfactory for processing bulky hydrocarbon molecules, such as those present in atmospheric petroleum residues. The HY zeolite showed a density of acidic sites equivalent to 2.30 mmol/g, whose value is due to the presence of structural aluminum in the zeolite, which after calcination in the NH4-Y form generates Bronsted acid centers inserted in the micropores of the zeolite. In the case of MCM-41, due to the absence of aluminum, this material showed negligible acidity. The physicochemical properties of the catalysts are given in Table 1.

Table 1. Structural and acidic properties of the microporous HY zeolites and mesoporous MCM-41 catalysts.

Sample	Dp (nm)	Wt (nm)	Vp (cm^3/g)	Si/Al	SA (m^2/g)	Acidity (mmol/g)
HY (Zeolyst)	1.6	-	-	32	660	2.30
MCM-41	3.4	2.99	0.13	-	998	0.01

Dp: pore diameter; Wt: wall thickness; Vp: pore volume; SA: surface area.

2.2. Characteristics of the ATR

The atmospheric petroleum residue was submitted to specific mass tests (°API), rheological parameters, and pour points. The results obtained are given in Table 2.

Table 2. Results of the physicochemical characterization of the ATR.

Physicochemical Property	Result
Specific mass at 20 °C, kg/m^3	935.9
Degree API at 20 °C	19.7
Pour point	23
Dynamic viscosity at 40 °C, mPas	679.9
Kinematic viscosity at 40 °C, mm^2/s	737.6

Since the specific mass depends on the mass of the individual compounds that form the atmospheric residue and on the size of the hydrocarbon chains, the value measured for the ATR indicates that this atmospheric residue is constituted by fractions rich in aromatic and naphthenic compounds. Its API degree is equal to 19.7, being classified as a heavy oil.

The pour point is an important petroleum parameter, as the cooling of the residue causes the formation of paraffin crystals. During this process, the heaviest components, which have the highest melting point, precipitate first, indicating that they are present in the fluid in a higher proportion. At a temperature of 23 °C, the cessation of ATR flow was observed; that is, the viscous behavior of the fluid ceased to be predominant. Considering the result obtained, the atmospheric petroleum residue is classified as a low pour point fluid. Although paraffins interfere directly with the pour point, the presence of aromatics (asphaltenes) is responsible for this parameter in heavy oils, as is the case with ATR [35].

2.3. Standard Distillation of ATR

Figure 3 shows the distillation curves obtained for the pure residue and its mixtures with the catalyst.

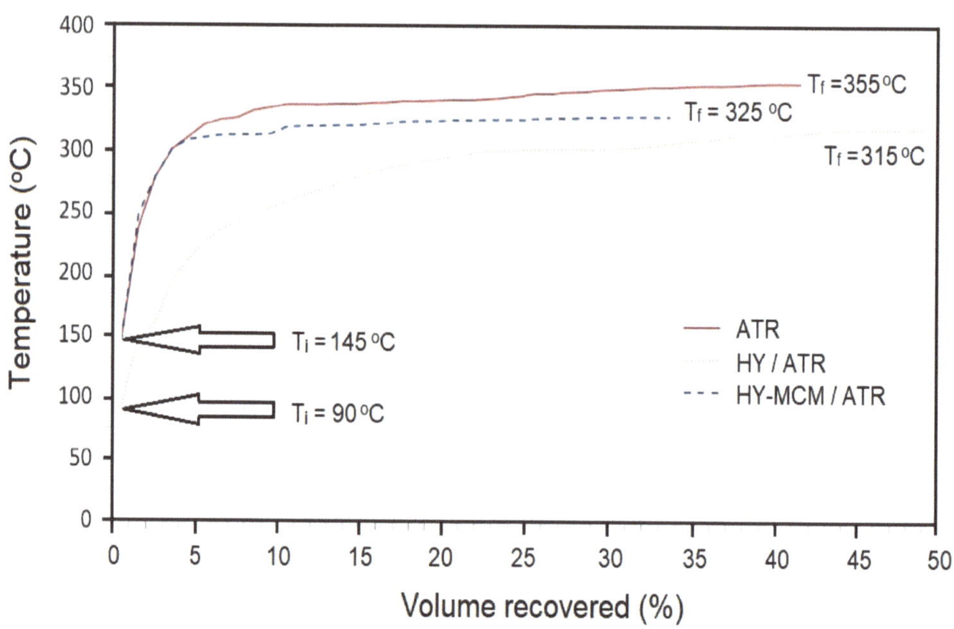

Figure 3. Standard distillation curve, as measured by the ASTM D86 method for ATR, and the catalytic distillation for HY/ATR and HY-MCM-41/ATR.

In general, in a refinery, the atmospheric distillation process is carried out at a temperature of up to about 350 °C. For ATR samples, with and without a catalyst, automatic distillation performed according to the ASTM D86 standard started after about 15 min, with higher initial boiling points for ATR and HY-MCM-41/ATR, equivalent to 149 °C, and for HY/ATR, the initial temperature dropped to 90 °C. The equipment used reaches a maximum heating temperature of 800 °C; however, it was verified that the distillation ended at the final temperatures of 355, 315, and 325 °C for ATR, HY/ATR, and HY-MCM-41/ATR, respectively. During the entire distillation, the HY/ATR sample had lower temperatures, indicating a greater efficiency in the breakdown of molecules from the atmospheric residue to smaller molecules, with a higher percentage of recovered volume, which reached about 50% vol, when compared to the distillation of the pure residue and HY-MCM-41/ATR, which obtained a recovery of 35% and 42% vol, respectively.

2.4. Analysis of the Distilled Products by Chromatography

The distillate products were collected and analyzed by gas chromatography to estimate composition as a function of hydrocarbon fractions. Firstly, a Supelco standard of n-paraffins (C_7 to C_{44}) was used to help identify the products, according to the retention times of the chromatogram peaks. From the analysis of this pattern, it was possible to determine the retention times of specific compounds and to determine the compositional profile of the sample of petroleum derivatives. Considering that all components of the injected sample reach the detector, the quantification could be performed by internal normalization, using FID response factors for the different compounds [36]. Figure 4 shows the chromatograms, with the identification of each n-paraffin from their respective retention times, for the pure ATR distillates, and with the HY and HY-MCM-41 catalysts.

By evaluating the chromatograms, a wide range of hydrocarbons is verified in the ATR, mainly between C_{10} and C_{32}. For the HY/ATR sample distillate, a greater number of peaks were identified in the first 7 min of retention time. This period corresponds to the elution of the components in the gasoline range, due to the acidity of the protonated zeolite associated with the presence of micropores in the zeolite HY. From this time on, no major changes were observed in relation to the pure ATR. However, when the HY-MCM-41 hybrid catalyst was used, the formation of hydrocarbon fractions in the C_{11}-C_{12} and C_{13}-C_{18} ranges, corresponding to kerosene and diesel, respectively, was observed. This is an indication that these fractions are present in a larger amount in the aforementioned distillate sample. From this point, heavier compounds, such as residual gas oil, are eluted.

Using the area normalization method, the composition of each of the distillates was determined in terms of petroleum derivative fractions. The graph in Figure 5 illustrates the yield of the products obtained.

In relation to the distillates with a catalyst addition, the ATR produced higher concentrations of diesel and heavy gas oil, which correspond to 56.5% of the fractions obtained in this condition. The use of HY in the distillation of the atmospheric residue was decisive for the production of lighter fractions, mainly gasoline. The elution of HY/ATR components in the first minutes of chromatographic analysis corresponds to the C_7-C_{18} range, including kerosene and diesel. These three fuels account for 75.4% of the products obtained by catalytic distillation with zeolite HY. In turn, the catalytic distillation of ATR with the mixture of HY-MCM-41 catalysts produced more diesel and heavy gas oil (C_{11}-C_{18}), which corresponds to 61.1% of the fractions obtained. From the data obtained by gas chromatography, when comparing the efficiency of HY and the HY/MCM-41 mixture, it is observed that the presence of the mesoporous material made the process more selective for diesel, while HY made it more selective for gasoline. Regarding the formation of lubrificans (C_{26}-C_{38}), the formation percentage was higher without the use of the catalyst.

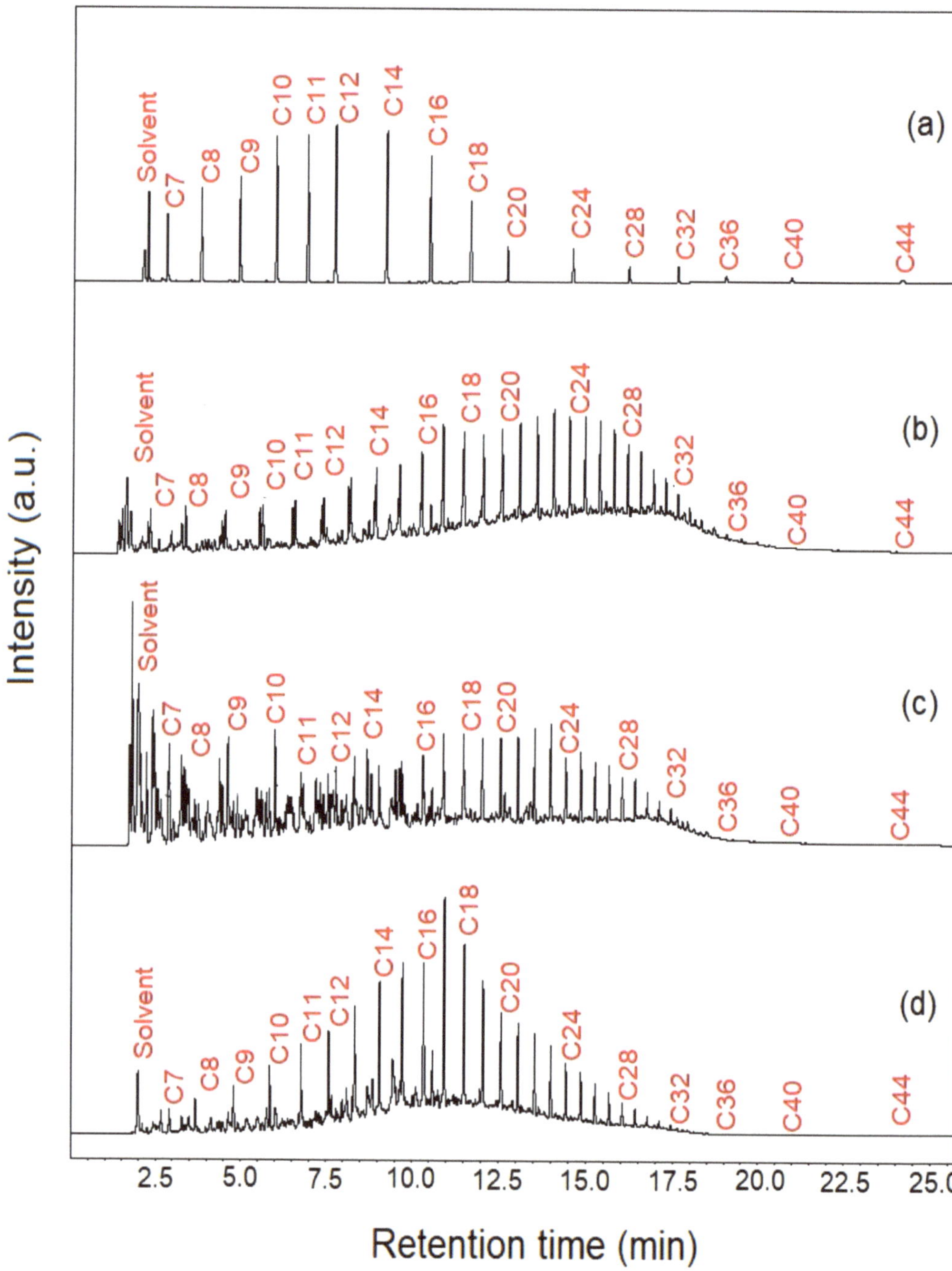

Figure 4. Chromatograms of hydrocarbons from: (**a**) Supelco standand for C_7-C_{44} hydrocarbons; (**b**) distilled products from ATR; and (**c**) distilled products from HY/ATR and (**d**) HY-MCM-41/ATR.

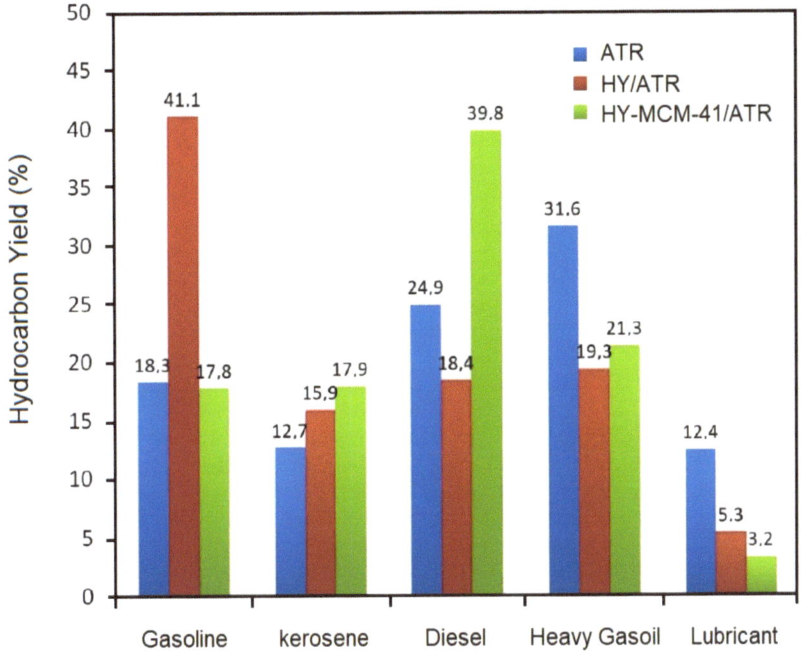

Figure 5. Hydrocarbon distribution obtained from the distillation of pure ATR and in presence of HY and hybrid HY-MCM-41 catalysts.

2.5. Thermal Analysis

Through thermogravimetry, the behavior of the thermal decomposition of the pure atmospheric residue, containing about 10% m of the catalyst, was evaluated. Thus, TG curves were obtained for ATR, HY/ATR, and HY-MCM-41/ATR. Figure 6 shows the overlapping of the TG curves obtained at a heating rate of 20 °C/min.

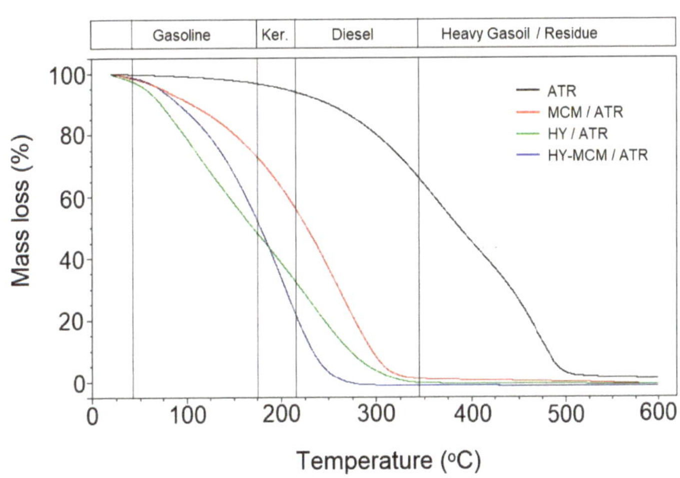

Figure 6. TG curves for ATR samples without catalyst; and ATR in the presence of porous materials, showing the temperature ranges for hydrocarbons.

The data from the TG analyses corroborate the results obtained by gas chromatography. In the TG curves, the percentages of mass losses related to the processes of separation and degradation of hydrocarbons are observed. The percentages were separated by the temperature ranges of gasoline, kerosene, diesel, and diesel residue (see Table 3). In Brazil, the hycrocarbon fraction and the values of boiling points are suggested by the National Agency of Petroleum, Natural Gas, and Biofuels (ANP).

Table 3. Approximate boiling ranges of different hydrocarbon fractions.

Fraction	Number of Carbons	Approximate Boiling Point (°C)
(i) Gasoline	C_5–C_{10}	40–175
(ii) Kerosene	C_{11}–C_{12}	175–215
(iii) Diesel	C_{13}–C_{18}	215–340
(iv) Heavy Gasoil	C_{18}–C_{25}	340–390

The Figure 6 shows the temperature ranges at which the catalytic conversion process associated with product separation in the presence of the zeolites began. The mass losses at temperature ranges of 40–175, 175–215, 215–340, and 340–390 °C are indicated in the TG curves. For the ATR sample, a significant mass loss occurred at a temperature range of 215–520 °C, relative to the non-catalyzed process.

The ATR samples containing catalysts showed mass losses in the ambient temperature range up to about 340 °C, which is the temperature range normally used for the atmospheric distillation of petroleum, and in the presence of a catalyst, it was called catalytic distillation. However, the existence of smaller carbon chains, such as those of gasoline, explain the appearance of mass losses observed in the first stage of thermal degradation for distillates for the HY/ATR and HY-MCM-41/ATR samples. Using these catalysts, the concentration of lighter compounds (C_5-C_{10}) was higher compared to the MCM-41/ATR sample, due to the presence of protonic acidity in the reaction medium due to the HY zeolite. The catalysts used showed that an appreciable amount of kerosene (C_{11}-C_{12}) can be obtained in the process. Regarding the diesel fraction (C_{13}-C_{18}), a greater selectivity was observed for the mesoporous material MCM-41, showing that the presence of mesopores associated with the accessibility of larger molecules favors the formation of diesel.

In the catalytic distillation steps, the ATR and the solid zeolite catalyst are physically mixed and heated simultaneously. Then, the products immediately begin to vaporize and are separated. By catalyzing and heating the reactants at the same time, the obtained products are quickly boiled out of the system. As the products are being continuously evaporated, the system does not reach equilibrium, and the formation of by-products causes the reaction to end. The products obtained from catalytic distillation, are more volatile than the initial reactants [37]. The proposed reactions for catalytic distillation are shown in Figure 7.

According to Figure 7, in the presence of the protons of the HY zeolite, or even the HY-MCM-41, the catalytic distillation reactions should occur according to the following steps: (1) primary cracking, forming a molecule from two species with weak or strong bonds; (2) isomerization, changing the structure of a molecule without changing its individual elements and their respective quantities; (3) alkylation, resulting in a formation of isoparaffins; and (4) the recombination of radicals to obtain large paraffins molecules, in the diesel range. The aromatics present in the ATR suffer thermal degradation, and should reacts with some free radicals, producing alkyl-aromatics.

The thermal degradation of oil and its products involves quite complex reactions that are difficult to interpret due to the presence of numerous components. These reactions may even vary their behavior at different heating rates. In this work, therefore, we opted for a kinetic model that uses three heating ratios.

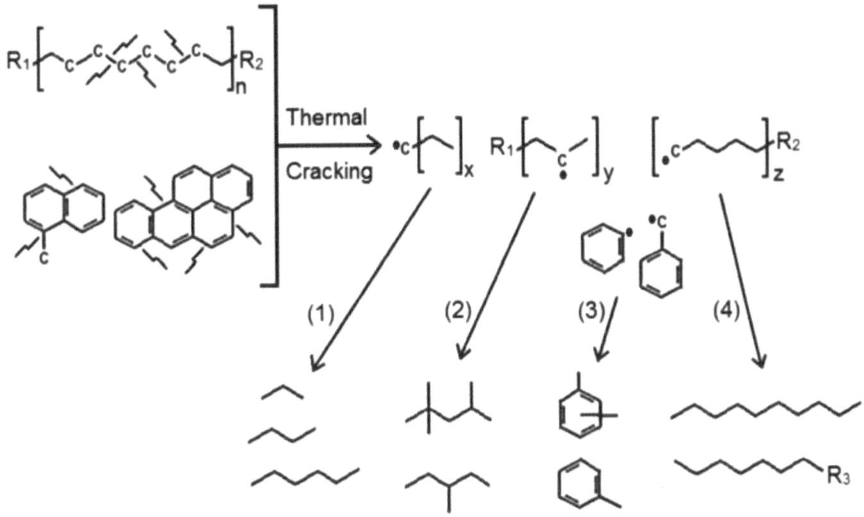

Figure 7. Proposed steps for thermal and catalytic distillation for the atmospheric residue of petroleum, where the steps are represented by (1) catalytic cracking, (2) isomerization, (3) alkylation, and (4) recombination.

Figure 8, shows the conversion curves vs. the temperature of the ATR, HY/ATR, and HY-MCM-41/ATR, using heating rates of 5, 10, and 20 °C/min. The conversion was determined by considering the initial and final masses isothermally. The ambient temperature range up to about 500 °C was used, and the conversion from 10% to 95% of conversion was considered.

The activation energies for the thermal and catalytic processes were determined using the Ozawa–Flynn–Wall kinetic (OFW) model from TG curves at multiple heating ratios. In this procedure, the conversion rate of the samples was initially determined as a function of temperature, time and heating ratio, in order to determine the activation energy (*Ea*), applying the for the OFW [38,39]. The thermal degradation of ATR is a complex reaction, considering that this decomposition follows a reaction of the nth order, where the phenomena of collision and energy barrier are generally correlated with Arrhenius constants. These phenomena can be represented by the apparent activation energy calculated from thermogravimetric data. The isoconversional methods are the most suitable to calculate the activation energy of the reactions, as they use different degrees of conversions (α) as a function of temperature and heating rate, without the knowledge degradation mechanism. The model proposed by Ozawa–Flynn–Wall is relatively simple and follows the Arrhenius equation for multiple heating ratios. One of the most important considerations of this model is that the conversion function $f(\alpha)$ is independent of the variation in the heating rate (β) for different values of the degree of conversion. The application of dynamic thermogravimetric methods is interesting for evaluating the physical and chemical processes that occur during the thermal degradation of ATR. This process can be represented by Equation (2).

$$\text{Heavy Oil} \rightarrow [\text{Gases/Naphta/Gasoil}] + [\text{Residue}] \quad (1)$$

The degree of decomposition or ATR conversion (α) is defined as the ratio of the mass lost in a time t with the total mass lost, as follows:

$$\alpha = \frac{w_0 - w_t}{w_0 - w_f} \quad (2)$$

where α is the extent of the decomposition; w_t, w_0, w_f are the actual, initial, and final mass of the sample, respectively. The reaction rate can be expressed by:

$$\frac{d\alpha}{dt} = kf(\alpha) \tag{3}$$

where k is the velocity constant, expressed as a function of temperature; $f(\alpha)$ is a conversion function and depends on the degradation mechanism. Assuming that the heating rate is constant (β), then:

$$T = T_0 + \beta T \tag{4}$$

$$\frac{dT}{dt} = \beta \tag{5}$$

The ATR conversion can be expressed as a function of the temperature variation, which depends on time:

$$\frac{d\alpha}{dt} = \frac{d\alpha}{dT}\cdot\frac{dT}{dt} = \beta\frac{d\alpha}{dT} \tag{6}$$

Thus, from the Arrhenius equation and Equations (2) and (5), we have:

$$\frac{d\alpha}{dT} = \frac{A}{\beta}f(\alpha)e^{-\frac{E_a}{RT}} \tag{7}$$

Rearranging and integrating Equation (7), we have Equations (8) and (9), in which $g(\alpha)$ is the integral function of α and $x = (-E_a/RT)$.

$$g(\alpha) = \int_0^\alpha \frac{d\alpha}{f(\alpha)} = \frac{A}{\beta}\int_{T_0}^{T} e^{-\frac{E_a}{RT}}dT \rightarrow g(\alpha) = \frac{AE_a}{\beta R}\cdot\left(-\frac{e^x}{x} + \int_{-\infty}^{x}\frac{e^x}{x}dx\right) \tag{8}$$

$$g(\alpha) = \frac{AE_a}{\beta R}\cdot p(x) \rightarrow log(g(\alpha)) = log\left(\frac{AE_a}{R}\right) - log(\beta) + log(p(x)) \tag{9}$$

The function $p(x)$, in Equation (8), does not present an exact analytical solution, requiring an approximation. The Ozawa-Flynn-Wall method is based on the approximation proposed by Doyle [40], which is presented in Equation (10) and used for values of $-60 \leq x \leq -20$, so that the errors due to the approximation are $\leq$5%. Thus, replacing Equation (10) in (9), in logarithmic form, the Equation (11) is obtained.

$$log(p(x)) \cong -2.315 + 0.457x \tag{10}$$

$$log(g(\alpha)) \cong log\left(\frac{AE_a}{R}\right) - log(\beta) - 2.315 - 0.457\left(\frac{E_a}{RT}\right) \tag{11}$$

Isolating the $log(\beta)$ term and deriving the equation with respect to $1/T$, we obtain Equation (12), where R = 8.314 $Jmol^{-1}K^{-1}$. Thus, the Ozawa-Flynn-Wall method provides the activation energy (Ea) in J/mol through a graph of $log(\beta)$ vs $1/T$.

$$\frac{\partial(log(\beta))}{\partial(1/T)} \cong -\left(\frac{0.457}{R}\right)\cdot E_a \rightarrow E_a \cong -18.192\frac{\partial(log(\beta))}{\partial(1/T)} \tag{12}$$

where β is the heating rate (K/min), T is the absolute temperature (K), and Ea is the activation energy (kJ/mol), for different degrees of conversion, from 10% to 95%, as shown in Figure 9. Thus, applying Equation (12), the Ea values can be easily calculated over a wide temperature range during the sample decomposition process, without knowing the reaction order.

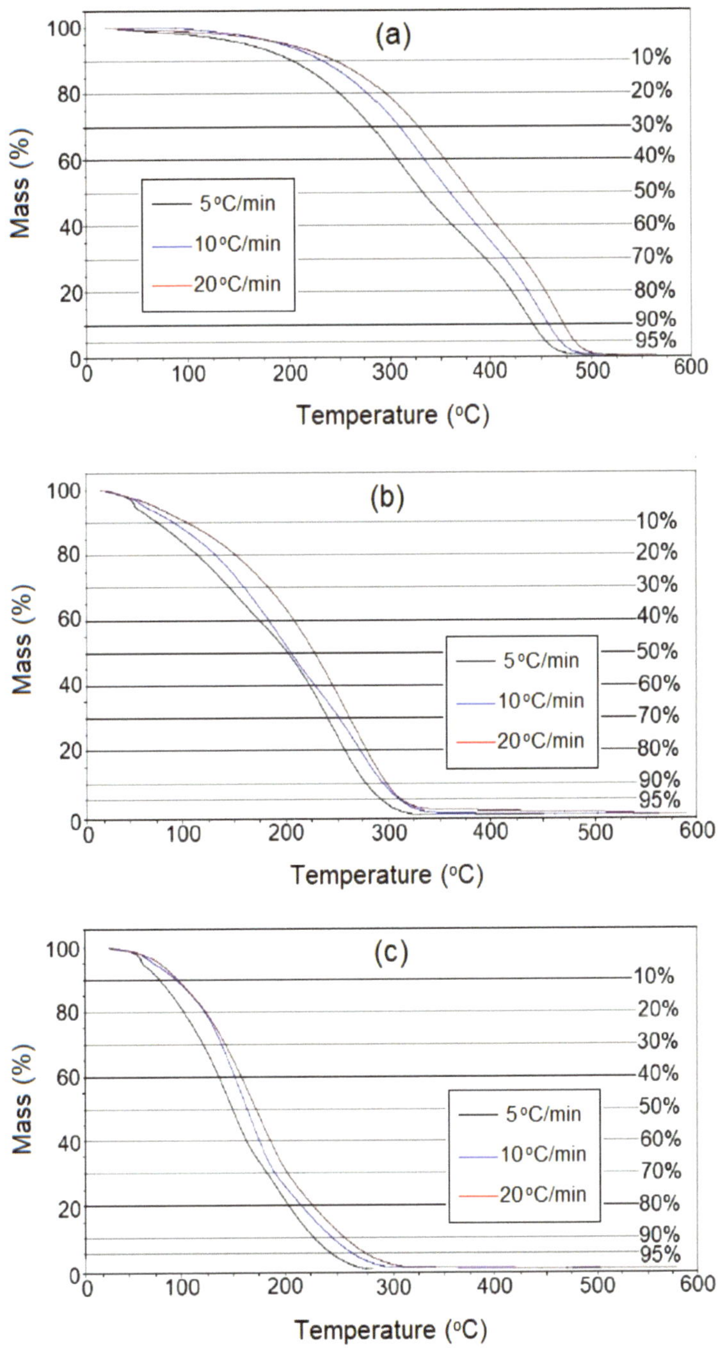

Figure 8. TG curves for thermal degradation ATR (**a**); catalytic distillation of HY/ATR (**b**); and HY-MCM-41 (**c**) samples at different heating rates, along with the degree of conversions.

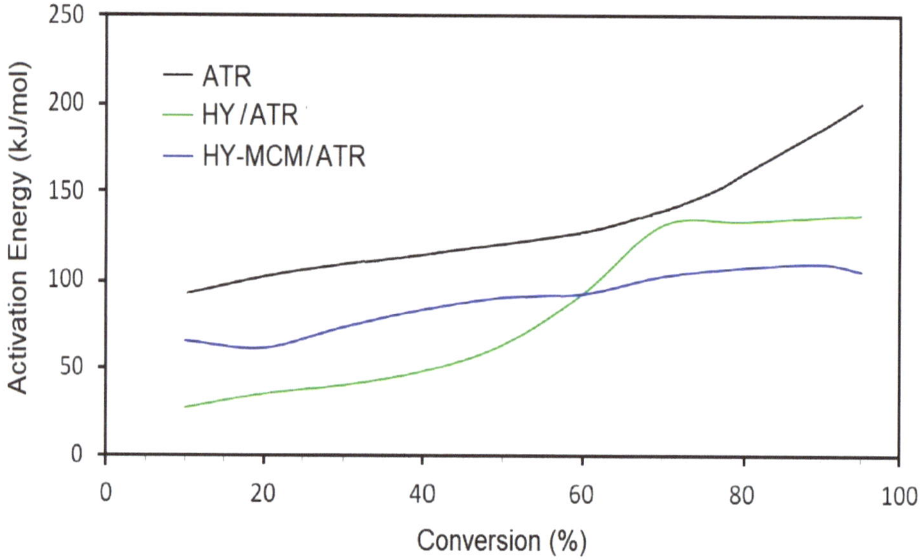

Figure 9. Activation energy as a function of degree of convention for the thermal and catalytic distillation of ATR, HY/ATR, and HY-MCM-41/ATR.

In the process of the thermal decomposition of the residue and distillates, the activation energy varied along the conversions, as illustrated in Figure 9. From the graph above, an increase in activation energy values is observed as the conversion increases. This behavior is similar to that reported in [41], since the curves of activation energy (*Ea*), as a function of conversion, express the characteristic of complex processes. This increase in energy and degree of conversion suggests the existence of parallel and secondary reactions in the decomposition process of petroleum residue.

As expected, pure ATR showed the highest *Ea*, with values in the range of 100 to 200 kJ/mol throughout the decomposition range. When ATR was used with the catalysts, it was observed that the HY zeolite had a lower *Ea* than the hybrid material up to 60% conversion, with values in the range of 25 to 90 kJ/mol, due to the acidity and microporosity of the zeolite HY. Up to this percentage, the *Ea* values for the HY-MCM-41 catalyst were observed in the range of 60 to 85 kJ/mol. After this percentage, there was an inversion in *Ea* values for 98 to 142 kJ/mol, and 95 to 105 kJ/mol, for HY and HY-MCM-41, respectively. From 60% conversion, the thermal decomposition process of the distillates with catalysts occurs with an *Ea* lower than the decomposition of the distillate without catalysts, with an increase in *Ea* being observed, probably due to the formation of carbonaceous residues inside the micropores of zeolite HY. However, the HY-MCM-41/ATR sample showed the smallest variation in activation energy along the conversions. The HY/ATR sample showed the lowest activation energies in the 10% to 60% range. From this point on, the HY-MCM-41/ATR presented the lowest *Ea* values, evidencing a greater external diffusion of the products formed, due to the existence of the mesopores of MCM-41. Compared with the results of distillation by the ASTM D86 method, these variations may be related to the composition of the distillates. Thus, it was verified that the catalytic distillation using the HY-MCM-41 mixture did not reach such a high recovery, so the energy required for the thermal degradation of HY-MCM-41/ATR becomes lower than that of HY/ATR.

3. Materials and Methods

3.1. Atmospheric Residue of Petroleum

The atmospheric residue of the petroleum sample was supplied by the Potiguar Clara Camarão Refinery, located at Guamaré City, Rio Grande do Norte State, Brazil. The sample was nominated as ATR, which was characterized by density (°API), viscosity, pour point, and thermogravimetry.

3.2. HY Zeolite and MCM-41 Catalysts

The commercial HY zeolite was obtained from the calcination of one NH_4-Y zeolite sample supplied by Zeolyst International (280 Cedar Grove Rd, Conshohocken, PA 19428, USA). For the calcinations process, the sample was heated from room temperature up to 500 °C, at a heating rate of 10 °C/min under nitrogen flowing at 100 mL/min. At this temperature, it was kept for 4 h for the decomposition of NH_4-Y to NH_3 and the obtaining of the HY acid form of the zeolites.

The MCM-41 was prepared according to the proposed methodology [42], with silica gel (Aldrich) as the source of silicon, cetyltrimethylammonium bromide (CTMABr), as the organic template, and sodium hydroxide, and distilled water as the solvent. These reactants were added in order to obtain a reactive hydrogel with the molar composition of $1CTMABr{:}2NaOH{:}4SiO_2{:}200H_2O$. The synthesis was carried out at in a Teflon-lined autoclave and heated at 100 °C, for a period of 4 d, with daily correction of pH in the range of 9–10 using a 30% solution of acetic acid. After the hydrothermal synthesis, the obtained material was filtered and washed with distilled water to remove the bromine ions and sodium waste. Then, the material was dried in an oven at 100 °C for 5 h. For removal of the organic template, the material was calcined at 450 °C, under nitrogen gas flowing at 100 mL/min, and a heating rate of 10 °C/min for 1 h. After that time, the gas was replaced with synthetic air, and the sample remained an additional 1 h at the same conditions.

The characterization of the materials was performed by X-ray diffraction (Shimadzu model XRD-6000 with CuKα radiation, Kyoto, Japan), scanning electron microscopy for characterization of the HY zeolites, and transmission electron microscopy for the characterization of MCM-41. The chemical composition was determined from X-ray fluorescency (XRF). The nitrogen adsorption and desorption isotherms were measured at a temperature of 77 K using a Quantachrome device, NOVA-2000 model (Boynton Beach, FL, USA), to determine the surface area from the BET method and the pore volume. Prior to adsorption measurement, the sample was degassed at a temperature of 300 °C for 2 h.

3.3. Catalytic Distillation of ATR

The thermal and catalytic distillation of the ATR was carried out using an automatic OptiDist–PAC apparatus as the reactor, as shown in Figure 10, according to the ASTM D86 methodology according to the Standard Test Method for Distillation of Petroleum Products and Liquid Fuels at Atmospheric Pressure. In the procedure, 100 mL of the sample was transferred into a 125 mL distillation flask. This balloon was placed in the equipment, jointly with the PT-100 thermocouple, which has been previously calibrated. At the end of the distillation, the observed steam temperatures were corrected to a barometric pressure of 101.3 kPa. For catalytic distillation, the ATR was previously mixed with the catalysts and then subjected to the reaction process. For the catalytic distillation, the HY and MCM-41 materials were physically added to the ATR at a concentration of 10% by mass, producing mixtures called HY/ATR and HY-MCM-41/ATR.

The calculation process to assess volume recovered consisted of monitoring the percentages evaporated at prescribed thermometer readings. The observed percentage loss of each of the observed values was reported according to Equation (13):

$$Vr = L - Pe \tag{13}$$

where Vr = percentage of volume recovered; Pe = percentage evaporated; and L = observed loss.

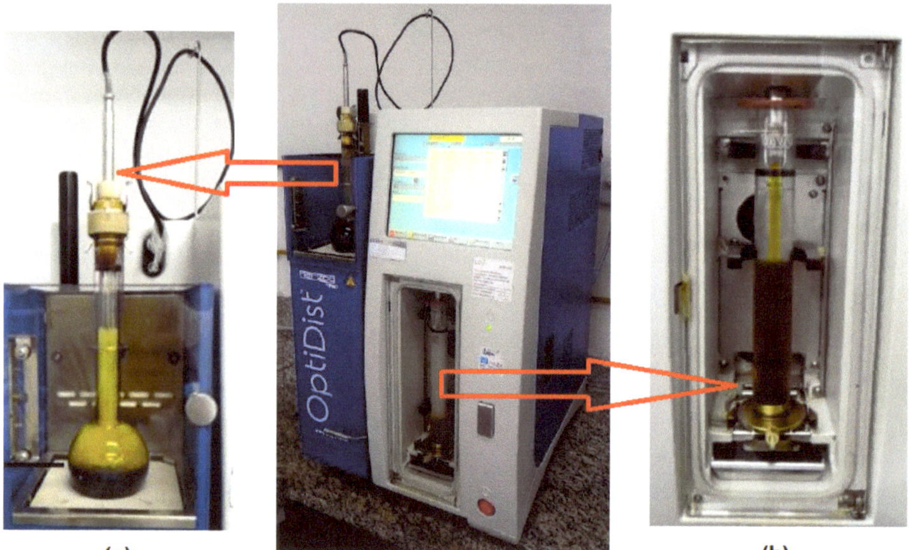

(a) (b)

Figure 10. Equipment used for automatic distillation according to ASTM D86: (**a**) the reactor containing the atmospheric residue of the petroleum and catalyst; (**b**) glass measuring cylinders with 100 mL capacity to collect the distilled products.

The gas chromatography analysis was carried out to estimate the composition of the distillates in terms of hydrocarbon fractions. A GC2010 chromatograph, from Shimadzu (Kyoto, Japan), was used with a split/splitless injector (SPL) and a hydrogen flame ionization detector. The separation column used was an HT5 SGE (nonpolar), with a 5% phenyl polysiloxane-carborane stationary phase, 25 m × 0.32 mm, and a film thickness of 0.1 µm. The results were processed using the GC Solution® software (SHIMADZU SCIENTIFIC INSTRUMENTS, INC., 7102 Riverwood Drive, Columbia, MD 21046, USA). The program used is shown in Table 4.

Table 4. Parameters used for gas chromatography analysis.

Oven Temperature Program	
-Initial temperature	35 °C for 2 min
-Heating rate; end temperature	20 °C/min up to 350 °C for 30 min
Detector parameters	
-Type	Flame ionization detector (FID)
-Temperature	350 °C
-Hydrogen flow	30 mL/min
-Sintetic air	300 mL/min
-Nitrogen (makeup)	30 mL/min
Gas flow to GC column	
-Carrier gas	Hydrogen
-Gas flow	12 mL/min
Injector parameters	
-Type	Split/Splitless
-Temperature of injector	300 °C
-Split ratio	10/1
-Volume of injection	0.5 µL
Time of analysis:	28 min

3.4. Thermal Analysis

In order to evaluate the distillation process of the petroleum residue, the thermal and catalytic degradation of the heavy oil was performed with ca. 10 mg of the sample, without a catalyst (ATR) and containing a catalyst, using thermogravimetry (TG) and differential thermogravimetry (DTG), in a TA Instruments thermobalance, model SQT600. For the experiments, the ATR was physically mixed with the HY and HY-MCM-41 materials, at concentration of ca. 10% mass, resulting in the HY/ATR and HY-MCM-41/ATR samples. The carrier gas was nitrogen, flowing at a rate of 50 mL/min. The mixtures were heated at heating rates of 5, 10, and 20 °C/min, from room temperature to 600 °C. The TG/DTG data were analyzed in the TA Universal Analysis® software, and the kinetic parameters were calculated in the TA Specialty Library® (TA Instruments, 159 Lukens Dr, New Castle, DE 19720, USA).

4. Conclusions

Catalytic distillation is equivalent to a reactive distillation that combines the thermal distillation process with catalysis using HY acid zeolite and MCM-41 to separate the hydrocarbon mixtures according to temperature and carbon ranges. The main function of the catalyst in the process was to maximize the yield of the organic catalytic reactions, such as gasoline and diesel. The zeolite catalysts were used in the form of a finely divided powder in order to increase the contact area with the ART, and consequently, to access to the micropores and internal acidity. In this way, it was possible to increase catalytic activity and selectivity, thus improving the efficiency of the process. In the presence of an active zeolite catalyst, these fractions are cracked and separated in specific temperature ranges, resulting in the catalytic distillation. In the process, the heat fractionates the molecular structure of the residue, releasing carbon compounds in gaseous and liquid forms, which may then be used as fuel. The use of hybrid HY-MCM-41 as catalysts revealed a promising path to increase yield and selectivity for desirable hydrocarbons.

For the catalytic distillation process, two factors were observed: heating, which promotes intense heat exchange, favoring solid–liquid reactions to obtain the desired products; and the porosities of the catalyst, which allowed bulky compounds present in the ATR to access their active reaction sites and become cracked, increasing the yield of the reactions. The distillation of ATR without catalysts converted this residue into diesel and heavy gas oil. However, the use of the HY catalyst was more efficient in the cracking and conversion of ATR molecules, as it achieved a higher percentage of conversion and a greater variety of products, in the gasoline and kerosene range, and showed lower selectivity for hydrocarbons in the C_{13}-C_{18} range, which is the diesel range. This occurred due to the characteristics of this catalyst, which has the ability to yield a fluid rich in lower molecular weight hydrocarbons, due to its protonic acidity and microporosity. The mixture of hybrid HY-MCM-41 micro-mesoporous catalysts was selective for the formation of hydrocarbons in the diesel range, due to the greater accessibility to the mesopores, with a subsequent reduction in coke formation.

The main advantage of catalytic distillation is the possibility of mixing the reactant, solid catalyst, and heat simultaneously. The products are continuously formed and are more volatile than the initial reactants, favoring immediate separation. In the presence of the acid sites of the zeolites, the short contact time of the hydrocarbons thermally cracked at low pressures favored secondary cracking reactions, increasing the selectivity to lower weigh molecular hydrocarbons, in the range of liquid gases, gasoline, diesel, and lubricants. The obtained results showed that catalytic distillation, combining micro and mesoporosity with protonic acidity, emerges as a promising technology for the valorization of industrial residues such as those generated in the initial stages of the refinery.

Author Contributions: Conceptualization, C.G.D.P.M. and J.B.S.; methodology, J.B.S. and J.S.A.; formal analysis, R.R.O. and G.J.T.F.; investigation, R.C.O.B.D.; data curation, A.C.F.C.; writing—original draft preparation, C.G.D.P.M. and A.S.A.; writing—review and editing, M.D.S.A., A.S.A. and C.G.D.P.M.;

project administration, A.S.A. and V.J.F.J.; funding acquisition, A.S.A. All authors have read and agreed to the published version of the manuscript.

Funding: This research was funded by CNPq, grant number 306780/2018-6.

Data Availability Statement: Not applicable.

Acknowledgments: The authors thank the Brazilian Agency of Petroleum, Natural Gas, and Biofuel (ANP) and the National Council for Scientific and Technological Development (CNPq Brazil, Grant number 306780/2018-6), for supporting this research.

Conflicts of Interest: The authors declare no conflict of interest.

References

1. Speight, J.G. Crude Oil–Distillation. In *Rules of Thumb for Petroleum Engineers*; Scrivener Publishing LLC: New York, NY, USA, 2017; pp. 185–186.
2. Jin, Q.; Li, Z.; Yan, Z.; Wang, B.; Wang, Z. Optimization Study on Enhancing Deep-Cut Effect of the Vacuum Distillation Unit (VDU). *Processes* **2022**, *10*, 359. [CrossRef]
3. Bartholomew, C.H.; Farrauto, R.J. Petroleum Refining and Processing. In *Fundamentals of Industrial Catalytic Processes*; John Wiley & Sons, Inc.: Hoboken, NJ, USA, 2005; pp. 635–704.
4. Zhao, R.F. *Progress and Application of Atmospheric and Vacuum Distillation Technology*; China Petrochemical Press: Beijing, China, 2020.
5. Corma, A. From Microporous to Mesoporous Molecular Sieve Materials and Their Use in Catalysis. *Chem. Rev.* **1997**, *97*, 2373–2420. [CrossRef] [PubMed]
6. de Jong, K.P.; Zecevic, J.; Friedrich, H.; de Jongh, P.E.; Bulut, M.; van Donk, S.; Kenmogne, R.; Finiels, A.; Hulea, V.; Fajula, F. Zeolite Y with trimodal porosity as ideal hydrocracking catalysts. *Angew. Chem.* **2010**, *49*, 10074–10078. [CrossRef] [PubMed]
7. Ruifeng, L.; Fan, W.; Ma, J.; Xie, K. Preparation of Y/MCM-41 composite materials. *Stud. Surf. Sci. Catal.* **2000**, *129*, 117–120.
8. Koch, H.; Liepold, A.; Roos, K.; Stöcker, M.; Reschetilowski, W. Comparative Study of the Acidic and Catalytic Properties of the Mesoporous Material H-MCM-41 and Zeolite H-Y. *Chem. Eng. Technol.* **1999**, *22*, 807–811. [CrossRef]
9. Serrano, D.P.; Escola, J.M.; Sanz, R.; Garcia, R.A.; Peral, A.; Moreno, I.; Linares, M. Hierarchical ZSM-5 zeolite with uniform mesopores and improved catalytic properties. *New J. Chem.* **2016**, *40*, 4206–4216. [CrossRef]
10. van Donk, S.; Janssen, A.H.; Bitter, J.H.; de Jong, K.P. Generation, Characterization, and Impact of Mesopores in Zeolite Catalysts. *Catal. Rev.* **2003**, *45*, 297–319. [CrossRef]
11. Kazakov, M.O.; Nadeina, K.A.; Danilova, I.G.; Dik, P.P.; Klimov, O.V.; Pereyna, V.; Yu Gerasimov, E.; Yu Dobryakova, I.V.; Knyazeva, E.E.; Ivanova, I.I.; et al. Hydrocracking of Vacuum Gas Oil over NiMo/γ-Al_2O_3: Effect of Mesoporosity Introduced by Zeolite Y Recrystallization. *Catal. Today* **2018**, *305*, 117–125. [CrossRef]
12. Ivanova, I.I.; Knyazeva, E.E. Micro-mesoporous materials obtained by zeolite recrystallization: Synthesis, characterization and catalytic applications. *Chem. Soc. Rev.* **2013**, *42*, 3671–3688. [CrossRef]
13. Theodore, L.; Dupont, R.R.; Ganesan, K. Distillation. In *Unit Operations in Environmental Engineering*; Scrivener Publishing LLC: New York, NY, USA, 2017; pp. 493–502.
14. Harmsen, G.J. Reactive distillation: The front-runner of industrial process intensification: A full review of commercial applications, research, scale-up, design and operation. *Chem. Eng. Process.* **2007**, *46*, 774–780. [CrossRef]
15. Stuart, F. Distillation in Refining. In *Distillation Operation and Applications*; Academic Press: New York, NY, USA, 2014; pp. 155–190.
16. Götze, L.; Bailer, O.; Moritz, P.; von Scala, C. Reactive distillation with KATAPAK®. *Catal. Today* **2001**, *69*, 201–208. [CrossRef]
17. Kiss, A.A. Novel Catalytic Reactive Distillation Processes for a Sustainable Chemical Industry. *Top. Catal.* **2019**, *62*, 1132–1148. [CrossRef]
18. Kiss, A.A. Distillation technology—Still young and full of breakthrough opportunities. *J. Chem. Technol. Biotechnol.* **2014**, *89*, 479–498. [CrossRef]
19. Prajapati, R.; Kohli, K.; Maity, S.K.; Garg, M.O. Coking propensity during hydroprocessing of vacuum residues, deasphalted oils, and asphaltenes. *Fuel* **2017**, *203*, 514–521. [CrossRef]
20. Grushova, E.I.; Sharif, A.S. Effect of polar organic substances on the distillates yield in atmospheric and vacuum distillation of crude oil. *Pet. Chem.* **2014**, *54*, 225–228. [CrossRef]
21. Sawarkar, A.; Aniruddha, P.; Shriniwas, S.; Jyeshtharaj, J. Petroleum residue upgrading via delayed cocking: A review. *Can. J. Chem. Eng.* **2007**, *85*, 1–24. [CrossRef]
22. Ramzan, N.; Faheem, M.; Gani, R.; Witt, W. Multiple steady states detection in a packed-bed reactive distillation column using bifurcation analysis. *Comput. Chem. Eng.* **2010**, *34*, 460–466. [CrossRef]
23. Naranov, E.; Dementev, K.; Gerzeliev, I.; Kolesnichenko, N.; Roldugina, E.; Maksimov, A. The role of zeolite catalysis in modern petroleum refining: Contribution from domestic technologies. *Pet. Chem.* **2019**, *59*, 247–261. [CrossRef]
24. Yan, L.; Ma, H.; Wang, B.; Mao, W.; Chen, Y. Advanced purification of petroleum refinery wastewater by catalytic vacuum distillation. *J. Hazard. Mater.* **2010**, *178*, 1120–1124. [CrossRef]
25. Zou, Z.; Dai, C.; Li, Q.; Chen, B. Synthesis of dimethyl ether (DME) by catalytic distillation. *Chem. Eng. Sci.* **2011**, *66*, 3195–3203.

26. Hu, S.; Liu, D.; Li, L.; Borgna, A.; Yang, Y. A non-sodium synthesis of highly ordered V-MCM-41 and its catalytic application in isomerization. *Catal. Lett.* **2009**, *129*, 478–485. [CrossRef]
27. Zhang, Y.P.; Xue, K.C.; Zhang, W.P.; Song, C.; Zhang, R.; Zhao, J. Synthesis and catalytic performance of MCM-41 modified with tetracarboxylphthalocyanine. *Chem. Pap.* **2013**, *67*, 372–379. [CrossRef]
28. Castro, K.K.V.; Paulino, A.A.D.; Silva, E.F.B.; Chellappa, T.; Lago, M.B.D.L.; Fernandes, V.J.; Araujo, A.S. Effect of the Al-MCM-41 catalyst on the catalytic pyrolisis of atmospheric petroleum residue (ATR). *J. Therm. Anal. Calorim.* **2011**, *106*, 759–762. [CrossRef]
29. Castro, K.K.V.; Figueiredo, A.L.; Gondim, A.D.; Coriolano, A.C.F.; Alves, A.P.M.; Fernandes, V.J., Jr.; Araujo, A.S. Pyrolisis of atmospheric residue of petroleum (ATR) using AlSBA-15 mesoporous material by TG and Py-GC/MS. *J. Therm. Anal. Calorim.* **2014**, *117*, 953–959. [CrossRef]
30. Coriolano, A.C.F.; Barbosa, G.F.S.; Alberto, C.K.D.; Delgado, R.C.O.B.; Castro, K.K.V.; Araujo, A.S. Catalytic processing of atmospheric residue of petroleum over AlSBA-15 nanomaterials with different acidity. *Pet. Sci. Technol.* **2016**, *34*, 627–632. [CrossRef]
31. Silva, E.F.B.; Ribeiro, M.P.; Coriolano, A.C.F.; Melo, A.C.R.; Santos, A.G.D.; Fernandes, V.J.; Araujo, A.S. Kinetic study of degradation of heavy oil over MCM-41. *J. Therm. Anal. Calorim.* **2011**, *106*, 793–797. [CrossRef]
32. Coriolano, A.C.F.; Silva, C.G.C.; Costa, M.J.F.; Pergher, S.B.C.; Caldeira, V.P.S.; Araujo, A.S. Development of HZSM-5/AlMCM-41 hybridmicro-mesoporous material and application for pyrolysis of vacuum gasoil. *Microporous Mesoporous Mater.* **2013**, *172*, 206–212. [CrossRef]
33. Daou, T.J.; Dos Santos, T.; Nouali, H.; Josien, L.; Michelin, L.; Pieuchot, L.; Dutournie, P. Synthesis of FAU-Type Zeolite Membranes with Antimicrobial Activity. *Molecules* **2020**, *25*, 3414. [CrossRef]
34. Beck, J.S.; Vartuli, J.C.; Roth, W.J.; Leonowicz, M.E.; Kresge, C.T.; Schmitt, K.D.; Chu, C.T.W.; Olson, D.H.; Sheppard, E.W.; McCullen, S.B.; et al. A new family of mesoporous molecular sieves prepared with liquid crystal templates. *J. Am. Chem. Soc.* **1992**, *114*, 10834–10843. [CrossRef]
35. Sad, C.M.S.; Lacerda, V., Jr.; Filgueiras, P.R.; Rigoni, V.S.; Bassane, J.P.F.; Castro, E.V.R.; Pereira, K.S.; Santos, M.P.F. Limitations of the pour point measurement and the influence of the oil composition on its detection using principal component analysis. *Energy Fuels* **2014**, *28*, 1686–1691. [CrossRef]
36. Beens, J.; Brinkman, U.A. The role of gas chromatography in compositional analyses in the petroleum industry. *Trends Anal. Chem.* **2000**, *19*, 260–275. [CrossRef]
37. Silva, J.M.R.; Oliveira, M.H.R.; Nosman, T.; Coriolano, A.C.F.; Fernandes, G.J.T.; Fernandes, V.J.; Araujo, A.S. Catalytic distillation of an atmospheric petroleum resid using HZSM-5 and HY zeolites. *Pet. Sci. Technol.* **2017**, *35*, 1938–1943. [CrossRef]
38. Flynn, J.H. Early papers by Takeo Ozawa and their continuing relevance. *Thermochim. Acta* **1996**, *283*, 35–42. [CrossRef]
39. Flynn, J.H. A function to aid in the fitting of kinetic data to a rate equation. *J. Phys. Chem.* **1957**, *61*, 110. [CrossRef]
40. Doyle, C. Kinetic analysis of thermogravimetric data. *J. Appl. Polym. Sci.* **1961**, *5*, 285–292. [CrossRef]
41. Kök, M.V. Effect of metal oxide on light oil combustion. *Therm. Anal. Calorim.* **2003**, *73*, 241–246. [CrossRef]
42. Araujo, A.S.; Jaroniec, M. Determination of the surface area and mesopore volume for lanthanide-incorporated MCM-41 materials by using high resolution thermogravimetry. *Thermochim. Acta* **2000**, *345*, 173–177. [CrossRef]

catalysts

Review

Homogeneous and Heterogeneous Catalytic Ozonation of Textile Wastewater: Application and Mechanism

Magdalena Bilińska [1,2,*], Lucyna Bilińska [1,2] and Marta Gmurek [1]

[1] Department of Molecular Engineering, Lodz University of Technology, Wolczanska 213, 90-005 Lodz, Poland
[2] Bilinski Factory of Colour, Mickiewicza 29, 95-050 Konstantynow Lodzki, Poland
* Correspondence: magdalena.bilinska@dokt.p.lodz.pl; Tel.: +697826296

Abstract: This paper presents an overview of textile wastewater treatment by catalytic ozonation, highlighting the parameters of the process and accompanying mechanisms. Since more than 800,000 tons of dyes are produced annually and thousands of cubic meters of highly polluted textile wastewater have been emitted into the environment every day, this issue has become an environmental concern. Due to the high oxidative potential of ozone (2.08 V) and hydroxyl radical (2.80 V), the main reactive species in catalytic ozonation, the burdensome organic pollutants, including textile dyes, can be successfully decomposed. The paper shows the main groups of catalysts, emphasizing novel structural, nano-structured, and functionalized materials. The examples of catalytic ozonation in the industrial application for real textile wastewater were specially highlighted.

Keywords: textile wastewater treatment; advanced oxidation processes; catalytic ozonation; reaction mechanism; parameters of the ozonation catalytic process

check for updates

Citation: Bilińska, M.; Bilińska, L.; Gmurek, M. Homogeneous and Heterogeneous Catalytic Ozonation of Textile Wastewater: Application and Mechanism. *Catalysts* **2023**, *13*, 6. https://doi.org/10.3390/catal13010006

Academic Editor: Narendra Kumar

Received: 30 November 2022
Revised: 16 December 2022
Accepted: 19 December 2022
Published: 21 December 2022

1. Introduction

Thousands of cubic meters of highly polluted textile wastewater are emitted into the environment every day, making it a severe environmental threat. The average volume between 150 and 200 L per 1 kg of textiles produced in chemical processing is assumed [1]. However, according to Ghaly and co-workers, this value can be as high as 933 L per 1 kg in specific cases, e.g., in felt production [2]. Considering the ever-growing demand for textile goods, it is no wonder that textile wastewater has a significant impact on global pollution. China, India, Pakistan, Indochina countries, and some of EU countries are the major exporters of textiles [2]. Excluding Europe, where environmental protection is at a high technological level, most of these regions suffer from massive pollution. Australia's Perth region has an interesting approach to wastewater treatment for recycling purposes to meet long-term water needs. They use, among others, the method of oxidation ditches [3]. Membranes for closed water circulation at an on-site textile plant (Osmonics, USA) were also investigated to prevent clean water consumption [4]. There are several textile industrial zones in Bangladesh (such as Dhaka, Narayanganj, and Gazipur) reported being afflicted by untreated effluent discharge into water bodies [5]. The next region severely affected by the textile industry is the Southcoast area in China [6]. Heavy metals from dyes are reported to be the most burdensome pollution in local rivers [7]. The most shocking example of how the local landscape can be deformed by irresponsible wastewater management was shown in the Punjab region in India. Bhatia and co-workers reported the effects of direct textile wastewater discharge into surface water bodies, which made it a sewerage-like quality. What is more, the local community uses this water for living purposes, dramatically decreasing their hygienic standards [8]. Conclusively, numerous local legal acts uphold the limiting values of wastewater discharge [9]. At the same time, valid global institutions, such as the European Parliament [10], Organization for Economic Co-operation and Development (OECD) [11], or even public responsibility organizations such as Greenpeace [12], took up efforts to decrease textile waste. This opens a wide field for investigation in the area of textile wastewater treatment. Especially when the priority is to make this wastewater a source of recyclable water for technological purposes as it stands in EU and OECD reports [10,11].

As far as textile wastewater is concerned, numerous treatments were used [13–15], and color removal was the most investigated. The color is the most vivid manifestation of pollution emitted by the textile industry into the water environment. Besides visual disgust, when emitted into natural water bodies, dyes' ability to light absorption disturbs living conditions by impairing water organisms' photosynthesis. Moreover, dyes' bioaccumulation and genotoxicity, exhibited in chromosomal aberrations, were proven for testing organisms [16]. Some azo dyes, such as Acid Black 077 [17], are confirmed as cancerogenic, and those were listed under the register of legally forbidden substances in EU countries [18]. However, environmental and health concerns are not the only pollution brought by textile wastewater. It has to be kept in mind that chemical textile processing on an industrial scale is a chain of many technological stages. Operations such as washing, scouring, bleaching, rinsing, dyeing, printing, fixing, and padding, following one by another, using a vast spectrum of chemicals. Therefore, numerous detergents, soda, acids, salts, urea, silicones, perfluorinated carbonates, enzymes, biologically active agents (e.g., silver, permethrin), and flame retardants can be used as textile auxiliaries in the production process, and finally occur in wastewater [9,19–21]. This wastewater is characterized by a high mineral content, resulting in low biodegradability (biodegradability index (BI) below 0.4) [22]. Consequently, bio-treatment is limited [13]. Textile wastewater is characterized by large ranges of chemical oxygen demand (COD): 150–12,000 (mg/L), biochemical oxygen demand (BOD): 80–6000 (mg/L) [23], and total organic carbon (TOC): 50–40,000 (mg/L) [13]. The textile wastewater purification is extremely difficult for additional reasons, as the chemical composition of the wastewater changes depending on the production settings. The settings of textile processing are imposed primarily by the composition of the textile garment. Each fiber type demands its dying technology and chemicals, which are presented in Table 1. This variety in the production system results in diversified wastewater characteristics changing over time.

Table 1. Examples of the most common textile fibers processing chemically in dye houses (own study of a different kind of textile treatment from Bilinski Factory sewage).

Fiber Type	Type of Dye	Auxiliaries	Temperature (Celsius) of Wastewater	pH of Wastewater
Cellulosic (cotton, viscose)	Reactive Direct	dye salt (NaCl, Na_2SO_4) soda (NaOH, Na_2CO_3) leveling agent the enzyme (against H_2O_2)	60, 80	10–12
Woll, Silk, Polyamide	Acid	Dye acid (H_2SO_4, formic, acetic) salt (Na_2SO_4) ammonium sulfate	100	2–7
Polyester	Disperse	Dye Dispergator acid (formic, acetic)	130 (under pressure)	5
Acrylic	Cationic	Dye acid (formic, acetic) salt (Na_2SO_4, sodium acetate)	90	4–5

This short characteristic of the textile processing system shows the complexity of the issue. The variety of pollutants which can occur besides dyes in textile wastewater, high temperature, low BI, and instability in time make it a challenge to find a proper treatment. This opens the discussion to novel enhanced strategies of textile-source pollutants removal from water, making it recyclable. One of the last willingly presented approaches is combining chemical and biological treatments used as a sequence. This idea's motivation is to promote biodegradability and decrease toxicity at the first stage of chemical treatment to make further bio-treatment more efficient [13,14,24]. The preferable group of chemical treatments is advanced oxidation processes (AOPs). Ozone and reactive oxygen species (ROS) can decompose organics, including dyes, so ozone-based AOPs have been widely used in textile wastewater treatment [25–32]. Simultaneously, some traditional AOPs, H_2O_2/UV or Fenton, were insufficient for real textile wastewater purification because, despite discoloration, they were still characterized by high impurities [33–35].

Moreover, the application of hydrogen peroxide and UV-based technologies like O_3/H_2O_2, O_3/UV, and $O_3/H_2O_2/UV$ for textile wastewater treatment is limited [33]. Since H_2O_2 is an expensive compound and the use of UV provides an additional energy cost of around EUR, USD $2/m^3$. The new trend in the treatment of textile wastewater is catalytic ozonation. The main objective of this review paper was to show perspectives on the use of catalytic ozonation for textile wastewater treatment. The presentation of vast catalysts is used for enhanced reactive oxygen species (ROS) production. The discussion on the possible mechanisms of catalytical action includes the main groups of catalysts. Particular emphasis was put on novel structural, nano-structured, and functionalized materials used as catalysts. Efforts to expose developments in reactor construction, ensuring proper contact between the catalyst, ozone, gas, and wastewater, were undertaken. The authors aimed to highlight examples of catalytic ozonation in the industrial application for real textile wastewater, showing its practical use.

2. Basic of Ozonation

Some facts from ozone chemistry must be brought up to understand the ozone-based catalytic processes. Ozone (O_3) is a strong oxidant, which can decompose numerous pollutants. Nevertheless, molecular ozone as an oxidant is selective, and some groups of pollutants, like carboxylic acids, naphthol, and phenol derivatives, were referred to as hardly degradable via single ozonation [36]. Moreover, it should be kept in mind that ozonation is usually carried out in the liquid phase, but ozone is applied in this process in its gaseous form. What is more, the O_3 molecule is unstable in the gas phase as well as in the liquid phase. As a consequence, two more issues arise: limited ozone absorption from the gaseous into the liquid phase and ozone self-decomposition [37,38].

The O_3 self-decomposition is one of the most relevant factors which define the treatment process. It was recognized as a chain of initiation, propagation, and termination steps. There are two of the most popular mechanisms which describe the O_3 self-decomposition. First, SBH (Staehelin, Buhler, and Hoigne), given by reactions (1)–(10) [39,40]:

Initiation:

$$O_3 + OH^- \rightarrow HO_2^\bullet + O_2^- \quad (1)$$

Propagation:

$$HO_2^\bullet \rightarrow H^+ + O_2^{\bullet-} \quad (2)$$

$$O_3 + O_2^{\bullet-} \rightarrow O_3^{\bullet-} + O_2 \quad (3)$$

$$O_3^{\bullet-} + H^+ \rightarrow HO_3^\bullet \quad (4)$$

$$HO_3^\bullet \rightarrow O_3^{\bullet-} + H^+ \quad (5)$$

$$HO_3^\bullet \rightarrow HO^\bullet + O_2 \quad (6)$$

$$HO^\bullet + O_3 \rightarrow HO_4^\bullet \quad (7)$$

$$HO_4^\bullet \rightarrow HO_2^\bullet + O_2 \quad (8)$$

Termination:

$$HO_4^\bullet + HO_4^\bullet \rightarrow H_2O_2 + 2O_3 \quad (9)$$

$$HO_4^\bullet + HO_3^\bullet \rightarrow H_2O_2 + O_3 + O_2 \quad (10)$$

The second most popular pathway of ozone decomposition is the TFG (Tomiyasu, Fukutomi, and Gordon) mechanism, in which the initial step starts by OH^- ions, as in the SBH mechanism, or by OH_2^- ions. This second mechanism was more relevant in an alkaline reaction medium [39,40].

Because the ozone molecule is non-polar, its solubility in water is limited. Therefore, O_3 concentration in the liquid phase is the next important factor during the ozonation process. When a bubble column is used for ozonation, the mass balance of ozone in the liquid phase can be described by Equation (11) [41]:

$$\frac{d\,C_L}{dt} = (k_L a)(C_L^* - C_L) - r_D \tag{11}$$

where $k_L a$ is the volumetric mass transfer coefficient in the liquid phase (s^{-1}), C_L^* is the equilibrium molar concentration of ozone in the liquid phase (mol m^{-3}), C_L is the molar concentration of ozone in the liquid phase (mol m^{-3}), and r_D is the ozone decomposition rate (1000 mol m^{-3} s^{-1}).

All the above-mentioned ozone properties form a few consequences which should be taken into consideration when ozone-based treatment is planned. Firstly, according to the SBH model or TFG model, and other models as well, the decomposition of O_3 results in the formation of free radicals. This phenomenon is enhanced by alkaline pH. Therefore, two separate pathways of pollutants decomposition can be observed during ozonation: direct oxidation by the O_3 molecule and indirect reaction with the $HO^\bullet$ radical. In contrast to oxidation by the $HO^\bullet$, the reactions with O_3 are selective. O_3 can react with molecules at locations of high electron density by either cycloaddition, electrophilic substitution, or nucleophilic reactions [39]. Due to the presence of numerous double bonds in the structure of dye molecules, they can therefore likely be decomposed by O_3 molecules. However, low-molecular-weight by-products of dye decomposition are more likely decomposed by $HO^\bullet$ radicals, not O_3. Consequently, the accumulation of the by-products after basic ozone treatment is highly probable [42]. Secondly, the transferred ozone dose should be monitored during ozone-based treatment because this factor was found to be sensitive to any process modification, e.g., catalyst addition [43].

3. Catalytic Ozonation

Catalytic ozonation is widely concerned as an AOPs because of the production of hydroxyl radicals in the process [44]. The ozone self-decomposition resulting in hydroxyl radicals ($HO^\bullet$) production is referred to as a region of the catalytic action of catalysts most often [45]. However, this phenomenon is not a sufficient explanation of the crux of this process. Consequently, a deeper discussion on catalytic ozonation mechanisms should be raised, depending on the catalyst type. Generally, two systems of catalysis, homogeneous and heterogenous, are possible for catalytic ozonation, and both are discussed below (Figure 1). The first studies catalytic ozonation focused on metals and their various forms, such as salts, oxides, or deposits on the support [46]. The latest research focuses mainly on modern materials that combine the capabilities of catalysts from different groups. In a homogeneous system, ozone decomposition is initiated by transition metal ions, while in a heterogeneous system—by solid catalysts [44].

There are two mechanisms in homogeneous catalytic ozonation. The first leads to the formation of free radicals through the decomposition of ozone by metal ions [47]. The second is by the formation of complexes between the catalyst and the organic molecule, followed by oxidation of the resulting complex [48].

In heterogeneous catalytic ozonation, the catalytic activity of the catalyst is important, and for the catalyst to show it, there must be adsorption of ozone and an organic molecule on its surface. We have several catalytic ozonation decomposition mechanisms. The first is the decomposition of ozone on a reduced or oxidized form of metal deposited on the surface of the supporting material [49]. When metal oxides are used as catalysts, ozone decomposition can occur on undissociated hydroxyl groups [50]. Ozone decomposition in the presence of activated carbon takes place at the primary centers of the catalyst [51].

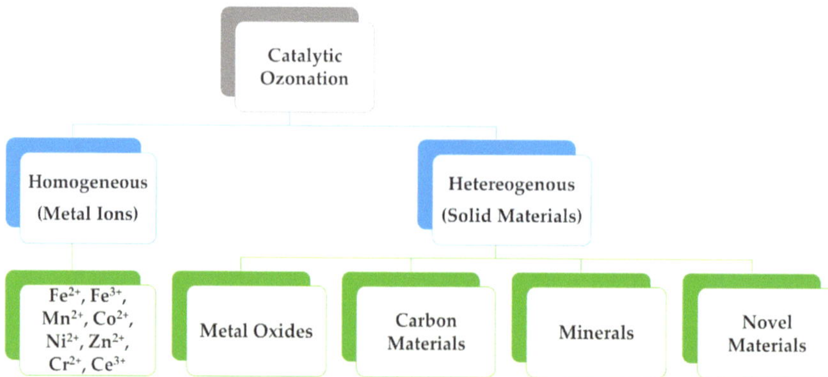

Figure 1. Classification of catalysts used in catalytic ozonation based on Nawrocki, 2010 [44].

The difference between the two methods is as follows [52]:

- Recovery of homogeneous catalysts is difficult and expensive, while in heterogeneous, it is easy and cheaper.
- Homogeneous shows poor thermal stability compared to heterogeneous.
- During homogeneous catalyst application, the selectivity is very good and focused on a single active site, while in heterogeneous, it is weaker, but more functioning for active centers.

Catalytic ozonation stands out from other purification methods in that it does not require additional costs (no UV lamps, often pH adjustment or removal of the resulting sludge) for the industry, and we often obtain a higher degree of mineralization [53]. In addition, the presence of the catalyst increases the number of hydroxyl radicals, i.e., it increases the speed of purification of the dye solution or textile wastewater than in the case of ozonation alone [54].

4. Homogeneous Catalytic Ozonation

The homogeneous system of catalytic ozonation can be catalyzed by transition metal cations, such as Fe^{2+}, Fe^{3+}, Mn^{2+}, Zn^{2+}, Co^{2+} and Ni^{2+} [55–58]. The possible catalytic activity of metal cations was referred to as indirect ozone decomposition of the complex of metal-pollutant [37].

The accompanying mechanisms of using various combinations of metallic ions with ozone to decolorize Reactive Red 2 have been studied by Wu et al. [55]. Table 2 shows the most effective parameters for the catalytic ozonation of a dye or wastewater. Low pH is most often used when using metal cations. In this group of catalysts, the most common system is the one based on the Fenton reaction Fe(II)/O^3 and Fe(III)/O^3 [55–57]:

$$Fe^{2+} + O_3 \rightarrow FeO^{2+} + O_2 \tag{12}$$

$$FeO^{2+} + H_2O \rightarrow Fe^{3+} + HO^{\bullet} + HO^{-} \tag{13}$$

$$Fe^{3+} + O_3 + H_2O \rightarrow FeO^{2+} + H^{+} + HO^{\bullet} + O_2 \tag{14}$$

Using this method, it is important not to exceed the optimal dose of the catalyst, because FeO^{2+} also oxidizes Fe(II) ions. The formation of radicals then occurs at low Fe(II) concentrations, reactions (15) and (16) start to work above the optimal dose and the reaction is inhibited [55]:

$$Fe^{2+} + HO^{\bullet} \rightarrow Fe^{3+} + HO^{-} \tag{15}$$

$$Fe^{2+} + FeO^{2+} + 2H^{+} \rightarrow 2Fe^{3+} + H_2O \quad (16)$$

Table 2. Results of homogeneous catalytic ozonation.

Wastewater Type/Pollutant	Catalyst Type	pH	C_0 (mg/L)	Conditions O_3 Dose (g/L)	Catal. Dose (g/L)	Color (%)	Removal COD (%)	TOC (%)	k (min^{-1})	Year, Ref
Dye solution Reactive Red 2	Fe(II), Fe(III), Mn(II), Zn(II), Co(II), Ni(II)	2	100	-	0.0335	-	-	13–23 (after 5 min)	1.299 1.278 3.296 1.015 0.843 0.822	2008, [55]
Textile Effluent	Fe(II), nZVI	-	-	0.05–0.2	0.7	87 (after 40 min)	73.5 (after 40 min)	-	0.000751 0.000948	2018, [56]
Reactive Red 120	Fe(III)	3	100	-	0.00558	-	40 (after 30 min)	-	-	2012, [57]
Reactive Red 2	Mn(II)	2	100	-	0.1	95 (after 5 min)	-	17–21 (after 5 min)	-	2008, [58]

Studies by Malik et al. have shown that the use of Fe^{2+}/O_3 not only increases the process decolorization efficiency for dye and wastewater but will also increase the BI index from 0.26 to 0.4 for textile wastewater because catalytic ozonation accelerated the conversion of high molecular weight compounds into more biodegradable forms [56]. In their work, they also checked how zero valent iron nanoparticles (nZVI) work in the treatment of textile wastewater.

Another catalyst from the homogeneous group is Mn^{2+}/O_3, which also showed an efficient process of decolorizing the Reactive Red 2 dye solution [55,58]. Mn(II) reacts with ozone to form hydrated manganese oxide. The resulting oxide reacts with the dye, accelerating the discoloration process. Wu et al. have proposed the path of the reaction that may follow [58]:

$$Mn^{2+} + O_3 + 2H^{+} \rightarrow Mn^{4+} + O_2 + H_2O \quad (17)$$

$$Mn^{4+} + 1.5O_3 + 3H^{+} \rightarrow Mn^{7+} + 1.5O_2 + 1.5H_2O \quad (18)$$

$$Dye + Mn^{7+} + 1.5H_2O \rightarrow Mn^{4+} + 3H^{+} + products \quad (19)$$

$$Mn^{2+} + Mn^{4+} \rightarrow 2Mn^{3+} \quad (20)$$

$$Mn^{2+} + O_3 + H^{+} \rightarrow Mn^{3+} + O_2 + HO^{\bullet} \quad (21)$$

$$Mn^{3+} + O_3 + (Dye)^{2-} + H^{+} \rightarrow Mn^{2+} + O_2 + HO^{\bullet} + products \quad (22)$$

$$Dye + HO^{\bullet} \rightarrow products \quad (23)$$

In the case of Zn(II), Co(II), and Ni(II), the lowest efficiency was achieved by nickel, regardless of the catalyst dose used [55]. The reaction mechanism for cobalt and nickel can be represented by the following reaction (24):

$$M^{(n-1)+} + O_3 + H^{+} \rightarrow M^{n+} + HO^{\bullet} + O_2 \quad (24)$$

The use of homogeneous catalysts for the decolorization of dye solutions or treatment of textile wastewater is not very frequent and it may be due to the lack of a developed method of removing ions from the solution after the process. Often homogeneous catalysts are used for preliminary research into later heterogeneous catalysts [48,58–60].

5. Heterogeneous Catalytic Ozonation

Heterogeneous systems in catalytic ozonation are solids, so they can be removed from the solution in a simpler way than homogeneous systems. There are three probable mechanisms of operation of heterogeneous catalytic ozonation, as shown in Figure 2 [44]. In this system, the adsorption capacity of the catalyst is critical as at least one of the reagents must be adsorbed onto the catalyst surface. Correspondingly, the process's catalytic action is possible when one of the presented activities occurs: ozone is adsorbed on the catalyst surface. A pollutant molecule is adsorbed on the catalyst, and both ozone and pollutants are adsorbed on the surface.

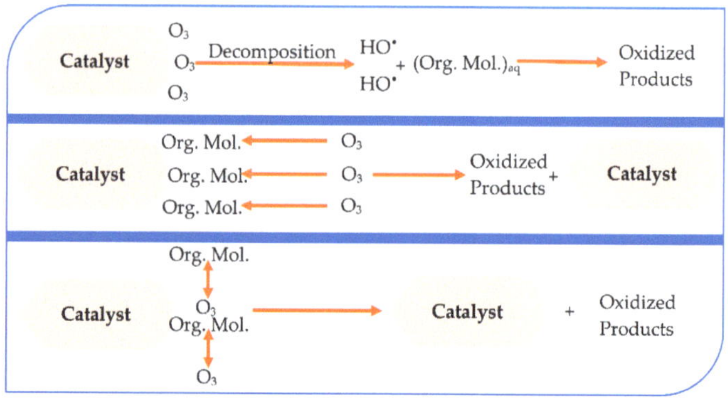

Figure 2. Different mechanisms of heterogeneous catalytic ozonation are based on Nawrocki, 2010 [44].

The main groups of substances used as heterogeneous catalysts are metal oxides, carbon species, minerals, and new materials. The further part of the article will present the use of particular groups for the discoloration of the dye solution as well as textile wastewater, along with the test parameters for which the best results were obtained, along with the test parameters for which the best results were obtained.

5.1. Metal Oxides

For the treatment of textile wastewater or dye solutions, metal oxides such as Al_2O_3 [61,62], CeO_2 [63], MgO [64–66], MnO_2 [58,67], CoO [67–69], FeO_3 [70,71], $Ca(OH)_2$ [72] were used. When using metal oxides, it must be remembered that the pH value has a great influence on the properties of their surface, and hence on their efficiency in removing contaminants. Surface properties that can change are the isoelectric point, the specific surface area, and the Lewis acid sites (accept electrons) [44].

Polat et al. showed that the effectiveness of catalytic ozonation with the use of alumina depends on the surface of the catalyst and the pH of the solution [61,62]. The dye molecules are absorbed on the Al_2O_3 surface, therefore the larger the surface area, the greater the adsorption capacity [62]. They are degraded by radicals on the catalyst surface and by ozone in the solution. Conducting the process at acidic pH resulted in a low concentration of hydroxide ions and significantly reduced the share of ozone decomposition into $HO^•$ [61].

CeO_2 was another catalyst used for dye discoloration [63]. Efficient dye removal and COD reduction was increased significantly with the addition of the catalyst to the process, a phenomenon that can be explained by the fact that CeO_2 promotes the decomposition of ozone to form reactive oxygen species (such as $HO^•$). Qiu et al. analyzed the mechanism of organic matter removal using CeO_2. First, the dye and ozone are adsorbed on the surface of the catalyst, then the catalyst initiates the decomposition of ozone into hydroxyl radicals. The resulting radicals attack organic substances, creating H_2O and CO_2 [63].

MgO has also been used to decolorize dye solutions [64–66]. Moussavi et al. obtained a maximum COD and color removal at pH 8, hydroxyl radicals are mainly generated at pH above 10, and the obtained results are explained by the presence of MgO which could have contributed to lowering the optimal pH or led to the generation of other radicals [65]. Increasing the pH did not lead to an increase in the efficiency of the process. The results obtained at an alkaline pH can be ascribed to an increase in the availability of hydroxide ions and an accelerated breakdown of ozone. This translates into the formation of hydroxyl radicals and MgO-hydroxyl radicals. The mechanism is adopted that the catalyst is also the initiator of radical formation, similar to CeO_2, and it adsorbs ozone on its surface. Similar results were obtained by Chokshi et al. where the best results were obtained for pH 7 [64]. The use of MgO also has advantages for environmental reasons, it is environmentally friendly, recyclable, and easy to separate [66].

Wu et al. used MnO_2 in their research on the discoloration of the dye solution. They presented the mechanism of operation in this process (25)–(27). Ozone reacts with the hydroxide ions on the catalyst surface according to the following reactions to form hydroxyl radicals on the catalyst surface [58,67]:

$$MnO_2 + H_2O \rightarrow MnO_2 - OH^- + H^+ \tag{25}$$

$$MnO_2 - OH^- + O_3 \rightarrow MnO_2 - OH^\bullet + O_3^- \tag{26}$$

$$MnO_2 - OH^\bullet + dye \rightarrow MnO_2 + products \tag{27}$$

Cobalt oxide occurs mainly in the form of Co_3O_4. According to Faria et al., the increased mineralization of pollutants achieved by catalytic ozonation is mainly due to the presence of $HO^\bullet$ formed as a result of catalytic ozone decomposition [67]. On the other hand, Chokshi et al. put forward the theory that cobalt oxide works according to the third mechanism, where the catalyst absorbs not only ozone but also pollutions [69].

Fe-shaving-based catalyst was used for the catalytic ozonation of industrial wastewater in the research conducted by Li et al. In both studies, an analysis of the existing mechanisms was carried out, where the formation of hydroxyl radicals was confirmed. In this process, the absorption of water and ozone takes place on the catalyst surface, where water is dissociated and ozone is decomposed [70,71].

The last representative of metal oxides used in the purification of the dye solution is $Ca(OH)_2$. Quan et al. proposed a mechanism to improve the catalytic ozonation process. During the mineralization of organic pollutants, HCO^{3-} and ${CO_3}^{2-}$ ions are formed, which act as radical scavengers that negatively affect the efficiency of the process. $Ca(OH)_2$ dissociates into Ca^{2+} and HO^- in an aqueous solution. The resulting cation reacts with HCO^{3-} and ${CO_3}^{2-}$ ions to form $CaCO_3$. The hydroxide anion reacts with ozone in the O_3/HO^- system. These two processes taking place simultaneously lead to an increase in the degradation efficiency and the mineralization of organic pollutants [72].

Table 3 summarizes the best process parameters using metal oxides for catalytic ozonation.

The use of metal oxide catalysts, due to their easy removal from the solution, is more common than their homogeneous counterparts. Comparing the test results from both groups, it can be noticed that by using metal oxides, better mineralization of organic pollutants can be obtained, which is associated with more complex mechanisms accompanying the purification process.

5.2. Carbon Materials

Another group of heterogeneous catalysts used in catalytic ozonation, distinguished by a large specific surface area (from 500 m^2/g to 1500 m^2/g), adsorption capacity, chemical resistance, no metal leaching [51,73,74], relatively low price, and the possibility of recycling [66,75,76] is activated carbon.

In studies on catalytic ozonation with the use of activated carbon, it has been proven that the presence of this catalyst in solution initiates the radical decomposition of ozone in water, leading to the formation of hydroxyl radicals. Research has shown that metal centers, basal plane electrons, and functional groups, such as chromene, pyrene, and pyrrole, are active centers in the production of $HO^\bullet$ radicals from ozone decomposition [75,77,78]. Mostly, the catalytic decomposition of ozone with the use of activated carbon takes place in the alkaline pH range (6–10) [51,74,77,79,80].

Wang et al. discovered that granular activated carbon (GAC) used for catalytic ozonation of textile wastewater can be successfully regenerated and reused for 20 days of a continuous process. The obtained results indicated more than twice greater efficiency after the regeneration process [76].

Table 3. Examples of catalytic ozonation with metal oxides.

Wastewater Type/Pollutant	Catalyst Type	Conditions pH	C_0 (mg/L)	O_3 Dose (g/L)	Catal. Dose (g/L)	Removal Color (%)	COD (%)	TOC (%)	k (min^{-1})	Year, Ref
Textile Wastewaters (Basic Blue 41, Basic Yellow 28, Basic Red 18.1)	Al_2O_3	4	-	0.0918	5	-	Blue Dye—45.3; Red/Yellow Dyes—100	-	-	2015, [61]
Acid Red 151;	Al_2O_3	2.5	200	0.0147	5	98.4 (after 30 min)	78.7 (after 30 min)	-	0.136	2008, [62]
Remazol Blue R						98.3 (after 30 min)	82.6 (after 30 min)	-	0.132	
Lemon Yellow	CeO_2	6	-	-	0.5	-	-	97	-	2014, [63]
Reactive Black 5	MgO	7	-	-	0.05–2	99.9	-	38.8	-	2016, [64]
Reactive Red 198	MgO	8	-	-	1–6	100	69	-	-	2009, [65]
Methylene Blue	MgO	9	500	-	-	77 (after 60 min)	12.15 (after 60 min)	-	0.025	2016, [66]
Reactive Red 2	MnO_2	2	100	-	0.8	95 (after 5 min)	-	17–21 (after 5 min)	-	2008, [58]
CI Acid Blue 113, aniline	Mn-O, Co-O, Ce-O	3, 5.5, 6	-	0.05	0.35	-	-	~80	-	2009, [67]
Naphtol Blue Black	Co_3O_4,	5	-	-	0.1			10–30	-	2007, [68]
Reactive Black 5	Co_3O_4	7	-	-	0.2	99.99	80	-	-	2017, [69]
Industrial Wastewater	Fe-shaving based	6.81 ± 0.84	-	0.065	3	-	-	98.5	-	2018, [70] 2019, [71]
Acid Red 18	$Ca(OH)_2$		-	0.065	3	~100		~100	-	2017, [72]

To increase the efficiency of the process, attempts were made to increase the surface of activated carbon and change the morphology of the catalyst surface [80]. The more developed surface area (BET surface area and pore size) contributed to an increase in the absorption of the dye on the catalyst surface and an increase in discoloration.

Mahmoodi et al. investigated the performance of multi-walled carbon nanotubes (MWCNTs) in catalytic ozonation. This material, thanks to its unique mechanical, thermal, chemical, and electrical properties can be an interesting alternative to activated carbon. The mechanism of catalytic ozonation should be similar to that of activated carbon due to the similarity in bonding and structure [81]. Qu et al. also used carbon nanotubes for catalytic ozonation of the Indigo. The nanotubes have been enriched with -COOH functional groups. The presence of the catalyst contributed to the increase in mineralization. In these studies, it was concluded that the mechanism of catalytic ozonation may be different, however. The special nanostructure of carbon nanotubes can lead to the storage of ozone in a microstate and lead to an increase in direct electron transfer, translating into an increase in the efficiency of the process [82].

For catalytic ozonation, Faria et al. used a composite of activated carbon and cerium oxide. In this way, they obtained a catalyst that can be used three times without any change in the efficiency of the process. It has been noted that the mechanism of catalytic ozonation using an activated carbon-cerium oxide composite includes both surface reactions that occur when using only activated carbon and that are aqueous solution reactions involving $HO^•$ that occurred during catalytic ozonation with cerium oxide [83].

The next carbon-based catalyst is carbon aerogel, which has a very extensive surface and high porosity [84,85]. The mechanism proposed by Hu et al. contains an increase in ozone mass transfer to the water phase due to the porosity of the catalyst. It also promotes the breakdown of ozone into reactive forms [84]. According to Niu et al., the addition of MnO_2 to the graphene aerogel leads to an increase in the efficiency of the decomposition of organic pollutants by increasing the density of the electron cloud of Mn atoms deposited on the aerogel, it favors the decomposition of ozone and the generation of $HO^•$ radicals [85].

Copper(II)-doped carbon dots were used for catalytic ozonation of dyes as well as real textile wastewater [86]. The use of polyethylene glycol (PEG) in the catalyst synthesis made it possible to adsorb the dye on the surface of the dots. The addition of copper increased the active surface of the carbon dots and the ozone adsorbed on the surface is catalytically transformed into hydroxyl radicals, which then react with the adsorbed dye.

Table 4 summarizes the best process parameters using carbon materials as catalysts.

Table 4. Examples of catalytic ozonation with the application of carbon materials.

Wastewater Type/Pollutant	Catalyst Type	Conditions pH				Removal Color (%)				Year, Ref
		pH	C_0 (mg/L)	O_3 Dose (g/L)	Catal. Dose (g/L)	Color (%)	COD (%)	TOC (%)	k (min^{-1})	
Reactive Black 5	Activated carbon	11.29	-	0.9	0.005	-	40 (after 60 min)	35 (after 60 min)	-	2020, [75]
Aniline	Activated Carbon	3 7 9	13.2	50	0.35	-	-	-	0.188, 0.290, 0.233	2007, [79]
Real textile effluent	Activated carbon	8.5	100	0.00754	0.3	~100 (15 min)	-	20.7 (after 60 min)	0.47	2005, [87]
Reactive Blue 194	Granular activated carbon	7	200	0.179	-	100 (after 20 min)	-	-		2020, [51]
Basic Blue 9	Granular activated carbon	10	-	-	2	80 (after 5 min)	68 (after 5 min)	-	0.72	2013, [77]
Reactive Red 195	Granular activated carbon	11	100	-	1	90.4 (after 2 min)	-		-	2012, [80]
Acid Red 3R	Granular Activated Carbon	7	100	0.0417	2	-	-	78.2 (after 60 min)	-	2018, [74]
C.I. Reactive Red 194, C.I. Reactive Yellow 145	Granular Activated Carbon	6.3 5.9	100	0.035	10	~100 (after 30 min)	80 (after 30 min)	50 (after 30 min)	-	2009, [78]
Methylene Blue	Granular Activated Carbon	9	500	-	-	63 (after 60 min)	25.36 (after 60 min)	-	0.016	2016, [66]
Bio-treated dyeing finishing wastewater	Regenerated granular activated carbon	-	-	0.0185	2	81.7 (after 25 min)	71 (after 25 min)	-	-	2019, [76]
Reactive Red 198, Direct Green 6	Multiwalled carbon nanotube	3	150	-	0.03	100 (after 16 min), 100 (after 20 min)	-	-	-	2013, [81]
Indigo	Carbon nanotubes functionalized with carboxyl groups	4	100	0.141	0.005	~99 (after 20 min)	-	35.1 (after 2 h)	−0.219	2015, [82]
CI Acid Blue 113, CI Reactive Yellow 3 CI Reactive Blue 5	Activated carbon–cerium oxide	5.6	50	50	0.35	97 88 98 (after 5 min)	-	88 (after 120 min)	-	2009, [83]
Dyeing effluents	Carbon aerogel-Co_3O_4	7, 10	-	0.008	3	100 (after 10 min)	80 (after 30 min)	-	-	2016, [84]

Table 4. *Cont.*

Wastewater Type/Pollutant	Catalyst Type	Conditions pH	C_0 (mg/L)	O_3 Dose (g/L)	Catal. Dose (g/L)	Removal Color (%)	COD (%)	TOC (%)	k (min^{-1})	Year, Ref
Rhodamine B	Graphene/α-MnO_2 nanocrystals hybrid aerogel (GMA)	-	50	0.035	-	100 (after 60 min)	89.02 (after 15 min)	-	0.2859	2019, [85]
Methyl Orange	Copper(II)-doped carbon dots	7	-	1.98	-	99.8 (after 6 min)	-	-	1.184	2022, [86]
Real textile effluent						41 (after 60 min)			0.012	

Combining ozonation with carbonaceous materials is effective in removing color and some organic matter from highly colored wastewater. Despite the popularity of, for example, carbon nanotubes and their wide application, most research focuses on the use of granulated activated carbon, which is easily available and achieves good performance in purifying dye solutions.

5.3. Minerals

The use of natural minerals as catalysts for catalytic ozonation is still not widely used in the treatment of textile wastewater. The use of natural-based materials may limit the application of synthetic catalysts, which leads to decreased production often harmful to the environment and associated with high costs [88]. Natural minerals can be used as catalyst supports [89,90] or catalysts [91,92]. Due to their functionality and abundance of resources, they are relatively cheap equivalents and may in the future become substitutes for currently used catalysts in environmental protection [92]. Moreover, due to their high porosity and specific surface area, they can show greater adsorption and thus higher efficiency in the catalytic ozonation process [89].

Montmorillonite (Mt) is an aluminosilicate that Boudissa et al. modified to give acid-base and hydrophobic properties. In the first modification, Na was deposited on the Mt surface, and in this variant, the dye molecules were absorbed by hydrophobic interactions. In the second modification, Fe(II) was used, and cation exchanges and mobility of Fe^{2+} to catalytic activity took place here. The last modification was the activation of the surface with acid, and the surface showed the lowest hydrophilic character, increasing the mineralization of the dye [89].

Valdes et al. used volcanic sand [91] and natural zeolite [93] for catalytic ozonation of the dye solution. Metal oxides in volcanic sand generate hydroxyl groups on their surface, which are involved in the decomposition of ozone. Volcanic sand accelerates the self-decomposition of ozone, which means it accelerates the generation of active forms [91]. The mechanism of catalytic ozonation with zeolite is the same as with volcanic sand because hydroxyl groups are also formed on its surface [93].

Taseidifar et al. first treated natural magnetite (Fe_3O_4) with oxygen plasma (chemical etching) and argon plasma (sputtering effect), resulting in greater surface roughness. Thanks to this modification, more active sites, and better mass exchange were obtained. On the magnetite surface, hydroxyl groups are formed, which later take part in ozone decomposition, as in the case of earlier minerals [94].

For the catalytic ozonation of the dye solution, Moussavi et al. used raw (27 m^2/g) and calcine (34 m^2/g) magnetite. Calcination led to the destruction of pyrite and the transformation of magnetite (Fe_3O_4) into hematite (Fe_2O_3). The mechanism of the catalytic ozonation reaction with the participation of magnetite is similar to the previous minerals. The greater efficiency of the calcined catalyst can be attributed to Fe_2O_3, which is its main component and has a higher catalytic potential than Fe_3O_4 [88].

Dong et al. used natural brucite for catalytic ozonation of a dye solution, which consists mainly of $Mg(OH)_2$ and dopants of Si, Ca, and Fe oxides. An increase in pH occurs during the reaction due to the dissociation of magnesium oxide, therefore the possibility that the reaction mechanism is homogeneous here due to OH^-, although there is solid brucite powder in the solution, was highlighted [92].

The last presented mineral used for catalytic ozonation of the dye solution is the aluminosilicate Montanit300®, and although it is not very active by itself, it gains greater activity after modification with H_2SO_4 [90]. Inchaurrondo et al. found that there are active sites (Mn) on the surface of the mineral, acid treatment increased the Si: Al ratio, the specific surface of the catalyst increased, and the percentage of iron and manganese decreased. The improved catalytic properties of the modified zeolite can be explained by an increase in the Si: Al ratio and an increase in the hydrophobicity, which in such minerals is responsible for the reactions between ozone adsorption and organic pollutants on the catalyst surface.

Table 5 summarizes the best process parameters using minerals for catalytic ozonation.

Table 5. Examples of catalytic ozonation with minerals.

Wastewater Type/Pollutant	Catalyst Type	Conditions pH	C_0 (mg/L)	O_3 Dose (g/L)	Catal. Dose (g/L)	Removal Color (%)	COD (%)	TOC (%)	k (min^{-1})	Year, Ref
Methylene Blue, Methyl Green, Methyl Orange, Methylthymol Blue	Ion-exchanged montmorillonite, (NaMt and Fe(II)Mt), crude bentonite, and acid-activated counterparts (HMt)	-	200	0.0083	0.04	-	-	-	-	2019, [89]
Methylene Blue	Volcanic sand	8	30	0.006	50	70 (after 50 min)	-	-	0.09	2010, [91]
Methylene Blue	Zeolite Volcanic sand	2		-	15	-	-	-	0.054 0.12	2009, [93]
Basic Blue 3	Natural Magnetite modified with argon plasma	6.7	90	0.003	0.6	93.47 (after 15 min)	-	-	0.1814	2015, [94]
Reactive Red-120	Raw and Calcined magnetite	11	100	-	3	-	-	96.1 (after 120 min)	0.082	2012, [88]
Active Brilliant Red X-3B	Brucite	-	500	-	0.5	89 (after 15 min)	32.5 (after 15 min)	-	-	2007, [92]
Orange II	Aluminosilicate, Montanit300	6	100	-	1	100 (after 240 min)	-	91 (after 240 min)	-	2021, [90]

Cheap solid materials such as zeolites, aluminosilicates, sand, brucite, and magnetite have been used for the heterogeneous ozonation of dyes. The use of natural catalysts is a great advantage from an ecological point of view because production processes that can deliver toxic compounds to the environment are avoided and costs are reduced. However, the use of mineral catalysts for the treatment of textile wastewater is still a little-explored area, which in time may turn out to be one of the main paths that scientists will follow. The disadvantage of using natural minerals is their wearability during the catalytic ozonation process, as the oxides on their surfaces often dissociate [88,92].

5.4. Novel Materials

Modern catalytic materials strive to increase the effectivity of catalytic ozonation by changing the properties of the catalyst surface, such as porosity, pore-volume, or mechanical strength. When creating such catalysts, these already proven materials are often combined to obtain a product that exhibits even better properties: composites [95,96], metals, or metal oxides on support [62,97–102]. Increasing the mesoporosity of the material or its delamination provides more active sites where catalytic reactions can take place [66,103,104]. Then, such surfaces can be modified, thanks to which we obtain a larger catalytic surface [105–110].

Chokshi et al. used the composites La-Co-O (La_2O_3/Co_3O_4) and Ag-La-Co to discolor the Reactive Black 5 solution [95,96]. The obtained Ag-La-Co composite is characterized by the simplicity of preparation by the co-precipitation method, which is quick and simple and does not require the use of any organic solvent. The mechanism of catalytic ozonation with the use of these composites can be described in several simultaneous reactions (Figure 3). First, ozone passes from the gas to the liquid phase, and there is adsorption of ozone and organic substances on the catalyst surface and chemical reactions between them, reactions between the adsorbed substances and $HO^\bullet$ radicals, and reactions of by-products [96].

Alumina was used as a carrier for perfluorooctylic acid (PFO), and the resulting catalyst was used to decolorize the Acid Red 151 (AR-151) and Remazol Blue R (RBBR) solution in the catalytic ozonation process [62]. Eriol et al. proved that the percentage by weight of PFO on the support did not have a large impact on the degree of discoloration, but it contributed to the increase in COD reduction concerning the alumina alone. These results were explained by the presence of a significant amount of PFO acid on the surface which produced longer alkyl chains that increased perfluorooctyl alumina (PFOA) activity. The adsorption capacity of PFOA also improved over alumina alone due to the increased hydrophobicity of the catalyst surface. The formation of perfluorinated alkyl chains on the surface promotes the change of character from acidic/basic to hydrophobic. This phenomenon is attributed to the easier adsorption of organic molecules to the catalyst surface. The obtained tests showed that the efficiency of this catalyst largely depends on the pH of the solution and the surface properties as well as the nature of the dye. Therefore, for 100% PFOA, the best results were obtained with RBBR at pH 13.

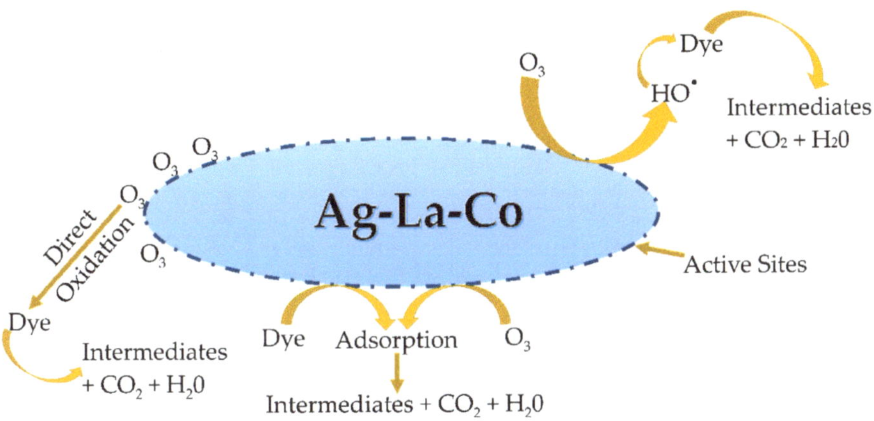

Figure 3. Proposed mechanisms of action of Ag-La-Co catalyst based on Chokshi, 2021 [96].

Asgari et al. produced carbon-doped magnesium oxide that was additionally doped with the powdered, acid-free eggshell membrane (C-MgO-EMP). The obtained catalyst was used to decolorize textile wastewater. The doping of activated carbon will allow an increase in the degradation of pollutants in catalytic ozonation by increasing the number of radicals. From the obtained results, it was concluded that the performance of the catalyst improved the degree of mineralization of the byproducts, and the improvement of the results was attributed to better interactions and reduced resistance to mass transfer. Due to the extensive surface of the catalyst, organic pollutants and ozone are absorbed on the surface undergoing mineralization [97].

Ghuge et al. investigated the process of ozonation of the catalytic dye solution Reactive Orange 4 and textile wastewater, using two catalysts Cu/SBA-15 and Ru-Cu/SBA-15. SBA-15 is a mesoporous material with a highly ordered structure, characterized by uniform pore size and a parallel arrangement. Its catalytic activity was improved by immobilizing metallic copper on its surface, creating new active sites [98]. Using the t-butanol radical scavenger, it was found that the decomposition of ozone corresponds to the removal of color from the solution, while the demineralization takes place using hydroxyl radicals which are formed by the reaction of ozone with hydroxyl groups formed on the catalyst surface [99]. Another example of bimetallic nanostructures used to treat dye wastewater is $Fe/Mn@\gamma-Al_2O_3$ produced by Liang et al. Through studies of the mechanism, they found that the use of a catalyst increases the constant reaction rate of removing dye wastewater compared to ozone alone, due to the increased efficiency of $HO^\bullet$ and $^1O_2$ formation [102].

The chip catalyst, i.e., porous copper fibers (PCFSS) sintered together and then loaded with Cu/Zn/Al/Zr metal oxide catalysts, was used to decolorize Basic Yellow 87 in the catalytic ozonation process [100]. The obtained catalyst can be used many times without losing its catalytic activity. After 10 uses, 99.2% dye discoloration and 58.9% COD removal were achieved. PCFSS has a three-dimensional lattice structure with high porosity and a specific surface area, facilitating mass transfer and promoting multiphase reactions. Zhu et al. noticed a slight loss of the catalytic chip mass, the reason for this may be the dissolution of metal oxides on the support surface in water. This phenomenon may contribute to receiving H^+ ions from the solution. The mechanism of chip operation was proposed by adsorbing ozone and organic matter on its surface, and then the interaction of ozone with metal oxides. This leads to obtaining both the radicals and the reactants on the surface of the catalyst, facilitating the reaction between them [100]. A similar adsorption/catalytic effect of a ceramic membrane with deposited Mg, Ce, and Mn oxides was used by Li et al. The membrane uses two techniques: filtration and catalytic ozonation. Coating the membrane with metal oxides increased the efficiency of pollutant removal from 30% to 80% [101].

The Mg-O-$SC_{CA\text{-}Zn}$ granular catalyst was made of synthesized porous sludge-derived char and magnesium hydroxide and was used for catalytic ozonation of an aqueous solution of Methylene Blue [66]. Kong et al. attributed the catalytic ability of the resulting catalyst to the combination of MgO nanoparticles deposited on the surface and $SC_{CA\text{-}Zn}$ porosity.

Lu et al. prepared a magnetic mesoporous catalyst for the catalytic ozonation of Acid Orange II [103]. Spinel ferrites are soft magnetic materials with a stable structure and the general formula $MeFe_2O_4$. Due to their magnetic property, they can be easily removed after the process. The decomposition of ozone into active radicals is controlled by electron transfer between ozone on the catalyst surface and $MgFe_2O_4$. The obtained results indicate the presence of the $HO^•$ radical mechanism and the adsorption of both dye and ozone on the catalyst surface along with the occurrence of surface reactions [103].

El Hassani et al. looked at Ni-based double hydroxide (Ni-LDH) layered nanomaterials and used them for the catalytic ozonation of Methyl Orange. The mechanism with the use of the Ni-LDH catalyst consists in adsorbing ozone on its surface and creating weak bonds with the hydroxyl groups present in each layer of the catalyst. Nickel, on the other hand, acts as an active site for ozone decomposition. Additionally, dye absorption takes place on the catalyst surface [104].

Another example of a catalyst used to decolorize the Methyl Orange dye solution is nanohybrid $NiFe_2O_4$-NiO growing on porous nickel 3D (NF) foam [105]. In heterogeneous catalytic ozonation, it has shown high stability and activity, and most importantly the catalyst is recycled. Numerous active sites are created on the catalyst surface, which enables the formation of $HO^•$ and $O_2^{•-}$ radicals, accelerating the decomposition of O_3. Due to the porous structure of the catalyst, there is a rapid liquid flow and mass transfer during catalytic ozonation with its use. The used support material is flexible, which allows it to be shaped for the needs of future reactors and easier to remove from them.

Hien et al. used ground zinc slag (Zn-S) in the process of catalytic ozonation of the Direct Black 22 (DB22) dye solution. Zinc slag consists of calcium and zinc which increase the degradation of O_3. The degradation of the dye occurs through the mechanism of reaction with hydroxyl radicals and adsorption on the catalyst surface. The structure of the catalyst is characterized by large pores and an extensive surface, thanks to which its use has resulted in better mineralization and discoloration of the dye. DB22 is an anionic dye with sulfo groups that can form a complex with the Ca^{2+} ($CaCO_3$) ions present. This further increases the catalyst's ability to absorb the dye from the solution [106].

An example of improving the already existing catalyst parameters is the work of Pereira et al. where calcium alginate beads incorporated with nZVI were made [107]. Using nZVI is associated with self-aggregation and passivation of nanoparticles, reducing their reactivity and reaction area. Immobilizing them in an alginate shell overcomes these limitations. However, the use of this catalyst involves the use of pneumatic agitation, due to their poor shear resistance, and the use of conventional mixers could lead to breakage. Figure 4 shows the proposed catalyst mechanism in the catalytic ozonation of Reactive Red 195. Sone et al. focused on the green synthesis of ZnO nanoparticles, using an aqueous extract of the palm tree Doum. Their research enables more ecological catalyst production with equally good yields [111].

Pervez et al. developed hydrophilic electrospun membranes based on polyvinyl alcohol (PVA) using low-temperature synthesis. The obtained catalyst was used for the catalytic ozonation of Methylene Blue (MB). PVA membranes are characterized by chemical and thermal stability and have numerous hydroxyl groups on their surface. The produced membrane has a hydrophilic surface, which results in strong electrostatic interactions between it and the cationic Methylene Blue (greater adsorption on the catalyst surface) [108].

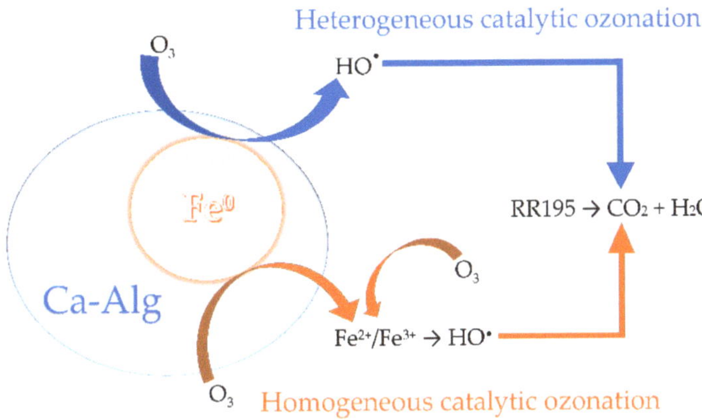

Figure 4. Proposed mechanism of catalytic ozonation of Reactive Red 195 dye using nZVI based on Pereira, 2021 [107].

Chen et al. prepared bimetallic sulfur-doped yttrium copper and yttrium oxide by the co-precipitation method [109]. The obtained S-CuYO catalyst was used for the catalytic ozonation of aniline solution. Due to the synergy between the forms ≡Cu(I) and ≡Y(III), the obtained catalyst is characterized by a large number of hydroxyl groups on its surface, a large specific surface area, and a large pore size. The dominant reactive oxygen species are $HO^{\bullet}$ radicals, in addition, electrons are transferred from S-CuYO to dissolved oxygen, which creates $O_2^{\bullet -}$. The research also discovered that during the process of catalytic ozonation, H_2O_2 is produced, which will turn into $HO^{\bullet}$ [109].

Metal-organic frameworks are characterized by very good adsorption properties. Obtained by Yu et al., the MIL-53(Fe) catalyst showed high catalytic activity due to its porous surface (mass-transfer properties) and active surface area of 372.65 m^2/g. The study of the mechanism accompanying the use of the MIL-53(Fe)/O_3 system proved that the active site is conducive to ozone decomposition. The main ROS here are $HO^{\bullet}_{ads}$, $O_2^{\bullet -}$, and O_2^1. Additionally, the reaction rate using this catalyst increases almost 13 times [112].

Faghihinezhad et al. created an O@g-C_3N_4/Al_2O_3 nanocatalyst for catalytic ozonation of textile wastewater. The catalyst consists of magnetic oxidized g-C_3O_4 nanoparticles and Al_2O_3 immobilized on their surface. Within 60 min of the process, color reductions of 99% and COD reductions of 77% were obtained. Moreover, the regenerative capabilities of the obtained catalyst were checked and after 5 cycles, the COD efficiency decreased by only about 8% [110]. The mechanism of catalytic ozonation here is as follows: oxidation by O_3 on both surfaces, oxidation by free radicals $HO^{\bullet}$ generated from the decomposition of O_3.

Javed et al. used broken laboratory-grade borosilicate glass and coated it with cobalt (Co-BSG). The resulting catalyst decolorized the Methylene Blue solution in 8 min to 92% and removed the COD in 40 min to 93% [113]. Thanks to the use of recycled glass, this method is not only more environmentally friendly but also economically beneficial. The reaction mechanism using this catalyst takes place through the interaction of O_3 with the surface of the catalyst. The radical mechanism is dominant here.

The Table 6 summarizes the best process parameters with the use of a given catalyst.

Table 6. Examples of catalytic ozonation with novel materials.

Wastewater Type/Pollutant	Catalyst Type	Conditions				Removal				Year, Ref
		pH	C_0 (mg/L)	O_3 Dose (g/L)	Catal. Dose (g/L)	Color (%)	COD (%)	TOC (%)	k (min^{-1})	
Reactive Black 5	La-Co-O	3	100	-	0.005	99 (after 30 min)	-	-	-	2020, [95]
Reactive Black 5	Ag-La-Co	7	100	-	0.5	95 (after 10 min)	-	60 (after 80 min)	0.0727	2021, [96]
Acid Red 151; Remazol Blue R	100% perfluorooctyl alumina	13	200	0.0147	5	98.8 (after 30 min), 97.4 (after 30 min)	75.7 (after 30 min), 96.6 (after 30 min)	- -	0.132, 0.158	2008, [62]
Textile wastewater	C-doped MgO eggshell C-MgO-EMP	-	-	0.08	0.23	93 (after 10 min)	-	78 (after 10 min)	1.545	2019, [97]
Reactive Orange 4	Cu/SBA-15	9	100	0.005	2	100 (after 21 min)	-	86 (after 60 min)	0.031	2018, [98]
Reactive Orange 4	Mesoporous	9								
Dye Industrial Effluent	Ru-Cu/SBA-15		100	0.005	2	100 (after 21 min)	70.4 (after 60 min), 90 (after 4 h)	-	-	2018, [99]
		7.5								
Dye wastewater	Fe/Mn@γ–Al_2O_3	7	-	0.006	0.2	-	-	-	0.132	2022, [102]
Basic Yellow 87	Porous copper fiber sintered sheet	6.6	216	0.5	-	100 (after 120 min)	60 (after 120 min)	-		2014, [100]
Azo Dye 4BS	Mg-Ce membrane, Mg-Mn membrane	8.5	12	0.012	-	85 (after 5 min) 88 (after 5 min)	-	>75 (after 5 min) >75 (after 5 min)	-	2021, [101]
							-		-	
Methylene Blue	G-MgO-$SC_{Ca\text{-}Zn}$	9	500	-	-	98 (after 60 min)	50 (after 60 min)	-	0.04	2016, [66]
Acid Orange II	$MgFe_2O_4$	5	50	0.005	0.1	-	-	48.1 (after 40 min)	-	2016, [103]
Methyl Orange (MO)	Ni-LDHs	9	500	-	1	96 (after 60 min)	72 (after 60 min)	-	0.053	2019, [104]
Methyl Orange	$NiFe_2O_4$-NiO/NF	6.84	-	0.0041	-	96 (after 60 min)	-	71 (after 60 min)	-	2021, [105]
Direct Black 22	Zn-S	3–11	100	3.38	0.75	99 (after 20 min)	82 (after 25 min)	-	-	2020, [106]
Reactive Red 195	nZVI-Alg	3.0–6.5	25	0.008	50	100 (after 30 min)	98 (after 90 min)	-	-	2021, [107]
Coralene Rubine-S2G	ZnO-400 nanoparticles	6.8–8.4	130	-	0.050	100 (after 35 min)	-	-	0.359	2021, [111]
Methylene Blue	Polyvinylalcohol (PVA) nanofibrous membranes	3.02	20	-	0.03	94 (after 60 min)	-	-	-	2020, [108]
Aniline	S-CuYO	3–11	10	-	0–2.0	96 (after 15 min)	-	57.7 (after 15 min)	-	2020, [109]
Rhodamine B	Fe-MOFs	7	40	-	0.2	100 (after 2.5 min)	-	40 (after 30 min)	5.76	2019, [112]
Real textile wastewater	O@g-C_3N_4/ Al_2O_3	7.1	-	-	0.5	99 (after 60 min)	77 (after 60 min)	-	0.155	2022, [110]
Methylene Blue	Co-BSG	6.8	30	-	5	92 (after 8 min)	93 (after 40 min)	-	-	2022, [113]

Modern materials can over time replace the most commonly used catalysts, such as TiO_2 or activated carbon. They are gaining popularity because by modifying their structure [100,111], surfaces [104,110,112], or composition [62,66], it is possible to improve their catalytic abilities, obtaining better and better pollutant removal abilities. The modifications also overcome the limitations of already existing catalysts [107].

6. Conclusions

Despite many studies, it is still unclear how ozone adsorbs on the catalyst surface. Many articles claim that the decomposition of ozone takes place on the catalyst surface, leading to the formation of free radicals. It is also questionable whether the adsorption of organic substances on the catalyst surface plays a key role in the catalytic process itself. It is certain, however, that the parameters of the process have a great influence on the catalytic efficiency of the degradation of limited pollutants. Therefore, it is important to study the mechanisms of catalytic ozonation to better understand the phenomena that occur during this process and the main factors influencing its efficiency.

The important advantages of using catalytic ozonation to purify colored solutions and textile wastewater are:

- Increasing the degradation of pollutants in water, mainly organic.
- Supporting the mineralization of organic compounds.
- Reduces ozone consumption compared to the ozonation process itself. When reviewing the parameters used in the studies with the best results, some relationships can be noticed:
- The pH of the solution affects the charge of active centers located on the catalyst surface and the ionic charge of organic molecules. This parameter is responsible for the interaction between the catalyst and the impurities. Low pH slows down the ozone decomposition process, which contributes to the longer contact time of ozone with pollutants, but from the industrial point of view, it is less profitable.
- An important parameter due to the increased interest in environmental protection and costs for the company is the stability of the catalyst and the possibility of its reuse.
- By increasing the amount of the catalyst, we provide more active sites, contributing to the decomposition of ozone, i.e., increasing the reactive radicals in the solution.
- Increasing the ozone flow rate also increases the generation of reactive radicals. This parameter is limited by the number of active sites on the catalyst surface.
- Increasing the ozone dose increases gas permeation into the dye solution/sewage, thus improving its availability to react with pollutants. Increasing the ozone dose is also associated with higher production costs.

Despite many studies conducted on the use of catalytic ozonation to purify dye solutions, there is still a lack of research focused on textile wastewater. Textile wastewater is more complex and contains additional substances which makes it more difficult to clean. Additionally, it is important to pay attention to the financial aspect so that the developed method of textile wastewater treatment is beneficial for the industry.

Author Contributions: Conceptualization, M.B. and L.B.; methodology, M.B., L.B. and M.G.; validation, M.B., L.B. and M.G.; investigation, M.B., L.B. and M.G.; data curation, M.B.; writing—original draft preparation, M.B.; writing—review and editing, L.B. and M.G.; supervision, M.G.; project administration, L.B. All authors have read and agreed to the published version of the manuscript.

Funding: The research was funded by the Norway Grants 2014–2021 via the National Centre for Research and Development under grant number NOR/SGS/TEX-WATER-REC/0026/2020-00.

Acknowledgments: Special thanks to Textile Company Biliński, Konstantynów Łódzki, Poland for their cooperation.

Conflicts of Interest: The authors declare no conflict of interest.

References

1. Bisschops, I.; Spanjers, H. Literature Review on Textile Wastewater Characterisation. *Environ. Technol.* **2008**, *24*, 1399–1411. [CrossRef] [PubMed]
2. Ghaly, A.; Ananthashankar, R.; Alhattab, M.; Ramakrishnan, V. Production, Characterization and Treatment of Textile Effluents: A Critical Review. *J. Chem. Eng. Process Technol.* **2013**, *5*, 1–19. [CrossRef]
3. McFarlane, D. *Wastewater as a Potential Source of Recycling in The-Peel Region*; Water Corporation: Perth, Australia, 2019.
4. Sojka-Ledakowicz, J.; Zylla, R.; Mrozinska, Z.; Pazdzior, K.; Klepacz-Smolka, A.; Ledakowicz, S. Application of Membrane Processes in Closing of Water Cycle in a Textile Dye-House. *Desalination* **2010**, *250*, 634–638. [CrossRef]
5. Hossain, L.; Sarker, S.K.; Khan, M.S. Evaluation of Present and Future Wastewater Impacts of Textile Dyeing Industries in Bangladesh. *Environ. Dev.* **2018**, *26*, 23–33. [CrossRef]
6. Reddy, A.S.; Kalla, S.; Murthy, Z.V.P. Textile Wastewater Treatment via Membrane Distillation. *Environ. Eng. Res.* **2022**, *27*, 210228. [CrossRef]
7. Li, F.; Zhong, Z.; Gu, C.; Shen, C.; Ma, C.; Liu, Y.; Yin, S.; Xu, C. Metals Pollution from Textile Production Wastewater in Chinese Southeastern Coastal Area: Occurrence, Source Identification, and Associated Risk Assessment. *Environ. Sci. Pollut. Res.* **2021**, *28*, 38689–38697. [CrossRef]
8. Bhatia, D.; Sharma, N.R.; Kanwar, R.; Singh, J. Physicochemical Assessment of Industrial Textile Effluents of Punjab (India). *Appl. Water Sci.* **2018**, *8*, 83. [CrossRef]
9. Senthil Kumar, P.; Saravanan, A. *Sustainable Wastewater Treatments in Textile Sector*; Elsevier Ltd.: Amsterdam, The Netherlands, 2017; ISBN 9780081020425.
10. The European Commission, Integrated Pollution Prevention and Control (IPPC). *Reference Document on Best Available Techniques for the Textiles Industry*; EU Publ.: Brussels, Belgium, 2003.
11. OECD. *OECD Due Diligence Guidance for Responsible Supply Chains in the Garment and Footwear Sector*; OECD Publishing: Paris, France, 2018.
12. Ortega-Egea, J.M.; García-de-Frutos, N. *Greenpeace's Detox Campaign: Towards a More Sustainable Textile Industry*; Springer International Publishing: Cham, Switzerland, 2019; pp. 37–47. [CrossRef]
13. Paździor, K.; Bilińska, L.; Ledakowicz, S. A Review of the Existing and Emerging Technologies in the Combination of AOPs and Biological Processes in Industrial Textile Wastewater Treatment. *Chem. Eng. J.* **2019**, *376*, 120597. [CrossRef]
14. Oller, I.; Malato, S.; Sánchez-Pérez, J.A. Combination of Advanced Oxidation Processes and Biological Treatments for Wastewater Decontamination-A Review. *Sci. Total Environ.* **2011**, *409*, 4141–4166. [CrossRef]
15. Brillas, E.; Martínez-Huitle, C.A. Decontamination of Wastewaters Containing Synthetic Organic Dyes by Electrochemical Methods. An Updated Review. *Appl. Catal. B Environ.* **2015**, *166–167*, 603–643. [CrossRef]
16. Lellis, B.; Fávaro-Polonio, C.Z.; Pamphile, J.A.; Polonio, J.C. Effects of Textile Dyes on Health and the Environment and Bioremediation Potential of Living Organisms. *Biotechnol. Res. Innov.* **2019**, *3*, 275–290. [CrossRef]
17. Ahlström, L.H.; Sparr Eskilsson, C.; Björklund, E. Determination of Banned Azo Dyes in Consumer Goods. *TrAC-Trends Anal. Chem.* **2005**, *24*, 49–56. [CrossRef]
18. European Parliament. *Directive 2002/61/EC of the European Parliament and of the Council of 19 July 2002 Amending for the Nineteenth Time Council Directive 76/769/EEC Relating to Restrictions on the Marketing and Use of Certain Dangerous Substances and Pre*; EU Publ.: Brussels, Belgium, 19 July 2002.
19. Allègre, C.; Moulin, P.; Maisseu, M.; Charbit, F. Treatment and Reuse of Reactive Dyeing Effluents. *J. Memb. Sci.* **2006**, *269*, 15–34. [CrossRef]
20. Cay, A.; Tarakçioğlu, I.; Hepbasli, A. Assessment of Finishing Processes by Exhaustion Principle for Textile Fabrics: An Exergetic Approach. *Appl. Therm. Eng.* **2009**, *29*, 2554–2561. [CrossRef]
21. Lu, X.; Liu, L.; Liu, R.; Chen, J. Textile Wastewater Reuse as an Alternative Water Source for Dyeing and Finishing Processes: A Case Study. *Desalination* **2010**, *258*, 229–232. [CrossRef]
22. Al-Momani, F.; Touraud, E.; Degorce-Dumas, J.R.; Roussy, J.; Thomas, O. Biodegradability Enhancement of Textile Dyes and Textile Wastewater by VUV Photolysis. *J. Photochem. Photobiol. A Chem.* **2002**, *153*, 191–197. [CrossRef]
23. Wang, X.; Jiang, J.; Gao, W. Reviewing Textile Wastewater Produced by Industries: Characteristics, Environmental Impacts, and Treatment Strategies. *Water Sci. Technol.* **2022**, *85*, 2076–2096. [CrossRef]
24. Ledakowicz, S.; Paździor, K. Recent Achievements in Dyes Removal Focused on Advanced Oxidation Processes Integrated with Biological Methods. *Molecules* **2021**, *26*, 870. [CrossRef]
25. Chung, J.; Kim, J.-O.O. Application of Advanced Oxidation Processes to Remove Refractory Compounds from Dye Wastewater. *Desalin. Water Treat.* **2011**, *25*, 233–240. [CrossRef]
26. Azbar, N.; Yonar, T.; Kestioglu, K. Comparison of Various Advanced Oxidation Processes and Chemical Treatment Methods for COD and Color Removal from a Polyester and Acetate Fiber Dyeing Effluent. *Chemosphere* **2004**, *55*, 35–43. [CrossRef]
27. Emami-Meibodi, M.; Parsaeian, M.R.M.-R.; Banaei, M. Ozonation and Electron Beam Irradiation of Dyes Mixture. *Ozone Sci. Eng.* **2013**, *35*, 49–54. [CrossRef]
28. Hsing, H.J.; Chiang, P.C.; Chang, E.E.; Chen, M.Y. The Decolorization and Mineralization of Acid Orange 6 Azo Dye in Aqueous Solution by Advanced Oxidation Processes: A Comparative Study. *J. Hazard. Mater.* **2007**, *141*, 8–16. [CrossRef] [PubMed]

29. Oguz, E.; Keskinler, B. Comparison among O_3, PAC Adsorption, O_3/HCO_3^-, O_3/H_2O_2 and O_3/PAC Processes for the Removal of Bomaplex Red CR-L Dye from Aqueous Solution. *Dye. Pigment.* **2007**, *74*, 329–334. [CrossRef]
30. Muthukumar, M.; Sargunamani, D.; Selvakumar, N. Statistical Analysis of the Effect of Aromatic, Azo and Sulphonic Acid Groups on Decolouration of Acid Dye Effluents Using Advanced Oxidation Processes. *Dye. Pigment.* **2005**, *65*, 151–158. [CrossRef]
31. Arslan, I.; Balcioglu, I.A.; Tuhkanen, T. Oxidative Treatment of Simulated Dyehouse Effluent by UV and Near-UV Light Assisted Fenton's Reagent. *Chemosphere* **1999**, *39*, 2767–2783. [CrossRef]
32. Alaton, I.; Balcioglu, I.A.; Bahnemann, D. Advanced Oxidation of a Reactive Dyebath Effluent:Comparison of O_3,H_2O_2/UV-C and TiO_2/UV-A Processes. *Water Res.* **2002**, *36*, 1143–1154. [CrossRef]
33. Bilińska, L.; Gmurek, M.; Ledakowicz, S. Comparison between Industrial and Simulated Textile Wastewater Treatment by AOPs—Biodegradability, Toxicity and Cost Assessment. *Chem. Eng. J.* **2016**, *306*, 550–559. [CrossRef]
34. Sreeja, P.H.; Sosamony, K.J. A Comparative Study of Homogeneous and Heterogeneous Photo-Fenton Process for Textile Wastewater Treatment. *Procedia Technol.* **2016**, *24*, 217–223. [CrossRef]
35. Sahunin, C.; Kaewboran, J.; Hunsom, M. Treatment of Textile Dyeing Wastewater by Photo Oxidation Using UV/H_2O_2/Fe^{2+} Reagents. *ScienceAsia* **2006**, *32*, 181–186. [CrossRef]
36. Rejman, A. Bioindykacja i Oddziaływanie Na Organizmy Żywe Ścieków Przemysłu Włókienniczego. *Probl. Ekol.* **2007**, *11*, 41–46.
37. Siddique, K.; Rizwan, M.; Shahid, M.J.; Ali, S.; Ahmad, R.; Rizvi, H. *Textile Wastewater Treatment Options: A Critical Review;* Enhancing Cleanup of Environmental Pollutants; Springer International Publishing: Cham, Switzerland, 2017; Volume 2, pp. 183–207. ISBN 978-3-319-55422-8.
38. Kasprzyk-Hordern, B.; Ziółek, M.; Nawrocki, J. Catalytic Ozonation and Methods of Enhancing Molecular Ozone Reactions in Water Treatment. *Appl. Catal. B Environ.* **2003**, *46*, 639–669. [CrossRef]
39. Fernando, J. *Beltran Ozone Reaction Kinetics for Water and Wastewater Systems;* CRC Press: Boca Raton, FL, USA, 2003.
40. Ikehata, K.; Li, Y. Ozone-Based Processes. *Adv. Oxid. Process. Wastewater Treat. Emerg. Green Chem. Technol.* **2018**, *5*, 115–134. [CrossRef]
41. Biń, A.K. Ozone Dissolution in Aqueous Systems Treatment of the Experimental Data. *Exp. Therm. Fluid Sci.* **2004**, *28*, 395–405. [CrossRef]
42. Bilińska, L.; Blus, K.; Gmurek, M.; Ledakowicz, S. Brine Recycling from Industrial Textilewastewater Treated by Ozone. By-Products Accumulation. Part 1: Multi Recycling Loop. *Water* **2019**, *11*, 460. [CrossRef]
43. Gomes, J.F.; Bednarczyk, K.; Gmurek, M.; Stelmachowski, M.; Zaleska-Medynska, A.; Bastos, F.C.; Quinta-Ferreira, M.E.; Costa, R.; Quinta-Ferreira, R.M.; Martins, R.C. Noble Metal–TiO_2 Supported Catalysts for the Catalytic Ozonation of Parabens Mixtures. *Process Saf. Environ. Prot.* **2017**, *111*, 148–159. [CrossRef]
44. Nawrocki, J.; Kasprzyk-Hordern, B. The Efficiency and Mechanisms of Catalytic Ozonation. *Appl. Catal. B Environ.* **2010**, *99*, 27–42. [CrossRef]
45. Nawrocki, J. Catalytic Ozonation in Water: Controversies and Questions. Discussion Paper. *Appl. Catal. B Environ.* **2013**, *142–143*, 465–471. [CrossRef]
46. Legube, B.; Karpel Vel Leitner, N. Catalytic Ozonation: A Promising Advanced Oxidation Technology for Water Treatment. *Catal. Today* **1999**, *53*, 61–72. [CrossRef]
47. Gracia, R.; Aragües, J.L.; Ovelleiro, J.L. Study of the Catalytic Ozonation of Humic Substances in Water and Their Ozonation Byproducts. *Ozone Sci. Eng.* **1996**, *18*, 195–208. [CrossRef]
48. Pines, D.S.; Reckhow, D.A. Effect of Dissolved Cobalt(II) on the Ozonation of Oxalic Acid. *Environ. Sci. Technol.* **2002**, *36*, 4046–4051. [CrossRef]
49. Pines, D.S.; Reckhow, D.A. Solid Phase Catalytic Ozonation Process for the Destruction of a Model Pollutant. *Ozone Sci. Eng.* **2003**, *25*, 25–39. [CrossRef]
50. Zhang, T.; Lu, J.; Ma, J.; Qiang, Z. Fluorescence Spectroscopic Characterization of DOM Fractions Isolated from a Filtered River Water after Ozonation and Catalytic Ozonation. *Chemosphere* **2008**, *71*, 911–921. [CrossRef] [PubMed]
51. Bakht Shokouhi, S.; Dehghanzadeh, R.; Aslani, H.; Shahmahdi, N. Activated Carbon Catalyzed Ozonation (ACCO) of Reactive Blue 194 Azo Dye in Aqueous Saline Solution: Experimental Parameters, Kinetic and Analysis of Activated Carbon Properties. *J. Water Process Eng.* **2020**, *35*, 101188. [CrossRef]
52. Bertini, I. *Inorganic and Bio-Inorganic Chemistry—Volume II—Homogeneous and Heterogeneous Catalysis;* Eolss: Paris, France, 2009; pp. 50–56.
53. Wang, J.; Chen, H. Catalytic Ozonation for Water and Wastewater Treatment: Recent Advances and Perspective. *Sci. Total Environ.* **2020**, *704*, 135249. [CrossRef]
54. Rame, R. Potential of Catalytic Ozonation in Treatment of Industrial Textile Wastewater in Indonesia: Review. *J. Ris. Teknol. Pencegah. Pencemaran Ind.* **2020**, *11*, 1–11. [CrossRef]
55. Wu, C.H.; Kuo, C.Y.; Chang, C.L. Homogeneous Catalytic Ozonation of C.I. Reactive Red 2 by Metallic Ions in a Bubble Column Reactor. *J. Hazard. Mater.* **2008**, *154*, 748–755. [CrossRef]
56. Malik, S.N.; Ghosh, P.C.; Vaidya, A.N.; Mudliar, S.N. Catalytic Ozone Pretreatment of Complex Textile Effluent Using Fe^{2+} and Zero Valent Iron Nanoparticles. *J. Hazard. Mater.* **2018**, *357*, 363–375. [CrossRef]

57. Vergara-Sánchez, J.; Pérez-Orozco, J.P.; Suárez-Parra, R.; Hernández-Pérez, I. Degradation of Reactive Red 120 Azo Dye in Aqueous Solutions Using Homogeneous/Heterogeneous Iron Systems Degradac on Del Colorante Azo Rojo Reactivo 120 En Soluciones Acuosas Usando Sistemas Homo Eneos/Hetero Eneos De Hierro. *Abril Rev. Mex. Ing. Química* **2009**, *8*, 121–131.
58. Wu, C.H.; Kuo, C.Y.; Chang, C.L. Decolorization of C.I. Reactive Red 2 by Catalytic Ozonation Processes. *J. Hazard. Mater.* **2008**, *153*, 1052–1058. [CrossRef]
59. Trapido, M.; Veressinina, Y.; Munter, R.; Kallas, J. Catalytic Ozonation of M-Dinitrobenzene. *Ozone Sci. Eng.* **2005**, *27*, 359–363. [CrossRef]
60. Andreozzi, R.; Marotta, R.; Sanchirico, R. Manganese-Catalysed Ozonation of Glyoxalic Acid in Aqueous Solutions. *J. Chem. Technol. Biotechnol.* **2000**, *75*, 59–65. [CrossRef]
61. Polat, D.; Balci, I.; Özbelge, T.A. Catalytic Ozonation of an Industrial Textile Wastewater in a Heterogeneous Continuous Reactor. *J. Environ. Chem. Eng.* **2015**, *3*, 1860–1871. [CrossRef]
62. Erol, F.; Özbelge, T.A. Catalytic Ozonation with Non-Polar Bonded Alumina Phases for Treatment of Aqueous Dye Solutions in a Semi-Batch Reactor. *Chem. Eng. J.* **2008**, *139*, 272–283. [CrossRef]
63. Qiu, X.H.; Su, X.Y.; Li, X.J.; Li, N. Preparation of CeO2 with Different Morphologies and Its Application to Catalytic Ozonation of Lemon Yellow Solutions. *Adv. Mater. Res.* **2014**, *997*, 3–8. [CrossRef]
64. Chokshi, N.P.; Ruparelia, J.P. Degradation of Reactive Black 5 Azo Dye Using Catalytic Ozonation with MgO. **2016**, *III*, 2–6.
65. Moussavi, G.; Mahmoudi, M. Degradation and Biodegradability Improvement of the Reactive Red 198 Azo Dye Using Catalytic Ozonation with MgO Nanocrystals. *Chem. Eng. J.* **2009**, *152*, 1–7. [CrossRef]
66. Kong, L.; Diao, Z.; Chang, X.; Xiong, Y.; Chen, D. Synthesis of Recoverable and Reusable Granular MgO-SCCA-Zn Hybrid Ozonation Catalyst for Degradation of Methylene Blue. *J. Environ. Chem. Eng.* **2016**, *4*, 4385–4391. [CrossRef]
67. Faria, P.C.C.; Monteiro, D.C.M.; Órfão, J.J.M.; Pereira, M.F.R. Cerium, Manganese and Cobalt Oxides as Catalysts for the Ozonation of Selected Organic Compounds. *Chemosphere* **2009**, *74*, 818–824. [CrossRef]
68. Gruttadauria, M.; Liotta, L.F.; Di Carlo, G.; Pantaleo, G.; Deganello, G.; Lo Meo, P.; Aprile, C.; Noto, R. Oxidative Degradation Properties of Co-Based Catalysts in the Presence of Ozone. *Appl. Catal. B Environ.* **2007**, *75*, 281–289. [CrossRef]
69. Chokshi, N.P.; Bhutia, H.; Chotalia, A.; Kadiwala, A.; Ruparelia, J.P. Heterogeneous Catalytic Ozonation of Reactive Black 5 with Cobalt Oxide. *Int. J. ChemTech Res.* **2017**, *10*, 402–409.
70. Li, X.; Chen, W.; Ma, L.; Wang, H.; Fan, J. Industrial Wastewater Advanced Treatment via Catalytic Ozonation with an Fe-Based Catalyst. *Chemosphere* **2018**, *195*, 336–343. [CrossRef] [PubMed]
71. Li, X.; Chen, W.; Ma, L.; Huang, Y.; Wang, H. Characteristics and Mechanisms of Catalytic Ozonation with Fe-Shaving-Based Catalyst in Industrial Wastewater Advanced Treatment. *J. Clean. Prod.* **2019**, *222*, 174–181. [CrossRef]
72. Quan, X.; Luo, D.; Wu, J.; Li, R.; Cheng, W.; Ge, S. Ozonation of Acid Red 18 Wastewater Using O_3/$Ca(OH)_2$ System in a Micro Bubble Gas-Liquid Reactor. *J. Environ. Chem. Eng.* **2017**, *5*, 283–291. [CrossRef]
73. Wu, G.; Wei, W.; Cui, L. Adsorption and Catalytic Ozonation Performance of Activated Carbon and Cobalt-Supported Activated Carbon Derived from Brewing Yeast. *Can. J. Chem. Eng.* **2014**, *92*, 36–40. [CrossRef]
74. Zhang, J.; Huang, G.Q.; Liu, C.; Zhang, R.N.; Chen, X.X.; Zhang, L. Synergistic Effect of Microbubbles and Activated Carbon on the Ozonation Treatment of Synthetic Dyeing Wastewater. *Sep. Purif. Technol.* **2018**, *201*, 10–18. [CrossRef]
75. Bilińska, L.; Blus, K.; Bilińska, M.; Gmurek, M. Industrial Textile Wastewater Ozone Treatment: Catalyst Selection. *Catalysts* **2020**, *10*, 611. [CrossRef]
76. Wang, W.L.; Hu, H.Y.; Liu, X.; Shi, H.X.; Zhou, T.H.; Wang, C.; Huo, Z.Y.; Wu, Q.Y. Combination of Catalytic Ozonation by Regenerated Granular Activated Carbon (RGAC) and Biological Activated Carbon in the Advanced Treatment of Textile Wastewater for Reclamation. *Chemosphere* **2019**, *231*, 369–377. [CrossRef]
77. Gholami-Borujeni, F.; Naddafi, K.; Nejatzade-Barandozi, F. Application of Catalytic Ozonation in Treatment of Dye from Aquatic Solutions. *Desalin. Water Treat.* **2013**, *51*, 6545–6551. [CrossRef]
78. Gül, Ş.; Özcan, Ö.; Erbatur, O. Ozonation of C.I. Reactive Red 194 and C.I. Reactive Yellow 145 in Aqueous Solution in the Presence of Granular Activated Carbon. *Dye. Pigment.* **2007**, *75*, 426–431. [CrossRef]
79. Faria, P.C.C.; Órfão, J.J.M.; Pereira, M.F.R. Ozonation of Aniline Promoted by Activated Carbon. *Chemosphere* **2007**, *67*, 809–815. [CrossRef]
80. Gül, Ş.; Eren, O.; Kir, Ş.; Önal, Y. A Comparison of Different Activated Carbon Performances on Catalytic Ozonation of a Model Azo Reactive Dye. *Water Sci. Technol.* **2012**, *66*, 179–184. [CrossRef] [PubMed]
81. Mahmoodi, N.M. Photocatalytic Ozonation of Dyes Using Multiwalled Carbon Nanotube. *J. Mol. Catal. A Chem.* **2013**, *366*, 254–260. [CrossRef]
82. Qu, R.; Xu, B.; Meng, L.; Wang, L.; Wang, Z. Ozonation of Indigo Enhanced by Carboxylated Carbon Nanotubes: Performance Optimization, Degradation Products, Reaction Mechanism and Toxicity Evaluation. *Water Res.* **2015**, *68*, 316–327. [CrossRef] [PubMed]
83. Faria, P.C.C.; Órfão, J.J.M.; Pereira, M.F.R. Activated Carbon and Ceria Catalysts Applied to the Catalytic Ozonation of Dyes and Textile Effluents. *Appl. Catal. B Environ.* **2009**, *88*, 341–350. [CrossRef]
84. Hu, E.; Shang, S.; Tao, X.M.; Jiang, S.; Chiu, K. lok Regeneration and Reuse of Highly Polluting Textile Dyeing Effluents through Catalytic Ozonation with Carbon Aerogel Catalysts. *J. Clean. Prod.* **2016**, *137*, 1055–1065. [CrossRef]

85. Niu, Z.; Yue, T.; Hu, W.; Sun, W.; Hu, Y.; Xu, Z. Covalent Bonding of MnO_2 onto Graphene Aerogel Forwards: Efficiently Catalytic Degradation of Organic Wastewater. *Appl. Surf. Sci.* **2019**, *496*, 1–9. [CrossRef]
86. Cardoso, R.M.F.; Cardoso, I.M.F.; da Silva, L.P.; da Silva, J.C.G.E. Copper(II)-Doped Carbon Dots as Catalyst for Ozone Degradation of Textile Dyes. *Nanomaterials* **2022**, *12*, 1211. [CrossRef]
87. Faria, P.C.C.; Órfão, J.J.M.; Pereira, M.F.R. Mineralisation of Coloured Aqueous Solutions by Ozonation in the Presence of Activated Carbon. *Water Res.* **2005**, *39*, 1461–1470. [CrossRef]
88. Moussavi, G.; Khosravi, R.; Omran, N.R. Development of an Efficient Catalyst from Magnetite Ore: Characterization and Catalytic Potential in the Ozonation of Water Toxic Contaminants. *Appl. Catal. A Gen.* **2012**, *445–446*, 42–49. [CrossRef]
89. Boudissa, F.; Mirilà, D.; Arus, V.A.; Terkmani, T.; Semaan, S.; Proulx, M.; Nistor, I.D.; Roy, R.; Azzouz, A. Acid-Treated Clay Catalysts for Organic Dye Ozonation—Thorough Mineralization through Optimum Catalyst Basicity and Hydrophilic Character. *J. Hazard. Mater.* **2019**, *364*, 356–366. [CrossRef]
90. Inchaurrondo, N.; di Luca, C.; Žerjav, G.; Grau, J.M.; Pintar, A.; Haure, P. Catalytic Ozonation of an Azo-Dye Using a Natural Aluminosilicate. *Catal. Today* **2021**, *361*, 24–29. [CrossRef]
91. Valdés, H.; Godoy, H.P.; Zaror, C.A. Heterogeneous Catalytic Ozonation of Cationic Dyes Using Volcanic Sand. *Water Sci. Technol.* **2010**, *61*, 2973–2978. [CrossRef] [PubMed]
92. Dong, Y.; He, K.; Zhao, B.; Yin, Y.; Yin, L.; Zhang, A. Catalytic Ozonation of Azo Dye Active Brilliant Red X-3B in Water with Natural Mineral Brucite. *Catal. Commun.* **2007**, *8*, 1599–1603. [CrossRef]
93. Valdés, H.; Farfán, V.J.; Manoli, J.A.; Zaror, C.A. Catalytic Ozone Aqueous Decomposition Promoted by Natural Zeolite and Volcanic Sand. *J. Hazard. Mater.* **2009**, *165*, 915–922. [CrossRef] [PubMed]
94. Taseidifar, M.; Khataee, A.; Vahid, B.; Khorram, S.; Joo, S.W. Production of Nanocatalyst from Natural Magnetite by Glow Discharge Plasma for Enhanced Catalytic Ozonation of an Oxazine Dye in Aqueous Solution. *J. Mol. Catal. A Chem.* **2015**, *404–405*, 218–226. [CrossRef]
95. Chokshi, N.P.; Patel, D.; Atkotiya, R.; Ruparelia, J.P. Catalytic Ozonation of Reactive Black 5 in Aqueous Solution over a La-Co-O Catalyst. *J. Indian Chem. Soc.* **2020**, *97*, 373–378.
96. Chokshi, N.P.; Ruparelia, J.P. Synthesis of Nano Ag-La-Co Composite Metal Oxide for Degradation of RB 5 Dye Using Catalytic Ozonation Process. *Ozone Sci. Eng.* **2021**, *44*, 182–195. [CrossRef]
97. Asgari, G.; Faradmal, J.; Nasab, H.Z.; Ehsani, H. Catalytic Ozonation of Industrial Textile Wastewater Using Modified C-Doped MgO Eggshell Membrane Powder. *Adv. Powder Technol.* **2019**, *30*, 1297–1311. [CrossRef]
98. Ghuge, S.P.; Saroha, A.K. Ozonation of Reactive Orange 4 Dye Aqueous Solution Using Mesoporous Cu/SBA-15 Catalytic Material. *J. Water Process Eng.* **2018**, *23*, 217–229. [CrossRef]
99. Ghuge, S.P.; Saroha, A.K. Catalytic Ozonation of Dye Industry Effluent Using Mesoporous Bimetallic Ru-Cu/SBA-15 Catalyst. *Process Saf. Environ. Prot.* **2018**, *118*, 125–132. [CrossRef]
100. Zhu, S.N.; Hui, K.S.N.; Hong, X.; Hui, K.S.N. Catalytic Ozonation of Basic Yellow 87 with a Reusable Catalyst Chip. *Chem. Eng. J.* **2014**, *242*, 180–186. [CrossRef]
101. Li, P.; Miao, R.; Wang, P.; Sun, F.; Li, X. yan Bi-Metal Oxide-Modified Flat-Sheet Ceramic Membranes for Catalytic Ozonation of Organic Pollutants in Wastewater Treatment. *Chem. Eng. J.* **2021**, *426*, 131263. [CrossRef]
102. Liang, C.; Luo, X.; Hu, Y. Enhanced Ozone Oxidation by a Novel Fe/Mn@$\gamma-Al_2O_3$ Nanocatalyst: The Role of Hydroxyl Radical and Singlet Oxygen. *Water* **2022**, *14*, 19. [CrossRef]
103. Lu, J.; Wei, X.; Chang, Y.; Tian, S.; Xiong, Y. Role of Mg in Mesoporous $MgFe_2O_4$ for Efficient Catalytic Ozonation of Acid Orange II. *J. Chem. Technol. Biotechnol.* **2016**, *91*, 985–993. [CrossRef]
104. El Hassani, K.; Kalnina, D.; Turks, M.; Beakou, B.H.; Anouar, A. Enhanced Degradation of an Azo Dye by Catalytic Ozonation over Ni-Containing Layered Double Hydroxide Nanocatalyst. *Sep. Purif. Technol.* **2019**, *210*, 764–774. [CrossRef]
105. Feng, C.; Diao, P. Nickel Foam Supported NiFe2O4-NiO Hybrid: A Novel 3D Porous Catalyst for Efficient Heterogeneous Catalytic Ozonation of Azo Dye and Nitrobenzene. *Appl. Surf. Sci.* **2021**, *541*, 148683. [CrossRef]
106. Hien, N.T.; Nguyen, L.H.; Van, H.T.; Nguyen, T.D.; Nguyen, T.H.V.; Chu, T.H.H.; Nguyen, T.V.; Trinh, V.T.; Vu, X.H.; Aziz, K.H.H. Heterogeneous Catalyst Ozonation of Direct Black 22 from Aqueous Solution in the Presence of Metal Slags Originating from Industrial Solid Wastes. *Sep. Purif. Technol.* **2020**, *233*, 115961. [CrossRef]
107. Pereira, C.A.A.; Nava, M.R.; Walter, J.B.; Scherer, C.E.; Dominique Kupfer Dalfovo, A.; Barreto-Rodrigues, M. Application of Zero Valent Iron (ZVI) Immobilized in Ca-Alginate Beads for C.I. Reactive Red 195 Catalytic Degradation in an Air Lift Reactor Operated with Ozone. *J. Hazard. Mater.* **2021**, *401*, 123275. [CrossRef]
108. Pervez, M.N.; Stylios, G.K.; Liang, Y.; Ouyang, F.; Cai, Y. Low-Temperature Synthesis of Novel Polyvinylalcohol (PVA) Nanofibrous Membranes for Catalytic Dye Degradation. *J. Clean. Prod.* **2020**, *262*, 121301. [CrossRef]
109. Chen, C.; Jia, N.; Song, K.; Zheng, X.; Lan, Y.; Li, Y. Sulfur-Doped Copper-Yttrium Bimetallic Oxides: A Novel and Efficient Ozonation Catalyst for the Degradation of Aniline. *Sep. Purif. Technol.* **2020**, *236*, 116248. [CrossRef]
110. Faghihinezhad, M.; Baghdadi, M.; Shahin, M.S.; Torabian, A. Catalytic Ozonation of Real Textile Wastewater by Magnetic Oxidized G-C_3N_4 Modified with Al2O3 Nanoparticles as a Novel Catalyst. *Sep. Purif. Technol.* **2022**, *283*, 120208. [CrossRef]
111. Sone, B.T.; Makamu, E.; Mohamed, H.E.A.; Oputu, O.; Fester, V. Green-Synthesized ZnO via Hyphaene Thebaica Fruit Extracts: Structure & Catalytic Effect on the Ozonation of Coralene Rubine-S2G Azo Disperse Dye. *Environ. Nanotechnol. Monit. Manag.* **2021**, *16*, 100515. [CrossRef]

112. Yu, D.; Wu, M.; Hu, Q.; Wang, L.; Lv, C.; Zhang, L. Iron-Based Metal-Organic Frameworks as Novel Platforms for Catalytic Ozonation of Organic Pollutant: Efficiency and Mechanism. *J. Hazard. Mater.* **2019**, *367*, 456–464. [CrossRef] [PubMed]
113. Javed, F.; Tariq, A.; Ikhlaq, A.; Rizvi, O.S.; Ikhlaq, U.; Masood, Z.; Qazi, U.Y.; Qi, F. Application of Laboratory-Grade Recycled Borosilicate Glass Coated with Iron and Cobalt for the Removal of Methylene Blue by Catalytic Ozonation Process. *Arab. J. Sci. Eng.* **2022**, 1–16. [CrossRef]

catalysts

MDPI

Article

Effect of Basic Promoters on Porous Supported Alumina Catalysts for Acetins Production

Rita de Cássia F. Bezerra [1], Gabriela Mota [1], Ruth Maria B. Vidal [1], Jose Vitor do Carmo [1], Gilberto D. Saraiva [2], Adriana Campos [3], Alcineia C. Oliveira [1,*], Rossano Lang [4], Larissa Otubo [5], José Jiménez Jiménez [6] and Enrique Rodríguez-Castellón [6,*]

1 Departamento de Química Analítica e Físico-Química, Universidade Federal do Ceará, Campus do Pici, Bloco 940, Fortaleza 60455-760, Brazil
2 Faculdade de Educação Ciências e Letras do Sertão Central, Universidade Estadual do Ceará, Quixadá 63902-098, Brazil
3 CETENE, Cidade Universitária, Recife 50740-545, Brazil
4 Instituto de Ciência e Tecnologia, Universidade Federal de São Paulo—UNIFESP, São José dos Campos 12231-280, Brazil
5 Centro de Ciência e Tecnologia de Materiais—CCTM, Instituto de Pesquisas Energéticas e Nucleares—IPEN, São Paulo 05508-000, Brazil
6 Departamento de Química Inorgánica, Facultad de Ciencias, Universidad de Málaga, 29071 Málaga, Spain
* Correspondence: alcineia@ufc.br (A.C.O.); castellon@uma.es (E.R.-C.)

Abstract: A facile strategy for the design of porous supports was obtained by modifying the sol-gel method followed by the wet impregnation technique. In this respect, herein, the acidity of the γ-Al_2O_3 phase was modulated by adding basic MgO, La_2O_3 or ZnO promoters to form binary supported catalysts. The Ni and Co dispersion on the supports associated with their tunable acidity and morphologies resulted in highly porous supported alumina-based catalysts. The physicochemical properties of the solids were comprehensively investigated by XRD, textural properties, Raman and FTIR spectroscopy, SEM-EDS, TEM, EPR and XPS analyses. The catalytic performances in the esterification of glycerol in the presence of acetic acid (EG) for the acetins production were evaluated. The highly dispersed NiO and Co_3O_4 active species on binary porous supports produced synergistic effects appearing to be the reason for the activity of the solids in the EG reaction. Under the optimized reaction conditions, NiCo/MgO-Al_2O_3 was found to be a robust solid with superior catalytic performance and improved stability in four reaction cycles with 65.0% of glycerol conversion with an exclusive selectivity of 53% for triacetin. The presence of Co^{2+}/Co^{3+} and Ni^{2+} strongly interacting with the spinel γ-Al_2O_3 and $MgAl_2O_4$ phases, the latter having a large number of lattice oxygen species, was considered another active component besides those of Ni and Co in the esterification of glycerol.

Keywords: porous alumina; support; esterification; glycerol; basic promoters

check for updates

Citation: Bezerra, R.d.C.F.; Mota, G.; Vidal, R.M.B.; Carmo, J.V.d.; Saraiva, G.D.; Campos, A.; Oliveira, A.C.; Lang, R.; Otubo, L.; Jiménez Jiménez, J.; et al. Effect of Basic Promoters on Porous Supported Alumina Catalysts for Acetins Production. *Catalysts* **2022**, *12*, 1616. https://doi.org/10.3390/catal12121616

Academic Editor: Narendra Kumar

Received: 29 October 2022
Accepted: 5 December 2022
Published: 9 December 2022

1. Introduction

In recent decades, glycerol valorization has received prominent research interest because of escalating biodiesel production [1–3]. The large amounts of crude glycerol as the main by-product of the biodiesel industries have impelled the scientific community to look for alternatives to convert the trialcohol into value-added chemicals [1–6].

To date, many catalytic routes, including dehydration, esterification, reforming and acetalization, among others, have been developed to utilize processes capable of consuming raw glycerol for industrial applications [4–8]. Particularly, the esterification reaction of glycerol in the presence of acetic acid (EG) becomes increasingly important for the direct use of glycerol to obtain more valuable products (Figure 1).

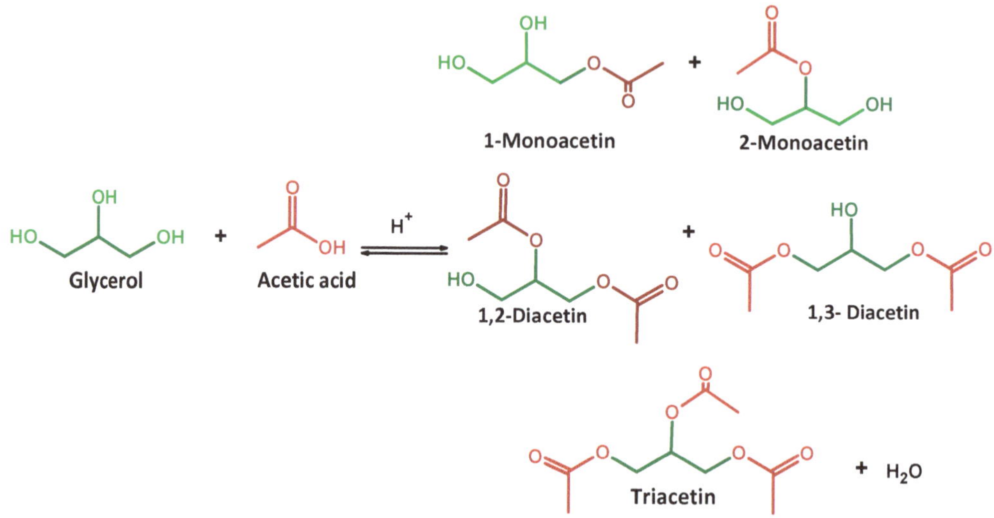

Figure 1. Schematic illustration of the EG reaction in the presence of acetic acid.

Thus, the transformation of glycerol into glycerol esters (acetins) through the EG reaction is of commercial importance, since the obtained acetins are applied as emulsifiers, cosmetics, food additives, stabilizers, biofuels additives and pharmaceuticals [9–11]. This has boosted progress in the design of a variety of advanced catalytic materials towards the EG reaction, such as zeolites, resins, graphene oxide, metal oxides, amorphous and mesoporous sulfated silicas, hydroxyapatites and a diversity of supported oxides [1–7,11–13]. Among all types of catalysts, acid-based solids have been intensively investigated to produce acetins because of their good prospect for improving catalytic activities and yields during EG reaction [2,4,11–16].

In this sense, the exceptional properties of alumina-based catalysts, such as porosity, large specific surface area, chemical stability and their addressable Lewis and Brønsted acid sites for glycerol conversion, have made them of particular interest in the EG reaction [3,13,17–20]. However, the main disadvantages of these types of solid acid catalysts for glycerol esterification are their low water tolerance, leaching of active sites and low selectivities to the acetins, which significantly restrict their performance in the reaction [5,17–20]. Overall, strategies for acidity modulation of the catalysts that permit the control of acid strength, types of acid sites and surface acidity to minimize the problem on water deactivation and subsequently improving the selectivity of diacetin and triacetin are vastly documented [6,11,15,20].

It is particularly interesting to note that the current efforts for enhancing the performance of the solids in the EG reaction are moving towards the use of catalysts possessing simultaneous acid-based and redox active sites to the title reaction [5]. Such a bifunctional catalyst represents an emerging strategy to have a robust catalyst for glycerol esterification that at least limits the rapid deactivation by water of the active sites. Among the many attempts to overcome these limitations, synthetic methodologies to improve the accessibility of acid sites and increase the reactivity of surfaces are reasonably mature [2,10,21]. Nonetheless, little attention has been focused on the influence of the acidity modulation of the alumina-based catalysts and the consequent modification of their properties to achieve good catalytic performances [5,21,22].

In the present study, the effect of the basic promoters addition to supported alumina catalyst was investigated in the EG reaction for acetins production. This type of bifunctional catalyst involves the combination of Ni and Co species representing the active sites dispersed on the alumina support, besides tuning their acidic properties provided by the incorporation of MgO, ZnO and La_2O_3. Because the decline in acidity of the alumina is known to be strongly influenced by the type of basic promoter [22–28], the suitable modification of the solid by MgO, ZnO or La_2O_3 seems to be well suited for solid textural properties and stability preservation, which would further alleviate the water deactivation in EG reaction and thus to improve the overall selectivity to the acetins.

Another important aspect of the synthesis of the catalysts under study is the efficiency of the sol-gel method to prevent particle aggregation and generate the porosity and accessibility of the acid sites to the reactant molecules [22–24]. It can be expected that a significant enhancement in catalytic properties can be achieved by combining the advantages of well dispersed Ni and Co sites on porous modified alumina possessing weak to medium acid site strengths. The physicochemical properties of the solids were intensively investigated by XRD, N_2-physisorption, Raman and FTIR spectroscopy, SEM-EDS, TEM, EPR and XPS measurements.

2. Results and Discussion

2.1. Structural Characterizations

XRD patterns are collected to illustrate the structural features of the aluminas, after the incorporation of the basic promoters. A typical XRD pattern of the semi-crystalline γ-Al_2O_3 phase is clearly recognized by the low intensity and broad peaks (Figure 2a). Accordingly, small reflections at 2θ values of 19.9, 31.2, 37.0, 39.8, 45.4, 60.7 and 67.1° are assigned to the (111), (220), (311), (222), (400), (511) and (440) crystallographic planes of the face centered-cubic Fd-3m planes for γ-Al_2O_3 phase (JCPDS 10-425). Notably, the most intense peaks of cubic γ-Al_2O_3 are depicted at 2θ values of ca. 45.4 and 67.1° having quite similar intensities, except to MA sample. This allows us to deduce that the prevalence of γ-Al_2O_3 phase in all samples is related to the successful introduction of foreign oxides, such as ZnO, MgO and La_2O_3 into alumina lattice structure to form intermingled mixed-metal oxides.

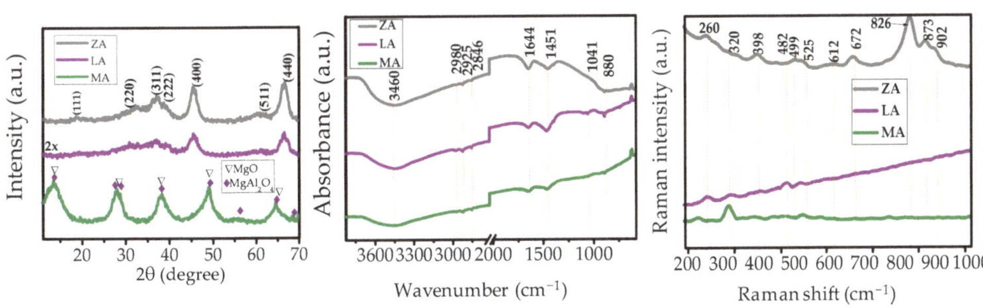

(**a**) (**b**) (**c**)

Figure 2. Physicochemical characterizations of the supports: (**a**) XRD, (**b**) FTIR and (**c**) Raman measurements. The letters LA, ZA, MA represent the La_2O_3–Al_2O_3, ZnO–Al_2O_3 and MgO–Al_2O_3 supports, respectively. The XRD pattern of LA sample is amplified 2× in Figure 2a.

To prove this hypothesis, XRD patterns of the unsupported samples (Figure 2a) reveal no detectable differences for a γ-Al_2O_3 sample, as found elsewhere [10,13,21]. It is supposed that the incorporation of the aforesaid basic oxides occurs because of the use of single-source alkoxy precursors, which form pre-existing Me–O–Al bonds via sol–gel chemistry process, where Al^{3+} (50 pm) cations are replaced by larger La^{3+} (105 pm), Zn^{2+} (75 pm) or Mg^{2+} (78 pm) during the synthesis [21,23,24]. Some others but very similar preparations of alumina are used with basic promoters resulting in weaker XRD peaks, which suggests the inclusion of the basic promoters into the cubic structure [25–28]. Obviously, weak diffraction peaks for LA indicate poor crystallinity due to either small particles or amorphous structure related to La_2O_3 oxides, but reflections of the γ-Al_2O_3 phase prevail in the solid. On the contrary, the diffraction peaks of ZA have a higher intensity than those of LA owing to the existence of the inherently larger particles in the former. For MA, a particular composition of both cubic MgO and spinel $MgAl_2O_4$ phases is seen, probably due to the high calcination temperatures above 800 °C favoring the spinel phase formation besides that of γ-Al_2O_3. The findings also state that the diffraction peaks are shifted to small 2θ values because of the lattice parameters of γ-Al_2O_3 increase caused by cell expansion, which evidences that some divalent Mg ions with larger ionic radius enter into the spinel skeleton by isomorphic replacement of Al^{3+} ions [29].

FTIR measurements are used to further characterize the structural features of the solids. Figure 2b shows the FTIR spectra of unsupported samples. A broad absorption band centered at 3460 cm^{-1} is visible with low intensity for all samples, which corresponds to the hydrogen-bonded O–H stretching ν(O-H) of physisorbed water. Another possible assignment of this band could be the structural hydroxyl groups present in the oxides, which is consistent with previous reports [13,30]. Moreover, weak absorption bands emerged at approximately 1644 cm^{-1} are assigned to the bending vibration δ(O-H) of hydroxyl groups (Figure 2b). Moreover, the stretchings of the alkyl groups ν(C–H) are located at 2980, 2925, and 2846 cm^{-1}. Furthermore, the band observed at 1041 cm^{-1} for the unsupported solids could be due to C-O stretching of the carbonyl groups, which are not completely removed upon calcination.

Raman spectroscopy is helpful to complement the structural investigation of solids. Raman spectra of the unsupported solids are characterized by the presence of four weak bands located in the low-frequency region at approximately 260, 320, 482 and 672 cm^{-1} (Figure 2c). According to the findings, the corundum-type structure, e.g., γ-Al_2O_3 phase has a $D^6{}_{3d}$ symmetry with seven Raman active phonon modes and the signals assigned as those of 2A1g+5Eg modes are observed at 378, 418, 432, 451, 578, 645 and 751 cm^{-1} [30,31]. This is in agreement with the bands found in the FTIR spectra. The Raman band at 260 cm^{-1} is slightly shifted to lower frequencies within 2–20 cm^{-1} range possibly due to Mg-O bonds from free MgO and $MgAl_2O_4$ species on solid surface of MA support, as evidenced by XRD.

Additionally, the corundum structure may also depict a strong fluorescence background without visible Raman bands for the γ-Al_2O_3 phase, depending on the calcination temperature and laser excitation and power of the source used [30,31]. Interestingly, long exposure times, such as 50 scans applied to the samples under study allow the observed weak Raman bands for the unsupported samples. This clearly suggests that the aforesaid signals are consistent with the Al-O-Al vibrational modes in AlO_6 octahedra [30,31]. Additional signals appearing at approximately 398, 499, 525, 612, 826, 873 and 902 cm^{-1} are clearly detectable for ZA support (Figure 2c). Such type of signals could be a hexagonal wurtzite structure of ZnO belonging to the *P63mc* (C_{6v}) space group [32], mostly coming from some ZnO on the solid surface. Although ZnO contributions are not observable by XRD results, Raman spectroscopy is very sensitive to detect surface species. Instead, XRD measurements are devoted to observing the bulk species. In addition, other contributions of the MgO or $MgAl_2O_4$ and La_2O_3 promoters to the Raman spectra are not observable, which indicates the inclusion of these oxides in the alumina host, in line with XRD measurements (Figure 2a). These results are later confirmed by TEM measurements.

It seems that the XRD peak positions and intensities remain unchanged, after the dispersion of relatively low Ni and Co amounts on the surface of binary supports (Figure 3a). The NiCo/MA is an exception owing to the reflections of cubic MgO (JCPDS 89-7746) or spinel cubic $MgAl_2O_4$ (JCPDS9-1627) appearing in the diffractograms. Furthermore, NiO nanoparticles are initially formed in the first impregnation step, whereas the subsequent Co_3O_4 nanoparticles co-impregnation process results in a high dispersion the aforesaid oxide species (Figure 3a). Hence, the peaks of the supported solids have similar intensities to those of the unsupported catalysts (Figure 2a). Additionally, it is not possible to rule out that the relatively low Ni and Co amounts on the support has little influence on the XRD diffractograms, due to the detection limit of the technique. Accordingly, the existence of the $NiAl_2O_4$ and $CoAl_2O_4$ spinel-like phases in the form of nanoparticles cannot be neglected due to the high dispersed and isolated Ni^{2+} and Co^{2+} interacting with alumina [30–33].

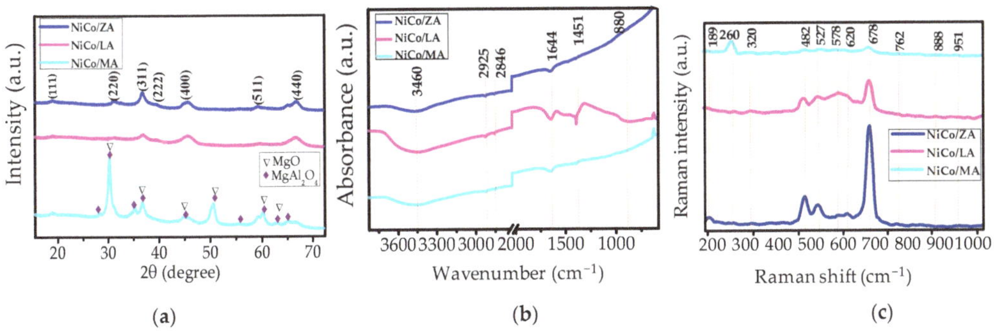

Figure 3. Physicochemical characterizations of the supported samples: (**a**) XRD, (**b**) FTIR and (**c**) Raman measurements. The letters NiCo/LA, NiCo/ZA and NiCo/MA represent the NiCo/La_2O_3–Al_2O_3, NiCo/ZnO–Al_2O_3 and NiCo/MgO–Al_2O_3 supported catalysts, respectively.

Interestingly, a noticeable increase in the FTIR band intensity at 1644 cm^{-1} implies that a large fraction of the hydroxyl groups is intensified from the samples, after impregnating the metal oxides on the supports (Figure 3b). It could indicate the presence of more surface OH groups, after the support of the active components on alumina. As found elsewhere, dehydroxylation of supports after consecutive impregnation and calcination steps is a common phenomenon found in the preparation of the supported oxide catalysts [30]. This apparent discrepancy is explained based on the fact the above FTIR results confirm a significant enhancement of hydrophobicity through the generation of surface acid Brønsted sites, owing to the successful synthesis of porous materials.

The medium intensity band at 1451 cm^{-1} for the supported samples is assigned the bending δ(C–H) vibrations of the organic compounds, such as residual trisec-butoxy aluminum [13]. This means that the precursors are not completely removed when calcinating the supports (Figure 3b). Further EDS and XPS results assign the presence of these residual carbon species on the solid surface. On the contrary, the disappearance or intensity attenuation of the residual carbon absorption bands in the spectra may be ascribable to the double calcination step of the metal oxides, after the impregnation process for supported samples (Figure 3b).

FTIR spectra of the supported solids (Figure 3b) do not exhibit representative bands at approximately 1041 cm^{-1}, which suggests the remaining trisec butoxyl groups removal from the solids after various calcination steps. At low frequency regions, the bands below 880 cm^{-1} are assigned to Me-O lattice vibrations from stretching Me–OH, Me–O–Me or even O–Me–O bonds [22,30,33].

Raman spectra of the supported samples display distinct features compared with those of the unsupported solids (Figure 3c). Low frequencies bands positioned at approximately 260, 320, 482 and 678 cm^{-1} attributable to the γ-Al_2O_3 phase remain unperturbed, but intensity change of other bands occurred simultaneously with the appearance of a broad band with little signals located at 480, 527, 578 and 620 cm^{-1}. Literature reports reveal that crystalline NiO has a defect rocksalt cubic structure (space group *Fm3m*) with Ni^{2+} cations in octahedral sites with Raman bands 460–600 cm^{-1} [34–36]. Therefore, the observed broad band between 480–620 cm^{-1} is attributed to the NiO nanoparticles dispersed on the supports. These observations are also consistent with the FTIR measurements, which reveal the presence of Me-O vibrations in the high frequencies region. In addition, Raman spectroscopy is capable of detecting surface nickel oxide nanoparticles in comparison with the XRD technique, which is sensitive to bulk species [34]. Bands found at approximately 189, 482, 527, 620 and 678 cm^{-1} may also be attributed to the Raman active modes of Co_3O_4. Accordingly, cubic Co_3O_4 crystallizes in the normal spinel structure $Co^{2+}(Co^{3+})_2O_4{}^{2-}$ belonging to the *Fd3m* space group ($O^7{}_h$) [36]. Hence, the cubic spinel lattice has Co^{2+} and Co^{3+} ions placed at tetrahedral and octahedral sites, respectively. The active *A1g*, *Eg* and

3*F2g* Raman modes for the spinel structure are active in opposite to the 4*F1u* mode, which is infrared active. Accordingly, the Raman mode positioned at 189 cm^{-1} is ascribable to *F2g* phonon modes, meanwhile those at 482 and 678 cm^{-1} correspond to the *Eg* and *A1g* phonon modes, respectively [36].

In particular, the NiCo/MA sample has Raman bands at approximately 250 (*F2g*), 307 (*F2g*), 410 (*F2g*), 492 (*F2g*), 670 (*F2g*) and 762 (*A1g*) cm^{-1} that can be attributed to lattice vibrations of $MgA1_2O_4$ along with bands at approximately 410, 670 and 762 cm^{-1} that may also be attributed to MgO. However, the low signal/noise ratio impedes the exact assignment of these vibrational modes. Although XRD and FTIR techniques do not allow detection of the $NiA1_2O_4$ spinel phase, in the present case, the aforesaid spinel structure depicts Raman bands at approximately 200,370 and 612 cm^{-1}, in line with an earlier report [31]. These Raman bands appear to be superimposed with those of the NiO, $MgAl_2O_4$, MgO and Co_3O_4 phases and thus the presence of the spinel structure cannot be ruled out. When comparing the Raman spectra of the supported samples, no distinct changes in the spectra are observed compared to the unsupported solids.

Summarizing, XRD, FTIR and Raman measurements demonstrate either good dispersion of Ni and Co nanoparticles or their interactions with supports, as further seen by textural and morphological properties.

2.2. Textural and Morphological Properties

The textural properties of the solids are examined via N_2 physisorption isotherms and the corresponding pore size distributions (Figure 4). As expected, the isotherms exhibit a hysteresis loop at high relative p/p_o values indicating a type IV isotherm, as defined by the IUPAC classification [13,30,37]. Such features suggest that the sol-gel method leads to the presence of uniform mesopores arrangements. The porosity of the unsupported solids is evident by their large nitrogen adsorption uptake at saturation, which has a hysteresis very similar to the H_4 and H_1 types (Figure 4A$_1$). In other words, LA and ZA supports are formed by particles crossed by nearly cylindrical channels or agglomerates of particles [38,39], as further seen by SEM-EDS analyses.

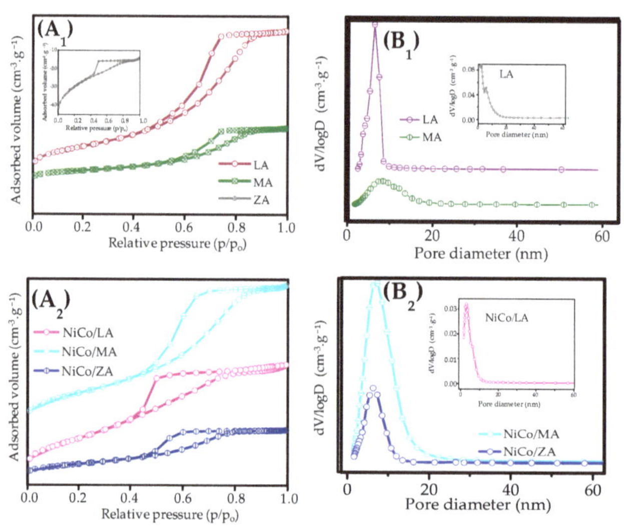

Figure 4. (**A**) N_2 physisorption analyses and the corresponding (**B**) pore size distributions of the supported and unsupported solids. The inset figure (**A$_1$**) is the isotherm of the ZA sample, whereas insets (**B$_1$**,**B$_2$**) are the pore size distribution curves of ZA and NiCo/LA samples, respectively. The figure (**A$_2$**) corresponds to the isotherms of the supported solids.

Regarding the MA support, a steady rise in p/p_0 in the range of 0.1–0.6 is followed by a gradual increase in nitrogen uptake till the saturated adsorption plateau at p/p_0 values is nearly 1. This is likely because of the capillary condensation phenomena in mesopores larger than 4 nm [39].

The textural properties of the solids are presented in Table 1. The BET surface area of the γ-Al_2O_3 sample is 182 $m^2 \cdot g^{-1}$, with a pore volume of 0.29 $cm^3 \cdot g^{-1}$, which confirms that the sol–gel method is efficient to obtain porous metal oxides [13,21,35]. The BET surface area of unsupported LA is nearly 3 folds higher than that of bare γ-Al_2O_3 with similar trends followed by the pore volume. The textural properties differ significantly with ZA and MA supports having lower surface areas and pore volumes compared with LA and γ-Al_2O_3 counterparts as well (Table 1). In this regard, the expansion of the alumina lattice due to La^{3+} cations seems to be well in agreement with the highest textural parameters of the LA sample. Meanwhile, the lattice contraction with smaller cations such as Mg^{2+} and Zn^{2+} accounts for the lower textural parameters of MA and ZA.

Table 1. Textural properties of the solids in study.

Samples	Surface Area (m^2 g^{-1})			[d] Mesopore Volume (cm^3 g^{-1})	[e] Total Pore Volume (cm^3 g^{-1})	[f] Pore Diameter (nm)
	[a] BET	[b] External	[c] *t*-Plot			
γ-Al_2O_3	182	87	23	0.29	0.29	6.1
LA	476	445	31	0.85	0.86	6.7
MA	162	152	9	0.30	0.31	7.5
ZA	109	82	27	0.04	0.05	3.1
NiCo/LA	148	152	0	0.20	0.20	4.1
NiCo/MA	130	125	5	0.17	0.17	5.6
NiCo/ZA	50	48	1	0.08	0.15	5.0

[a] Brunauer–Emmett–Teller surface areas. [b] External surface area obtained from *t*-plot method. [c] Micropore surface area obtained by the *t*-plot method. [d] Mesopore volume and [e] Total pore volume taken by Barrett–Joyner–Halenda method. [f] Adsorption average pore diameter derived from Barrett–Joyner–Halenda method (4 V/A).

This clearly illustrates the rapid diffusion of divalent cations to be included in the alumina framework during the peptization and co-condensation steps that occurred in the synthesis, besides alumina avoiding the particle coarsening, as shown by earlier reports [13,37,38].

It is worth to note that the micropores volumes of the unsupported solids have quite similar values, all being much lower than the total pore volumes, which confirms the supports are mesoporous. In addition, the *t*-plot surface areas for micropores are much lower than the mesoporous parameters suggesting a marginal microporous contribution from binary solids LA and MA (Table 1). Additionally, the external surface areas corresponding to the mesopores and macropores and yet the crystal void spaces in pore structures are listed in Table 1. The ZA support holds the lowest external surface area evidencing the presence of micro and meso porosity, while other solids have mainly mesoporous structures. Meanwhile, the corresponding pore size curves (Figure 4(B_1)) show a monomodal distribution of pore diameters centered between 3.1 and 7.5 nm (Table 1), which confirms the prevalence of the mesoporous structure in the unsupported samples. Despite ZA has the lowest textural parameters among the unsupported samples, the average pore diameter of ca. 3.1 nm confirms the mesopore structure (Figure 4(A_1) inset) along with some micropores (Figure 4(B_1) inset).

The isotherms of the supported samples have similar features to the unsupported solids, e.g., type IV isotherms, although H_1 and H_2 hysteresis loops are observed (Figure 4(B_1)).

In addition, an unavoidable decrease in surface areas and the same, if not slightly lower, pore volumes are observed compared with the unsupported solids (Table 1). These evident changes are illustrated by the BET surface areas of NiCo/MA and NiCo/ZA, which drop strikingly by approximately 15% comparable to their supports, along with the total pore volumes declining to well below 20%. These results can be attributed to the cooperative interactions between the metal oxides dispersed on alumina-based solids by reducing the surface areas and pore volumes during the consecutive steps of calcination of the supported solids. Such effects also result in the enlargement of the surface coverage, as shown by the external area values of supported solids, in general, being lower than those of unsupported solids. Remarkably, micropore areas and mesopore volumes decrease are a result of the absence of pore blockage by Ni and Co nanoparticles, as later seen by TEM. The NiCo/ZA is an exception, since the micropore volume augment suggests the smaller mesopores transformation into micropores upon dispersion of the metal oxides followed by two consecutive calcination steps of the sample. Correspondingly, a steep nitrogen uptake in NiCo/ZA at low relative pressure regions suggests the occurrence of some micropores, whereas the extent of adsorption in micropores partially disappears in NiCo/LA and NiCo/MA. The pore size distribution curves broadened and mean pore diameters range from 4.1 to 5.6 nm, compared with the unsupported solids. For instance, the pore diameter of NiCo/LA is ca. 4.1 nm appearing slightly smaller compared to those of NiCo/ZA and NiCo/MA. This closure of the pores of the former sample can be a result of the preferential deposition of the metal oxide particles in the mesopores or micro macropores with little effect coming from the Ni and Co nanoparticle sintering phenomenon.

The high porosity of the solids is consistent with the expected features for sol-gel based-solids. It is clear that the alumina support restrains the growth of metal oxide nanoparticles in the supported solids owing to the strong metal-support interaction.

In addition, the obtained values are all close to the nominal content of 1.0 wt.%, which suggests the samples successfully synthesized.

The transmission electron microcopy images depict the structural features of the supports (Figure 5). It is evident from the top of Figure 5(A_1) that LA consists of disordered particles, most of them agglomerated. The included Figure 5A_1 shows that these particles are 1–10 nm in size. Furthermore, the clear amorphous regions along with crystalline regions (square area in Figure 5(A_2)) generally show the features observed in the γ-Al_2O_3-based samples [21,22]. The arrow in Figure 5(A_3) depicts a border between two regions suggesting the grain boundary. The HRTEM image suggests lattice spacings of ca. 0.280 and 0.456 nm, which are indexed to be in the (022) and (111) planes of γ-Al_2O_3, as found elsewhere [40,41]. Additionally, inset Figure 5(A_3) illustrated a dark particle too large to be discerned, suggesting segregation of La_2O_3, in agreement with EDS results (Figure S1 in Supplementary Materials). In case of the MA support, similar particle agglomeration is seen in the low magnification TEM image with particle sizes of approximately 16 nm (Figure 5(B_1)). In the same vein, the crystalline regions appear evidently more than the LA counterparts (highlighted square and spherical regions in Figure 5(B_2)) but differ only in discontinuity of the pattern. This is indeed due to the superposition of the formed phases. The XRD and Raman results indicate supportive evidences for the MgO and $MgAl_2O_4$ formation, in agreement with Figure 5(B_3) through their lattice fringes.

Interestingly, ZA is composed of smaller particles with diameters within the range 1–7 nm being less agglomerated than those of LA and MA (Figure 5(C_1)). Besides, crystalline domains are also found with a d-spacing of ca. 0.247 nm (Figure 5(C_2)), which is associated with the (101) for cubic ZnO [42]. Moreover, the d-spacing of ca.0.280 nm (022) is associated with the γ-Al_2O_3 besides the presence of amorphous regions (the inset of Figure 5(C_2)). In addition, the particles of the binary ZA support appear to be significantly crystalline to generate the lattice fringes (Figure 5(C_3)) as those of MA.

Further information on the structural features of the supported samples is obtained by direct imaging of the structure through TEM. After dispersing the Ni and Co on LA, particles remain disordered with some degree of agglomeration (Figure 6(A_1)). This is in line with EDS results in Figure S2 in Supplementary Materials. The magnified trapeze highlighted area illustrates that the rippled aggregated particles contain notable nanoparticles dispersed on it (top left, Figure 6(A_1)). In these aggregates, the presence of lattice fringes indicates the crystalline domains of NiCo/LA (top middle, Figure 6(A_2)).

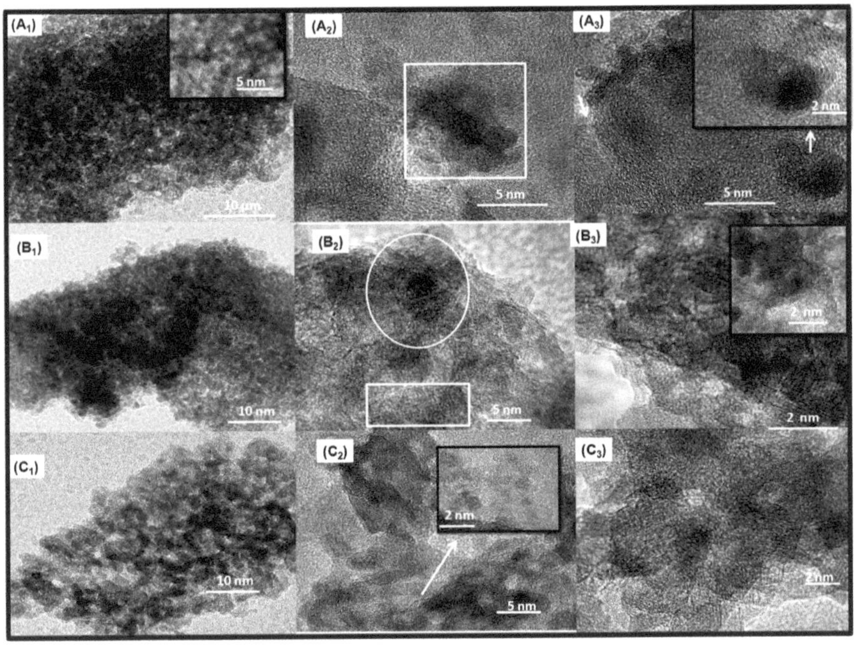

Figure 5. TEM images of the supports: (**A**) LA, (**B**) MA and (**C**) ZA samples. Insets illustrate parts of the figure at higher magnifications.

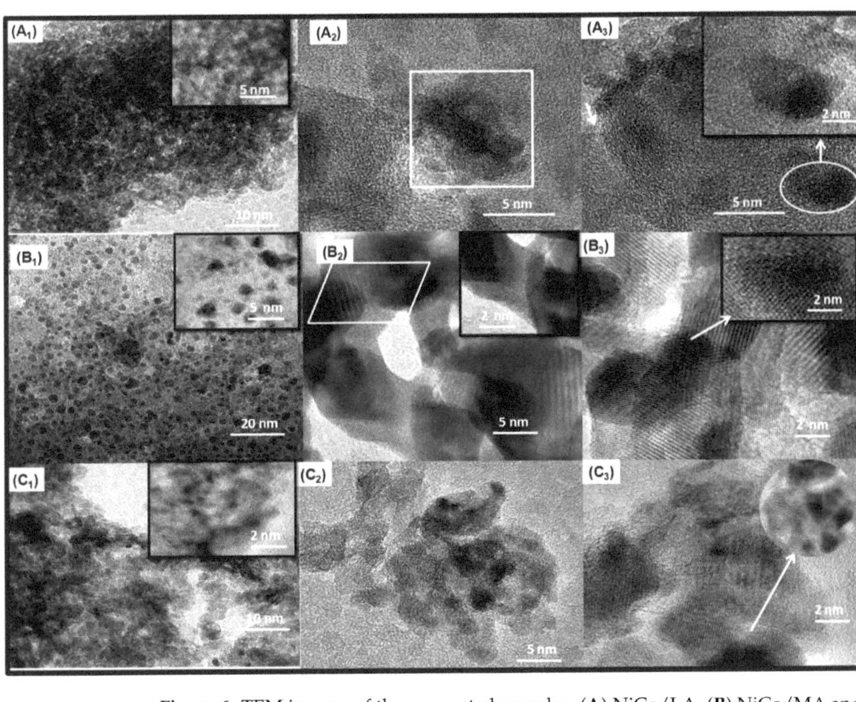

Figure 6. TEM images of the supported samples: (**A**) NiCo/LA, (**B**) NiCo/MA and (**C**) NiCo/ZA samples. Insets illustrate parts of the figure at higher magnifications.

The included high magnification image shows particle intergrowth predominantly appearing in the crystalline regions (top middle, inset Figure 6(A_2)). The periodic planes correspond to the d-spacing of 0.280, 0.290 and 0.241 nm with (022), (220) and (111) lines, respectively. These planes are indicative of a regular arrangement of γ-Al_2O_3, spinel Co_3O_4 and cubic NiO, in agreement with the findings [41,43]. The nanoparticles have somewhat small sizes ranging from 1 to 5 nm, being finely dispersed throughout the bulk (top right, Figure 6(A_3)) some of them included in the large pores of the sample, as shown by the selected area (inset top right, Figure 6(A_3)).

This is well-matched with the textural properties that depict a large surface area of ca. 400 m^2 g^{-1} for NiCo/LA and graded meso and macroporosity. Comparatively, well-dispersed NiO or Co_3O_4 nanoparticles are clearly visible in the low magnification TEM image of NiCo/MA with a higher dispersion of these entities, e.g., more than 83.2% detected for NiCo/MA (center right, Figure 6(B_1)) against 21% for NiCo/LA. Thus, there is an obvious uniform distribution of the nanoparticles on the surface of MA support with sizes extending from 1 to 13 nm (inset, Figure 6(B_1)). Evidence for the porosity of the MA support through the mesoporous and macroporous structure is given in Figure 6(B_2), top middle. An illustrative example of interparticle growth is shown by the highlighted trapeze, which is amplified in the inset of Figure 6(B_2). Importantly, the lattice spacings of 0.280 (022), 0.290 (220), 0.241 (111), 0.210 (200) and 0.281 nm (103) correspond to the Co_3O_4, NiO, Al_2O_3, MgO and $MgAl_2O_4$, respectively. Furthermore, NiCo/ZA exhibits a platelet of particles (top bottom, Figure 6(C_1)) with a uniform distribution compared to NiCo/LA. From the magnified image, the smaller particles are arranged in a perfect manner inside the crystalline ZA support (inset, Figure 6(C_1)). The sizes of NiO and Co_3O_4 nanoparticles are within 5–19 nm range, which is slightly lower than those of NiCo/LA. The high magnification image in the bottom center of Figure 6(C_2) shows a well-organized lattice arrangement of crystalline Al_2O_3, Co_3O_4 and NiO and a d-spacing of ca. 0.247 nm (101) is attributed to cubic ZnO. Figure 6(C_3) illustrates the regular porous structure along with the crystalline structure of the particles by the lattice fringes. Additionally, the magnified view of the figure depicts the presence of some nanoparticles inside the pores (inset of Figure 6(C_3)).

2.3. *Acidity of the Catalysts*

Acidity is measured by NH_3-TPD and the results are summarized in Table 2. A theoretically confirmed rule illustrates that at temperatures lower than 250 °C, the acid sites measured by NH_3-TPD are of weak strength, whereas those of medium strength retain ammonia showing desorption peaks close to the 250–350 °C range [13,28,44]. On the contrary, the characteristic acid sites having strong strength present broad peaks located at 500–700 °C range [13,44]. The NH_3-TPD curves of the pure alumina sample comprise peaks located at 150–250 °C corresponding to the acid of weak acidity and those of medium strengths in the 250–400 °C range (Table 2), in very close agreement with the findings [13,43]. Hence, the γ-Al_2O_3 sample displays a total acidity of ca. 0.187 mmol·NH_3·g^{-1} [13]. These acid sites are positioned below 300 °C, being associated with Lewis acid sites whilst the high temperature desorption peaks superior to 500 °C is attributable to either solely Brønsted acid sites or both Brønsted and Lewis acid sites [28,45].

With the addition of the basic promoters to the alumina, the sol–gel prepared binary support oxides decrease their amounts of surface acid sites, prevailing very few weak to medium acid sites. Based on the reports, MgO, ZnO and La_2O_3 are assumed to be basic oxides and therefore, the absence of acid sites is not expected [41–44,46]. In contrast, alumina itself has both acid and base sites characteristic of an amphoteric oxide [42,44]. At the opposite, the acidity can be tuned by altering the alumina chemical composition through doping, and yet aluminas are a typical solid base, when promoted by alkali-metals, alkali-fluorides, alkali-amides, and/or alkali-hydroxide forming as super or strong base catalysts [42,44].

Table 2. Acidity measurements by NH_3-TPD of the solids. The amount of ammonia desorbed per gram of catalyst was obtained from the temperature ranges to calculate the total concentration of acid sites.

Catalyst	Acid Amount mmol NH_3 g_{cat}^{-1} at 150–250 °C	Acid Amount mmol NH_3 g_{cat}^{-1} at 250–400 °C	Acid Amount mmol NH_3 g_{cat}^{-1} at T > 500 °C	Total Amount mmol NH_3 g_{cat}^{-1}
NiCo/MA	0.13	0.10	0.02	0.25
NiCo/LA	0.14	0.12	-	0.26
NiCo/ZA	0.07	0.11	0.01	0.19

Thus, the acidity of the supports is adjustable depending on the promoter added. For instance, the Zn addition on alumina gives a considerable drop in total acidity to 0.19 mmol·NH_3·g^{-1} while the modification by Mg increases the total acidity to 0.25 mmol·NH_3·g^{-1} and the addition of La slightly increases the total acidity to 0.26 mmol·NH_3·g^{-1}. Upon examination of the strength distributions, almost all samples appear in regions of weak to mild acidity with the absence of acid sites of strong strength, as expected [28,45]. Furthermore, the amount of acid sites for the supported samples has distinctive shifts for higher temperatures owing to the NiO and Co_3O_4 phases dispersed on the supports, making the total acidity considerably higher than that of γ-Al_2O_3, except for NiCo/ZA. Remarkably, the good dispersion of NiCo on the supports provides an evident enhancement of acid sites strengths with similarity in terms of distributions, in spite of large differences in surface acidity due to Ni and Co acting as Lewis acid sites. Therefore, the supported solids have much higher total acidity values than the unsupported ones, following the order: NiCo/LA ≅ NiCo/MA > NiCo/ZA.

All these properties suggest that the promoters are included in the alumina structure, modulating its textural properties, morphology and acidity to favor the interaction of Ni and Co in close contact with the support. This will result in a much high number of active sites possessing weak to mild acidity for promoting AG reaction, as further shown.

2.4. Electronic Properties and Surface Compositional Characterizations

In an attempt to further characterize the valence states and the presence of possible defects in the supported solids, EPR measurements are performed. As expected, γ-Al_2O_3 does not exhibit EPR signals owing to the absence of paramagnetic impurities, as found elsewhere [47]. On the contrary, the EPR spectra of the supported solids depict asymmetric resonance signals in two distinct regions (Figure 7a). At the magnetic field, the strong resonance located in the 2400–5000 G range is attributable to the paramagnetic Ni^{2+} species from the small NiO particles [48,49]. The g value close to 2.2 is assigned to either small ferromagnetic Ni clusters or the substitutional cubic Ni^{2+} ions in oxides matrix [48,49]. These Ni^{2+} species in such a low NiO amount on the surface could be strongly interacting with the alumina, as further demonstrated by the XPS results.

It is noteworthy that the asymmetrical peak in the 2400–5000 G range is superimposed on a broad anisotropic peak with a g value of 2.10, which is attributed to Co^{2+} ions from Co_3O_4 [50]. This result agrees with the TEM analyses that demonstrate the presence of the spinel phase in all the solids. According to the findings, the g value of ca. 2.00–2.12 can also be attributed to the electrons trapped by oxygen vacancies [29]. This also indicates that the samples have the oxygen atoms at the surface appearing to be interacting with the vacancies and these oxygen species could mask their own vacancies present or even interact with them, generating the ferromagnetic signals at 2900 G [51].

Additionally, it cannot be excluded the Mg, La and Zn species incorporated in the alumina structure contribute to some extend to the background of the EPR spectra. However, the supported samples have similar EPR curves, independently of the MA, LA and MA support and thereby the influence of the aforesaid ions is not detectable because of their low concentrations or masking under the EPR spectra of the Ni and Co species.

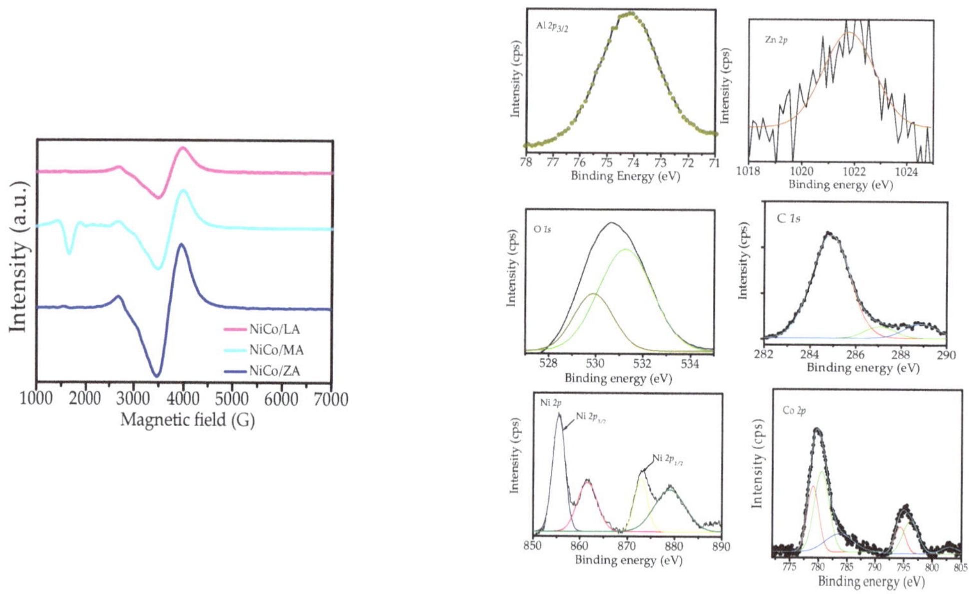

(a) (b)

Figure 7. (**a**) EPR measurements for supported solids (**b**) Representative XPS spectra for the NiCo/ZA supported solid.

The chemical states and surface compositions of the supported solids are evaluated by XPS analyses. A full scan range XPS survey spectrum depicts the presence of the Al 2*p*, Zn 2*p*, O 1*s*, C 1*s*, Ni 2*p*, and Co 2*p* signals for the NiCo/ZA sample (Figure 7b). Besides the aforesaid elements, the XPS spectra of NiCo/MA and NiCo/LA samples have additional Mg 1*s* and La $3d_{5/2}$ core levels signals. For all samples, the dominant peak at 73.8–74.1 eV is assigned to Al 2*p* signal (Table 3). This suggests the presence of Al(III) in Al-O and Al-OH bonds on the solid surface existing purely in the form of γ-Al_2O_3, in agreement with the findings [2]. To prove that the binary oxides supports are formed, the Zn $2p_{3/2}$ core level spectrum appears in NiCo/ZA as a very weak fitted peak (Figure 7b) corresponding to Zn (II) species from ZnO [52]. It is worth noting that the surface Zn content is quite low ca. 0.46 wt.% whereas the Al content is 28.59 wt.%, which suggests that most of ZnO oxide is on the bulk forming ZnO-Al_2O_3 support. This result agrees with the XRD, Raman and TEM results that indicate the existence of the ZnO-Al_2O_3 binary support. In the case of NiCo/LA, the high resolution La $3d_{5/2}$ core level spectrum reveals an intense characteristic peak that emerges at 834.9 eV matching well with La(III) from La_2O_3 [21]. Moreover, the surface La content is 0.39 wt.% with respect to that of 25.66 wt.% of Al, which indicates that lanthanum is mostly included in the bulk of La_2O_3-Al_2O_3 support. For NiCo/MA, the peak of Mg 1*s* core level occurred at 1303.4 eV, which stands for the Mg(II) from $MgAl_2O_4$ and MgO phases [29].

A relatively small amount of Mg of approximately 0.35 wt.% is detected together with 22.44 wt.% of Al because of the formation of bulk phase. This is in line with the XRD results that suggest the presence of these phases besides alumina. Based on the fact that the Mg, Zn and La contents in the bulk are 12.0 wt.% each, it can be inferred that the surface contributions of these species are too low being these species mostly included in the bulk. In addition, the curve fitting of the high resolution C 1*s* core level spectrum shows three contributions in all solids, with binding energies values of 284.8, 286.8 and 288.9 eV, which are assigned to adventitious carbon/C-C/-C=C- bonds, C-OH bonds and C=O bonds, respectively [2,21]. These carbon species arise probably due to the presence of some organic contaminants adsorbed on solid surfaces being in small amount of ca. 6–7 wt.%.

Table 3. Binding energies values (eV) and Ni/Al and Co/Al ratios obtained from XPS spectra of the supported samples.

Sample	Al 2*p*	Mg 1*s*	Zn $2p_{3/2}$	La $3d_{5/2}$	C 1*s*	Co $2p_{3/2}$	Ni $2p_{3/2}$	O 1*s*	Ni/Al	Co/Al
NiCo/LA	73.8	-	-	834.9	284.9 286.9 288.9	780.3 795.0 802.3	855.6 861.1 872.6 878.2	530.6 531.2	0.44	0.57
NiCo/MA	74.2	1303.4	-	-	284.7 286.8 288.8	780.6 795.1 802.4	855.8 861.1 872.3 878.5	530.4 531.3	0.78	0.45
NiCo/ZA	74.1	-	1022.1	-	284.8 286.4 288.7	780.2 795.0 802.3	855.4 861.5 872.5 878.8	530.4 531.0	0.31	0.47

The analysis of the high resolution O 1*s* core level spectrum illustrates two distinguished oxygen species, including 530.6 and 531.3 eV, which is consistent with the chemisorbed surface oxygen and water surface adsorbed OH and/or oxygen vacancy and lattice O^{2-} species in Me-O bonds [2,21,52]. Indeed, the bulk oxygen species possessing a binding energy of 530.6 eV is accompanied by hydroxyl groups with a binding energy of 532.3 eV with a relatively high amount for NiCo/ZA whereas NiCo/LA and NiCo/MA have minor amounts. Thus, NiCo/MA holds the most oxygen vacancies among these solids, which is consistent with the abundant oxygen vacancies found by the EPR results.

The high resolution Co 2*p* spectrum is deconvoluted into two contributions with binding energies at 780.3 and 795.0 eV, which are associated with the doublet Co $2p_{3/2}$ -Co $2p_{1/2}$. In addition, a weak satellite at approximately 802.1 eV appears in all spectra. The peaks at 780.3 and 795.0 eV are ascribed to the simultaneous presence of Co(III) and Co(II) from Co_3O_4 on the solid surface [22,52]. The EDS analyses found a large amount of Co species on the solid surface. The findings state that the absence of strong shake-up satellites could be associated with the presence of Co_3O_4 [36,52]. The Co/Al ratio is calculated to be 0.57, 0.45 and 0.47 for NiCo/ZA, NiCo/MA and NiCo/LA, suggesting a lower amount of these species on the latter solids. The doublet Ni $2p_{3/2}$ -Ni $2p_{1/2}$ appears at 855.6 and 878.2 eV and two shake up satellites at 861.1 and 872.6 eV for NiCo/LA (Table 3). The Ni 2*p* spectra for samples NiCo/LA for NiCo/MA have similar binding energy values. These components are attributed to Ni(II) from surface NiO [22,52]. The Ni/Al ratios are 0.44. 0.78 and 0.31 for NiCo/LA, NiCo/MA and NiCo/ZA, respectively. This suggests a higher coverage of the NiCo/MA surface by the NiO, as observed by the TEM measurements and EDS analyses (Supplementary Materials).

2.5. Catalytic Performance in EG Reaction

The esterification of glycerol in the presence of acetic acid is conducted to evaluate the catalytic properties of the solids. In preliminary investigations, the reaction is carried out in the presence of various catalysts (Table 4). It is notable that the glycerol conversions on unsupported solids are too low with MA holding a conversion of 4.5%, while those of ZA and LA are just 3.6 and 2.3%, respectively.

Table 4. Glycerol conversions, reaction rates and product selectivities for the catalysts evaluated in 1h of EG reaction. Reaction conditions: glycerol to acetic acid (molar ratio) = 1:3, reaction temperature = 80 °C, catalyst mass = 0.50 g.

Catalysts	X (%)	Rate ($mmol_{gly}$ g^{-1} cat h^{-1})	Selectivity (%)		
			Monoacetin	Diacetin	Triacetin
Blank	1.7	0.01	-	-	-
MA	4.5	0.05	-	9.0	23.0
ZA	3.6	0.04	-	12.0	12.1
LA	2.3	0.02	-	15.0	17.4
NiCo/MA	11.0	0.31	-	7	18.0
NiCo/LA	7.0	0.17	-	-	14.0
NiCo/ZA	5.0	0.14	-	-	16.0
[a] γ-Al_2O_3	80	0.80	84	10	0.8

[a] Reaction conditions: glycerol/acetic acid (molar ratio) = 1:9, reaction temperature = 100 °C, reaction time = 6 h, catalyst weight = 0.50 g.

On the other hand, glycerol conversion on γ-Al_2O_3 is higher than those of the most active binary supports. The acidity of the alumina decreases significantly upon incorporating the basic promoters appearing to be, at first sight, disadvantageous to convert glycerol. As shown previously, alumina posses acid sites of weak to medium strength [13]. Hence, it is expectable that conversions and selectivities would be high under the very mild reaction conditions tested. However, alumina catalyst requires high temperature, e.g., 100 °C and long reaction times, e.g., 6 h to transform glycerol, besides the selectivity of the triacetin is very poor. Importantly, a NiCo/Al_2O_3 sample prepared in this study has similar conversion and selectivity within 1h comparing with γ-Al_2O_3, but Ni and Co leaching over the course of the reaction deactivated the solid. Particularly, studies on EG reaction using basic promoters added to alumina, for example, CaO and MgO as well as Ni or Co addition to Al-based catalysts with different metal loadings have shown substantially lower activities than the those of the supported catalysts in the present work [53–56]. This is due to the low water tolerance of the catalysts during the esterification of glycerol causing the deactivation of the solids over the course of the reaction.

According to mechanistic considerations [12], the EG reaction with acetic acid involves the protonation of the carbonyl group of the acetic acid molecule over Brønsted acid catalysts possessing strong strengths [5]. Subsequently, the activated carbonyl group formed in the previous step reacts with a hydroxyl group of glycerol through nucleophilic attack to give an intermediate C–O bond [5,12]. The reaction proceeds via the loss of a water molecule from the intermediate and the acetin isomers formation, e.g., 1-monoacetin and 2-monoacetin [2,5]. The serial mechanism of monoacetins reaction and acetic acid molecules results in diacetins, e.g., 1,2-diacetin and 1,3-diacetin, and further reaction of these latter molecules with acetic acid produces triacetin [1–5,53].

One thing that deserves to be mentioned is that the initial ratios somehow achieve better values for the supported solids (Table 4) compared to those of the unsupported catalysts due to the low amount of acid sites present in the latter (Table 2). In the case of supported solids, glycerol conversions are also similar in terms of the trends for the kind of support with slightly higher values than those of binary catalysts. Such an effect is found prevalently with the presence of Ni and Co nanoparticle dispersed on the supports, since the nanoparticles themselves exert a dual role of chemisorb glycerol and work synergistically with basic promoters to adsorb acetic acid. This would influence the catalytic performance to some degree. In these systems, the nanoparticles dispersed on the supports are not affected by the water formation that leads to the common acid sites deactivation during the EG reaction, as found elsewhere [5,9]. Selectivities to triacetin do not change significantly over the supported solids whilst mono and diacetin are not produced. This is due to the large amount of by-products observed, after conducting the reaction in the harsh reaction conditions shown in Table 4.

Further studies on the activity of solids on supported solids are being carried out to examine the effects of the structure on their catalytic performances. Figure 8A clearly depicts a gradual increase in the glycerol conversion with increasing reaction time for supported catalysts. Such an effect is generally ascribed to the concentrations of the components being far from equilibrium at the initial stages of the reaction and then, the system approaches equilibrium with the progress of the reaction. The trend of the activities summarized in Figure 8A shows that the binary supports contribute to convert glycerol with conversions greater than 11% in short times. Raising reaction times within 4 h gives stable conversion of NiCo/ZA, while NiCo/LA conversion slightly increased. This may cause aggregation of the support particles and, subsequently, a lower stability of the catalyst up to 4 h, as shown later in the recyclability experiments. Notably, NiCo/MA also exhibits a large boost at longer reaction times and still maintains more than 28% glycerol conversion after 6 h owing to the solid possessing plenty of acid sites and existence of more exposed Ni and Co sites on the support surface, resulting in a high availability of the active sites to enhance the catalytic performance. Thus, Ni and Co nanoparticles dispersion on MA and the low acidity and probably also to its synergistic effect may contribute to the catalytic activity of the solid. Additionally, the better catalytic performance of NiCo/MA may be due to the porosity.

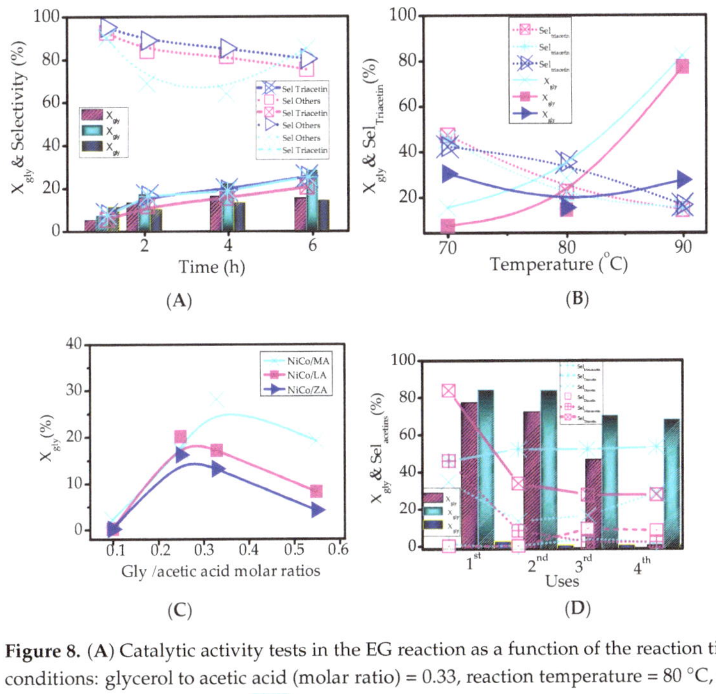

Figure 8. (**A**) Catalytic activity tests in the EG reaction as a function of the reaction time. Reaction conditions: glycerol to acetic acid (molar ratio) = 0.33, reaction temperature = 80 °C, catalyst mass = 0.50 g. The symbols for ( ) NiCo/MA, ( ) NiCo/LA and ( ) NiCo/ZA samples represent glycerol conversions, selectivity to triacetin and selectivity to others by-products. (**B**) Influence of the temperature on the catalytic properties of the samples. The symbols for ( ) NiCo/MA, ( ) NiCo/LA and ( ) NiCo/ZA samples represented glycerol conversions and selectivity to triacetin taken in 6 h of reaction. (**C**) Glycerol to acetic acid molar ratios studied at 80 °C using a catalyst mass of 0.50 g in 6 h of reaction. (**D**) Reusability studies. Reaction conditions: glycerol to acetic acid (molar ratio) = 0.33, reaction temperature = 80 °C, catalyst mass = 0.50 g for each cycle of 24 h. The symbols for ( ) NiCo/LA ( ) and NiCo/MA samples represent glycerol conversions, selectivity to triacetin, diacetin and monoacetin.

In spite of the high porosity of NiCo/LA in comparison to NiCo/ZA and NiCo/MA, the catalytic performance of the former is attributable to the low interaction between Ni and Co nanoparticles and the LA support. This facilitates the leaching of the particles during the reaction, thereby causing loss of activity after 6 h, as demonstrated by the observed tendency to activity decay (Figure 8A). For NiCo/ZA possessing lower acidity, the deactivation of the acid sites by water is possibly the reason why the solid exhibit lower glycerol conversion compared with NiCo/MA. Moreover, the selectivities to triacetin increases as the reaction proceeds, but it hardly changes within 6 h for all solids (Figure 8A). Contrary, selectivities to other by-products significantly enhanced during the whole time intervals reaching values nearly 70% in 6 h due to the triacetin oligomers formation, namely other by-products (Figure 8A).

Figure 8B illustrates the dependence of the temperature of glycerol conversion and selectivity to triacetin. The catalysts show a glycerol conversion initially below 10% at 70 °C, reflecting the need for heating to enhance catalytic performance. The evolution of the glycerol conversion at temperature of 80 °C shows an increment in activities due to the effect of the furnishment of heat to the reaction [54,56]. For NiCo/ZA, a sudden decrease of glycerol conversion suggests that by-products are formed, while the oligomers are irreversible adsorbed to the active sites at 90 °C achieving 27% of conversion and 16% of triacetin selectivity (Figure 8B). Such temperature effects are typical of the drop in glycerol conversion with rising temperature due to the overall exothermicity of the reaction [12]. In contrast, the glycerol conversion behaviors of the NiCo/MA and NiCo/LA samples are apparently different from that of NiCo/ZA. For instance, the glycerol conversion of NiCo/MA and NiCo/LA is found to be ca. 3 times greater than that of NiCo/ZA at 90 °C (Figure 8B). That is, the conversion of glycerol of NiCo/ZA remains almost constant in the temperature range investigated. In that case, the temperature causes variations of the specific heats, when the hydroxyl group of glycerol is substituted by acetyl groups from acetic acid until reaching a plateau [12]. The calculated energy of activation for NiCo/MA, assuming pseudo-second-order dependence on glycerol concentration, is found to be 18.0 kJ mol^{-1}, which is lower to that observed for sulfated alumina e.g., 70 kJ mol^{-1} [54].

In addition, the results indicate a trivial decrease in monoacetin and diacetin selectivities with the rising temperature and thus the catalysts offer high selectivity for triacetin and by-products in the range of the studied temperature (Figure 8B). The conversions steeply increase for NiCo/MA and NiCo/LA over the entire temperature range evaluated and selectivities to triacetin finally approach low values due to the oligomers production at high temperatures. Accordingly, increasing the temperature from 70 to 90 °C results in a decay in selectivity to triacetin from 47% to 13% on all solids. These results are hardly confirmed by the fact that the furnishment of heat to the system contributes to the esterification of monoacetin to triacetin as this reaction is a highly endothermic process [12]. In contrast, this effect is also observable for NiCo/MA, which reveals differences in acidity and porosity in comparison with those of the other samples, and thus, high temperatures improve the triacetin production.

Notably, samples possess remarkable catalytic performances at 80 °C and, thereby the catalytic activity studies are further continued at this selected temperature. The typical glycerol conversion curve dependence of the molar ratio is shown in Figure 8C. A considerable increase in the catalytic activity of the solids is found with increasing glycerol to acetic acid molar ratios from 0.10 to 0.33, providing a subsequent drop in glycerol conversion at glycerol to acetic acid molar ratio of 0.55 for all solids (Figure 8C). Reaction mixture containing high concentrations of viscous glycerol reactant, such as glycerol to acetic acid molar ratio of 0.55, limits the access of reactants to either active NiCo on the surface of the support or the acid sites of the support might be restricted. For instance, NiCo/MA seems to reach glycerol conversions of approximately 2 to 28%, when the molar ratio of glycerol to acetic acid is increased from 0.10 to 0.33 affording 16 and 22% of triacetin. A closer inspection of the behavior of NiCo/ZA shows evidence of obvious deactivation of the solid by leaching of the active sites, which nicely explains its poor catalytic activity varying the molar ratios between glycerol and acetic acid. It is also apparent that increasing the molar ratios from 0.25 to 0.55, the glycerol conversions little decays from 20% to 8% over NiCo/LA. No obvious change in triacetin selectivity is shown over all supported solids with molar ratios above 0.55. In this sense, the findings illustrate that the modulation of the acidity and porosity of the catalysts has an important role in determining the efficiency of diffusion of reactants and products to achieve good selectivities to triacetin [3,5]. In view of these findings, triacetin possesses a molecular diameter of ca. 4.5 nm and thus, it requires space to diffuse into the catalyst pores and, if the active sites are on the surface, this is not a limiting factor for triacetin formation over the solids under investigation.

As the supported solids exhibit improved performance in terms of glycerol conversion and triacetin selectivity at the molar ratio of glycerol to acetic acid of 0.33 and temperature of 80 °C, reusability studies are conducted under optimized conditions for 1–4 cycles of uses of 24 h each. As described in Figure 8D, an activity loss of 8% from the 1st cycle to the 2nd cycle use is observable for NiCo/LA, possibly because of the simultaneous increase in the concentration of the by-products and leaching of the active sites. Accordingly, the values for the triacetin selectivity presented by NiCo/LA decline from 84% to 35%, whereas monoacetin and diacetin follow similar trends of decay. Amongst the catalysts tested herein, the NiCo/ZA appears to be less active and selective to acetins due to the inevitable nanoparticles leaching and strong by-products adsorption reaching complete deactivation especially in the first use. When the EG reaction lasted for 48 h, e.g., 2nd use, glycerol conversion of NiCo/LA reached a slightly continuous drop achieving glycerol conversion and monoacetin selectivity of 77 and 10%, respectively. Again, NiCo/MA appears to be very stable during repeated uses with glycerol conversions ranging between 84 and 70%, in the second and third cycles of uses. This could be an effect of the presence of the NiCo active sites helped by the stable spinel $MgAl_2O_4$ phase, acting as an additional active component to improve the catalytic performance. The production of the acetins consisting of ca. 53% of triacetin, 29% of diacetin and 4% of monoacetin besides 14% of by-products of its condensation, follows the same trends with quite constant glycerol conversions. When the reaction time is extended above 78 h in the fourth cycle, the glycerol conversion of NiCo/LA is null, considering the nanoparticles leaching as the most important deactivation factor for its poor reusability. After three repeated uses, NiCo/MA retains more than 67% of its initial activity with 53% of triacetin selectivity remaining in a state of slow down for another five cycles. Previous studies on EG reaction carried out with supported Ni and Co-based catalysts have demonstrated that the strong by-products adsorption thought to be mainly on the surface of the acidic sites of supporting oxides and the metal nanoparticles leaching are among the leading causes of catalyst deactivation [5].

From the obtained results, the activity of NiCo/MA is nicely preserved for up to four cycles of the reusability test and thereby, the solid is more resistant to deactivation compared with the other NiCo supported catalysts counterparts due to the intrinsic synergistic effect between support and nanoparticles. The presence of the stable spinel $MgAl_2O_4$ phase possessing lattice oxygen species acts as additional active component to improve the NiCo/MA activity and selectivity in the esterification of glycerol. The catalytic performances of the supported solids definitely demonstrate the crucial role of the structure, porosity, acidity of the catalysts in the activity.

3. Materials and Methods

3.1. Materials

Aluminum tri-sec-butoxide ($Al(O\text{-}sBu)_3$, 99%), ethanol (99.5%), lanthanum(III) nitrate hexahydrate ($La(NO_3)_3{\cdot}6H_2O$, 99.9%) magnesium(II) nitrate hexahydrate ($Mg(NO_3)_2{\cdot}6H_2O$, zinc(II) acetate dihydrate ($Zn(CH_3CO_2)_2{\cdot}2H_2O$, nickel(II) nitrate hexahydrate ($Ni(NO_3)_2{\cdot}6H_2O$) and cobalt(II) nitrate hexahydrate ($Co(NO_3)_2{\cdot}6H_2O$) were purchased from Sigma-Aldrich (Sigma-Aldrich, St. Louis, MO, USA). All chemicals were used as received without further purification.

3.2. Synthesis of the Alumina Support

The alumina support was synthesized by a sol-gel route based on previous studies [21,22]. Then, Co and Ni active components were dispersed on the support by the wet impregnation method. Approximately 103.5 mmol of aluminum tri-sec-butoxide and 3.25 mol of absolute ethanol were added to 4.2 mmol of lanthanum nitrate hexahydrate under stirring. Then, the formed suspension was refluxed for 1 h at 100 °C. Afterwards, a solution of nitric acid (0.05 mol·L^{-1}) was added dropwise into the previous mixture forming a slurry under continuous stirring. Then, the resulting mixture was kept again at 100 °C under reflux followed by stirring for 14 h. The pH of the medium was 5.0. Subsequently, the obtained gel was washed five times by centrifugation, before standing overnight. The xerogel was dried at room temperature for further calcination at 850 °C for 6 h under flowing air. The obtained solid was denoted as LA, representing the La_2O_3–Al_2O_3 support with 12.0 mol% of La.

In a similar procedure, the MgO–Al_2O_3 support was obtained by adding a desired amount of magnesium nitrate salt into the synthesis mixture. Briefly, 103.5 mmol of aluminum trisec-butoxide was dissolved into 3.25 mol of absolute ethanol for homogenization. The magnesium nitrate salt was then added to the suspension, which was immediately refluxed at 100 °C after adding the nitric acid solution. In the next step, the gel was washed, dried and calcined 850 °C to obtain the MA representing the MgO–Al_2O_3 support with 12.0 mol% of Mg. A summary of the preparation of the support is shown in Figure S3 in Supplementary Materials.

The ZnO–Al_2O_3 support was synthesized in a similar fashion as described above. In short, aluminum tri-sec-butoxide, absolute ethanol and zinc acetate were mixed simultaneously to the zinc acetate solution. Subsequently, the pH of the mixture was adjusted with the dropwise addition of nitric acid. After that, the mixture was refluxed at 100 °C for 14 h and calcined at 850 °C for 8 h. The support obtained was denoted as ZA referring to the ZnO–Al_2O_3 support with 12.0 mol% of Zn.

The incipient wetness impregnation method was used to prepare the NiCo catalysts supported on LA, ZA and MA (Figure S3b). Briefly, 1wt.% mmol of nickel nitrate aqueous solution was impregnated on 1 g of LA support in a rotatory evaporator at 70 °C for 2 h. After removing the excess solvent, the solid was dried at 120 °C and then calcined in air at 350 °C for 2 h. Another sequential step by impregnation 1wt.% of cobalt nitrate aqueous solution on the previous solid in a rotatory evaporator was performed. Afterwards, catalysts were dried overnight and calcined in air at 350 °C for 4 h. The resulting NiCo/La_2O_3–Al_2O_3 catalyst, namely NiCo/LA, had 1 wt.% of each metal.

For the preparation of the NiCo/MgO–Al_2O_3 and NiCo/ZnO–Al_2O_3, Ni and Co were introduced in two sequential steps of impregnation-evaporation methods similar to the abovementioned procedure. Then, the solids were dried and calcined to obtain the NiCo/MgO–Al_2O_3 and NiCo/ZnO–Al_2O_3 designated as NiCo/MA and NiCo/ZA, respectively.

The pure alumina was prepared as reference material and the catalytic results compared with those of NiCo catalysts supported on the modified alumina in study. Details about the synthesis are given elsewhere [21].

3.3. Characterizations

The powder X-ray diffraction (XRD) patterns were recorded in a Shimadzu XRD6000 (Shimadzu, Kyoto, Japan) diffractometer using a Cu-Kα monochromatized radiation source at 40 kV and 30 mA (λ = 0.154 nm). The data were collected in the 2θ range of 10–70° in a step-scan mode of 0.02° s^{-1}. The Joint Committee on Powder Diffraction files was used as reference to compare the obtained XRD patterns.

Fourier transform infrared spectroscopy (FTIR) spectra were collected in a Bruker equipment (Bruker, Rheinstetten, Germany) in the range of 400–4000 cm^{-1} with a resolution of 2 cm^{-1}. Before measurements, the self-supporting KBr disks were prepared by dilution of 1 wt.% of the samples in KBr.

Raman measurements were recorded using a LabRAM HORIBA HR Evolution model, which was equipped with a CCD detector (Horiba, Gloucestershire, UK). Room-temperature Raman spectra were excited at 532 nm with 600 gr·mm^{-1} grating using a He–Ne laser. The laser power was set at 2 mW on the sample surface. All spectra have been recorded in the 100–1800 cm^{-1} range with a spectral resolution of 4 cm^{-1}.

The morphology of samples was investigated by Scanning electron microscopy (SEM) analyses in a Quanta-FEG FEI electron microscope (FEI Quanta, Hillsboro, OR, USA). The elemental distributions of the solids were obtained in an EDX Link Analytical QX-20000 system coupled to the SEM microscope at 2 kV acceleration voltage. Previously, the samples were sputtered with Ag to perform the analyses.

The surface area, pore volume and average pore sizes were determined by N_2 physisorption isotherms at −196 °C, using an ASAP 2000 Micromeritics equipment instrument. Prior to analyses, samples were treated at 250 °C under vacuum for 2 h. The specific surface areas were using the Brunauer-Emmet-Teller (BET) equation from the adsorption branch of the isotherms, whereas pore size distributions were determined by the Barret–Joyner–Halenda (BJH) method.

Electron paramagnetic resonance (EPR) spectra were performed on a Bruker spectrometer (Bruker, Rheinstetten, Germany) with frequency modulation of 100 kHz. The EPR spectra were obtained at the X-band microwave frequencies at 9.5 GHz. The values of g were obtained by the EPR marker from module ER031, which was adjustable to mark g = 2.0040.

Acid properties of the catalysts were analyzed using Temperature programmed of ammonia desorption (TPD-NH_3). The curves were recorded Chembet-3000 Quantachrome. Approximately 100 mg of the solids were placed on U-tube and then heated under flowing helium at 120 °C for 2 h. The catalyst was cooled to 100 °C and subsequently, a 5% of NH_3 diluted in He was placed into the tube. Thereafter, the physically adsorbed ammonia was flushed with helium for 1 h. The ammonia desorption was conducted from 50 a 350 °C to obtain the TPD curves.

Transmission electron microscopy (TEM) micrographs were collected on a JEOL JEM 2010F microscope with an accelerating voltage of 200 kV (Texas, TX, USA). Before the analyses, samples were treated in ethanol, sonicated and deposited on carbon-coated copper.

X-ray photoelectron spectra (XPS) were conducted on a Physical Electronics Versa Probe II Scanning XPS Microprobe (Minneapolis, MN, USA) spectrometer equipped with a monochromatic Al Kα radiation source. Spectra were fitted by using the Gaussian–Lorentzian. The spectra were corrected to the adventitious carbon component (284.8 eV) of the C 1*s* region. The Multipack software version 9.6.0.15 was employed to obtain the XPS spectra.

3.4. Catalyst Testing

A typical esterification of glycerol in the presence of acetic acid reaction was carried out in a jacketed 30 mL three-necked round bottom flask reactor, which was equipped with a condenser. The composition of the reaction mixture was glycerol and acetic acid to have a glycerol to acetic acid molar ratio of 0.33. The catalyst (50 mg) was placed in the batch reactor followed by addition of the reaction mixture under vigorous stirring at 1000 rpm for a certain period of time.

Unless otherwise specified, the reaction temperature measured by a thermocouple placed into the solution was 80 °C. The analysis system consisted in a Shimadzu gas-chromatography equipped with a capillary column and a flame ionization detector (FID). The liquid products of the reaction mixture were withdrawn periodically during the reaction course at each 60 min intervals and analyzed in the GC chromatograph. The EG esterification products were also identified by gas chromatography–mass spectroscopy (GC–MS).

The effects of the reaction time, temperature and glycerol to acetic acid were also examined. The catalyst reusability was carried out over the most active solids by recovering the solid from the reaction mixture and adding a fresh substrate at 80 °C. Each catalyst recycle experiment was performed in 24 h intervals at least 1–4 times. Before each reuse, catalysts were separated by centrifugation, washed with ethanol and dried at room temperature. An additional loading of the solid was added to the runs to ensure 3–5% of the catalyst in the reaction media.

The catalytic performance of solids was estimated in terms of the glycerol conversion ratio (X_{gly}, Equation (1)), selectivity to the products (S, Equation (2)), the initial reaction rate is calculated for all catalysts after 0.5 h of reaction (a, Equation (3)), as follows:

$$\%\,X(t) = \frac{[\%glyin - \%glyout]}{\%glyin} \times 100 \quad (1)$$

where %glycerol in is the percentage by mass of inlet glycerol concentration, and %glycerol out is the percentage by mass of outlet glycerol concentration time t (h).

Selectivity was calculated according to the equation:

$$\%\,S = \frac{\text{Percentage of desired product}}{\text{Sum of the percentages of products formed}} \times 100 \quad (2)$$

$$\%\text{Initial rate} = \frac{\text{mmol glycerol converted}}{\text{catalysts mass} \times \text{time}} \quad (3)$$

4. Conclusions

Porous $MgO\text{-}Al_2O_3$, $La_2O_3\text{-}Al_2O_3$ or $ZnO\text{-}Al_2O_3$ supports with controllable acidity were modified with well dispersed Ni and Co nanoparticles. XRD, Raman and FTIR results demonstrated the prevalence of $\gamma\text{-}Al_2O_3$ with modulated porosity and acidity. EPR and XPS analyses also confirmed the existence of lattice oxygen species as well as Ni^{2+} and Co^{3+}/Co^{2+} active species, enhancing the catalytic performance. This was attributed to a synergistic effect between the nanoparticles and the porous support, rather than acidity improving the activity of the solids in the EG reaction. The effects of temperature, glycerol to acetic acid molar ratios, and reusability studies surely demonstrated that the Ni and Co nanoparticles interacted strongly and well with the support and promoted the best performance of $NiCo/MgO\text{-}Al_2O_3$ and $NiCo/La_2O_3\text{-}Al_2O_3$ compared to $NiCo/ZnO\text{-}Al_2O_3$. In this regard, the presence of Ni and Co active sites helped by the spinel $MgAl_2O_4$ phase possessing high stability and a large amount of lattice oxygen species provided $NiCo/MgO\text{-}Al_2O_3$ to achieve superior catalytic performance in all reaction conditions. The catalyst had a good durability after being reused four times, revealing an excellent catalytic activity for the acetins production.

Supplementary Materials: The following supporting information can be downloaded at: https://www.mdpi.com/article/10.3390/catal12121616/s1, Figure S1: (A) SEM images, (B) EDS images and (C) EDS mapping of the unsupported solids. The numbers 1, 2 and 3 at the right side of the letters represent the LA, MA and ZA samples. Figure S2: (A) SEM images, (B) EDS images and (C) EDS mapping of the supported solids. The numbers 1, 2 and 3 at the right side of the letters represent the LA, MA and ZA samples. Figure S3: Schematic illustration of the (a) synthesis of the supports and (b) NiCo impregnation on the supports.

Author Contributions: R.d.C.F.B., G.M., J.V.d.C., R.M.B.V., R.L., L.O. and J.J.J. performed the experiments; A.C.O. and E.R.-C. analyzed the data and wrote the manuscript, funding acquisition and conceptualization; R.M.B.V. analyzed the data; G.D.S. and A.C. designed and performed the experiments and analyzed the data, as well. All authors have read and agreed to the published version of the manuscript.

Funding: This work is supported by Funcap (Grant n° PS1-0186-00346.01.00/21). Financial assistance received from Ministerio de Ciencia e Innovación, Junta de Andalucía and FEDER is also thankfully acknowledged for funding project n° PID2021-126235OB-C32, UMA18-FEDERJA-126 and P20_00375.

Acknowledgments: The Central Analítica da Universidade Federal do Ceará facility is gratefully acknowledged. R.C.F.B. would like to acknowledge support received from the CAPES for her PhD scholarship.

Conflicts of Interest: The authors declare no conflict of interest.

References

1. Koranian, P.; Huang, Q.; Dalai, A.K.; Sammynaiken, R. Chemicals Production from Glycerol through Heterogeneous Catalysis: A Review. *Catalysts* **2022**, *12*, 897. [CrossRef]
2. Neto, A.B.S.; Oliveira, A.C.; Rodriguez-Castellón, E.; Campos, A.F.; Freire, P.T.C.; Sousa, F.F.F.; Filho, J.M.; Araujo, J.C.S.; Lang, R. A comparative study on porous solid acid oxides as catalysts in the esterification of glycerol with acetic acid. *Catal. Today* **2020**, *349*, 57–67. [CrossRef]
3. Perez, F.M.; Gatti, M.N.; Nichio, N.N.; Pompeo, F. Bio-additives from glycerol acetylation with acetic acid: Chemical equilibrium model. *Results Eng.* **2022**, *15*, 100502. [CrossRef]
4. Checa, M.; Nogales-Delgado, S.; Montes, V.; Encinar, J.M. Recent Advances in Glycerol Catalytic Valorization: A Review. *Catalysts* **2020**, *10*, 1279. [CrossRef]
5. Mota, G.; do Carmo, J.V.C.; Paz, C.B.; Saraiva, G.D.; Campos, A.; Duarte, G.; Filho, E.C.S.; Oliveira, A.C.; Soares, J.M.; Rodríguez-Castellón, E.; et al. Influence of the Metal Incorporation into Hydroxyapatites on the Deactivation Behavior of the Solids in the Esterification of Glycerol. *Catalysts* **2022**, *12*, 10. [CrossRef]
6. Abida, K.; Ali, A. A review on catalytic role of heterogeneous acidic catalysts during glycerol acetylation to yield acetins. *J. Indian Chem. Soc.* **2022**, *99*, 100459. [CrossRef]
7. Carvalho, D.C.; Pinheiro, L.G.; Oliveira, A.C.; Millet, E.R.C.; de Sousa, F.F.; Saraiva, G.D.; da Silva Filho, E.C.; Fonseca, M.G. Characterization and catalytic performances of copper and cobalt-exchanged hydroxyapatite in glycerol conversion for 1-hydroxyacetone production. *Appl. Catal. A Gen.* **2014**, *471*, 39–49. [CrossRef]
8. Carvalho, D.C.; Souza, H.S.A.; Filho, J.M.; Assaf, E.M.; Thyssen, V.V.; Campos, A.; Padron Hernandez, E.; Raudel, R.; Oliveira, A.C. Nanosized Pt-containing Al_2O_3 as an efficient catalyst to avoid coking and sintering in steam reforming of glycerol. *RSC Adv.* **2014**, *4*, 61771–61780. [CrossRef]
9. Zada, B.; Kwon, M.; Kim, S.-W. Current Trends in Acetins Production: Green versus Non-Green Synthesis. *Molecules* **2022**, *27*, 2255. [CrossRef]
10. Alves, N.F.; Neto, A.B.S.; Bessa, B.D.S.; Oliveira, A.C.; Mendes Filho, J.; Campos, A.F. Binary Oxides with Defined Hierarchy of Pores in the Esterification of Glycerol. *Catalysts* **2016**, *6*, 151. [CrossRef]
11. Testa, M.L.; La Parola, V.; Liotta, L.F.; Venezia, A.M. Screening of different solid acid catalysts for glycerol acetylation. *J. Mol. Catal. A Chem.* **2013**, *367*, 69–76. [CrossRef]
12. Mou, R.; Wang, X.; Wang, Z.; Zhang, D.; Yin, Z.; Lv, Y.; Wei, Z. Synthesis of fuel bioadditive by esterification of glycerol with acetic acid over hydrophobic polymer-based solid acid. *Fuel* **2021**, *302*, 121175. [CrossRef]
13. Carmo, J.V.C.; Oliveira, A.C.; Araújo, J.C.S.; Campos, A.; Duarte, G.C.S. Synthesis of highly porous alumina-based oxides with tailored catalytic properties in the esterification of glycerol. *J. Mater. Res.* **2018**, *33*, 3625–3633. [CrossRef]
14. Zhou, L.; Al-Zaini, E.; Adesina, A.A. Catalytic characteristics and parameters optimization of the glycerol acetylation over solid acid catalysts. *Fuel* **2013**, *103*, 617–625. [CrossRef]
15. Kong, P.S.; Aroua, M.K.; Daud, W.M.A.W.; Lee, H.V.; Cognet, P.; Perez, Y. Catalytic role of solid acid catalysts in glycerol acetylation for the production of bio-additives: A review. *RSC Adv.* **2016**, *6*, 68885. [CrossRef]
16. Gao, X.; Zhu, S.; Li, Y. Graphene oxide as a facile solid acid catalyst bioadditives from glycerol esterification. *Catal. Commun.* **2015**, *62*, 48–51. [CrossRef]
17. Ferreira, F.B.; Silva, M.J.; Rodrigues, F.Á.; Faria, W.L.S. Sulfated-Alumina-Catalyzed Triacetin Synthesis: An Optimization Study of Glycerol Esterification. *Ind. Eng. Chem. Res.* **2022**, *61*, 4235–4243. [CrossRef]
18. Pankajakshan, A.; Pudi, S.M.; Biswas, P. Acetylation of Glycerol over Highly Stable and Active Sulfated Alumina Catalyst: Reaction Mechanism, Kinetic Modeling and Estimation of Kinetic Parameters. *J. Chem. Technol. Biotechnol.* **2018**, *50*, 98–111. [CrossRef]
19. Ramalingam, R.J.; Radhika, T.; Adam, F.; Dolla, T.H. Acetylation of glycerol over bimetallic Ag–Cu doped rice husk silica based biomass catalyst for bio-fuel additives application. *Int. J. Ind. Chem.* **2016**, *7*, 187–194. [CrossRef]
20. Kale, S.S.; Armbruster, U.; Eckelt, R.; Bentrup, U.; Umbarkar, S.B.; Dongare, M.K.; Martin, A. Understanding the role of Keggin type heteropolyacid catalysts for glycerol acetylation using toluene as an entrainer. *Appl. Catal. A Gen.* **2016**, *527*, 9–18. [CrossRef]
21. Araújo, J.C.S.; Oton, L.F.; Oliveira, A.C.; Lang, R.; Otubo, L.; Bueno, J.M.C. On the role of size controlled Pt particles in nanostructured Pt-containing Al_2O_3 catalysts for partial oxidation of methane. *Int. J. Hydrogen Energy* **2019**, *447*, 27329–27342. [CrossRef]
22. Nascimento, J.P.S.; Oliveira, A.C.; Araujo, J.C.S.; Sousa, F.F.; Saraiva, G.D.; Rodríguez-Aguado, E.; Rodríguez-Castellón, E. Combined promoting effect of molybdenum on the bimetallic Al_2O_3-La_2O_3 catalysts for NO_x reduction by CO. *Fuel* **2020**, *275*, 117872. [CrossRef]

23. Kirszensztejn, P.; Przekop, R.; Szymkowiak, A.; MaćkowskabJerzy, E.; Gaca, J. Preparation of MgO–Al_2O_3 binary gel system with mesoporous structure. *Microporous Mesoporous Mater.* **2006**, *89*, 150–157. [CrossRef]
24. Tantirungrotechai, J.; Chotmongkolsap, P.; Pohmakotr, M. Synthesis, characterization, and activity in transesterification of mesoporous Mg–Al mixed-metal oxides. *Microporous Mesoporous Mater.* **2010**, *128*, 41–47. [CrossRef]
25. Gonullu, M.P. Design and characterization of single bilayer ZnO/Al_2O_3 film by ultrasonically spray pyrolysis and its application in photocatalysis. *Micro Nanostruct.* **2020**, *164*, 107113. [CrossRef]
26. Abdollahifar, A.; Haghighi, M.; Babaluo, A.A.; Talkhoncheh, S.K. Sono-synthesis and characterization of bimetallic Ni–Co/Al_2O_3–MgO nanocatalyst: Effects of metal content on catalytic properties and activity for hydrogen production via CO_2 reforming of CH_4. *Ultrason. Sonochem.* **2016**, *31*, 173–183. [CrossRef]
27. Hilli, Y.; Kinnunen, N.K.; Suvanto, M.; Savimäki, A.; Kallinen, K.; Pakkanen, T.A. Sulfur adsorption and release properties of bimetallic Pd–Ni supported catalysts. *J. Mol. Catal. A Chem.* **2015**, *408*, 161–171. [CrossRef]
28. Charisiou, N.D.; Italiano, C.; Pino, L.; Sebastian, V.; Vita, A.; Goula, M.A. Hydrogen production via steam reforming of glycerol over Rh/g-Al_2O_3 catalysts modified with CeO_2, MgO or La_2O_3. *Renew. Energy* **2020**, *162*, 908–925. [CrossRef]
29. Liang, D.; Hu, Y.; Xiao, C.; Wang, G.; Xie, J.; Zhu, X. Highly efficient catalytic ozonation for ammonium in water upon γ-Al_2O_3@Fe/Mg with acidic-basic sites and oxygen vacancies. *Sci. Total Environ.* **2022**, *834*, 155278. [CrossRef]
30. Silva, A.N.; Neto, A.B.S.; Oliveir, A.C.; Junior, M.C.; Junior, J.A.L.; Freire, P.T.C.; Filho, J.M.; Oliveira, A.C.; Lang, R. Raman studies of nanocomposites catalysts: Temperature and pressure effects of CeAl, CeMn and NiAl oxides. *Spectrochim. Acta Part A* **2018**, *198*, 160–167. [CrossRef]
31. Aminzadeh, A.; Sarikhani-fard, H. Raman spectroscopic study of Ni:Al_2O_3 catalyst. *Spectrochim. Acta Part A* **1999**, *55*, 1421–1425. [CrossRef]
32. Jindal, K.; Tomar, M.; Katiyar, R.S.; Gupta, V. Raman scattering and photoluminescence investigations of N doped ZnO thin films: Local vibrational modes and induced ferromagnetism. *J. Appl. Phys.* **2016**, *120*, 135305. [CrossRef]
33. Gonçalves, A.A.S.; Costa, M.J.F.; Zhang, L.; Ciesielczyk, F.; Jaroniec, M. One-Pot Synthesis of $MeAl_2O_4$ (Me = Ni, Co, or Cu) Supported on γ-Al_2O_3 with Ultralarge Mesopores: Enhancing Interfacial Defects in γ-Al_2O_3 To Facilitate the Formation of Spinel Structures at Lower Temperatures. *Chem. Mater.* **2018**, *30*, 436–446.
34. Luo, M.-F.; Fang, P.; He, M.; Xie, Y.-L. In situ XRD, Raman, and TPR studies of CuO/ Al_2O_3 catalysts for CO oxidation. *J. Mol. Catal. A Chem.* **2005**, *239*, 243–248. [CrossRef]
35. Chan, S.S.; Wachs, I.E. In Situ Laser Raman Spectroscopy of Nickel Oxide Supported on γ-Al_2O_3. *J. Catal.* **1987**, *103*, 224–227. [CrossRef]
36. Escudero, M.J.; Mendoza, L.; Cassir, M.; Gonzalez, T.; Daza, L. Porous nickel MCFC cathode coated by potentiostatically deposited cobalt oxide II. Structural and morphological behavior in molten carbonate. *J. Power Sources* **2006**, *160*, 775–781. [CrossRef]
37. Nadeina, K.A.; Kazakova, M.O.; Kovalskaya, A.A.; Danilova, I.G.; Cherepanova, S.V.; Danilevich, V.V.; Gerasimova, E.Y.; Prosvirin, I.P.; Kondrashev, D.O.; Kleimenov, A.V.K.; et al. Influence of alumina precursor on silicon capacity of NiMo/γ- Al_2O_3 guard bed catalysts for gas oil hydrotreating. *Catal. Today* **2020**, *353*, 53–62. [CrossRef]
38. Núnez, S.; Escobar, J.; Vázquez, A.; Reyes, J.A.; Hernández-Barrera, M. 4,6-Dimethyl-dibenzothiophene conversion over Al_2O_3–TiO_2-supported noble metal catalysts. *Mater. Chem. Phys.* **2011**, *126*, 237–247. [CrossRef]
39. Thommes, M.; Kaneko, K.; Neimark, A.V.; Olivier, J.P.; Rodríguez-Reinoso, F.; Rouquerol, J.; Sing, K.S.W. Physisorption of gases, with special reference to the evaluation of surface area and pore size distribution (IUPAC Technical Report). *Pure Appl. Chem.* **2015**, *87*, 1051–1069. [CrossRef]
40. Kunde, G.B.; Yadav, G.D. Green approach in the solegel synthesis of defect free unsupported mesoporous alumina films. *Microporous Mesoporous Mater.* **2016**, *224*, 43–50. [CrossRef]
41. Wang, L.; Liu, H.; Fan, Y.; Yuan, P.; Huang, D.; Oyama, T.; Wang, S.T.; Bao, X. Cooperative bimetallic catalyst for thio-etherification reaction prepared bycrystal-facet engineering of γ-Al_2O_3 support. *Catal. Today* **2021**, *377*, 196–204. [CrossRef]
42. Nguyen, T.T.; Tsai, H.D.-H. Low-temperature methanol synthesis via (CO_2 + CO) combined hydrogenation using Cu-ZnO/Al_2O_3 hybrid nanoparticle cluster. *Appl. Catal. A Gen.* **2022**, *645*, 118844. [CrossRef]
43. Jiang, T.; Kong, D.; Xu, K.; Cao, F. Hydrogenolysis of glycerol aqueous solution to glycols over Ni–Co bimetallic catalyst: Effect of ceria promoting. *Appl. Petrochem. Res.* **2016**, *6*, 135–144. [CrossRef]
44. Tsiotsias, A.I.; Charisiou, N.D.; Sebastian, V.; Gaber, S.; Hinder, S.J.; Baker, M.A.; Polychronopoulou, K.; Goula, M.A. A comparative study of Ni catalysts supported on Al_2O_3, MgO–CaO– Al_2O_3 and La_2O_3– Al_2O_3 for the dry reforming of ethane. *Int. J. Hydrogen Energy* **2022**, *14*, 5337–5353. [CrossRef]
45. Ju, F.; Liu, C.; Li, K.; Meng, C.; Gao, S.; Lin, H. Reactive Adsorption Desulfurization of Fluidized Catalytically Cracked (FCC) Gasoline over a Ca-Doped Ni-ZnO/ Al_2O_3 −SiO_2 Adsorbent. *Energy Fuels* **2016**, *30*, 6688–6697. [CrossRef]
46. Yamamoto, T.; Hatsui, T.; Matsuyama, T.; Tanaka, T.; Funabiki, T. Structures and Acid−Base Properties of La/Al_2O_3 Role of La Addition to Enhance Thermal Stability of γ-Al_2O_3. *Chem. Mater.* **2003**, *15*, 4830–4840. [CrossRef]
47. Nikolova, D.; Edreva-Kardjieva, R.; Gouliev, G.; Grozeva, T.; Tzvetkov, P. The state of (K)(Ni)Mo/g-Al_2O_3 catalysts after water–gas shift reaction in the presence of sulfur in the feed: XPS and EPR study. *Appl. Catal. A Gen.* **2006**, *297*, 135–144. [CrossRef]
48. Zhu, J.; Cannizzaro, F.; Liu, L.; Zhang, H.; Kosinov, N.; Filot, I.A.W.; Rabeah, J.; Brückner, A.; Hensen, E.J.M. Ni—In Synergy in CO_2 Hydrogenation to Methanol. *ACS Catal.* **2021**, *11*, 11371–11384. [CrossRef]

49. Zhang, Z.-H.; Wu, S.-Y.; Xu, P.; Li, L.-L. Theoretical studies of the EPR parameters for Ni^{2+} and Co^{2+} in MgO. *Braz. J. Phys.* **2010**, *40*, 361. [CrossRef]
50. Natte, K.; Jagadeesh, R.V.; Sharif, M.; Neumann, H.; Beller, M. Synthesis of Nitriles from Amines using Nanoscale Co_3O_4 -based Catalysts via Sustainable Aerobic Oxidation. *Org. Biomol. Chem.* **2016**, *14*, 3356–3359. [CrossRef]
51. Silva-Calpa, L.R.; Zonetti, C.; Rodrigues, C.P.; Alves, O.C.; Appel, L.G.; de Avillez, R.R. The $Zn_xZr_{1-x}O_{2-y}$ solid solution on m-ZrO_2: Creating O vacancies and improving the m-ZrO_2 redox properties. *J. Mol. Catal. A Chem.* **2016**, *425*, 166–173. [CrossRef]
52. Carmo, J.V.; Bezerra, R.C.F.; Tehuacanero-Cuapa, S.; Rodríguez-Aguado, E.; Lang, R.; Campos, A.F.; Duarte, G.; Saraiva, G.D.; Otubo, L.; Oliveira, A.C.; et al. Synthesis of tailored alumina supported Cu-based solids obtained from nanocomposites: Catalytic application for valuable aldehyde and ketones production. *Mater. Chem.* **2022**, *292*, 126800. [CrossRef]
53. Manríquez-Ramírez, M.E.; Elizalde, I.; Ramírez-López, R.; Trejo-Valdez, M.; Estrada-Flores, M. Acetylation of glycerol using MgO–CaO catalysts with different CaO loadings. *React. Kinet. Mech. Catal.* **2020**, *130*, 417–431. [CrossRef]
54. Rane, S.A.; Pudi, S.M.; Biswas, P. Esterification of Glycerol with Acetic Acid over Highly Active and Stable Alumina-based Catalysts: A Reaction Kinetics Study. *Chem. Biochem. Eng. Q.* **2016**, *30*, 33–45. [CrossRef]
55. Tentor, F.R.; Dias, D.B.; Gomes, M.R.; Vicente, J.G.; Cardozo-Filho, P.L.; Berezuk, M.E. Glycerol Acetylation with Propionic Acid Using Iron and Cobalt Oxides in Al-MCM-41 Catalysts. *Bull. Chem. React. Eng. Catal.* **2020**, *15*, 803. [CrossRef]
56. Malaik, A.; Ptaszyńsk, K.; Kozłowski, M. Conversion of renewable feedstock to bio-carbons dedicated for the production of green fuel additives from glycerol. *Fuel* **2021**, *288*, 119609. [CrossRef]

catalysts

MDPI

Article

Ni and Ce Grafted Ordered Mesoporous Silica KIT-6 for CO_2 Adsorption

Mariana Suba [1], Alexandru Popa [1,*], Orsina Verdeș [1], Silvana Borcănescu [1] and Paul Barvinschi [2]

[1] "Coriolan Drăgulescu" Institute of Chemistry, Bl. Mihai Viteazul, No. 24, 300223 Timisoara, Romania
[2] Faculty of Physics, West University of Timisoara, Vasile Parvan, No. 4, 300223 Timisoara, Romania
* Correspondence: alpopa_tim2003@yahoo.com; Tel.: +40-256-491818

Abstract: In this study, the Ni/KIT-6 and Ce/KIT-6 materials were prepared through the impregnation method and then amino-functionalized materials were obtained by the grafting of an amino-silane coupling agent 3-aminopropyl triethoxysilane (APTES). The samples were characterized by thermogravimetric analysis (TGA-DTA), Fourier transform infrared spectroscopy (FT-IR), X-ray diffraction, scanning electron microscopy (SEM) and nitrogen adsorption at 77 K. The study of CO_2 adsorption–desorption on prepared materials was investigated using thermogravimetric analysis (TGA-DTA) coupled with mass spectrometry (MS). The influence of metal oxides on the performance of CO_2 adsorption on functionalized mesoporous silica was presented. The results showed that doping the molecular sieve with cerium oxide can significantly increase the adsorption capacity of the amino-functionalized KIT-6. As the CO_2 adsorbents were prepared by functionalization through grafting with APTES, the amount of amine loading is one of the important factors which improves CO_2 adsorption capacity. Additionally, CO_2 adsorption performance depends on the textural properties and the temperature used for the adsorption process. The maximum adsorption capacity of Ce/KIT-6 Sil is 3.66 mmol/g, which is 2.4 times higher than Ni/KIT-6 Sil. After the nine cycles of cyclic CO_2 adsorption/desorption, the Ce/KIT-6 Sil still had higher adsorption capacities, indicating their good cyclical stability.

Keywords: molecular sieves; KIT-6; amine; CO_2; adsorption–desorption

check for updates

Citation: Suba, M.; Popa, A.; Verdeș, O.; Borcănescu, S.; Barvinschi, P. Ni and Ce Grafted Ordered Mesoporous Silica KIT-6 for CO_2 Adsorption. *Catalysts* **2022**, *12*, 1339. https://doi.org/10.3390/catal12111339

Academic Editor: Narendra Kumar

Received: 29 September 2022
Accepted: 30 October 2022
Published: 1 November 2022

1. Introduction

Today, global warming as a result of the unprecedented increase of carbon dioxide emissions, which are caused mainly through the combustion of fossil fuels, is a severe environmental issue. Considerable research efforts have been carried out to capture CO_2, so in the short term it can be considered a promising way to reduce CO_2 emissions. Thus, the research directions have been aimed at obtaining materials with improved features for the CO_2 adsorption–desorption process [1–3]. Since the capture of carbon dioxide on a global scale can result in large quantities of gas, its prevention has been attempted, and one of the more effective methods is synthetic CO_2 hydrogenation in order to obtain combusted hydrocarbons and the CO_2 reforming of methane [4,5].

In addition to the mesoporous materials MCM-41, MCM-48 and SBA-15, considerable attention has been given to the mesoporous KIT-6 sieve by researchers in the fields of catalysis, adsorption–desorption or ion exchange. The mesoporous KIT-6 sieve is characterized by an easy synthesis method, its symmetric three-dimensional cubic structures *Ia3d*, an average pore diameter of 4–12 nm and wall thickness of 4–6 nm [6,7].

The functionalization of the mesoporous sieves MCM-41, MCM-48, SBA-15 and KIT-6 with amines for CO_2 capture has been studied for a long time due to the high structural stability of the pore volume and adjustable pore dimensions [8–10].

The structure of KIT-6 is clearly superior to the other mesoporous supports due to the gyroid two-dimensional networks with pores that provide direct access to active sites, which facilitates insertion or diffusion inside the pores, thus allowing more reactants and precursor solutions through pore channels [11,12].

To improve the efficiency of the adsorption properties but also, implicitly, the catalytic properties of metal oxide materials that have applications in many industrial processes, researchers have attempted to immobilize metal oxide nanoparticles on porous supports with a large surface area and pore volume, such as zeolites or mesoporous sieves [13–18].

Sun et al. studied the method of increasing the adsorption capacity of CO_2 and cyclic stability for CaO by impregnating CaO into a mesoporous silica structure, namely KIT-6 [19]. The prepared CaO supported KIT-6 has a good cycling stability of over 10 cycles of carbonation/calcination. The performance increase for CO_2 capture is due to the interaction between the structure of KIT-6 and active CaO, through the formation of interfacial mixed oxides, which enhanced the sintering-resistant properties of CaO.

Puertolas et al. studied the cerium oxide supported on mesoporous KIT-6 prepared by the impregnation of $Ce(NO_3)_3$ on siliceous KIT-6. These cerium oxide doped materials Ce-KIT-6 have been tested as catalysts for the total oxidation of naphthalene and display excellent activity, remarkably higher than commercial CeO_2 [20].

The advantages of mesoporous CeO_2 materials for glycerol conversion were demonstrated by Ruiz et al. [21], showing that the acidic properties of CeO_2 proved to be a determining factor for the selectivity towards acetol and acrolein. Mahfouz et al. [22] investigated the effect of the promotion of cerium on Ni and/or Ru KIT-6 mesoporous catalysts for the CO_2 reformation of methane, with the catalysts being synthesized using the wet impregnation technique.

The low cost and wide range of applications have led to detailed studies of nickel-based catalysts [23,24]. In order to obtain a high catalytic activity, studies were conducted on the production of nickel nanoparticles, controlling the dimensions at the nanometer scale using various synthesis methods [25–29]. These catalysts have high catalytic activity and are prepared by simple methods, making them appropriate for different applications, such as the gas cleaning of polycyclic aromatic hydrocarbons from gas exhaust. Additionally, they could be used as supports for different metals and oxides which are catalytically active and in other applications for the protection of the environment from harmful emissions.

In the present study, materials containing Ni and Ce were prepared by impregnating the hydrothermally prepared mesoporous KIT-6 support with 30 wt% total metal oxide content. Then amino-functionalized materials were obtained by the grafting of an amino-silane coupling agent 3-aminopropyl triethoxysilane (APTES) on previously prepared samples.

The three-dimensional mesoporous structure of the KIT-6 support and Me/KIT-6 (Me= Ni and Ce) was confirmed by small angle X-ray diffraction (SAXRD) patterns. N_2 adsorption–desorption analysis results confirmed that the mesoporous structure of KIT-6 and Me/KIT-6 was preserved after metal oxide loading. The structural bonds of KIT-6 support and prepared amino-functionalized materials were determined by Fourier transform infrared (FT-IR) spectroscopy. As far as we know, there are no reports available in the literature regarding Ni/KIT-6 and Ce/KIT-6 composites modified with APTES used for CO_2 adsorption–desorption studies. Compared with other mesoporous silica composites, the CO_2 adsorption–desorption experiments clearly reveal that Ce/KIT-6 composites modified with APTES in particular are very efficient adsorbents.

The CO_2 adsorption capacity and amine efficiency for all prepared amino-functionalized materials were measured by CO_2 adsorption/desorption experiments at adsorption temperatures of 40, 50, 60 and 70 °C, using thermogravimetric analysis. Some representative samples from studied amino-functionalized materials were also tested for stability and recyclability during prolonged operation to justify their consistency for practical applications.

2. Results and Discussion

2.1. Physical–Chemical Characterization

The low angle XRD patterns of KIT-6, Ni/KIT-6 and Ce/KIT-6 are shown in Figure 1 and to large angle in Figure 2. It can be seen that KIT-6, Ni-/KIT-6 and Ce/KIT-6 exhibit similar characteristic peaks, which suggest and confirm that the cubic structure Ia3d specific to the mesoporous sieves of KIT-6 type was still maintained after the addition of Ni- and Ce-active metals. Anyway, in the case of Ni/KIT-6 and Ce/KIT-6, the peaks were also retained but the intensities decreased.

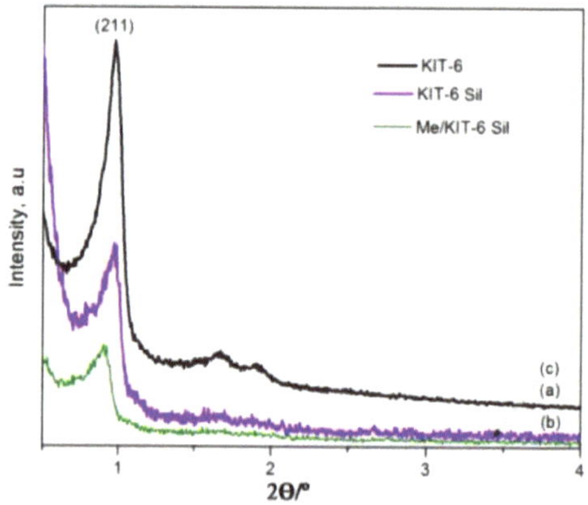

Figure 1. Small-angle XRD patterns of (**a**) KIT-6, (**b**) KIT-6 Sil and (**c**) Me/KIT-6.

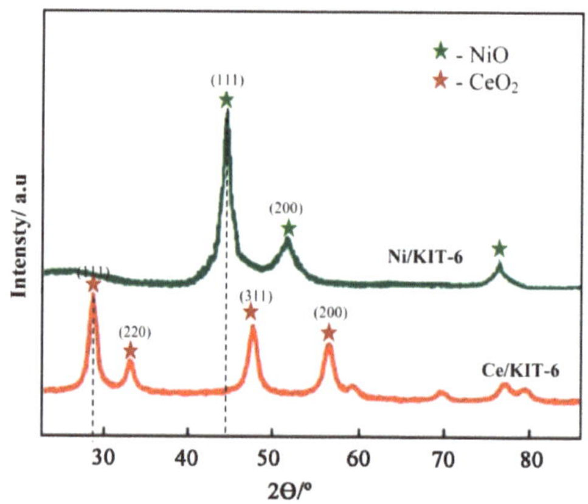

Figure 2. Large-angle XRD patterns of Ni/KIT-6 and Ce/KIT-6.

Figure 2 shows the large-angle XRD patterns of Ni/KIT-6 and Ce/ KIT-6 materials. The characteristic diffraction peaks observed at 2θ = 44.3° and 52.5° were indexed as the plane (111) and (220) of the face-centered cubic NiO (JCPDS No.04-0850) [22] and CeO_2 [21] at 2θ = 28.8 °, 33.1°, 47.5° and 56.4° peaks, corresponding to the (111), (220), (311) and (200) planes of cerium oxide spacegroup Fm3m (225), with cubic fluorite structure (JCPDS No. 34–0394).

The N_2 adsorption–desorption isotherms of KIT-6 support and metal oxide-impregnated materials are the type IV hysteresis loop, which, according to the IUPAC classification, is the indication of a mesoporous structure (Figure 3). It could be observed a sharp increase in volume adsorbed at p/p_0 = 0.5–0.8 is characteristic of the highly ordered pores of composites. Even after metal oxide loading, the preservation of mesoporous structure was confirmed by XRD and N_2 adsorption–desorption analyses.

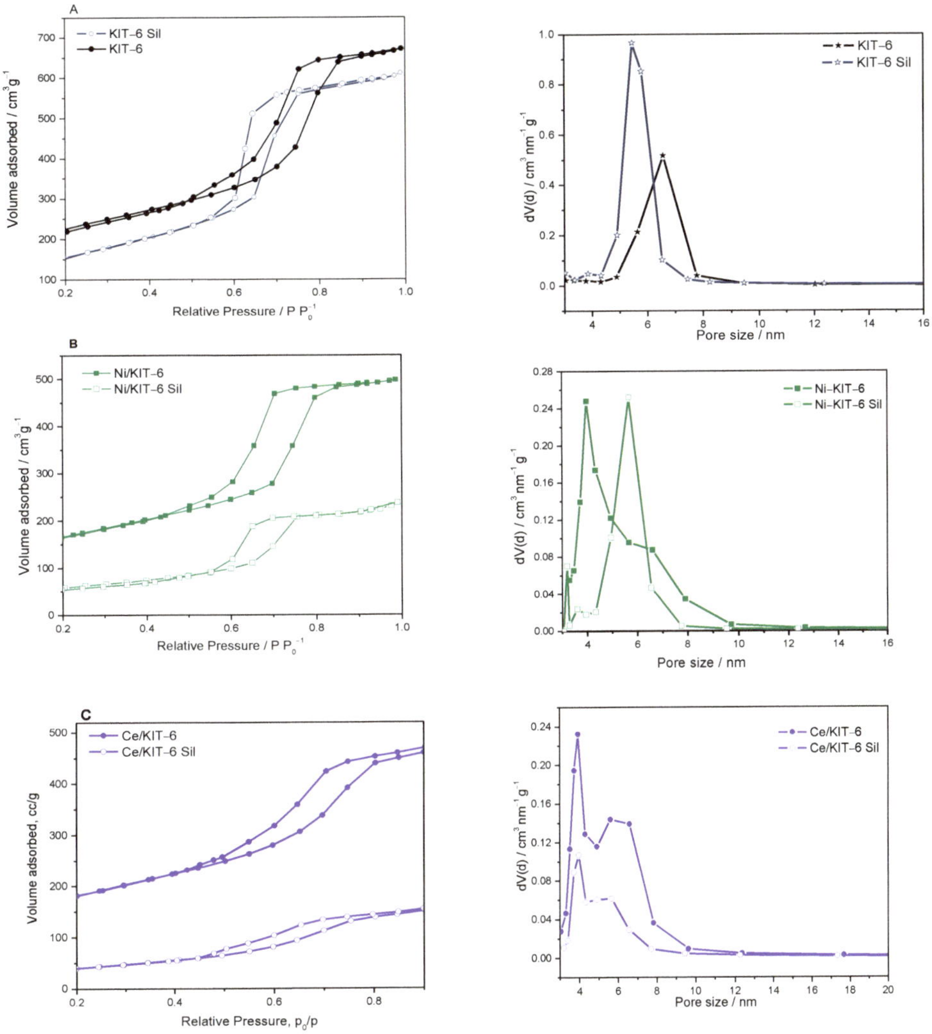

Figure 3. Nitrogen adsorption–desorption isotherms and the pore size distribution of KIT-6 and KIT-6 Sil (**A**), Ni/KIT-6 and Ni/KIT-6 Sil (**B**) and Ce/KIT-6 and Ce/KIT-6 Sil (**C**).

As can be seen from Table 1, the largest surface area (S_{BET} = 813 m^2/g) and pore volume (Vp = 0.99 cm^3/g) were recorded for the KIT-6 support. The specific surface areas and pore volumes decreased, respectively, to 649 m^2/g and 0.49 cm^3/g for Ce/KIT-6 and 551 m^2/g and 0.65 cm^3/g for Ni/KIT-6. This is an anticipated consequence of pore filling during the impregnation process.

The pore parameters of amino-functionalized molecular sieves reveal that the surface areas of KIT-6 Sil, Ni/KIT-6 Sil and Ce/KIT-6 Sil decreased after modification with amino-silane coupling agent 3-aminopropyl triethoxysilane. The reduction in surface areas can be attributed to the increase in the agglomeration of silica particles and/or the occupation of pores after modification with APTES.

Table 1. Textural properties of KIT-6, Me/KIT-6 and amine-grafted samples.

No.	Sample	Specific Surface Area (m^2/g)	Pore Volume BJH_{Des} (cc/g)	Average Pore Diameter BJH_{Des} (nm)
1	KIT-6	813	0.991	7.867
2	KIT-6 Sil	561	0.950	5.462
3	Ni/KIT-6	551	0.650	4.723
4	Ni/KIT-6 Sil	260	0.395	5.666
5	Ce/KIT-6	649	0.490	4.896

As can be seen in Figure 4, the FT-IR spectra of all the samples display three bands at 461, 803, and 1075 cm^{-s}, which are typical for Si-O-Si bands and are related to the presence of a condensed silica network. The adsorption peaks at 461 cm^{-m}, 803 cm^{-m8} and 1075 cm^{-md} are the bending vibration peak and the symmetric and asymmetric stretching vibration peaks of Si-O-Si bonds in the KIT-6 network [28].

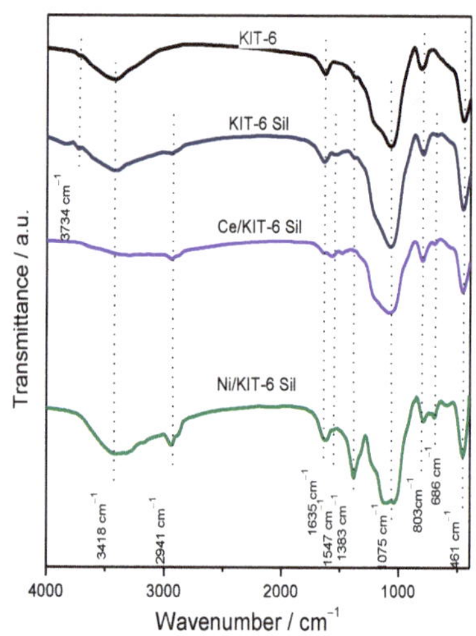

(a)

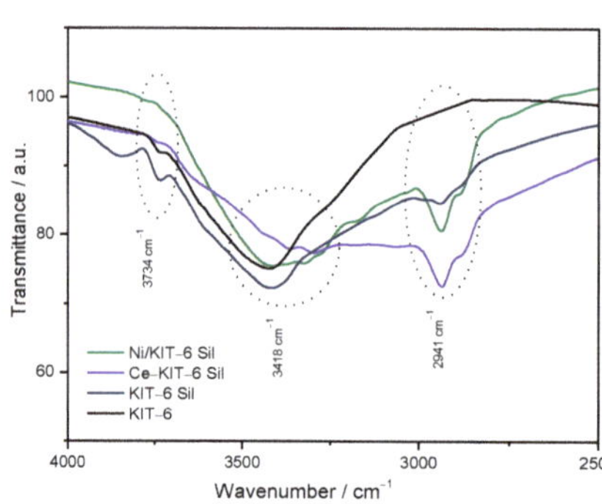

(b)

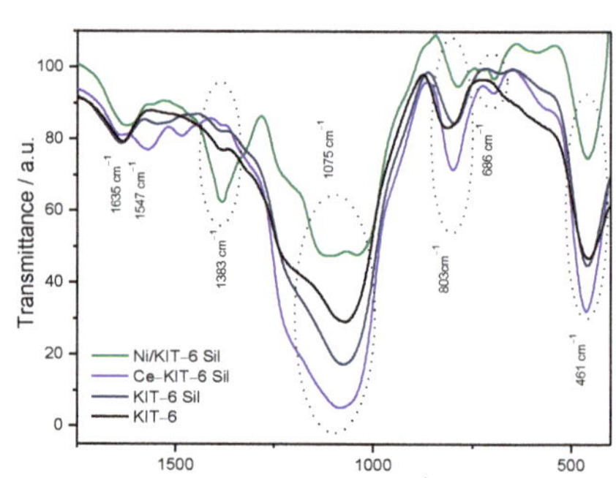

(c)

Figure 4. FTIR spectra of parent KIT-6, KIT-6 Sil, Ce/KIT-6 Sil and Ni/KIT-6 Sil (**a**) and detailed spectra in the intervals of interest (**b**,**c**).

In order to confirm the presence of the APTES in the amino-functionalized molecular sieves, the FT-IR spectra were recorded in the range 500–4000 cm^{-a} (Figure 4). In the APTES-grafted materials, the peaks at 1383 $cm^{-(}$ and 1547 cm^{-a}, representing the –NH_2 and –NH groups associated with the materials surface [16]. The broad peak at around 3734 represents the -OH stretching vibrations, and its intensity is lower for amino-functionalized materials than for the parent KIT-6. The other peaks at 2941 cm^{-o} and 1635 cm^{-a} represent the asymmetric deformation of CH_2 groups present in APTES and the physically adsorbed water [14].

The Si-O-Si symmetric stretching vibrations of Ni/KIT-6 and Ce/KIT-6 materials presented a slight shift toward lower wavenumbers in comparison with KIT-6, which was probably because Ni and Ce species were introduced into the 3D-mesopores of KIT-6.

Scanning electron micrographs images were used to analyze the morphology of the KIT-6, KIT-6 Sil and Ni/KIT-6 Sil particles (Figure 5). The picture of KIT-6 shows that the particles appear to be spherical due to the aggregation of the fine threads of KIT-6. It can be seen that the spherical morphology is modified by the introduction of APTES after amine functionalization, as the threads are dispersed and thus destroy the spherical morphology [14,30]. From the EDX spectrum it was confirmed that Ni and Ce elements have 30% loading.

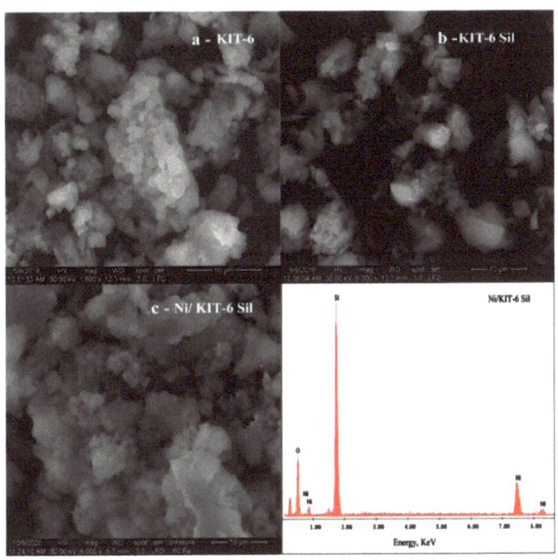

Figure 5. SEM micrographs images of the KIT-6 and KIT-6 Sil particles.

2.2. The Adsorption-Desorption Process CO_2

The adsorption–desorption process of CO_2 was studied for amine-grafted KIT-6 Sil, Ni/KIT-6 Sil and Ce/KIT-6 Sil adsorbents. It should be mentioned that non-functionalized materials were not active for CO_2 capture. The CO_2 adsorption capacity of the adsorbent in mmol of CO_2 per gram of the adsorbent was determined from the mass gain of the sample in the adsorption process step (Figures 6a and 7). The steps of pre-treated samples during adsorption–desorption processes are: first, each sample was pre-treated in flowing N_2 at 150 °C (1/2 h) and temperature is then decreased to the desired adsorption temperature (40, 50, 60, 70 °C) and held there for 30 min under N_2 flow. Following this, the sample was exposed to the adsorption gas mixture 30% CO_2/ N_2 (70 mL/min) and held for 90 min. After the adsorption process is ended, the samples are maintained for another 30 min in N_2 flow in order to remove the physisorbed CO_2 from the sample.

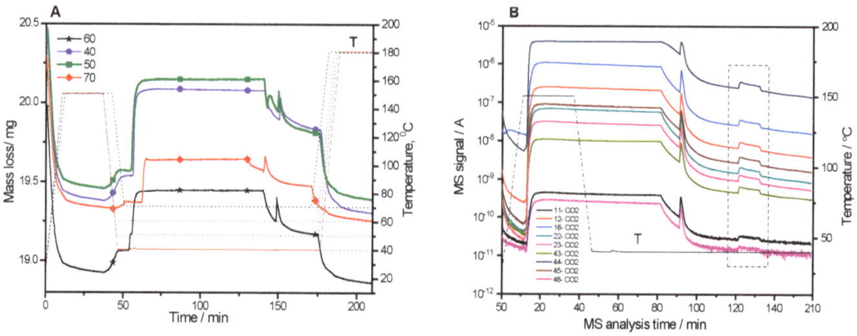

Figure 6. CO_2 adsorption–desorption steps of functionalized samples KIT-6 Sil with an adsorption isotherm at 40–70 °C (**A**) and MS for KIT-6 Sil composites at 40 °C (**B**).

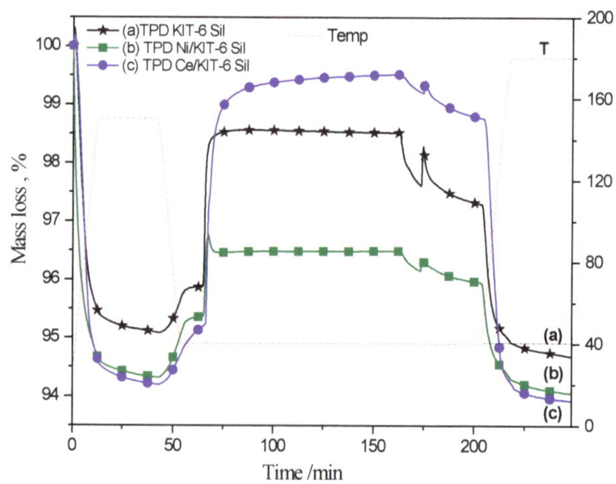

Figure 7. CO_2 adsorption–desorption steps of functionalized samples KIT-6 Sil (a), Ni/KIT-6 Sil (b) and Ce/KIT-6 Sil (c) with an adsorption isotherm at 40 °C.

The desorption step of the chemisorbed CO_2 from the amine-grafted materials was carried out from the adsorption temperature until 180 °C, with an increasing temperature rate of 10 °C/min and with an isotherm at 180 °C for 30 min. From the MS spectra (Figure 6b) the evolution of CO_2 (at a specific mass-to-charge ratio m/z = 46, 45, 44, 43, 23, 22, 16, 12 and 11) during all steps of the adsorption–desorption process was observed. During adsorption process, the MS signal of CO_2 increased and then remained relatively constant for 90 min when the sample was exposed to the adsorption gas mixture. After the CO_2/ N_2 mixture flow was stopped, the MS signal of CO_2 decreased when the samples were maintained for another 30 min in the N_2 atmosphere only (when the physisorbed CO_2 is removed from the sample). The CO_2 signal increased again during the desorption step of the chemisorbed CO_2 from the adsorption temperature until 180 °C (selected area in the Origin graphs of Figure 6b).

In Table 2, the amounts of the captured CO_2 on KIT-6 Sil adsorbent in the temperatures: 40, 50, 60 and 70 °C are shown. By increasing the temperature, the adsorption capacity and efficiently of amino groups decrease from 2.23 to 0.95 mmol CO_2/g SiO_2 and from 0.51 to 0.22 mmol CO_2/mmol NH_2, respectively. The best result was obtained for KIT-6 Sil at 40 °C, which means an adsorption capacity of 2.23 mmol CO_2/g SiO_2 and an efficiency of amino groups of 0.51 mmol CO_2/mmol NH_2. The levels of the capture CO_2 using KIT-6 Sil at different temperatures were calculated from mass gain during the adsorption step. The same calculation method was applied to KIT-6 Sil, Ni/KIT-6 Sil and Ce/KIT-6 Sil at 40 °C by using the following calculation formulas:

$$m_{KIT-6\ Sil(ini)} - m_{H_2O(mass\ loss\ total)} = m_{KIT-6\ Sil(dry)} \quad (1)$$

$$n_{SiO_2} = \frac{m_{KIT-6\ Sil(dry)}}{60} \quad (2)$$

$$n_{CO_2} = \frac{m_{CO_{2(ads)}}}{44} \quad (3)$$

$$\frac{n_{CO_2}}{n_{SiO_2}} * 60 = x\ mmoleCO_2/gSiO_2 \quad (4)$$

$$\frac{n_{CO_2}}{n_{NH_2}} = x\ mmole\ CO_2/mmole\ NH_2 \quad (5)$$

where $m_{KIT-6\ Sil(ini)}$ is the initial mass of KIT-6; $m_{H2O(mass\ loss\ total)}$ is water loss until 150 °C; $m_{KIT-6\ Sil(dry)}$ is the dried mass of KIT-6; and n_{CO_2}, n_{SiO_2} and n_{NH_2} are the number of mmole.

Table 2. The amounts of the capture CO_2 using KIT-6 Sil at different temperatures.

No.	Sample	Temp (°C)	$n_{CO2}/gSiO_2$ (mmol/gSiO$_2$)	n_{CO2}/n_{NH2} (mmol/mmol)
1	KIT-6-Sil	40	2.23	0.51
2	KIT-6-Sil	50	1.76	0.40
3	KIT-6-Sil	60	1.31	0.29
4	KIT-6-Sil	70	0.95	0.22

The results are promising and showed that both the adsorption capacity (mmol CO_2/g adsorbent) and the efficiency of amino groups (mmol CO_2/mmol NH_2) depend on the temperature at which the experiment is performed. The rate of CO_2 adsorption reaches maximum values at 40 °C and then decreases at higher temperatures. One of the external energy sources that can activate the CO_2 molecule is temperature, among others such as light or electricity. So, the CO_2 adsorption capacity is influenced by the temperature as a result of the activation of the CO_2 molecule by facilitating charge transfer (mainly electron injection) processes to the molecule from, for instance, a catalytic surface. It is probably a question of equilibrium between the processes of CO_2 adsorption and desorption processes, respectively. Until 40 °C, the adsorption process in favorably influenced by the temperature; however, with increasing temperature, desorption is the process favored by the temperature.

The evolution of the CO_2 uptake with time and also its time derivative—which is taken as a measure of adsorption rate—is shown in Figure 8a,b for KIT-6 Sil, Ni/KIT-6 Sil and Ce/KIT-6 Sil adsorbents at 40 °C. The rate of CO_2 adsorption has reached the maximum values for Ce/KIT-6 Sil and then decreases for the other adsorbents.

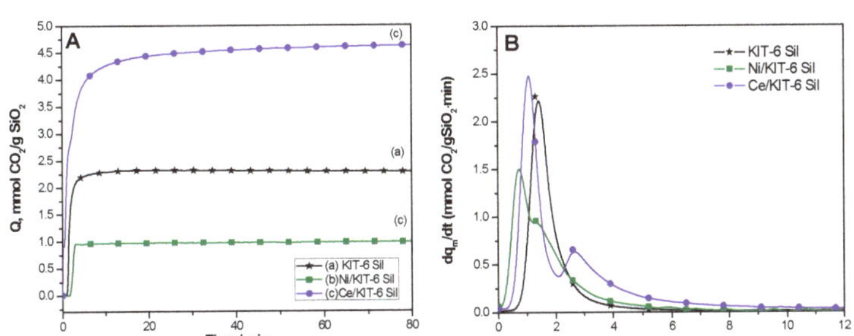

Figure 8. Carbon dioxide adsorption and (**A**) carbon dioxide adsorption rate of KIT-6 Sil, Ni/KIT-6 Sil and Ce/KIT-6 Sil at temperatures between 40 °C (**B**).

The efficiency of the adsorbent (measured in mmol CO_2/mmol NH_2), was calculated from the mass loss during the desorption step of CO_2 (Table 3). CO_2 adsorption isotherms of functionalized samples at 40 °C showed that the adsorption capacity (mmol CO_2/g ads) is higher for KIT-6 Sil, while the efficiency of the amino group is higher for Ce/KIT-6 Sil (Figure 9). The higher adsorption capacity of Ce/KIT-6 Sil compared to Ni/KIT-6 Sil can be explained by the fact that the anchoring of nickel oxide on the inner surface of the 3D mesopores was more uniform and in depth, owing to its lower volume ($r_{Ni\text{-}O}$ = 2.112 Å, Ni^{2+} =0.69 Å, O^{2-} = 1.40 Å and V_{NiO} = 74.97 Å^3,) in comparation with that of cerium oxide ($r_{Ce\text{-}O}$ = 2.37 Å, Ce^{3+} = 0.97 Å, O^{2-} = 1.40 Å and V_{CeO2} = 163.44 Å^3). Additionally, a higher decrease in specific surface area and pore volume can be observed in the case of Ni/KIT-6 Sil with APTES grafting, which suggests that a part of the pores was blocked, hindering CO_2 access to the active sites.

Table 3. The amount of the captured CO_2 at 40 °C using KIT-6 Sil, Ni/KIT-6 Sil and Ce/KIT-6 Sil.

No.	Sample	Temperature (°C)	n_{CO2}/g SiO_2 (mmol/g SiO_2)	n_{CO2}/n_{NH2} (mmol/mmol)
1	KIT-6 Sil	40	2.23	0.512
2	Ni/KIT-6 Sil	40	1.63	0.188
3	Ce/KIT-6 Sil	40	3.66	0.256

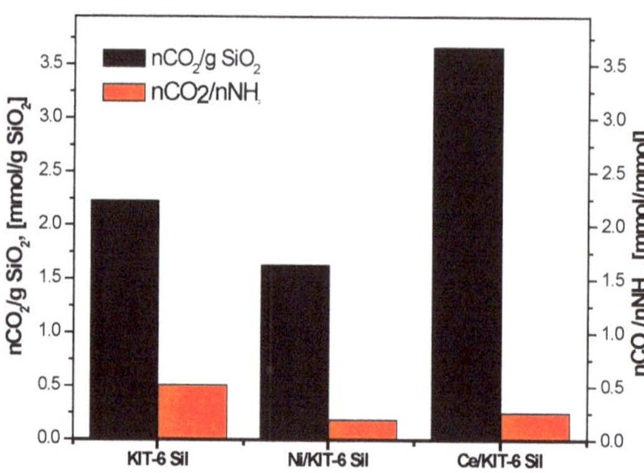

Figure 9. The adsorbed amounts of CO_2 by KIT-6 Sil, Ni/KIT-6 Sil and Ce/KIT-6 at 40 °C.

As we mentioned before, the modifications to the surface chemistry of the silica porous materials by incorporating basic organic group (various amines) and inorganic metal oxides increase CO_2 adsorption capacity. The interaction of CO_2 with amine groups in a water-free environment gives rise to carbamates, via the formation of the CO_2-amine zwitterion, such as R-NH_2 $^+$COO. Overall, this reaction requires two amine groups per CO_2 molecule, i.e., adsorption efficiency, defined as the CO_2/N molar ratio, equal to 0.5. The chemical interaction between CO_2 and amine (APTES) can be expressed as follows:

$$\text{R-NH}_2 + \text{CO}_2 \rightarrow \text{R-NH}_2{}^{+}\text{COO}^{-} \quad (6)$$

$$\text{R-NH}_2\text{+COO}^{-} + \text{R-NH}_2 \rightarrow \text{R-NHCOO}^{-} + \text{RNH}_3{}^{+} \quad (7)$$

While the desorption of CO_2 and regeneration of grafted mesoporous silica is expressed as follows:

$$\text{R-NH-COO}^{-} + \text{R-NH}_3{}^{+} + \text{Heat} \rightarrow \text{CO}_2 + 2\text{R-NH}_2 \quad (8)$$

The formation of the carbamate involves equilibrium and is, therefore, reversible. It is also important to point out that, while this mechanism corresponds to chemisorption, physisorption also takes place at higher partial pressures [31,32].

Kishor and Ghoshal [7] investigated KIT-6 performance as absorbent, for CO_2 adsorption using APTES as silanization agent. At different amine loadings quantities of 1.43, 1.50 and 1.72 mmol/g, they obtained CO_2 adsorption capacities of 0.79, 0.90 and 0.83 mmol CO_2/g, respectively, and a stable performance after 10 cycles of adsorption–desorption. Muchan et al. [33] concluded that high surface area does not represent the only important factor that influences the adsorption capacity, as pore-size diameter and the level of amine loading must also be taken into account. The presence of mesopores channels in the KIT-6 structure provides a better mass transfer of CO_2 throughout the adsorbent, and APTES addition increases the CO_2 adsorbed quantity to 1.30 mmol CO_2/g. Analyzing the adsorption–desorption cycle it was found that the regeneration efficiency at 100 °C for unmodified KIT-6 and modified KIT-6/APTES adsorbent reaches 95%, representing good regenerability and stability. Yilmaz S.M. [34] studied KIT-6 mesoporous silica via partitioned cooperative self-assembly method. CO_2 adsorption–desorption experiments were performed at the different temperatures of 25, 75, and 100 °C, and the maximum adsorption capacity, 0.65 mmol/g, was achieved at the 25 °C adsorption temperature. J.A.A. Gibson et al. [35] investigated how the different amine loadings influenced the CO_2 uptake at 0.1 bar partial pressure on two different carbon porous materials. In the case of micro-AC, the results show that for optimum amine loading, the absorbed amount of CO_2 is 1.05 mmol g^{-i}. After four cycles of the adsorption–desorption experiment, the CO_2 efficiency proved to be relatively constant in case of both micro-AC and meso-AC studied materials.

The cycles of CO_2 adsorption–desorption of the functionalized sample Ce/KIT-6 Sil are shown in Figure 10. Nine cycles of CO_2 adsorption/desorption measurements were performed on grafted samples with the highest CO_2 absorption at 40 °C, which means that Ce/KIT-6 Sil can be used to assess their potential for recycling. Every test was pretreated in N_2 flowing at 120 °C for 10 min, then cooled to a 40 °C adsorption temperature and exposed to 30% CO_2 in N_2 for 40 min. CO_2 desorption was performed by heating the sample to 120 °C with 10 °C/min. An examination of the adsorption/desorption cycle data shows that the performance of the Ce/KIT-6 Sil adsorbent is relatively stable, with a low decrease in the adsorption capacity after nine adsorption–desorption cycles.

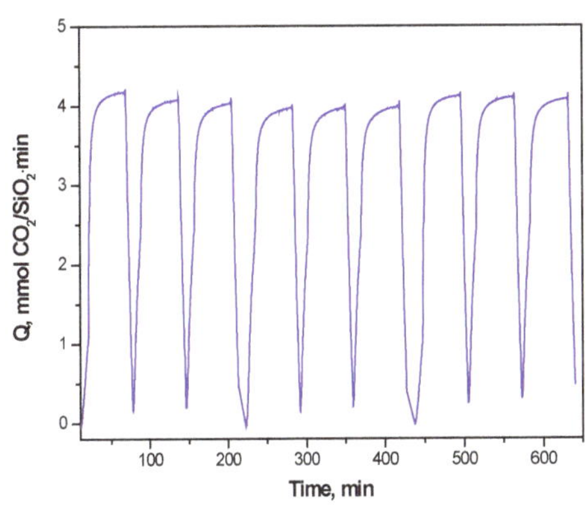

Figure 10. Cycles of CO_2 adsorption–desorption of functionalized samples Ce/KIT-6 Sil with adsorption at 40 °C.

3. Experimental

3.1. Preparation of sample

Mesoporous silica KIT-6 type was prepared according to the method described by Kleitz et al. [36]. First, 3.69 g triblock copolymer Pluronic P123 ($EO_{20}PO_{70}EO_{20}$) (Merck,

Darmstadt, Germany) was dissolved in 132 g distilled water and 6.75 g 37% HCl under vigorous stirring at 35 °C for 8 h. After total dissolution, 3.63 g n-butanol (≥95.0% Reactivul București, Romania) was added to initial solution. The mixture remained under stirring at 35 °C for 1 h. Then, 7.92 g tetraethyl orthosilicate (TEOS, ≥99% Merck, Darmstadt, Germany) was added to the homogeneous solution and stirred for 24 h at 35 °C. This mixture was transferred into a Teflon-lined stainless steel autoclave and heated at 100 °C for 24 h under static hydrothermal conditions. After hydrothermal treatment, the obtained solid material was separated by vacuum filtration and washed with a 2% HCl-ethanol solution. At the end, the samples were calcined in air atmosphere at 550 °C for 6 h to remove the organic template. The resulting sample was denoted as KIT-6.

The Ni/KIT-6 and Ce/KIT-6 materials with 30% Ni and Ce loading were prepared by the impregnation method. An adequate amount of $Ni(NO_3)_2{\cdot}6H_2O$ (Merck, Darmstadt, Germany) and $Ce(NO_3)_3{\cdot}6H_2O$ (Merck, Darmstadt, Germany), respectively, was dissolved in 0.6 mL deionized water. Then the support KIT-6 was added in the solution and the mixture was stirred for 2.0 h. Next, the formed slurry was further filtered to obtain dried powder samples, and the obtained powder samples were dried at 80 °C overnight. Finally, the powder samples were calcined at 250 °C for 4.0 h.

Modified KIT-6 and Me/KIT-6 denoted as KIT-6 Sil and Me / KIT-6 Sil (Me= Ni and Ce) were prepared as follows: 0.5 g of KIT-6 was dispersed in 50 mL toluene and 0.79 mL of 3-aminopropyl triethoxysilane (APTES, ≥98.0% Merck, Darmstadt, Germany) was later added to the solution. The grafting reaction was carried out at 110 °C for 5 h. After filtration and drying, the absorbents were obtained as white solids.

3.2. *Characterization of the Samples*

The thermal analysis was carried out using a thermoanalyzer system Mettler TGA/SDT 851/LF/1100 (Columbus, OH, USA) coupled with mass-spectrometry (MS). The measurements were conducted in dynamic atmosphere of air (50 mL/min), using the alumina plates crucibles of 150 µL.

The FTIR absorption spectra were recorded with a Jasco 430 spectrometer (Tokio, Japan) in the spectral range 4000–400 cm^{-1} range, 256 scans and resolution 2 cm^{-1}, using KBr pellets.

Powder X-ray diffraction data were obtained with a XD 8 Advanced Bruker diffractometer (Karlsruhe, Germany) using the CuKα radiation in the range $2\theta = 0.5 \div 50$ and $2\theta = 5 \div 600$. The specific surface areas of samples were calculated from the nitrogen adsorption–desorption isotherms using a Quantachrome instrument, Nova 2000 series (Boynton Beach, FL, USA).

Scanning electron micrographs (SEM) were recorded using Jeol JMS 6460 LV instrument (Tokio, Japan) equipped with an Oxford Instruments EDS (energy dispersive spectroscopy) analyzer.

The tests of CO_2 adsorption–desorption were obtained using the same thermogravimetric analyzer coupled with MS used for thermal analysis. High-purity CO_2 and 30% CO_2 in N_2 at 1 atm was used for the adsorption experiments, and N_2 was used as purge gas for CO_2 desorption. Each sample was pretreated in N_2 atmosphere at 150 °C, then was cooled to the desired adsorption temperature (40–70 °C) and finally exposed to 30% CO_2/N_2 (70 mL/min) for 90 min. The CO_2 adsorption capacity of the adsorbent in milligrams of CO_2 per gram of adsorbent was calculated from the weight gain of the sample in the adsorption process.

4. Conclusions

In this paper, the Ni/KIT-6 and Ce/KIT-6 materials with 30% Ni and Ce loading were prepared by the impregnation method. Then amino-functionalized materials were prepared by grafting of a silane coupling agent 3-aminopropyl triethoxysilane (APTES) on previous prepared samples.

The small angles of XRD showed that KIT-6 has a cubic structure, which indicates that the synthesis was successfully carried out. For the KIT-6-Sil sample a decrease in peak intensity appears because the APTES was successfully grafted onto KIT-6. The FT-IR indicated that the KIT-6-Sil is present all the characteristic bands of amino-functional groups, confirming the grafting of amines onto KIT-6.

The CO_2 adsorption/desorption of KIT-6-Sil showed that both the adsorption capacity (mmolCO_2/g adsorbent) and the efficiency of amino groups (molCO_2/mol NH_2) depend on the temperature. The best results were obtained for KIT-6-Sil at 40 °C. Good CO_2 adsorption capacities were obtained for all prepared adsorbents, especially for the Ce/KIT-6 Sil sample. After nine adsorption–desorption cycles, the performance of the Ce/KIT-6 Sil adsorbent is relatively stable, with a low decrease in the adsorption capacity. In the case of the Ni/KIT-6 Sil adsorbent, a lower adsorption capacity was displayed. These studies show the potential of mesoporous silica for carbon dioxide capture. The ability of KIT-6 mesoporous sieves to change their structure with metal ions and organic functional groups opens the door for CO_2 adsorption.

Author Contributions: Conceptualization, A.P. and M.S.; methodology, S.B., O.V. and A.P; formal analysis, M.S., S.B. and P.B.; investigation, S.B., M.S. and O.V.; resources, P.B. and A.P.; data curation, S.B. and M.S.; writing—original draft preparation, A.P., S.B. and M.S.; writing—review and editing, A.P., S.B. and M.S.; visualization, P.B. and O.V.; supervision, A.P. and S.B. All authors have read and agreed to the published version of the manuscript.

Funding: This research was funded by Romanian Academy Project No. 4.3.

Conflicts of Interest: The authors declare no conflict of interest. The funders had no role in the design of the study; in the collection, analyses, or interpretation of data; in the writing of the manuscript; or in the decision to publish the results.

References

1. Liu, Y.; Yu, X. Carbon dioxide adsorption properties and adsorption/desorption kinetics of amine-functionalized KIT-6. *Appl. Energy.* **2018**, *211*, 1080–1088. [CrossRef]
2. Belmabkhout, Y.; Serna-Guerrero, R.; Sayari, A. Adsorption of CO_2-Containing Gas Mixtures over Amine-Bearing Pore-Expanded MCM-41 Silica. *Ind. Eng. Chem. Res.* **2010**, *49*, 359–365. [CrossRef]
3. Ma, B.; Zhuang, L.; Chen, S. Rapid synthesis of tunable-structured short-pore SBA-15 and its application on CO_2 capture. *J. Porous Mater.* **2016**, *23*, 529–537. [CrossRef]
4. Alreshaidan, S.B.; Ibrahim, A.A.; Fakeeha, A.H.; Almutlaq, A.M.; Ali, F.A.A.; Al-Fatesh, A.S. Effect of Modified Alumina Support on the Performance of Ni-Based Catalysts for CO2 Reforming of Methane. *Catalysts* **2022**, *12*, 1066. [CrossRef]
5. Ye, R.-P.; Ding, J.; Gong, W.; Argyle, M.D.; Zhong, Q.; Wang, Y.; Russell, C.K.; Xu, Z.; Russell, A.G.; Li, Q.; et al. CO_2 hydrogenation to high-value products via heterogeneous catalysis. *Nat. Commun.* **2019**, *10*, 5698. [CrossRef]
6. Fernandes, F.; Pinto, F.; Lima, E.; Souza, L.; Caldeira, V.; Santos, A. Influence of Synthesis Parameters in Obtaining KIT-6. *Mesoporous Material. Appl. Sci.* **2018**, *8*, 725. [CrossRef]
7. Kishor, R.; Ghoshal, A.K. APTES grafted ordered mesoporous silica KIT-6 for CO_2 adsorption. *Chem. Eng. J.* **2015**, *262*, 882–890. [CrossRef]
8. Popa, A.; Sasca, V.; Verdes, O.; Suba, M.; Barvinschi, P. Effect of the amine type on thermal stability of modified mesoporous silica used for CO_2 adsorption. *J. Therm. Anal. Calorim.* **2018**, *134*, 269–279. [CrossRef]
9. Heydari-Gorji, A.; Yang, Y.; Sayari, A. Effect of the Pore Length on CO_2 Adsorption over Amine-Modified Mesoporous Silicas. *Energy Fuels* **2011**, *25*, 4206–4210. [CrossRef]
10. Liu, Q.; Shi, J.; Zheng, S.; Tao, M.; He, Y.; Shi, Y. Kinetics Studies of CO_2 Adsorption/Desorption on Amine-Functionalized Multiwalled Carbon Nanotubes. *Ind. Eng. Chem. Res.* **2014**, *53*, 11677–11683. [CrossRef]
11. Oveisi, H.; Anand, C.; Mano, A.; Al-Deyab, S.S.; Kalita, P.; Beitollahi, A.; Vinu, A. Inclusion of size controlled gallium oxide nanoparticles into highly ordered 3D mesoporous silica with tunable pore diameters and their unusual catalytic performance. *J. Mater. Chem. A.* **2010**, *20*, 10120–10129. [CrossRef]
12. Karthikeyan, G.; Pandurangan, A. Post synthesis alumination of KIT-6 materials with Ia3d symmetry and their catalytic efficiency towards multicomponent synthesis of 1H-pyrazolo[1,2-]phthalazine-5,10-dione carbonitriles and carboxylates. *J. Mol Catal A Chem.* **2012**, *361–362*, 58–67. [CrossRef]
13. Sun, J.; Kan, Q.; Li, Z.; Yu, G.; Liu, H.; Yang, X.; Guan, J. Different transition metal (Fe^{2+}, Co^{2+}, Ni^{2+}, Cu^{2+} or VO^{2+}) Schiff complexes immobilized onto three-dimensional mesoporous silica KIT-6 for the epoxidation of styrene. *RSC Adv.* **2014**, *4*, 2310–2317. [CrossRef]
14. Visuvamithiran, P.; Palanichamy, M.; Shanthi, K.; Murugesan, V. Selective epoxidation of olefins over Co(II)-Schiff immobilised on KIT-6. *Appl. Catal. A Gen.* **2013**, *462–463*, 31–38. [CrossRef]

15. Taghizadeh, M.; Akhoundzadeh, H.; Rezayan, A.; Sadeghian, M. Excellent catalytic performance of 3D-mesoporous KIT-6 supported Cu and Ce nanoparticles in methanol steam reforming. *Int. J. Hydrog. Energy* **2018**, *43*, 10926–11093. [CrossRef]
16. Kishor, R.; Singh, S.B.; Ghoshal, A.K. Role of metal type on mesoporous KIT-6 for hydrogen storage. *Int. J. Hydrog. Energy* **2018**, *43*, 10376–10385. [CrossRef]
17. Li, Z.; Miao, Z.; Wang, X.; Zhao, J.; Zhou, J.; Si, W.; Zhuo, S. One-pot synthesis of ZrMo-KIT-6 solid acid catalyst for solvent-free conversion of glycerol to solketal. *Fuel* **2018**, *233*, 377–387. [CrossRef]
18. Prabhu, A.; Kumaresan, L.; Palanichamy, M.; Murugesan, V. Synthesis and characterization of aluminium incorporated mesoporous KIT-6: Efficient catalyst for acylation of phenol. *Appl. Catal. A Gen.* **2009**, *360*, 59–65. [CrossRef]
19. Sun, H.; Parlett, C.M.A.; Isaacs, M.A.; Liu, X.; Adwek, G.; Wang, J.; Wu, C. Development of Ca/KIT-6 adsorbents for high temperature CO_2 capture. *Fuel* **2019**, *235*, 1070–1076. [CrossRef]
20. Puertolas, B.; Solsona, B.; Agouram, S.; Murillo, R.; Mastral, A.M.; Aranda, A.; Garcia, T. The catalytic performance of mesoporous cerium oxides prepared through a nanocasting route for the total oxidation of naphthalene. *Appl. Catal. B Environ.* **2010**, *93*, 395–405. [CrossRef]
21. Ruiz, M.G.; Pliego, J.A.; Noreña Franco, L.E.; Álvarez, C.M.; Pariente, J.P.; Martin Guaregua, N.C. Synthesis and characterization of a mesoporous cerium oxide catalyst for the conversion of glycerol. *J. Appl. Res. Technol.* **2018**, *16*, 511–523. [CrossRef]
22. Mahfouz, R.; Estephane, J.; Gennequin, C.; Tidahy, L.; Aouad, S.; Abi-Aad, E. CO_2 reforming of methane over Ni and/or Ru catalysts supported on mesoporous KIT-6: Effect of promotion with Ce. *J. Environ. Chem. Eng.* **2021**, *9*, 104662. [CrossRef]
23. Zhou, X.; Zhao, H.; Liu, S.; Ye, D.; Qu, R.; Zheng, C.; Gao, X. Engineering nano-ordered of Ni nanoparticles on KIT-6 for enhanced catalytic hydrogenation of nitrobenzene. *Appl. Surf. Sci.* **2020**, *525*, 146382. [CrossRef]
24. Świrk, K.; Gálvez, M.E.; Motak, M.; Grzybek, T.; Rønning, M.; Da Costa, P. Syngas production from dry methane reforming over yttrium-promoted nickel-KIT-6 catalysts. *Int. J. Hydrog. Energy* **2019**, *44*, 274–286. [CrossRef]
25. Koltypin, Y.; Fernandez, A.; Rojas, T.C.; Campora, J.; Palma, P.; Prozorov, R.; Gedanken, A. Encapsulation of Nickel Nanoparticles in Carbon Obtained by the Sonochemical Decomposition of $Ni(C_8H_{12})_2$. *Chem. Mater.* **1999**, *11*, 1331–1335. [CrossRef]
26. Moreau, L.M.; Ha, D.H.; Bealing, C.R.; Zhang, H.; Hennig, R.G.; Robinson, R.D. Unintended Phosphorus Doping of Nickel Nanoparticles during Synthesis with TOP: A Discovery through Structural Analysis. *Nano Lett.* **2012**, *12*, 4530–4539. [CrossRef]
27. Pan, Y.; Jia, R.; Zhao, J.; Liang, J.; Liu, Y.; Liu, C. Size-controlled synthesis of monodisperse nickel nanoparticles and investigation of their magnetic and catalytic properties. *Appl. Surf. Sci.* **2014**, *316*, 276–285. [CrossRef]
28. Jia, F.L.; Zhang, L.Z.; Shang, X.Y.; Yang, Y. Non-Aqueous Sol–Gel Approach towards the Controllable Synthesis of Nickel Nanospheres, Nanowires, and Nanoflowers. *Adv. Mater.* **2008**, *20*, 1050–1054. [CrossRef]
29. Liu, Z.C.; Zhou, J.; Cao, K.; Yang, W.M.; Gao, H.X.; Wang, Y.D.; Li, H.X. Highly dispersed nickel loaded on mesoporous silica: One-spot synthesis strategy and high performance as catalysts for methane reforming with carbon dioxide. *Appl. Catal. B* **2012**, *125*, 324–330. [CrossRef]
30. Cao, H.-X.; Zhang, J.; Guo, C.L.; Chen, J.G.; Ren, X.K. Highly dispersed Ni nanoparticles on 3D-mesoporous KIT-6 for CO methanation: Effect of promoter species on catalytic performance. *Chin. J. Catal.* **2017**, *38*, 1127–1137. [CrossRef]
31. Sreenivasulu, B.; Gayatri, D.V.; Sreedhar, I.; Raghavan, K.V. A journey into the process and engineering aspects of carbon capture technologies. *Renew. Sust. Energ. Rev.* **2015**, *41*, 1324–1350. [CrossRef]
32. Sanz-Perez, E.S.; Dantas, T.C.M.; Arencibia, A.; Calleja, G.; Guedes, A.P.M.A.; Araujo, A.S.; Sanz, R. Reuse and recycling of amine-functionalized silica materials for CO_2 adsorption. *Chem. Eng. J.* **2017**, *308*, 1021–1033. [CrossRef]
33. Muchan, P.; Saiwan., C.C.; Nithitanakul, M. Investigation of adsorption/desorption performance by aminopropyltriethoxysilane grafted onto different mesoporous silica for post-combustion CO_2 capture. *Clean Energy* **2020**, *4*, 120–131. [CrossRef]
34. Yılmaz, S.M. A study of CO_2 adsorption behaviour and kinetics on KIT-6. *EJOSAT* **2020**, *19*, 48–55.
35. Gibson, J.A.A.; Gromov, A.V.; Brandani, S.; Campbell, E.E.B. The effect of pore structure on the CO_2 adsorption efficiency of polyamine impregnated porous carbons. *Microporous Mesoporous Mater.* **2015**, *208*, 129–139.
36. Kleitz, F.; Choi, S.H.; Ryoo, R. Cubic *Ia3d* large mesoupouros silica: Synthesis and replication to palatinum nanowires, carbon nanorods and carbon nanotube. *Chem. Commun.* **2003**, *7*, 2136–2137.

catalysts

Article

Ordered Mesoporous nZVI/Zr-Ce-SBA-15 Catalysts Used for Nitrate Reduction: Synthesis, Optimization and Mechanism

Ruimin Zhang [1,*], Haixia Liu [1], Weili Jiang [2] and Weijing Liu [2]

[1] School of Environment and Ecology, The City Vocational College of Jiangsu, Nanjing 210019, China; liuhaixia198027@126.com

[2] Jiangsu Province Key Laboratory of Environmental Engineering, Nanjing 210019, China; jiangweili@yeah.net (W.J.); liuwj@jshb.gov.cn (W.L.)

* Correspondence: zhangruimin_04@163.com; Tel.: +86-159-5056-8032

Abstract: Excessive concentrations of nitrate (NO_3-N) in water lead to the deterioration of water quality, reducing biodiversity and destroying ecosystems. Therefore, the present study investigated NO_3-N removal from simulated wastewater by nanoscale zero-valent iron-supported ordered mesoporous Zr-Ce-SBA-15 composites (nZVI/Zr-Ce-SBA-15) assisted by response surface methodology (RSM), an artificial neural network combined with a genetic algorithm (ANN-GA) and a radial basis neural network (RBF). The successful support of nZVI on Zr-Ce-SBA-15 was confirmed using XRD, FTIR, TEM, SEM–EDS, N_2 adsorption and XPS, which indicated ordered mesoporous materials. The results showed that ANN-GA was better than the RSM for optimizing the conditions of NO_3-N removal and the RBF neural network further confirmed the reliability of the ANN-GA model. The removal rate of NO_3-N by the composites reached 95.71% under the optimized experimental conditions (initial pH of 4.89, contact time = of 62.27 min, initial NO_3-N concentration of 74.84 mg/L and temperature of 24.77 °C). The process of NO_3-N adsorption onto Zr-Ce-SBA-15 composites was followed by the Langmuir model (maximum adsorption capacity of 45.24 mg/g), pseudo-second-order kinetics, and was spontaneous, endothermic and entropy driven. The yield of N_2 can be improved after nZVI was supported on Zr-Ce-SBA-15, and the composites exhibited a strong renewability in the short term within three cycles. The resolution of Fe^{2+} experiments confirmed that nZVI/Zr-Ce-SBA-15 was simultaneously undergoing adsorption and catalysis in the process of NO_3-N removal. Our study suggests that the ordered mesoporous nZVI/Zr-Ce-SBA-15 composites are a promising material for simultaneously performing NO_3-N removal and improving the selectivity of N_2, which provides a theoretical reference for NO_3-N remediation from wastewater.

Keywords: nitrate; nZVI/Zr-Ce-SBA-15; response surface methodology; artificial neural network combined with genetic algorithm; radial basis neural network

Citation: Zhang, R.; Liu, H.; Jiang, W.; Liu, W. Ordered Mesoporous nZVI/Zr-Ce-SBA-15 Catalysts Used for Nitrate Reduction: Synthesis, Optimization and Mechanism. *Catalysts* **2022**, *12*, 797. https://doi.org/10.3390/catal12070797

Academic Editor: Narendra Kumar

Received: 27 May 2022
Accepted: 14 July 2022
Published: 19 July 2022

1. Introduction

Excessive concentrations of nitrate (NO_3-N) entering rivers and lakes can stimulate the growth of algae, resulting in the deterioration of water quality, reduction in biodiversity and degradation of the ecosystem [1,2]. Although many countries have made efforts to reduce NO_3-N emissions to the environment, it is still one of the most serious environmental issues faced across the world [3]. In particular, its concentration in approximately 35% of well water in Tuscany in Italy exceeded 50 mg/L [4]. According to a groundwater survey in Alabama (America), NO_3-N levels in most parts of the central and northeast areas of the state were greater than 63 mg/L, and even exceeded 112 mg/L in some areas [5]. Similarly, in the vast rural and suburban areas of China, many people drink seriously polluted well water for a long time without knowing its harm [6]. NO_3-N is of lower toxicity, while it can be reduced to nitrite after entering the human body, and the toxicity of nitrite can be 11 times higher than NO_3-N [7]. The main biological effect of nitrite is to oxidize normal hemoglobin into a type that does not have the ability to deliver oxygen, reducing the ability of hemoglobin to deliver oxygen to methemoglobin. When A concentration of methemoglobin of more than 10% of the normal hemoglobin concentration will lead to methemoglobin disease (such as Verticillium wilt). The clinical symptoms include skin cyanosis, dizziness, nausea, accelerated heartbeat, dyspnea, fatigue, abdominal pain, and diarrhea. Higher concentrations will cause asphyxia and even death [8]. Therefore, excessive NO_3-N concentrated in the water body will seriously threaten human health. Several effective measures must be taken to control and prevent NO_3-N pollution in water environments.

Hereto, the methods of removing NO_3-N from water mainly include chemical denitrification, catalytic denitrification, reverse osmosis, electrodialysis, ion exchange, biological denitrification, adsorption, etc. [9–11]. Among them, adsorption is a superior method to remove NO_3-N from wastewater due to its low cost and fast reaction [12]. In particular, nanoscale zero-valent iron (nZVI) is widely used in environmental remediation because of its superior adsorption performance and higher reduction activity. nZVI is excellent in the treatment of heavy metals and organics as well as NO_3-N in water. However, the easy agglomeration and poor antioxidation of nZVI limit its application in the field of wastewater treatment [13]. Zhou and Li [14] used nZVI and modified polyethylene carrier, a novel composite packing of tea polyphenol, to remove NO_3-N. Their results demonstrated a novel approach for the fast and eco-friendly preparation and efficient application of nZVI. Zhang et al. [15] fabricated the nZVI supported on pillared clay, which was used for NO_3-N removal. Their results showed that NO_3-N could be absolutely removed by nZVI/PILC within 120 min. This efficiency was not only much higher than that (62.3%) when using nZVI alone, but was also superior to the reduction (71.5%) seen with nZVI plus adsorption (9.19%). The end-products of NO_3-N reduction were identified as NH_4^+-N and N_2, implying that nZVI/PILC may help to cut down the total amount nitrogen in water. Obviously, it is urgent to modify nZVI to avoid easy agglomeration and poor antioxidation, and thus achieving an efficient NO_3-N removal from water.

SBA-15, as a mesoporous molecular sieve, has a high specific surface area, larger pore volume, regular pores and improved mechanical and hydrothermal stability [16]. It has attracted extensive interest in the fields of catalysis, separation, biology and nanomaterials. In particular, the short-channel SBA-15 mesoporous materials with regular morphology have potential applications in many fields. However, due to the inherent shortcomings, such as its low chemical reaction activity, its practical application range is greatly limited [17]. Traditional SBA-15 is rod-shaped or fibrous [17]. When traditional SBA-15 is used as a carrier in the fields of adsorption, separation, and catalysis, its long pores in the micron range are not conducive to the diffusion and transmission of substances into the pores. Therefore, the synthesis of mesoporous SBA-15 materials with small pores has been studied [18]. Chen et al. [19] synthesized organic functionalized short-channel plate SBA-15 with P123 as the template and trace Zr (IV) under strong acid conditions. For the adsorption of organic macromolecules, short-channel plate SBA-15 shows a better transmission capacity than traditional SBA-15. In addition, metal oxides generally exhibit excellent catalytic performances; the catalytic activity of Ce is slightly lower than that of the precious metals [20]. The cost of Ce is lower than precious metals, and it can replace precious metals as catalysts to catalyze some important reactions. Therefore, Zr-Ce-SBA-15 combined with nZVI may result in shortly ordered channels, a higher catalytic capacity, and stronger reduction ability.

Currently, several studies [20,21] have reported nZVI/SBA-15 used for environmental remediation. Tang et al. [21] successfully prepared nZVI/SBA-15 used for the effective degradation of p-nitrophenol. Their results showed that abundant ultrasmall nanoscale zero-valent iron particles were formed and well dispersed on mesoporous silica (SBA-15). A previous study [22] used short-channel hexagonal ordered mesoporous Zr-Ce-SBA-15 materials as a carrier and adopted double-solvent impregnation calcination. nZVI-confined composites in ordered mesoporous channels were synthesized through a reduction method and used to remove Trinitrotoluene. Overall, nZVI/Zr-Ce-SBA-15 may possess excellent potential for NO_3-N removal, and unfortunately, to the best of our knowledge, these works were not reported. Therefore, the objective of the present study is to (1) fabricate the ordered mesoporous nZVI/Zr-Ce-SBA-15 composites used for NO_3-N removal, and tentatively try to improve the selectivity of N_2; (2) characterize the as-prepared nZVI/Zr-Ce-SBA-15 composites using different approaches, including X-ray diffraction (XRD), scanning electron microscopy combined with energy dispersive spectrum (SEM-EDS), N_2-sorption, high-resolution transmission electron microscope (HRTEM), and X-ray photoelectron spectroscopy (XPS); (3) optimize/predict the parameters of NO_3-N removal and nitrogen generation from simulated wastewater by nZVI/Zr-Ce-SBA-15 composites with the aid of response surface methodology (RSM), a back-propagation neural network combined with genetic algorithm (ANN-GA) and a radial basis function neural network (RBFNN).

2. Materials and Methods

2.1. Materials and Chemicals

All chemical reagents were of analytical grade, and all solutions were prepared with high-purity water (18.25 M/cm) from a Milli-Q water purification system. Triblock copolymer P123 (A.R.) was purchased from BASF Corp. (Florham Park, NJ, USA) and Tetraethyl orthosilicate (TEOS) came from ChengDu Chron Chemicals Co., Ltd. (Chengdu, China). $ZrOCl_8H_2O$, $Ce(NO_3)_3{\cdot}6H_2O$, $CH_3(CH_2)_4CH_3$ was from Sinopharm Chemical Reagent Co., Ltd. (Shanghai, China), Shanghai Qingxi Chemical Technology Co., Ltd. (Shanghai, China), and Nanjing Chemical Reagent Co., Ltd. (Nanjing, China), respectively. High-purity nitrogen was purchased from Shanghai bio gas Co., Ltd. (Shanghai, China).

2.2. Preparation of nZVI and nZVI/Zr-Ce-SBA-15

A certain amount of triblock copolymer P123 was weighed and added to 80 mL deionized water, and then heated and stirred at 35 °C, so it dissolved and formed micelles. Immediately, appropriate amounts of $ZrOCl_2{\cdot}8H_2O$, $Ce(NO_3)_3{\cdot}6H_2O$, and TEOS were put in the above micellar solution, in which the molar ratio of reactants was 0.01P123:1TEOS: $170H_2O$: $0.05ZrOCl_2{\cdot}8H_2O$: 0.05 $Ce(NO_3)_3{\cdot}6H_2O$. The mixtures were continuously stirred at 35 °C for 20 h. The formed gel was transferred into an autoclave for crystallization at 24 h at 100 °C and then cooled to room temperature. After, the produced white materials were filtered, washed, dried and finally calcined in 550 °C air atmosphere for 6 h (heating rate is 1 °C/min) to remove the template. Finally, the white powder sample is Zr-Ce-SBA-15.

nZVI/Zr-Ce-SBA-15 composites were fabricated following the method of Tang et al. [21]. Firstly, with continuous stirring, the as-prepared Zr-Ce-SBA-15 of 1.0 g was mixed with 30 mL of n-pentane. Fe(II) aqueous solution prepared by 1.112 g $FeSO_4{\cdot}7H_2O$ was gradually added into the above mixtures. The above mixture was dried at 60 °C for 12 h. Then, with continuous N_2 protection and stirring, 8 mL $NaBH_4$ solution (2 M) was added into the mixtures from the previous step, reducing Fe(II) to metallic Fe^0. The obtained materials were separated from mixtures using a magnet, washed with methanol three times, and then dried in vacuum at 50 °C for 20 h. Meanwhile, 4.0 g of $FeSO_4{\cdot}7H_2O$ was dissolved in 200 mL of methanol and deoxygenated water to fabricate nZVI. Immediately, with continuous N_2 protection and stirring, 10mL of $NaBH_4$ solution (2.1 M) was added gradually to the above solution, lasting for 30 min. The black solid was successively washed with ethanol and deionized water, and the nZVI was dried in vacuum at 50 °C for 20 h. Figure 1 shows the flow chart for preparing nZVI/Zr-Ce-SBA-15.

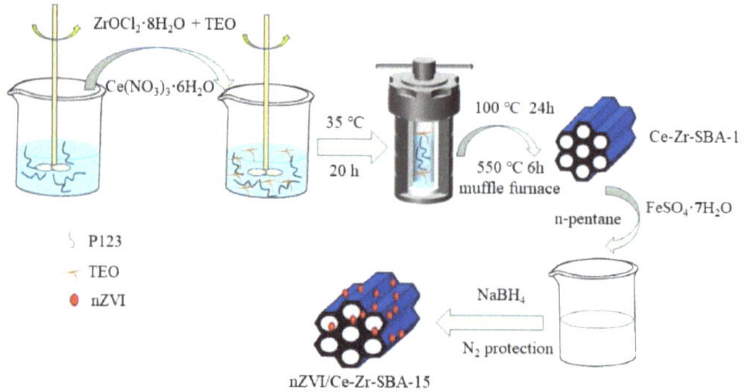

Figure 1. The flow chart for preparing nZVI/Zr-Ce-SBA-15.

2.3. Characterizations

The morphology of the samples was characterized by TEM (JEM-2100, Akishima, Tokyo, Japan) and SEM (Jsm-6490lv, JEOL, Tokyo, Japan). XRD (RIGAKUD/max 2500) was used to test the crystal structure of the sample with a test voltage of 40 kV, current of 20 mA, scanning rate of 2 (°)/min, and Cu target Kal radiation line (λ = 0.15405 nm). The N_2 adsorption–desorption isotherm, specific surface area and pore size distribution of the samples were measured by a 3Flex specific surface area pore analyzer (USA). The binding energy was tested by XPS (Thermoescalab 250Xi). The monochromatic alka hv = 1486.6 EV, power was 150 W, 500 um beam spots were used, and the binding energy was calibrated with C1s 284.8 eV. Nitrate and nitrite concentrations in the solution were determined by sulfamate spectrophotometry and N-(1-naphthyl)-ethylenediamine photometry, respectively. Nanoreagent spectrophotometry was used to test ammonia concentration in the solution. Fe^{2+} contents were determined by o-phenanthroline spectrophotometry (APHA, 2005). Total nitrogen (TN) was equal to the sum of NH_4^+-N, NO_3-N, and NO_2-N.

2.4. Batch Experiments for NO_3-N Removal Using nZVI/Zr-Ce-SBA-15 Composites

The stock solution of NO_3-N was prepared from $NaNO_3$ (analytical purity). The stock solution of 1000 mg /L NO_3-N was gradually diluted to 40, 60, and 80 mg/L. A total of 0.15 g of nZVI/Zr-Ce-SBA-15 composites was placed in a 250 mL glass bottles, and 50 mL of NO_3-N stock solution at 40, 60, and 80 mg/L concentrations (actual concentrations of 39.17, 60.82, and 79.65 mg/L) was, respectively, added into the above conical flask with a microinjector. The glass bottle was sealed and put in the thermostat water bath to remove NO_3-N. The samples were collected at certain intervals with a needle tube and immediately filtered with a 0.22 μm filter membrane. Immediately, the samples were analyzed by assessing the changes in NO_3-N, N_2 and NO_2-N, NH_4^+-N contents in the solution over time, thus obtaining the adsorption capabilities (q_e), removal efficiency (P), NO_2-N productivity ($S_{NO2\text{-}N}$), N_2 productivity (S_{N2}) and NH_4^+-N productivity ($S_{NH_4^+\text{-}N}$) (Equations (1)–(5)). The glass bottle and reaction solution were deoxidized before the reaction. The pH of the solution was adjusted by H_2SO_4 and $NaHCO_3$.

$$q_e = \frac{C_0 - C_{NO_3-N}}{m} \times v \quad (1)$$

$$P = \frac{C_0 - C_{NO_3-N}}{C_0} \times 100\% \quad (2)$$

$$S_{NH_4^+-N} = \frac{C_{NH_4^+-N}}{C_0 - C_{NO_3-N}} \quad (3)$$

$$S_{NO_2-N} = \frac{C_{NO_2-N}}{C_0 - C_{NO_3-N}} \quad (4)$$

$$S_{N_2} = \frac{C_0 - C_{NO_3-N} - C_{NO_2-N} - C_{NH_4^+-N}}{C_0 - C_{NO_3-N}} \tag{5}$$

where C_0 is the concentration of NO_3-N in the solution before adsorption, mg/L; $C_{NO3\text{-}N}$, C_{NO_2-N}, and $C_{NH_4^+-N}$ are the concentrations of NO_3-N, NO_2-N and NH_4^+-N in the solution after adsorption, mg/L; m is the mass of nZVI/Zr-Ce-SBA-15, g; q_e is the equilibrium adsorption quality of nZVI/Zr-Ce-SBA-15, mg/g; P is the adsorption efficiency; S_{NO_2-N}, S_{N_2}, and $S_{NH_4^+-N}$ represent the productivity of NO_2-N, N_2 and NH_4^+-N. The following steps were used for batch experiments using optimal nZVI/Zr-Ce-SBA-15. (1) Isothermal adsorption: the initial concentration gradients of NO_3-N were set to 20, 40, 60, 80 and 100 mg/L. (2) Adsorption kinetics: the adsorption times of NO_3-N were set to 1, 5, 10, 20, 30, 60, 120, and 180 min; other settings are the same (nZVI/Zr-Ce-SBA-15 = 0.15 g; initial pH = 4; temperature = 25 °C).

2.5. Modeling Methods of RSM, ANN-GA and RBFNN

The second-order model designed by Box–Behenken design (BBD) was used to study the relationship between response and variable values (Table 1). The selected variables are initial pH (x_1), contact time (x_2), temperature (x_3) and initial NO_3-N concentration (x_4). The response values are the removal rate of NO_3-N (y). The second-order model of BBD is shown in Equation (6):

$$y = \beta_0 + \sum_{i=1}^{k} \beta_i x_i + \sum_{i=1}^{k} \beta_{ii} (x_i)^2 + \sum_{i=1}^{k-1} \sum_{ij}^{k} \beta_{ij} x_i x_j + \varepsilon \tag{6}$$

where β_0, β_i, β_{ii} and β_{ij} are the intercept, primary term coefficient, secondary term coefficient and interaction coefficient, respectively, and ε is the test residual. The ANN-GA design was followed according to Xiang et al. [23] and ran in MATLAB 2016a. The modeling process of RBFNN is referred with [24,25].

Table 1. Parameter level of BBD experimental design.

Code	Parameters	Maximum	Middle	Minimum
x_1	Initial pH	5	4	3
x_2	Contact time (min)	70	60	50
x_3	Temperature (°C)	25	20	15
x_4	Initial NO_3-N concentration	80	60	40

3. Results and Discussion

3.1. Characterization of Zr-Ce-SBA-15 and nZVI/Zr-Ce-SBA-15 Composites

Figure 2 exhibits that the diffraction peak of nZVI is 44.88°, and its corresponding crystal plane is (100). After NO_3-N adsorption onto pure nZVI, the diffraction peak of nZVI was basically unchanged. A strong and broad-band satellite peak of Zr-Ce-SBA, nZVI/Zr-Ce-SBA, nZVI/Zr-Ce-SBA@NO_3-N composites was found in the range of 2θ = 20–30° due to the amorphous silica walls of mesoporous material [26]. In addition, a crystal plane (100) at 44.88° was also found in nZVI/Zr-Ce-SBA-15 and nZVI/Zr-Ce-SBA@NO_3-N composites, implying that nZVI was successfully supported onto Zr-Ce-SBA-15 and its valence state is α-Fe^0 [26].

The silica material exhibits two asymmetric stretching vibrational modes of siloxane moieties (–Si–O–Si–) at 1079.9 cm^{-1} and 814.2 cm^{-1}, and bending vibrational modes at 450.9 cm^{-1} [27,28] (Figure 3). The characteristic band of the hydroxyl group was found at 3455.8 cm^{-1} for the H_2O molecule, and the intense absorption band at 1637.2 cm^{-1} for the five materials is attributed to carbonyl groups, exhibiting that they are of the massive oxygen-containing unsaturated group. The peaks at 548 cm^{-1} and 621 cm^{-1} corresponding to Fe-O stretches of iron oxide were observed for the nZVI/Zr-Ce-SBA-15 and nZVI/Zr-

Ce-SBA-15 compositions, suggesting that nZVI was adsorbed onto Zr-Ce-SBA-15 and the oxidation occurred on the surface of nZVI particles. The peak for the Fe-O bond shifted to 636 cm^{-1} and 540 cm^{-1} after the adsorption of NO_3-N, indicating the transformation of different phases of iron oxides during aging. The characteristic band of nZVI and nZVI/Zr-Ce-SBA-15 at 1383.7 cm^{-1} is obviously enhanced after the adsorption of NO_3-N, which may be the characteristic nitro peak.

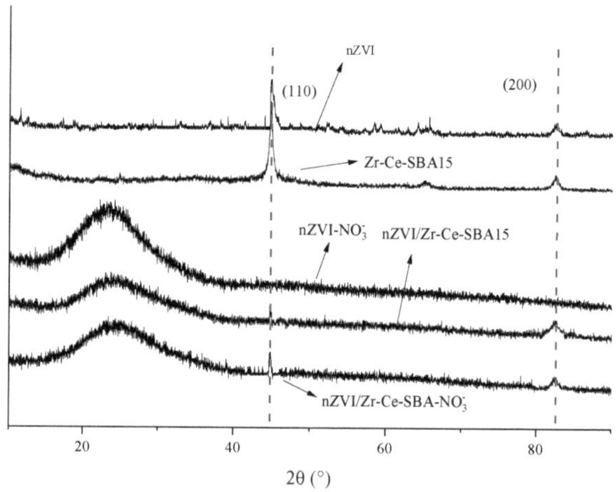

Figure 2. XRD characterization of nZVI/Zr-Ce-SBA-15 composites.

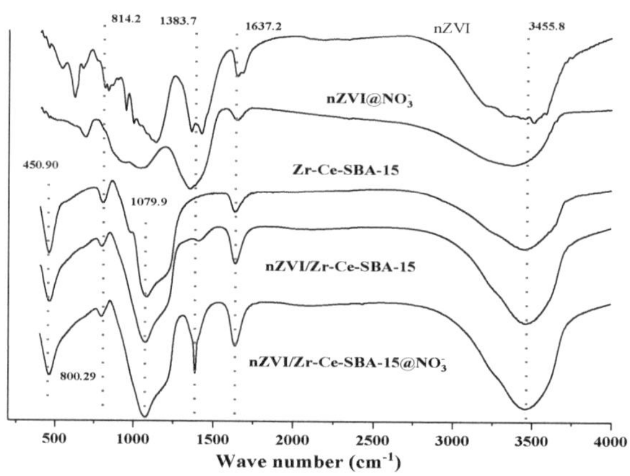

Figure 3. FTIR characterization of nZVI/Zr−Ce−SBA−15 composites.

Zr-Ce-SBA-15 is a hexagonal plate structure (Figure 4), with uniform size, a radial length of 1.5 μm and an axial length of 0.5 μm. There was no obvious change in the morphology of the two mesoporous materials after loading nZVI, indicating that the loading of nZVI does not damage the structure of mesoporous materials [29]. According to analysis of SEM-EDS, the element ratio of Zr-Ce-SBA-15 is 22.04% (C), 58.1% (O), 1.08% (N), 2.08% (Na), 13.79% (Si), 2.25% (Fe) and 0.66% (Zr). Apparently, the corresponding Si and Zr maps evidenced bright spots corresponding to the selected area and illustrated a homogeneous distribution of these elements in the field of view of the cross-section (Figure S1a,b).

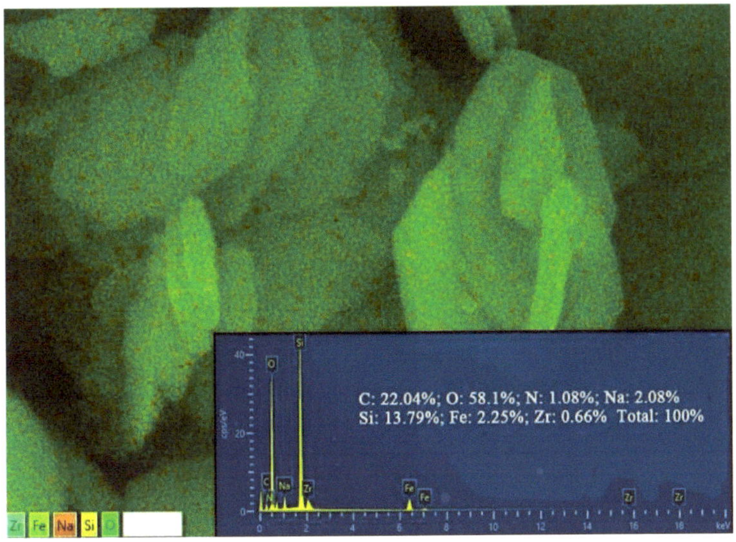

Figure 4. SEM characterization of nZVI/Zr-Ce-SBA-15 composites.

Figure 5a shows that Zr-Ce-SBA-15 has regular black-and-white stripes, with the hole wall and the mesoporous channel being black and white, respectively. The channel and the long axis direction are mutually parallel in the material. Figure 5b also exhibits that the mesoporous materials have long-range-ordered pores and a uniform pore size, which is consistent with the results of small-angle XRD in the previously published literature [22]. The channel direction of nZVI/Zr-Ce-SBA-15 is perpendicular to the hexagonal plate and parallel to the axial direction of the hexagonal plate. Figure 5b demonstrates that the composite material still has neat black-and-white stripes, indicating that the mesoporous structure is not damaged after the loading of nZVI particles; meanwhile, the material still maintains a certain order. There are a large number of black particles on the surface of the composite, which are nZVI particles on the carrier. These particles are of small size and evenly distributed on the carrier without obvious agglomeration. No nZVI particles dispersed outside were found in nZVI/Zr-Ce-SBA-15 composites, indicating that most of the nZVI particles in the composites synthesized entered the pores. The results further show that the method used in the present study can make it easier for metal particles to enter the carrier channel, limiting the chance of them remaining in the channel, and avoiding the growth and agglomeration of nZVI particles [30].

According to IUPAC, the curve of the nZVI/Zr-Ce-SBA-15 composites belongs to the type IV isotherm and had obvious pore condensation, indicating that the synthesized materials had a mesoporous structure (Figure S2). The existence of H1-type adsorption shows that the sample had a columnar pore structure; these characteristics are unique to mesoporous nZVI /Zr-Ce-SBA-15 materials [30]. In addition, compared to nZVI and Zr-Ce-SBA-15, the adsorption capacity of N_2 increased significantly. The specific surface areas of nZVI and nZVI/Zr-Ce-SBA-15 were 292 and 790 cm^2/g, respectively, and generally, the larger the specific surface areas of mesoporous materials, the higher the adsorption capabilities of pollutants are [30].

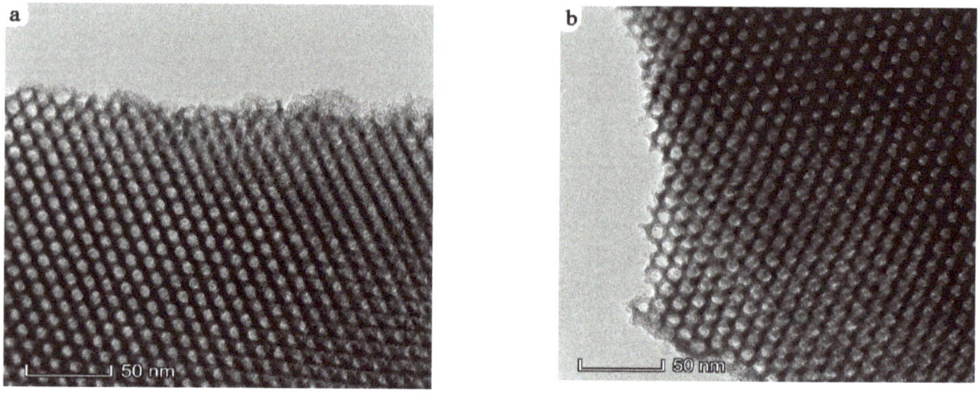

Figure 5. HRTEM characterization of Zr-Ce-SBA-15 composites (**a**) and nZVI/Zr-Ce-SBA-15 composites (**b**).

Figure 6 exhibits nZVI before and after loading onto Zr-Ce-SBA-15 composites. A weak peak for C1s and Si2p was found in the Zr-Ce-SBA-15 and nZVI/Zr-Ce-SBA-15 composites. Among them, the carbon that appears in the two samples results from the internal standard used to calibrate the binding energy of the other elements [31]. The existence of O1s confirmed that the two materials were of the massive oxygen-containing unsaturated groups, which are mainly from P123 and TEO. Obviously, before the nZVI loading onto Zr-Ce-SBA-15 composites, no peak was found in the range from 700 eV to 740 eV (Figure S3a). Further, after the nZVI loading onto Zr-Ce-SBA-15 composites, Fe(0), Fe(II) and Fe(III) were found at 710.5 eV, 711.8 eV, and 713.6 eV, suggesting that nZVI was loaded onto Zr-Ce-SBA-15 composites (Figure S3b). However, it was also found that nZVI had begun to be oxidized at this point.

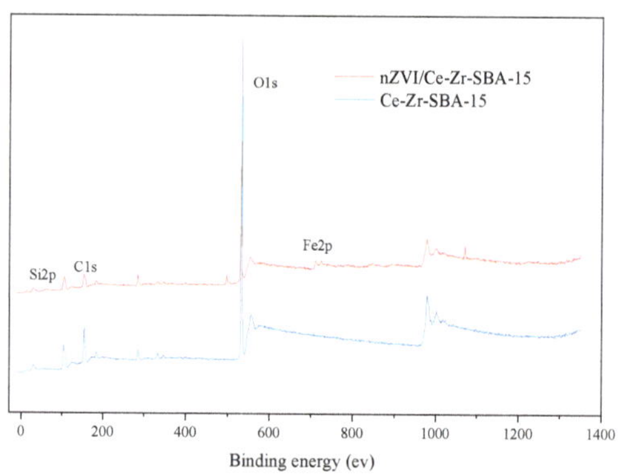

Figure 6. XPS characterization of Zr-Ce-SBA-15 and nZVI/Zr-Ce-SBA-15 composites.

3.2. Optimization of the NO_3-N Removal from Simulated Wastewater by RSM

Table 2 exhibits 29 groups of treatment designs using BBD. The experimental results are fitted by a quadratic polynomial, and the NO_3-N removal rate was selected as the response value (Equation (7)):

$$y = 87.18 + 4.82x_1 + 2.96x_2 - 1.45x_3 + 1.76x_4 - 1.81x_1x_2 + 2.17x_1x_3 + 6.61x_1x_4 + 4.89x_2x_3 - 1.01x_2x_4 - 1.22x_3x_4 - 8.52x_1^2 - 0.90x_1^2 - 2.22x_3^2 - 0.34x_4^2 \quad (7)$$

where y is the removal rate of NO_3-N, %; x_1, x_2, x_3 and x_4 are the coded values of the initial pH, contact time, temperature, and initial NO_3-N concentration, respectively. The positive sign (+) before the coefficient suggests a synergy between factors, and the negative sign (−) indicates the opposite relationship between factors. The absolute value of the coefficient can judge the effects of various factors on the adsorption effect of NO_3-N. Obviously, the order of the impact factors was initial pH > contact time > initial NO_3-N concentration > temperature. Yang et al. [32] optimized the nitrite-removal process from pickled meat by garlic by RSM. They found that the effect of extraction temperature on garlic clearance was the most significant, which showed that the surface was steep, followed by the amount of extraction solution, pH value of reaction solution and reaction time, and the relative effect of extraction time was the least. Rahdar et al. [33] used RSM to optimize the removal of NO_3-N by adsorption onto copper oxide nanoparticles. Their results suggest that the order of influence factors was NO_3-N concentration > initial pH > CuO-NPS dose > contact time. This is not inconsistent with this study, which shows that NO_3-N and nZVI were perhaps sensitive to pH in solutions.

Table 2. BBD test scheme and results.

Order	Contact Time (min)	Temperature (°C)	Initial NO_3-N Concentration(mg/L)	Initial pH	NO_3-N Removal Efficiency (%)
1	50	20	60	3	65.03
2	60	25	60	3	63.48
3	60	15	60	3	73.41
4	60	20	80	3	57.61
5	70	20	60	3	76.18
6	60	20	40	3	82.91
7	70	25	60	4	88.17
8	60	20	60	4	87.83
9	60	20	60	4	90.58
10	60	15	80	4	86.31
11	50	15	60	4	87.9
12	70	20	40	4	93.72
13	70	20	80	4	83.05
14	60	25	80	4	80.16
15	60	20	60	4	84.24
16	70	15	60	4	80.99
17	50	25	60	4	75.52
18	60	15	40	4	83.28
19	60	20	60	4	85.97
20	60	25	40	4	83.62
21	50	20	40	4	85.66
22	60	20	60	4	84.14
23	50	20	80	4	80.38
24	60	15	60	5	85.14
25	50	20	60	5	80.94
26	60	20	40	5	79.06
27	60	20	80	5	88.99
28	70	20	60	5	84.84
29	60	25	60	5	83.87

F and p values represent the significance of the developed model. According to the results of analysis of variance (Table 3), the adaptability of the developed model with NO_3-N removal is very significant (F = 1739.25, $p < 0.0001$). The mismatch term is less than 0.05, indicating that the model fitting is not significant, which also reflects the significant relationship between the factors described in the model and the response value. The determination coefficient ($R^2 - R^2adj = 0.0528 < 0.2$) is 0.9821, indicating that the experimental and predicted values are closed (Figure S4). The coefficient of variation (3.23%) is less than 10% and the precision of the analysis (17.28) is greater than 4, indicating that the reliability and accuracy of the model are high. Precision is the ratio of effective signal to noise, which is considered reasonable when its value is greater than 4. Overall, the quadratic models established by BBD can better fit the process of the NO_3-N removal using nZVI/Zr-Ce-SBA-15 composites, which can be used to optimize the experimental conditions.

Table 3. Response surface variance analysis results.

Source	Sum of Squares	Degree of Freedom	Mean Square	F Value	*p* Value	Source
Model	1739.25	14	124.23	17.95	<0.0001	significant
x	234.42	1	234.42	33.86	<0.0001	
x_1	88.75	1	88.75	12.82	0.003	
x_2	21.11	1	21.11	3.05	0.1027	
x_3	45	1	45	6.5	0.0231	
x_4	13.14	1	13.14	1.9	0.1899	
x_1x_2	18.75	1	18.75	2.71	0.1221	
x_1x_3	310.29	1	310.29	44.82	<0.0001	
x_1x_4	95.65	1	95.65	13.82	0.0023	
x_2x_3	7.26	1	7.26	1.05	0.3231	
x_2x_4	10.53	1	10.53	1.52	0.2378	
x_3x_4	470.78	1	470.78	68.01	<0.0001	
x_1^2	5.3	1	5.3	0.7663	0.3961	
x_2^2	31.98	1	31.98	4.62	0.0496	
x_3^2	2.4	1	2.4	0.3465	0.5655	
x_2^4	96.91	14	6.92			
Lack of Fit	67.55	10	6.76	0.9204	0.5862	not significant
Pure Error	29.36	4	7.34			
Cor Total	1836.16	28				

R^2 = 0.9472; Adj.R^2 value = 0.8944; Pre.R^2 value = 0.7631; adequate precision = 17.28; coefficient of variation = 3.23%.

According to the analysis of the quadratic fitting model, the optimized experimental conditions within the experimental range of this study are 24.35 °C (temperature), 69.98 min (contact time), 3.79 (initial pH), and 40.58 mg/L (initial NO_3-N concentration). In order to facilitate practical operation and analysis, the conditions can be adjusted to 24.4 °C, 70.00 min, pH = 3.8 and 40.6 mg/L, respectively, and the theoretical removal capacity under these conditions is 93.84%. The experiment was carried out under the above conditions and three groups were set up in parallel, and the experimental results were taken as the average. Finally, the optimized actual NO_3-N removal rate was 92.07%, and the absolute error between them was 1.77%. There are few studies on the removal of NO_3-N by RSM, and NO_2-N was therefore used for discussion in the present study. Rahdar et al. [33] utilized RSM to optimize NO_3-N removal with factors (input): initial NO_3-N concentration (20, 45, and 70 mg/L), CuO single-bond NP concentration (0.03, 0.06, and 0.08 g/L), pH (3, 7, and 11) and reaction time (40, 80, v120 min). The optimum conditions found were 70 mg/L for nanoparticles dose, 0.08 g/L for NO_3-N concentration, Ph = 3, and 120 min contact time. The above studies confirmed RSM as a reliable tool, and our results are not exactly the same as the two references, which may be due to the different subjects. Figure 7 demonstrates that the 3D response surface analysis concerning the interaction between the four factors affected the NO_3-N removal efficiency. The steeper the 3D surface, the greater the interaction between the two factors.

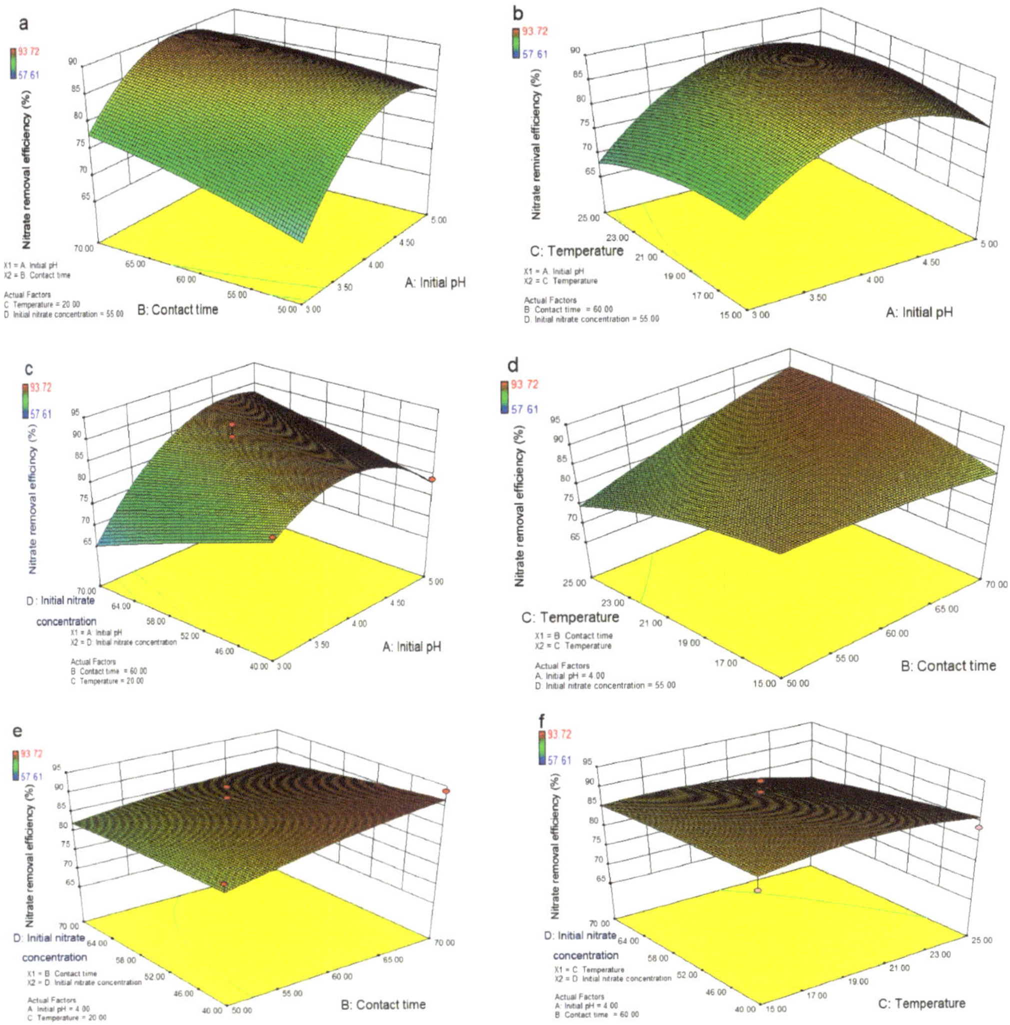

Figure 7. 3D response surface (**a**) of the effects of contact time and initial pH on removal efficiencies; 3D response surface (**b**) of the effects of temperature and initial pH on removal efficiencies; 3D response surface (**c**) of the effects of initial NO_3-N concentration and initial pH on removal efficiencies; 3D response surface (**d**) of the effects of temperature and contact time on removal efficiencies; 3D response surface (**e**) of the effects of initial NO_3-N concentration and contact time on removal efficiencies; 3D response surface (**f**) of the effects of initial NO_3-N concentration and temperature on removal efficiencies.

3.3. Optimization of the NO_3-N Removal from Simulated Wastewater by ANN-GA and RBF

The 29-group dataset designed by BBD in RSM was used as the source of the sample set. From 29 groups of BBD, 24 groups were randomly selected as the input vector of the training sample set, and the remaining 5 groups were used as the input vector of the prediction sample set. The error value between the actual and target outputs of the training sample in the neural network reflects the network training accuracy. A smaller value can mean that the number of network training steps is too large and time consuming. The higher error value will render the network incapable of evaluating the nonlinear mapping relationship accurately and cannot realize the simulation and prediction function. Therefore, referring to the work of Xiang et al. [23], the parameters of the developed ANN model are as the follows: epoch—2000; learning rate—0.1; goal—1×10^{-5}; and momentum factor—0.9 (Figure S5a). Most processing methods transform the original test data into 0 to 1 in a specific way to improve the network training effect and network simulation ability. In this study, to eliminate the numerical difference between the parameters of the input and the layers, the test data were first normalized using the mapminmax function covered by the Matlab toolbox before the network operation. The model developed by the ANN is a 'black box model', and its R^2 value is 0.99159, indicating that the experimental and predicted values of the model are highly consistent (Figure S6a) A lower or higher number of neurons in the hidden layer may lead to insufficient fitting or overfitting for the developed ANN model, respectively, reducing the prediction ability of model. The present study selected six neurons used in the hidden layer based on the lowest MSE value (Figure S6b).

GA was used to optimize the established BP neural network model and its basic parameters of GA by referring to Xiang et al. [23] (population size = 20; cross probability = 0.8; mutation probability = 0.01; genetic probability = 0.9; maxgen = 500). The network model is called fitness function. In the initial stage, the population search characteristic of GA has an obvious role, showing that the NO_3-N removal rate of the selected individuals has increased sharply. Then, GA was used to conduct a stable cross operation, making it so that the NO_3-N removal rate did not change. The optimization process of GA was used to carry out a selection operation, and the NO_3-N removal rate of the selected individual had a small range of positive changes, approaching the target step by step. The curve change process in Figure 8 shows that the NO_3-N removal rate converges at 93.45% at the 50th iterations. After a series of cyclic iterative operations, including crossover, selection, mutation, crossover, selection, the program of GA was terminated and obtained the highest NO_3-N removal rate with its optimization evolution iterations reaching 105 generations. The removal rate of NO_3-N by nZVI/Zr-Ce-SBA-15 composites reached 95.71% under the optimal parameters: initial pH = 4.89; contact time = 62.27 min; initial NO_3-N concentration = 74.84 mg/L; temperature = 24.77. Based on the above parameters, the experimental removal rate of NO_3-N is 94.64%, and the absolute error of removal efficiencies between the predicted and actual values is merely 1.07%. RBFNN is superior to the BP neural network in terms of approximation ability, classification ability and learning speed. It has a simple structure, is simple the train, and has a fast learning convergence speed, which can approximate any nonlinear function and overcome the problem of a local minimum. The reason for this is that its parameter initialization has a distinctive method, not a random initialization [34]. The parameters of the developed RBFNN model were set as the follows: spread = 0.8; err goal = 1×10^{-10}. After obtaining the predicted results of ANN-GA, the developed RBFNN model was used to verify the accuracy of them again. The training process was terminated after the MSE value reached 0.0226085 (Figure S5b), and the predicted and actual values were very consistent, confirming the reliability of the ANN-GA model (Figure S7). Cai et al. [35] reported that the prediction of the performance of simultaneous anaerobic sulfide and NO_3-N removal in an upflow anaerobic sludge bed reactor through an ANN. Their results showed that the ANN model predicted the simultaneous sulfide and NO_3-N removal process accurately. Lee et al. [36] studied NO_3-N sorption onto quaternary ammonium-functionalized poly(amidoamine) dendrimer, generation 2.0. Based on a pH experiment, these materials maintained a relatively stable NO_3-N sorption capacity in the pH range from 2 to 10. They also suggested that the developed ANN model (R^2 = 0.872) predicted more accurately than the RSM model (R^2 = 0.790) for the additional multi-parameter experimental data. Multi-parameter experiments and modeling for NO_3-N sorption were conducted on quaternary ammonium-functionalized poly(amidoamine) dendrimers in aqueous solutions. Due to its nonlinear fitting, the performance of ANN is obviously better than that of RSM. Liu et al. [37] introduced two widely used ANN models, the basic theories of BP and RBF neural networks, and expounded the learning process of the two algorithms from a mathematical point of view. The simulation results show that the generalization ability of RBF is better than the BP network in many aspects, but the structure of BP network is simpler than RBF network when solving the problem with the same accuracy requirements.

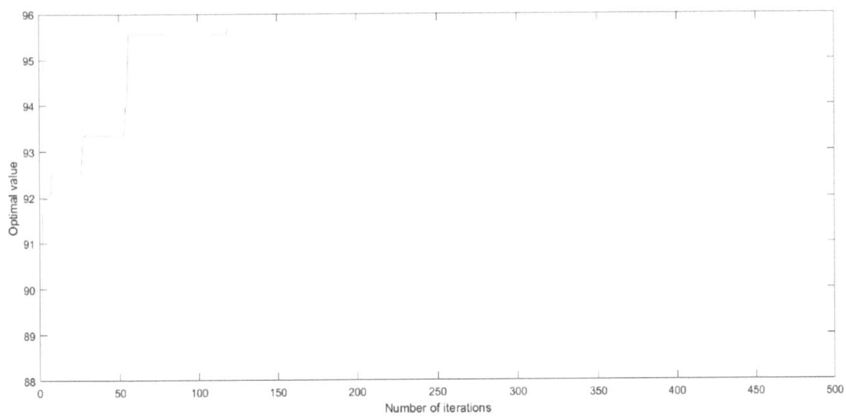

Figure 8. The predicted optimal NO_3-N removal efficiency by ANN-GA model.

3.4. *Kinetic Model, Isothermal Adsorption and Thermodynamic Parameters for NO_3-N Removal*

The kinetics process is mainly used to describe the rate of solute adsorption by adsorbent. The adsorption mechanism can be inferred by fitting the data to the kinetic model. The experimental data were fitted by two kinetic models (Equations (8) and (9)) [38].

$$\ln(q_e - q_t) = \ln q_e - k_1 t \quad (8)$$

$$\frac{t}{q_t} = \frac{1}{k_2 q_e^2} + \frac{t}{q_e} \quad (9)$$

where q_e and q_t are the adsorption capacity (mg/g) at equilibrium and time t, respectively; k_1 (1/min) is the rate constant of the pseudo-first-order kinetic model; k_2 is the rate constant for the pseudo-second-order kinetic model (g/mg·min^{-1}).

The capacities of NO_3-N adsorption onto nZVI/Zr-Ce-SBA-15 composites are shown in Figure 9a. The adsorption process with an initial NO_3-N concentration of 60 mg/L is divided into three stages, including rapid adsorption in the first 50 min, slow adsorption in the following 60 min and final apparent equilibrium adsorption. This is mainly because there are many adsorption sites on nZVI/Zr-Ce-SBA-15 composites in the initial stage. With the continuous reaction, the adsorption sites gradually decrease and reach the saturation state. The slow adsorption for 60 min may be because, with the continuous reaction, NO_3-N further diffuses into the internal pores of nZVI/Zr-Ce-SBA-15 composites and reacts with the active sites on the inner surface. A smaller internal pore size leads to difficult diffusion, making that the reaction slowed down. Compared with the pseudo first-order kinetic model, the pseudo second-order kinetic model can better fit the adsorption process ($R^2 > 0.99$), and its adsorption capacity (43.7) was nearer to the experimental value (42.6), indicating that the adsorption process was mainly controlled by chemical adsorption (Figure 9b). Zhao et al. [39] used green tea extract as a reducing agent to synthesize a nanoscale zero-valent iron–nickel bimetallic to remove NO_3-N from groundwater. The kinetic study showed that, under the optimum conditions, the removal process of NO_3-N by this material conforms to the pseudo-second-order adsorption kinetic model, which was dominated by adsorption and accompanied by the reduction reaction. Wang et al. [40] used ZVI supported on biochar composites to remove NO_3-N from groundwater. Their results demonstrated that the removal NO_3-N in water by these materials can meet the pseudo-first-order reaction only when the mass of the composite and pollutant is relatively large. When the second-order reaction is used to simulate the reduction reaction, the reaction rate decreases obviously. Therefore, the reaction process cannot be described by the reaction order alone.

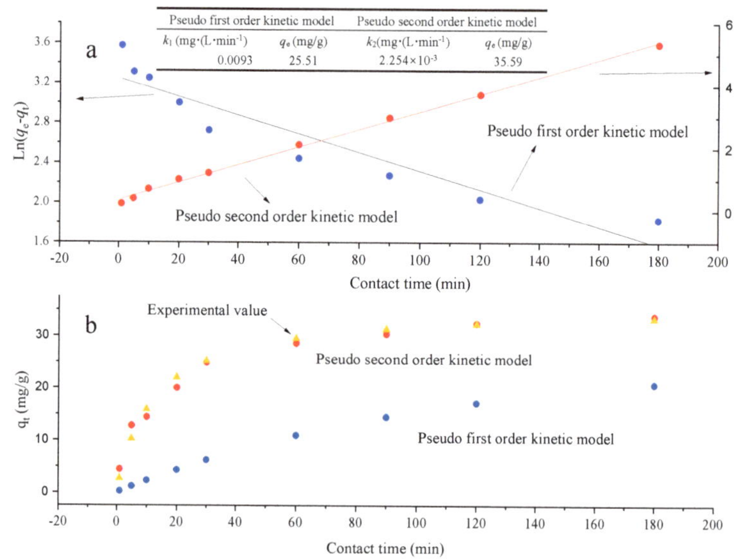

Figure 9. Comparison of pseudo-first and second-order kinetic models (**a**) and the capacities of NO_3-N adsorption onto nZVI/Zr−Ce−SBA−15 composites (**b**) (nZVI/nZVI/Zr−Ce−SBA−15 = 0.15 g; contact time = 1−180 min; initial NO_3-N concentration = 60 mg/L and initial pH = 4; temperature = 25 °C).

nZVI/Zr-Ce-SBA-15 was simultaneously adsorbed and catalyzed in the process of NO_3-N removal. Therefore, isothermal adsorption models were not suitable to describe the adsorption behavior of nZVI/Zr-Ce-SBA-15. The present study investigated the relationship between the equilibrium concentration of NO_3-N in the solution and the adsorption capacity of Zr-Ce-SBA-15 using isothermal adsorption models, and thus indirectly reflecting the adsorption capacity of NO_3-N on nZVI/Zr-Ce-SBA-15. The results show that the adsorption capacity increases with increasing NO_3-N in the solution. The adsorption capacity was not unchanged when the NO_3-N concentration in the solution reached a certain value. This may be because, with the increase in NO_3-N concentration, the adsorption sites on the surface of nZVI/Zr-Ce-SBA-15 were gradually occupied by NO_3-N under the constant doses of the adsorbents. The Langmuir and Freundlich isothermal adsorption models are as follows (Equations (10) and (11)) [41]:

$$\frac{C_e}{q_e} = \frac{1}{K_L q_{max}} + \frac{C_e}{q_{max}} \tag{10}$$

$$Lnq_e = LnK_F + \frac{1}{n}LnC_e \tag{11}$$

where C_e is the equilibrium concentration of NO_3-N in solution (mg/L), q_e is the amount of NO_3-N adsorbed (mg/g), q_{max} is q_e for a complete monolayer (maximum adsorption capacity) (mg/g) and K_L is the adsorption equilibrium constant (L/mg).

Figure 10a shows that the fitting correlation coefficient (R^2) of Langmuir and Freundlich isothermal adsorption models of NO_3-N adsorption on Zr-Ce-SBA-15 composites in simulated wastewater is greater than 0.900, indicating that their models can be better used to describe the adsorption process. A Langmuir adsorption isothermal model assumes that the adsorbent surface is uniform, the adsorption energy is the same everywhere, and the adsorption of the adsorbent to adsorbate belongs to monolayer adsorption [42]. Compared with the Freundlich model, the Langmuir model has a higher fitting correlation coefficient for the process of NO_3-N adsorption, implying that the NO_3-N adsorption onto Zr-Ce-SBA-15 is followed more closely by the Langmuir model and belongs to monolayer adsorption (Figure 10b). The calculated R_L value is between 0 and 1, indicating that the adsorption process is preferential adsorption. According to the Langmuir model, the maximum adsorption capacity of NO_3-N in simulated wastewater by Zr-Ce-SBA-15 composites is 45.25 mg/g, which is more than that of most adsorbents (Table 4), indirectly confirming that the maximum removal capability of nZVI/Zr-Ce-SBA-15 was higher than 45.24 mg/g, Figure 10a. The n value in the Freundlich model can be used as an indicator of the adsorption of NO_3-N by Zr-Ce-SBA-15 composites. The smaller the n value is, the greater the adsorption capacity is, and the larger the K_F value is, the greater the adsorption capacity is. Meng et al. [42] utilized the in situ growth synthesis of a CNTs@AC hybrid material for efficient nitrate–nitrogen adsorption. The isothermal analysis shows that NO_3-N adsorption is made up of multiple processes with a maximum adsorption capacity of 27.07 mg·g^{-1}. He et al. [43] studied the absorption characteristics of bamboo charcoal for NO_3-N. The Langmuir fitting results showed that K_L decreased linearly with increasing temperature, and the increase in temperature was not conducive to the adsorption of NO_3-N by bamboo charcoal. The pH value also affected the adsorption effect. When the pH value is 2.0, the adsorption effect of bamboo charcoal toward NO_3-N is at its best.

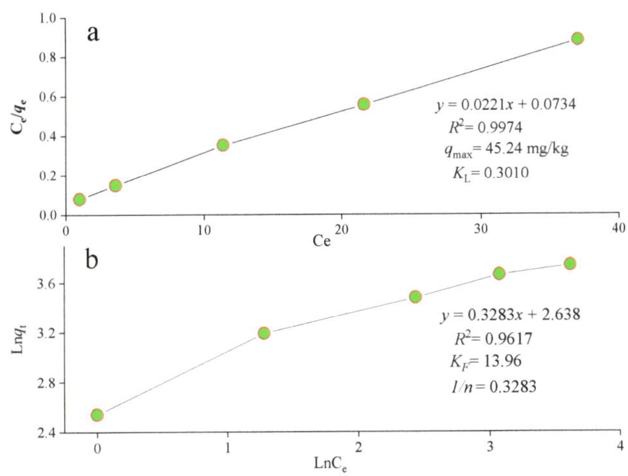

Figure 10. The fitting processes the Langmuir (**a**) and Freundlich (**b**) isothermal adsorption models (Zr-Ce-SBA-15 = 0.15 g; contact time = 180 min; temperature = 25 °C; initial pH = 4).

Adsorption thermodynamics are an important factor for the adsorption mechanism. The parameters, Gibbs-free energy change (ΔG^0), enthalpy change (ΔH^0) and entropy change (ΔS^0) are used to describe the effect of temperature on the adsorption equilibrium. ΔG^0 (J/mol) can judge the spontaneity of the adsorption reaction. The ΔH^0 (kJ/mol) can judge whether the adsorption reaction is an endothermic process or an exothermic process. The ΔS^0 (J/mol·K) represents the change of the degree of freedom of the system. Their calculated mean is found through the following equations, Equations (12)–(14) [23]:

$$\ln K_D = \frac{\Delta S^0}{R} - \frac{\Delta H^0}{RT} \tag{12}$$

$$K_D = \frac{C_0}{C_e} \tag{13}$$

$$\Delta G^0 = \Delta H^0 - T\Delta S^0 \tag{14}$$

where ΔH^0 is the standard enthalpy change (kJ/mol) and ΔS^0 is the standard entropy change ($kJ \cdot mol^{-1} \cdot K^{-1}$). The values of ΔH^0 and ΔS^0 can be obtained from the slope and intercept of a plot of lnK_0 against 1/T. ΔG^0 is the standard free energy change (kJ/mol), T is the temperature (K) and R is the gas constant (8.314 $J \cdot mol^{-1} \cdot K^{-1}$).

Table 4. Comparing the parameters of Langmuir and Freundlich isothermal adsorption models and the maximum adsorption capabilities with those of other studies [44–46].

Materials	Initial NO_3^--N Concentration (mg/L)	q_{max} (mg/g)/Removal Efficiency (%)
Trimethyl quaternary ammonium functionalized SBA-15 (C1Q-SBA1)	10–1000	89.4
Dimethylbutyl quaternary ammonium functionalized SBA-15 (C4Q-SBA1)	10–1000	129.9
Dimethyloctyl quaternary ammonium functionalized SBA-15 (C8Q-SBA1)	10–1000	136.4
Dimethyldodecyl quaternary ammonium functionalized SBA-15 (C12Q-SBA1)	10–1000	71.9
Dimethyloctadecyl quaternary ammonium functionalized SBA-15 (C18Q-SBA1)	10–1000	30.7
Propyl-ammonium SBA-15 G	100–700	55.24
Propyl-ammonium SBA-15 C	100–700	45.66
Propyl-N,N,N-trimethylammonium functionalized SBA-15 G	100–700	62.50
Propyl-N,N,N-trimethylammonium functionalized SBA-15 C	100–700	38.75
Fe-SBA-15	7.5 (15 min)	67.3%
Fe-SBA-15	12.5 (min)	50%
Fe-SBA-15	20 (min)	46.3%
Fe-SBA-15	30 (min)	43.3%
BC-Fe	20–1000 (24 h)	51.19
Zr-Ce-SBA-15	60	45.24 (In this study)

Figure S8 shows that the process of NO_3-N adsorption onto nZVI/Zr-Ce-SBA-15 composites is spontaneous. With the increase in temperature, ΔG^0 gradually decreased and the adsorption spontaneity gradually increased, indicating that the increase in temperature is more conducive to adsorption. Generally, when the ΔH^0 is less than 40 kJ/mol, adsorption is controlled by van der Waals' forces and is due to to physical adsorption. When 50 kJ/mol $< \Delta H^0 <$ 200 kJ/mol, the adsorption is controlled by chemical bonds and is due to physical adsorption [47]. The adsorption of nZVI/Zr-Ce-SBA-15 onto NO_3-N is mainly chemical adsorption, which is also in good agreement with the results of pseudo-second-order kinetic fitting. This may be due to the interaction between a small amount of organic functional groups on nZVI/Zr-Ce-SBA-15 composites or nZVI and NO_3-N. In addition, due to $\Delta H^0 > 0$, this adsorption process in the present study is endothermic, and heating is beneficial for the adsorption process. $\Delta S^0 > 0$ in the present study also indicates that the reaction process enhances the disorder degree of the solid–liquid interface, exhibiting that the adsorption process easily continues. Overall, this adsorption process was spontaneous, endothermic and entropy-driven.

3.5. The Balance of NO_3-N Removal, Ammonia Nitrogen and Nitrogen Generation

Figure S9a shows that nZVI/Zr-Ce-SBA-15 was simultaneously undergoing adsorption, reduction, and catalysis in the process of NO_3-N removal because the removal capabilities of NO_3-N using nZVI/Zr-Ce-SBA-15 are obviously higher than those when using Zr-Ce-SBA-15. Although the possible product, N_2, in the reaction process was not collected and detected, the part with insufficient total nitrogen balance is likely to be N_2 (g), and research results have showed that Fe^0 can be reduced to N_2 by NO_3-N, and a very small amount of NO_2^- is generated in the reaction process. With the progress of the reaction, NO_2^- is also reduced to N_2 or NH_4^+, and NO_2^- is the intermediate product of nZVI reducing NO_3^--N. In addition to the direct reduction function of Fe^0, it also can dissolve to produce Fe^{2+} in acidic water environment. Chen et al. [48] suggested that Fe^{2+} can significantly improve the removal rate of Fe^0 for NO_3^-, and the higher the concentration of Fe^{2+} was, the higher the removal efficiency of NO_3^- was. Because the non-acidified Fe^0 has an oxide film, the reduction rate of NO_3^- in the initial stage of the reaction was slow. Fe^{2+} converted the Fe_2O_3 oxide film on the surface of nZVI into Fe_3O_4, accelerating the transfer of electrons from Fe^0 to NO_3^-, and promoting the reduction of NO-3. Therefore, the dissolution of Fe^{2+} in the nZVI/Zr-Ce-SBA-15-NO_3-N system may indirectly accelerate or catalyze NO_3-N removal, which can be confirmed by the Fe^{2+} concentration increasing firstly and then decreasing in Figure S9b. In addition, cerium oxide is the most stable oxide among rare-earth elements. As an active site of surface, the concentration of oxygen vacancy is directly related to the catalytic performance of decorated catalysts. Based on defect chemistry, other metal ions are introduced into the oxide lattice to form solid solutions, which induce more abundant oxygen vacancies, thereby improving the migration rate of surface oxygen and lattice oxygen. Liu et al. [49] reported a promotion effect of cerium and lanthanum oxides on the Ni/SBA-15 catalyst for ammonia decomposition. Zhang et al. [50] exhibited a promoting effect of cerium doping on iron–titanium composite oxide catalysts for the selective catalytic reduction of NOx with NH_3. Combined with the published literature, the existence of Ce in nZVI/Zr-Ce-SBA-15 may act as a catalyst in the NO_3-N removal process. After NO_3-N adsorption, reduction and catalysis onto Zr-Ce-SBA-15 mesoporous materials, the proportion of N was significantly increased in comparison with before removal and pure Zr-Ce-SBA-15 mesoporous materials (Figure 11). This was indicted that the massive NO_3-N were adsorbed, reduced, and catalyzed onto Zr-Ce-SBA-15 mesoporous materials.

Figure 12 shows that the removal rate of NO_3-N by nZVI can reach 87%, and it cannot completely remove NO_3-N. This is probably due to the easy agglomeration of nZVI, as the effective active sites are reduced. nZVI/Zr-Ce-SBA-15 composites can improve the NO_3-N removal rate, as well as the NH_4^+-N and N_2 generation rate, which is because nZVI loaded onto Zr-Ce-SBA-15 as a reducing agent reduced agglomeration, providing more active sites, and thus improving the removal efficiency. In addition, Zr-Ce-SBA-15 improves the dispersion of nZVI, allowing it to better react with N_2 and enhance its selectivity. In particular, the selectivity of N_2 is 19.54% after NO_3-N adsorption onto Zr-Ce-SBA-15. Zha et al. [51] reported that NO_3-N from industrial wastewater was removed through Fe/Cu composites. A primary cell structure formed by iron–copper enhanced the effect of nZVI on NO_3-N removal. The removal NO_3-N efficiency and N_2 selectivity are 40.94% and 28.96%, respectively. There is little NO_2-N content in the process of NO_3-N removal using nZVI/Zr-Ce-SBA-15 composites. This may be because NO_2-N is only the intermediate product of NO_3-N being reduced to NH_4^+-N and will not exist stably in the reduction process. In the removal process, the removal rate of NO_3-N is greater than the generation rate of NH_4^+-N. This is because there is a certain order of reaction between the two substances. NO_3-N is first adsorbed on the surface of the composites, and then it is reduced by nZVI to NH_4^+-N. After the reaction, most of the NO_3-N removed is reduced to NH_4^+-N by nZVI, indicating that NH_4^+-N is the main product when nZVI reduces NO_3-N, which is in accordance with Song et al. [52] and Shi et al. [53]. Hwang et al. [54] found that with the increase in the pH of the reaction system, ammonium ions will discharge the solution in the form of ammonia with mechanical stirring. Chao et al. [55] found that the final product of NO_3-N reduction by nZVI is N_2, and there is no generation of NH_4^+-N. In addition, the yield of N_2 was been improved and the present study showed the two products were more likely to coexist in the adsorption process. Referring to the works of Wang et al. [40] and Pan et al. [56], the mechanism of this adsorption process in the present study is exhibited in Figure 12.

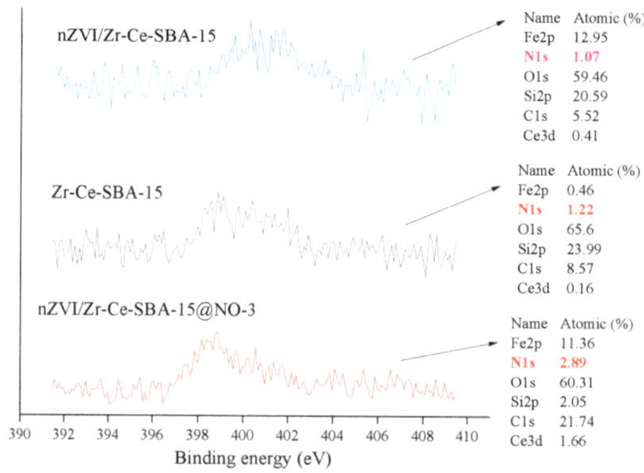

Figure 11. The XPS characterization of magnified N1S spectrogram of Zr-Ce-SBA-15, nZVI/Zr-Ce-SBA-15 and nZVI/Zr-Ce-SBA-15@NO_3-N composites.

3.6. *Stability and Regeneration of nZVI/Zr-Ce-SBA-15*

The desorption rate is an important index reflecting the economy of the adsorbent. Therefore, five adsorption/desorption cycle tests were conducted to study the reusability of nZVI/Zr-Ce-SBA-15 (Figure S10). The adsorption rate of nZVI/Zr-Ce-SBA-15 to NO_3-N decreased when desorbing with 0.5 mol/L HCI, while the effective removal of NO_3-N could still be maintained in the first three experiments (removal rate > 80%). After repeated desorption three times, the adsorption rate decreased significantly and tended to be stable, and the removal rate of NO_3-N was merely 65%. This may be because some H^+ occupies the surface binding sites during acid desorption. Although the treated adsorbent was washed many times to reach neutrality and for reuse, the effect of acid treatment is irreversible, resulting in the decrease in the removal rate. Another possible reason for this is that, due to incomplete acid desorption, some NO_3-N still occupied the adsorption sites, resulting in the reduction in adsorption sites. Five adsorption/desorption tests show that nZVI/Zr-Ce-SBA-15 composites have strong renewability in the short term, and repeated recycling will reduce its adsorption capacity.

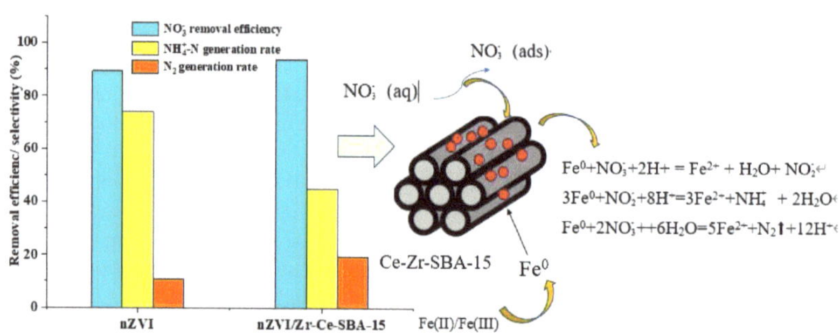

Figure 12. The efficiency/selectivity of nZVI/Zr−Ce−SBA−15 used for NO_3-N removal and its mechanism (nZVI/Zr−Ce−SBA−15 = 3 g/L; contact time = 0−8 h, temperature = 298 K and initial NO_3-N concentration = 60 mg/L).

4. Conclusions

In the present study, nZVI/Zr-Ce-SBA-15 composites were prepared and used to remove NO_3-N from simulated wastewater. The nZVI was successfully supported onto Zr-Ce-SBA-15, confirmed using XRD, FTIR, TEM, SEM–EDS, N_2 adsorption and XPS, which showed them to be ordered mesoporous materials. The ANN model was better than the RSM model. The developed ANN-GA model exhibited that the removal rate of NO_3-N by nZVI/Zr-Ce-SBA-15 reached 95.71% under the following optimal parameters: initial pH = 4.89, contact time = 62.27 min, initial NO_3-N concentration = 74.84 mg/L and temperature = 24.77. Based on the above parameters, the experimental removal rate of NO_3-N was 94.64%. Moreover, the RBF neural network further confirmed the reliability of the ANN-GA model. Pseudo-second-order kinetics can better describe the behavior of NO_3-N adsorption onto nZVI/Zr-Ce-SBA-15 composites, and this reaction was spontaneous, endothermic and entropy-driven. The process of NO_3-N adsorption onto Zr-Ce-SBA-15 composites was followed by the Langmuir model, and its maximum adsorption capacity was 47.17 mg/g. It was indirectly confirmed that the maximum removal capacity of nZVI/Zr-Ce-SBA-15 exceeded this value because the removal efficiency of NO_3-N using nZVI/Zr-Ce-SBA-15 was obviously higher than that when using Zr-Ce-SBA-15. The yield of N_2 can be improved after nZVI is supported on Zr-Ce-SBA-15, and the composites exhibited strong renewability in the short term within three cycles. The resolution of Fe ions experiments confirmed that nZVI/Zr-Ce-SBA-15 simultaneously underwent adsorption and catalysis in tprocess of NO_3-N removal. Although many methods have been used to remove NO_3-N from water, there are few studies on how to convert NO_3-N into versatile N_2. Several researchers found that the inclusion of precious metals in materials, such as Cu, Pd, Pt, etc., can significantly improve the selectivity of N_2, which is also the next direction to enhance the selectivity of nZVI/Zr-Ce-SBA-15 to N_2 in the next study.

Supplementary Materials: The following supporting information can be downloaded at: https://www.mdpi.com/article/10.3390/catal12070797/s1, Figure S1: SEM characterization of Si (a) and Zr (b) distribution of nZVI/Zr-Ce-SBA-15 composites; Figure S2 Characterization of N2 adsorption for nZVI/Zr-Ce-SBA-15 composites; Figure S3 XPS characterization of magnified Fe2p spectrogram of Zr-Ce-SBA-15 (a) and nZVI/Zr-Ce-SBA-15 (b); Figure S4. Comparison with the predicted and actual values; Figure S5 The training performance of ANN (a) and RBFNN (b); Figure S6 The correlation coefficient between target and output values (a) and the lowest MSE value in ANN model; Figure S7 The comparison of predicted and actual values by RBFNN; Figure S8 The fitting process and parameters of thermodynamics; Figure S9 The comparison of removal efficiency of NO_3-N using Zr-Ce-SBA-15 and nZVI/Zr-Ce-SBA-15 composites (a) and Fe^{2+} dissolution of nZVI/Zr-Ce-SBA-15 in solution (b) (Zr-Ce-SBA-15 or nZVI/Zr-Ce-SBA-15 = 0.15 g; initial NO_3-N concentration = 60 mg/L and initial pH = 4; temperature = 25 °C); Figure S10 The adsorption and desorption efficiencies of nZVI /Ce-Zr-SBA-15.

Author Contributions: Conceptualization, R.Z. and H.L.; methodology, H.L.; software, W.J. and W.L.; validation, R.Z., W.J. and W.L.; formal analysis, R.Z.; investigation, R.Z.; resources, R.Z.; data curation, R.Z. All authors have read and agreed to the published version of the manuscript.

Funding: This research was funded by Open Research Found of Jiangsu Province Key Laboratory of Environmental Engineering, [HX2017005] and [KF2015008].

Institutional Review Board Statement: Not applicable.

Informed Consent Statement: Not applicable.

Data Availability Statement: Not applicable.

Conflicts of Interest: The authors declare no conflict of interest.

Sample Availability: Not applicable.

References

1. Jiang, H.; Zhang, Q.; Liu, W.; Zhang, J.; Xu, Z. Isotopic compositions reveal the driving forces of high nitrate level in an urban river: Implications for pollution control. *J. Clean. Prod.* **2021**, *298*, 126693. [CrossRef]
2. Sendrowski, A.; Castaeda-Moya, E.; Twilley, R.; Passalacqua, P. Biogeochemical and hydrological variables synergistically influence nitrate variability in coastal deltaic wetlands. *J. Geophys. Res. Biogeo.* **2021**, *126*, e2020JG005737. [CrossRef]
3. Zhang, X.; Zhang, Y.; Shi, P.; Bi, Z.; Ren, L. The Deep Challenge of Nitrate Pollution in River Water of China. *Sci. Total Environ.* **2021**, *770*, 144674. [CrossRef] [PubMed]
4. Dinelli, E.; Testa, G.; Cortecci, G.; Barbieri, M. Stratigraphic and petrographic constraint to traced element and isotope geochemistry of Messinian sulphates of Tuscany. *Mem. Soc. Geol. Ital.* **1999**, *54*, 61–74.
5. Preetha, P.; Al-Hamdan, A. Integrating finite-element-model and remote-sensing data into SWAT to estimate transit times of nitrate in groundwater. *Hydrogeol. J.* **2020**, *28*, 2187–2205. [CrossRef]
6. Gu, B.; Ying, G.; Chang, S.; Luo, W.; Jie, C. Nitrate in groundwater of China: Sources and driving forces. *Glob. Environ. Chang.* **2013**, *23*, 1112–1121. [CrossRef]
7. Lundberg, J.; Weitzberg, E.; Gladwin, M. The nitrate-nitrite-nitric oxide pathway in physiology and therapeutics. *Nat. Rev. Drug Discov.* **2008**, *7*, 156–167. [CrossRef]
8. Cosby, K.; Partovi, K.; Crawford, J.; Patel, C.; Martyr, S.; Yang, B.; Waclawiw, M.; Zalos, G.; Xu, X. Nitrite reduction to nitric oxide by deoxyhemoglobin vasodilates the human circulation. *Nat. Med.* **2003**, *9*, 1498–1505. [CrossRef]
9. Zhang, F.; Yi, W.; Cao, J.; He, K.; Liu, Y.; Bai, X. Microstructure characteristics of tea seed dietary fibre and its effect on cholesterol, glucose and nitrite ion adsorption capacities in vitro: A comparison study among different modifications. *Int. J. Food Sci. Technol.* **2020**, *55*, 1781–1791. [CrossRef]
10. Babu, D.; Ravindhranath, K.; Mekala, S. Simple effective new bio-adsorbents for the removal of highly toxic nitrite ions from wastewater. *Biomass Convers. Biorefin.* **2021**, *12*, 1–13. [CrossRef]
11. Wang, Z.; Richards, D.; Singh, N. Recent discoveries in the reaction mechanism of heterogeneous electrocatalytic nitrate reduction. *Catal. Sci. Technol.* **2021**, *11*, 705–725. [CrossRef]
12. Dhakal, P.; Coyne, M.; Mcnear, D.; Wendroth, O.; Vandiviere, M.; D'Angelo, E.; Matocha, C. Reactions of Nitrite with Goethite and Surface Fe (II)-Goethite Complexes. *Sci. Total Environ.* **2021**, *782*, 146406. [CrossRef] [PubMed]
13. Crane, R.; Scott, T. Nanoscale zero-valent iron: Future prospects for an emerging water treatment technology. *J. Hazard. Mater.* **2012**, *211–212*, 112–125. [CrossRef] [PubMed]
14. Zhou, Y.; Li, X. Green synthesis of modified polyethylene packing supported tea polyphenols-NZVI for nitrate removal from wastewater: Characterization and mechanisms. *Sci. Total Environ.* **2022**, *806*, 150596. [CrossRef]
15. Zhang, Y.; Li, Y.; Li, J.; Hu, L.; Zheng, X. Enhanced removal of nitrate by a novel composite: Nanoscale zero valent iron supported on pillared clay. *Chem. Eng. J.* **2011**, *171*, 526–531. [CrossRef]
16. Song, L.; Wang, H.; Rui, C.; Liu, Q.; Zhang, Y.; Cheng, Y.; He, J. Preparation and properties of aflatoxins imprinted polymer grafted onto the surface of mesoporous silica SBA-15 functionalized with double bonds. *J. Sep. Sci.* **2021**, *44*, 4181–4189. [CrossRef]
17. Yuan, S.; Wang, M.; Liu, J.; Guo, B. Recent advances of SBA-15-based composites as the heterogeneous catalysts in water decontamination: A mini-review. *J. Environ. Manag.* **2020**, *254*, 109787. [CrossRef] [PubMed]
18. Voort, P. An Overview of the Challenges and Progress of Synthesis, Characterization and Applications of Plugged SBA-15 Materials for Heterogeneous Catalysis. *Materials* **2021**, *14*, 5082–5090.
19. Chen, S.; Jang, L.; Cheng, S. Synthesis of Zr-incorporated SBA-15 mesoporous materials in a self-generated acidic environment. *Chem. Mater.* **2004**, *16*, 4174–4180. [CrossRef]
20. Ruan, J.; Li, J.; Shen, Z.; Gu, J.; Li, H.; Sun, X.; Wang, L. Synthesis and Morphological Control of Zr-Ce-SBA-15 Mesoporous Materials. *Rare Met. Mater. Eng.* **2011**, *40*, 6–15.
21. Tang, L.; Tang, J.; Zeng, G.; Yang, G.; Xie, X.; Zhou, Y.; Pang, Y.; Fang, Y.; Wang, J.; Xiong, W. Rapid reductive degradation of aqueous p-nitrophenol using nanoscale zero-valent iron particles immobilized on mesoporous silica with enhanced antioxidation effect. *Appl. Surf. Sci.* **2015**, *333*, 220–228. [CrossRef]
22. Zhang, R.M.; Li, J.S.; Guan, Y.; Qi, L. Reduction of Trinitrotoluene by Nanoscale Zero-Valent Iron Confined in Mesoporous Channels. *Technol. Water Treat.* **2017**, *43*, 19–23. (In Chinese)
23. Xiang, G.; Long, S.; Liu, H.; Wu, X. Cd (II) removal from aqueous solutions by pomelo peel derived biochar in a permeable reactive barrier: Modelling, optimization and mechanism. *Mater. Res. Express* **2021**, *8*, 115508. [CrossRef]
24. Ayala, H.; Habineza, D.; Rakotondrabe, M.; Coelho, L. Nonlinear black-box system identification through coevolutionary algorithms and radial basis function artificial neural networks. *Appl. Soft. Comput.* **2019**, *87*, 105990. [CrossRef]
25. Hong, H.; Zhang, Z.; Guo, A.; Shen, L.; Sun, H.; Liang, Y.; Wu, F.; Lin, H. Radial basis function artificial neural network (RBF ANN) as well as the hybrid method of RBF ANN and grey relational analysis able to well predict trihalomethanes levels in tap water. *J. Hydrol.* **2020**, *591*, 125574. [CrossRef]
26. Lina, A.; Calzada, R.; García, A.; Tatiana, E. TiO_2, SnO_2 and ZnO catalysts supported on mesoporous SBA-15 versus unsupported nanopowders in photocatalytic degradation of methylene blue. *Micropor. Mesopor. Mat.* **2019**, *285*, 247–258.
27. Amr, A.; Es, B.; Mab, C.; Em, D.; Da, D. Mercury removal from aqueous solution via functionalized mesoporous silica nanoparticles with the amine compound. *Egypt. J. Pet.* **2019**, *28*, 289–296.

28. Betiha, M.; Moustafa, Y.; El-Shahat, M.; Rafik, E. Polyvinylpyrrolidone-Aminopropyl-SBA-15 schiff Base hybrid for efficient removal of divalent heavy metal cations from wastewater. *J. Hazard. Mater.* **2020**, *397*, 122675. [CrossRef]
29. Popa, A.; Sasca, V.; Verdes, O.; Barvinschi, P.; Holclajtner-Antunovic, I. Acidic and neutral caesium salts of 12-molybdophosphoric acid supported on SBA-15 mesoporous silica. The influence of Cs concentration and surface coverage on textural and structural properties. *Mater. Res. Bull.* **2014**, *50*, 312–322. [CrossRef]
30. Li, Q.; Wang, H.; Chen, Z.; He, X.; Wang, X. Adsorption-reduction strategy of U(VI) on NZVI-supported zeolite composites via batch, visual and XPS techniques. *J. Mol. Liq.* **2021**, *339*, 116719. [CrossRef]
31. Silva, F.; Silva, L.; Santos, A.; Caldeira, V.; Luz, G. Structural Refinement, Morphological Features, Optical Properties, and Adsorption Capacity of α-Ag_2WO_4 Nanocrystals/SBA-15 Mesoporous on Rhodamine B Dye. *J. Inorg. Organomet. Polym. Mater.* **2020**, *30*, 3626–3645. [CrossRef]
32. Yang, J.; Wang, S. Study on optimization of elimination effect of garlic extracts on nitrite by response surface method in salted meat. *China Condiment* **2009**, *34*, 47–51. (In Chinese)
33. Rahdar, S.; Pal, K.; Mohammadi, L.; Rahdar, A.; Kyzasg, G. Response surface methodology for the removal of nitrate ions by adsorption onto copper oxide nanoparticles. *J. Mol. Struct.* **2021**, *1231*, 129686. [CrossRef]
34. Chen, C.H.; Yang, S.H.; Chu, Y.J. Nitrite Scavenging Effect of Polyphenols from Major Small Grains in Yunnan Province. *Food Res. Dev.* **2019**, *40*, 17–23. (In Chinese)
35. Cai, J.; Zheng, P.; Qaisar, M. Prediction and quantifying parameter importance in simultaneous anaerobic sulfide and nitrate removal process using artificial neural network. *Environ. Sci. Pollut. Res.* **2015**, *22*, 8272–8279. [CrossRef]
36. Lee, S.C.; Kang, J.K.; Jang, H.Y. Multi-parameter experiments and modeling for nitrate sorption to quaternary ammonium-functionalized poly(amidoamine) dendrimers in aqueous solutions. *Int. J. Environ. Sci. Technol.* **2022**, *19*, 1–14. [CrossRef]
37. Liu, Y.; Zhang, L.Y. Implementation of BP and RBF neural network and their performance comparison. *Electron. Meas. Technol.* **2007**, *30*, 77–91. (In Chinese)
38. Yao, C.; Zhu, C. A new multi-mechanism adsorption kinetic model and its relation to mass transfer coefficients. *Surf. Interfaces* **2021**, *26*, 101422. [CrossRef]
39. Zhao, Y.; Zhu, F.; Ren, W.T. Kinetics of nitrate removal groundwater using green synthesized nanoscale zero valent iron-nickel. *Environ. Eng.* **2018**, *36*, 71–76. (In Chinese)
40. Wang, H.X.; Liao, B.; Lu, T.; Wang, J.Z.; Wei, S.M.; Liu, G. Nitrate removal from groundwater by zero-valent iron-biochar composites. *Chin. J. Environ. Eng.* **2020**, *14*, 3317–3328. (In Chinese)
41. Bhatnagar, A.; Ji, M.; Choi, Y.H.; Jung, W. Removal of nitrate from water by adsorption onto zinc chloride treated activated carbon. *Sep. Sci. Technol.* **2008**, *43*, 886–907. [CrossRef]
42. Meng, X.; Yao, L.; Jiang, W. In Situ Growth Synthesis of the CNTs@AC Hybrid Material for Efficient Nitrate-Nitrogen Adsorption. *ACS Omega* **2021**, *6*, 1612–1622. [CrossRef] [PubMed]
43. He, Z.P.; Han, M.D.; Pang, L.J. Adsorption behavior and mechanism of the bamboo-carbon for nitrate in aqueous solution. *Food Mach.* **2010**, *26*, 68–71. (In Chinese)
44. Dioum, A.; Hamoudi, S. Mono- and quaternary-ammonium functionalized mesoporous silica materials for nitrate adsorptive removal from water and wastewaters. *J. Porous Mat.* **2014**, *21*, 685–690. [CrossRef]
45. Kang, J.K.; Kin, S.B. Synthesis of quaternized mesoporous silica SBA-15 with different alkyl chain lengths for selective nitrate removal from aqueous solutions. *Micropor. Mesopor. Mat.* **2020**, *295*, 109967. [CrossRef]
46. Wang, H.; Wang, G.F.; Li, L. Study on the denitrification performance of Fe-SBA-15 mesoporous molecular sieve in the removal of nitrate nitrogen from Beijing Groundwater. *Resour. Conserv. Environ. Prot.* **2016**, *181*, 40–44. (In Chinese)
47. Aouaini, F.; Yahia, M.B.; Alanazi, M. Phenomenological statistical physics modeling of metalloporphyrins adsorption at the molecular level. *J. Mol. Liq.* **2021**, *340*, 117108. [CrossRef]
48. Chen, H.; Zhang, Y.P.; Ruan, X.H. Effect of Fe^{2+} on Nitrate Reduction Using Zero-valent Iron. *Geol. J. China Univ.* **2018**, *24*, 84–90. (In Chinese)
49. Liu, H.; Hua, W.; Shen, J. Promotion effect of cerium and lanthanum oxides on Ni/SBA-15 catalyst for ammonia decomposition. *Catal. Today* **2008**, *131*, 444–449. [CrossRef]
50. Zhang, W.; Shi, X.; Shan, Y. Promotion effect of cerium doping on iron–titanium composite oxide catalysts for selective catalytic reduction of NO_x with NH_3. *Catal. Sci. Technol.* **2020**, *10*, 648–657. [CrossRef]
51. Zha, X.S.; Feng, Z.L.; Jin, S.W. Removal of Nitrate in Water by Reduction of Iron-Based Bimetal. *Technol. Water Treat.* **2020**, *46*, 44–48. (In Chinese)
52. Song, Y.J.; Song, S.F. Preparation, characterization, and kinetics of nanoscale iron in nitrate nitrogen removal from polluted water. *Toxico. Environ. Chem.* **2015**, *97*, 379–387. [CrossRef]
53. Shi, J.; Yi, S.; He, H. Preparation of nanoscale zero-valent iron supported on chelating resin with nitrogen donor atoms for simultaneous reduction of Pb^{2+} and NO_3^-. *Chem. Eng. J.* **2013**, *230*, 166–171. [CrossRef]
54. Hwang, Y.H.; Kim, D.G.; Shin, H.S. Effects of synthesis conditions on the characteristics and reactivity of nano scale zero valent iron. *Appl. Catal. B-Environ.* **2011**, *105*, 144–150. [CrossRef]

55. Choe, S.; Chang, Y.Y.; Hwang, K.Y. Kinetics of reductive denitrification by nanoscale zero-valent iron. *Chemosphere* **2000**, *41* 1307–1311. [CrossRef]
56. Pan, W.L.; Wu, Q.Y.; Cao, Y.P.; Zhang, X.B.; Gu, L.; He, Q. Improvement of nitrate removal and nitrogen selectivity by the synergy of nZVI/BC and (Cu-Pd)/BC. *Chem. Ind. Eng. Prog.* **2022**, *41*, 981–989.

MDPI AG
Grosspeteranlage 5
4052 Basel
Switzerland
Tel.: +41 61 683 77 34

Catalysts Editorial Office
E-mail: catalysts@mdpi.com
www.mdpi.com/journal/catalysts

www.ingramcontent.com/pod-product-compliance
Lightning Source LLC
Chambersburg PA
CBHW041418210926
51850CB00013B/3304

* 9 7 8 3 7 2 5 8 3 5 9 8 0 *